Undergraduate Topics in Computer Science

Undergraduate Topics in Computer Science (UTiCS) delivers high-quality instructional content for undergraduates studying in all areas of computing and information science. From core foundational and theoretical material to final-year topics and applications, UTiCS books take a fresh, concise, and modern approach and are ideal for self-study or for a one- or two-semester course. The texts are all authored by established experts in their fields, reviewed by an international advisory board, and contain numerous examples and problems. Many include fully worked solutions.

More information about this series at http://www.springer.com/series/7592

Tom Jenkyns · Ben Stephenson

Fundamentals of Discrete Math for Computer Science

A Problem-Solving Primer

Second Edition

 Springer

Tom Jenkyns
Brock University
St. Catharines, ON
Canada

Ben Stephenson
University of Calgary
Calgary, AB
Canada

ISSN 1863-7310   ISSN 2197-1781   (electronic)
Undergraduate Topics in Computer Science
ISBN 978-3-319-70150-9   ISBN 978-3-319-70151-6   (eBook)
https://doi.org/10.1007/978-3-319-70151-6

Library of Congress Control Number: 2017959909

Printed on acid-free paper

This Springer imprint is published by the registered company Springer International Publishing AG part
of Springer Nature
The registered company address is: Gewerbestrasse 11, 6330 Cham, Switzerland

Preface

While this is a new edition of Fundamentals of Discrete Math for Computer Science, the goal of the book remains the same: To present discrete mathematics to computer science students in a form that is accessible to them, and in a way that will improve their programming competence. This edition improves upon its predecessor by introducing a new chapter on directed graphs, introducing a new section on drawing and coloring graphs, adding more than 100 new exercises, and providing solutions to selected exercises. We have also made numerous minor modifications in the second edition that make the text easier to read, improve clarity, or correct errata.

Most chapters begin with an example that sets the stage for its content. Chapter 1 begins by setting the stage for the whole book: How do you design algorithms to solve computing problems? How do you know your algorithm will work correctly for every suitable input? How long will your algorithm take to generate its output?

In our view, teaching is much more than presenting content, and we have written this text as *a design for the experience of students with our subject.* A students' first experience with a new subject can have a long-lasting impact on their perception it. We want that experience to be positive. Our text empowers students to think critically, to be effective problem solvers, to integrate theory and practice, and to recognize the importance of abstraction. It engages them with memorable, motivating examples. It challenges them with many new ideas, methods, and rigorous thinking, and we hope it entertains them like no other textbook with "Math" in its title.

This book introduces much of the "culture" of Computer Science and the common knowledge shared by all computer scientists (beyond programming). Much of it is devoted to the solutions to fundamental problems that all computer scientists have studied: how to search a list for a particular target; how to sort a list into a natural order; how to generate all objects, subsets, or sequences of some kind in such an order; how to traverse all of the nodes in a graph or digraph; and especially, how to compare the efficiency of algorithms and prove their correctness. Our constant theme is the relevance of the mathematics we present to Computer Science.

Perhaps the most distinguishing feature of this text is its informal and interactive nature. Detailed walkthroughs of algorithms appear from beginning to end. We motivate the material by inserting provocative questions and commentary into the prose, and we simulate a conversation with our reader, more like our lectures and less like other pedantic and ponderous mathematics texts. We employ the symbol "//" to denote "*comments*", to indicate *asides* (especially expanding and explaining mathematical arguments), as *signals* of what comes next, and to *prompt* questions we want to raise in the reader's mind. We've kept the text (words, sentences, and paragraphs) short and to the point, and we've used font changes, bold, italic and underlining to **capture**, *focus* and <u>recapture</u> our reader's attention.

But this book is a *mathematics text*. We expand on students' intuition with precise mathematical language and ideas. Although detailed proofs are delayed until Chap. 3, the fundamental nature of proof in mathematics is explained, maintained, and applied repeatedly to proving the correctness of algorithms. One of the purposes of the text is to provide a toolbox of useful algorithms solving standard problems in Computer Science. The other purpose is to provide a catalogue of useful concepts without which the underlying theory of Computer Science cannot be understood.

St. Catharines, Canada   Tom Jenkyns
Calgary, Canada   Ben Stephenson

Acknowledgements

We wish to acknowledge the support and encouragement of our friends and families throughout the long gestation of this book; especially Eleanor, Glenys, Janice, Flora and Jonathan. We are also grateful to the students and teachers that made this book possible, including the students at Brock University who indulged experimentation in both the form and content of this project over two decades of "class-testing", and every student and teacher that adopted, recommended, read or worked their way through the first edition. We also want to thank everyone who provided helpful comments and suggestions based on their experience using the first edition, and particularly acknowledge the comments submitted by Dominic Magnarelli and Carlo Tomasi. Finally, we are indebted to Eric R. Muller who initiated the proto-type course at Brock University.

Contents

Algorithms, Numbers, and Machines

We want to begin by illustrating the objectives of this book with two examples. First, an "algorithm" for multiplication you weren't taught in school.

Russian Peasant Multiplication

To find the product of integers M and N, both larger than one:

Step 1. Start two columns on a page, one labeled "A" and the other "B"; and put the value of M under A and the value of N under B.

Step 2. **Repeat**
    (a) calculate a new A value by multiplying the old A-value by 2; and
    (b) calculate a new B-value by dividing the old B-value by 2 and reducing the result by a half if necessary to obtain an integer;
    Until the B-value equals one.

Step 3. Go down the columns crossing out the A-value whenever the B-value is even.

Step 4. Add up the remaining A-values and "return" the sum.

To see how this works, let's "walk through" an example of the operation of this algorithm. Suppose that the input values are $M = 73$ and $N = 41$.

A	B	
73	41	
~~146~~	20	(20½ is reduced to 20)
~~292~~	10	
584	5	
~~1168~~	2	(2½ is reduced to 2)
2336	1	
2993		

© Springer International Publishing AG, part of Springer Nature 2018
T. Jenkyns and B. Stephenson, *Fundamentals of Discrete Math for Computer Science: A Problem-Solving Primer*, Undergraduate Topics in Computer Science,
https://doi.org/10.1007/978-3-319-70151-6_1

The algorithm ends by returning the value 2993. Is that equal to 73×41? What would happen if the initial values of A and B were both 100?

A	B
~~100~~	100
~~200~~	50
400	25
~~800~~	12
~~1600~~	6
3200	3
6400	1
10000	

Now <u>you</u> try a few more examples.

A	B		A	B		A	B
6	6		41	73		1000	1000
..	..		...	...		... ...	... ...
..	..		...	...		... ...	... ...
..			...	...		... ...	... ...
			...	...		... ...	... ...
			...	...		... ...	... ...
			...	...		... ...	... ...
			...			... ...	... ...
						... ...	... ...
						... ...	... ...
						... ...	

Do you believe that this algorithm is **correct**? (Will it produce the right answer for <u>every</u> possible pair of suitable input values – integers M and N larger than one?)

Is the "loop" in Step 2 sure to **terminate**? (Must B eventually equal one <u>exactly</u>?)

What is the **complexity** of this algorithm? (Do you think it is possible to predict, before applying the algorithm, how often B will be halved? That will determine how many rows the table will have, and give an upper bound on how many terms will be added in Step 4.)

We will mention this algorithm a number of times later (and answer these three questions), but we'll refer to it by its initials, RPM.

The second example is a problem involving a

Cake-Cutting Conundrum

Imagine you're cutting a circular cake with a large knife in the following manner. There are N "points" marked on the circumference of the cake, and you decide to make straight cuts joining <u>all pairs</u> of these points. How many pieces of cake, $P(N)$, will be produced?

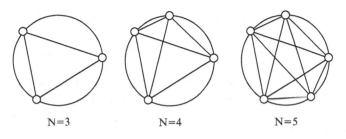

N=3   N=4   N=5

Tabulating the number of points and the number of pieces, we get the following.

N	P(N)
1	1
2	2
3	4
4	8
5	16
6	??

How many pieces will 6 points produce? Count the pieces in the diagram below.

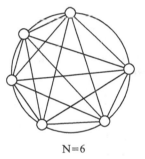

N=6

The number of pieces doubles each time as N goes from 1 to 5. This "pattern" would indicate that $P(6)$ should be 32. But it's not 32. The general pattern of doubling exists only in our minds. People seem programmed to find patterns of consistency, and we find it very disappointing when they are not correct; we feel we've been tricked. When students ask me "Why is $P(6)$ not 32?" I ask them "Which do you believe: what you actually saw when you counted the pieces for $N = 6$ or the pattern you imagined?"

Do you believe that RPM always works? On the basis of a few cases? Or could this be a "pattern" like the Cake-Cutting Conundrum that only applies some of the time?

We will look at RPM again later and prove that it is indeed correct (there are no cases where it does not return the product of N and M), it always terminates, and we can determine its complexity. Also, we'll see that while N must be an integer greater than one, M need not be greater than one, need not be an integer, and may even be negative!

And we will look again at the Cake-Cutting Conundrum and prove there is a formula for $P(N)$ that equals 2^{N-1} only for $N = 1, 2, 3, 4,$ and 5.

The two objectives of this book are to provide you with a portfolio of algorithms and a catalogue of mathematical ideas for designing and analyzing algorithms. The algorithms will be useful for solving standard problems. (Russian Peasant Multiplication is not used, even by Russian peasants, to multiply integers, but we used it to introduce the notion of an algorithm and will use it again later in several contexts. For our purposes, it is very useful.) We used the Cake-Cutting Conundrum to illustrate that *induction* – the method of reasoning that arrives at a general principle from several examples – cannot be trusted to establish the whole truth. We need better methods of reasoning, ones that will provide "mathematical" certainty.

1.1   What Is an Algorithm?

The word *algorithm* is a distortion of the name of a ninth century Persian mathematician, Al-Khowarizimi, who devised (or at least wrote down) a number of methods for doing arithmetic. Such methods were known long before Al-Khowarizimi's book and were known to you long before now. Much of your earlier mathematics education involved learning and applying algorithms for addition, multiplication, division, solving quadratic equations, and so on. But you probably didn't ask "Why does this procedure work?" or "Could this be done another way, that's easier or more efficient?" When using a computer, you must specify to the machine exactly which steps must be done and the order in which they must be done, making these questions important and interesting.

We shall use the word *algorithm* to mean

a step by step method

(like a recipe for a calculation divided up into sub-calculations). But there is always some intended purpose for an algorithm, whether it fulfils that intended purpose or not is another question. (A computer program that has bugs in it so it doesn't work correctly is still a computer program.) When the algorithm does accomplish the task or computation it was intended to do, the algorithm is *correct*. Also, to ever be of any practical value, an algorithm must *terminate*; that is, it must finish operation after a finite number of steps.

This description is very vague. What is a *step*? A *method*? A *task*? We will try to show you with a number of examples what we mean by these terms (even though we warned you about induction a few paragraphs ago, induction was the process we all used to learn our first language). In particular, a "step" is some *relatively easy subtask*. Computer programming is expressing a process as a sequence of statements in some formal language; here, a "step" roughly corresponds to a program statement. In an executable program, a "step" corresponds to execution of a machine instruction.

Throughout this book, we will present algorithms in terms we hope you will understand, not in a particular "computer language". However, we will often add "comments", set off by the symbols "//". These comments are not part of the list of steps in the algorithms. They are there to assist you in comprehending what's going on. In fact, we will use this convention to indicate commentary to the entire text of this book and will often use "//" to raise questions in imitation of a conversation with you. The symbols "//X" will indicate a question that will occur again among the exercises.

Example 1.1.1: Exponentiation

// Let us remind you that if n is an integer greater than one then x^n denotes the
// product of n x's; x^1 is x itself; x^0 equals 1; and, if $x \neq 0$ then x^{-n} denotes $1/x^n$.
// Then $(x^m) \times (x^n) = x^{m+n}$ and $(x^n)^m = x^{mn}$ for all integers m and n.

Evaluating x^{100} would seem to require 99 multiplications, but it can be done with far fewer as follows:

$$
\begin{array}{lll}
\text{multiply } x & \text{by } x & \text{to obtain } x^2 \\
\text{multiply } x^2 & \text{by } x^2 & \text{to obtain } x^4 \\
\text{multiply } x^4 & \text{by } x^4 & \text{to obtain } x^8 \\
\text{multiply } x^8 & \text{by } x^8 & \text{to obtain } x^{16} \\
\text{multiply } x^{16} & \text{by } x^{16} & \text{to obtain } x^{32} \\
\text{multiply } x^{32} & \text{by } x^{32} & \text{to obtain } x^{64}.
\end{array}
$$

Now multiply x^{64} by x^{32} to obtain x^{96} and multiply x^{96} by x^4 to obtain x^{100}. The total number of multiplications was only 8.

// How many multiplications would be required to evaluate x^{23}, x^{204}, or x^{2005}?

The first stage of the algorithm used 6 squaring operations to obtain x^{2^j} for $j = 1, 2, \ldots, 6$, and the second stage used the fact that

$$100 = 64 + 32 + 4 = 2^6 + 2^5 + 2^2.$$

Do you think that every positive integer can be expressed as a sum of certain, distinct powers of 2? If we were to use RPM to multiply 1 times 1,349, we would get

A	B
1	1349
~~2~~	674
4	337
~~8~~	168
~~16~~	84
~~32~~	42
64	21
~~128~~	10
256	5
~~512~~	2
1024	1
1349	

When A is given the initial value $1 = 2^0$, then the rest of the A-values are consecutive powers of 2, and the product returned by RPM is the initial value of B expressed as a sum of distinct powers of 2.

// Incidentally, several current cryptographic methods used to securely encode
// data raise certain numerical objects to very high powers and use variations of
// this "square and multiply" algorithm to do it.

Example 1.1.2: Three Subtraction Algorithms

Digit by digit subtraction was what we all learned in school along with some rule
to apply when the digit in the subtrahend was larger than the corresponding digit in
the minuend.

$$86 \quad \text{(the minuend)}$$
$$\underline{-38} \quad \text{(the subtrahend)}$$
$$?? \quad \text{(the difference)}$$

There are (at least) three ways to do subtraction.

(a) Borrowing: "Borrow" a 10 from the 80 so the problem becomes

$$70 + 16$$
$$\underline{-30 - 8}$$
$$40 + 8 = 48$$

(b) Carrying: Add 10 to the 6 and add 10 to the subtrahend by adding a 1 in the
     "tens place" so the problem becomes

$$80 + 16$$
$$\underline{-40 - 8}$$
$$40 + 8 = 48$$

(c) Complementation: No such rule would be needed if the minuend were all
     nines.

$$99$$
$$\underline{-38}$$
$$61$$

The difference, 61, is called "the 2-digit nines complement of 38". Adding 1 to
this produces 62 which is called "the 2-digit tens complement of 38".
Then

$$86 - 38 = 86 + \{\text{the 2-digit tens complement of } 38 - 10^2\}$$
$$= 86 + 62 \qquad\qquad\qquad\qquad\qquad - 100$$
$$= 148 - 100 \qquad\qquad\qquad // \text{ an easy subtraction}$$
$$= 48$$

$$// \, -38 = 100 \qquad - 38 \quad - 100$$
$$// \qquad\quad = 1 + (99 - 38) \quad - 100$$
$$// \qquad\quad = 1 \qquad\quad + 61 \quad - 100$$
$$// \qquad\quad = 62 \qquad\quad - 100$$

You probably feel that the best of these algorithms was the one you learned in elementary school. Which do you think machines use? Usually, there are many algorithms that accomplish the same task. *How do we decide that one algorithm is better than another?*

Example 1.1.3: Casting Out Nines

Is there a quick and easy way to decide whether or not 3 (or 9) divides evenly into a given positive integer n, without doing the division?

1. Let k start with the value of n.
2. While k has more than one digit,
       add the digits in k to produce a new value of k.
3. Return the answer to the question
       "Does 3 (or 9) divide evenly into k?"

For example, if n were 87 466, we would get

$$k_0 = 87\ 466 \qquad\qquad // = n, \text{the input}$$
$$k_1 = 31 \qquad\qquad\qquad // = 8+7+4+6+6$$
$$k_2 = 4$$

No. 3 does <u>not</u> divide evenly into 87 466.

// Is that correct?   Is $87\ 466 = 9(9\ 718) + 4 = 3(29\ 155) + 1$?

This process is known as "casting out nines" because at each stage (i.e., for each k we calculate)

$$n = 9q + k \qquad\qquad \text{for some integer } q.$$

So going from n to k, we have subtracted (cast out) a bunch of nines. Because 3 divides evenly into $9q$, 3 divides evenly into n (the left hand side of the equation, **LHS**) if and only if 3 divides evenly into k.

// But is $n - k$ always a multiple of 9?

If n were a 4-digit number "$d_3 d_2 d_1 d_0$," then for the first k-value

$$n = 1000d_3 + 100d_2 + 10d_1 + 1d_0$$
$$\underline{k = \qquad 1d_3 + \qquad 1d_2 + \quad 1d_1 + 1d_0}$$
$$n - k = \quad 999d_3 + \quad 99d_2 + \quad 9d_1$$

where each term on the right hand side (**RHS**) is clearly divisible by 9, and so the sum on the RHS is divisible by 9.

If we let $<n>$ denote the final k-value obtained, then   // as we'll prove later

$<n> = 9$ *precisely when n is divisible by 9, and in all other cases*
$<n>$ *is the remainder you get when you divide n by 9.*

Ancient accountants used "casting out nines" to check arithmetic.

If $A + B = C$, then $<<A> + <B>> = <C>$, and if $A \times B = D$, then
$<<A> \times <B>> = <D>$.

So if an apprentice-accountant claimed that

$$492 \times 61 + 2983 = 34995,$$

the master accountant would quickly determine

$$<<492> \times <61> + <2983>> = <6 \times 7 + 4> = <46> = 1$$

but $$<34995> = <30> = 3,$$

and conclude that the apprentice had made (at least one) mistake. This checking is
much simpler than redoing the original arithmetic (on a clay tablet or on an abacus)
because it uses relatively small numbers, but it's not foolproof. If it finds a mistake
there must be one, but there may be mistakes it doesn't find. For instance, the
correct answer in this example is 32995, but a "transposition error" giving 23995
would not be detected.

The Most Important Ideas in This Section.
An *algorithm* is *a step by step method* with some intended purpose. When it
does accomplish the task or computation it was intended to do, the algorithm
is *correct*. But to be of any practical value, an algorithm must *terminate* (after
a finite number of steps). Providing methods of proving termination and
correctness of algorithms is the "intended purpose" of this book.

1.2 Integer Algorithms and Complexity

This section gives more examples of algorithms. We will develop several related
algorithms to **test** a positive integer to see if it is prime or not, to **factor** a positive
integer into primes, and to find the **greatest common divisor** of two given positive
integers.

The first division algorithm I was taught in school was called "short division"
and consisted of finding the integer quotient of two numbers and the remainder.
I learned that 7 divided into 25 goes 3 times with a remainder of 4, or

$$25 = 7(3) + 4.$$

In fact,

if n is any integer and d is any positive integer,
then there are (unique) integers q and r where
$$n = d(q) + r \text{ and } 0 <= r < d.$$

Finding the integer quotient, q, and the remainder, r, is done so often that most high-level computer languages have built-in operations to do just that. In this book, we shall use DIV and MOD to denote these operations. That is,

$$q \quad \text{is} \quad n \,\mathrm{DIV}\, d \quad \text{and} \quad r \quad \text{is} \quad n \,\mathrm{MOD}\, d.$$

When $n \,\mathrm{MOD}\, d = 0$, d is said to **divide evenly into** n or to be a **factor** of n or to be a **divisor** of n. Furthermore it will be useful to let

$$\mathbf{d|n} \text{ denote the } \textit{statement } \text{``}d \text{ divides evenly into} n\text{''}.$$

Then $d|n$ is sometimes True and sometimes False depending on the values of d and n. That is, "$|$" is a *Boolean* operator on ordered pairs of integers.

$$2|6 \text{ is True} \quad \text{but} \quad 6|2 \text{ is False} \quad \text{and} \quad 7|25 \text{ is False}$$

The first operand, d, cannot be zero, but the second, n, may be negative, zero, or positive. For any integer $n \neq 0$

$$1|n, \quad n|n, \quad n|(-n) \quad \text{and} \quad n|0 \quad \text{are all True}.$$

When $d|n$ is True and $1 < d < n$, we'll say that d is a **proper divisor** of n. A **prime** *is an integer greater than one that has no proper divisors.* One is not a prime; the smallest prime is two.

1.2.1   Prime Testing

We can use the definition of a prime to construct an algorithm to test an input integer n to see if n is prime or not, by "trying" all the integers in the range from 2 to $n - 1$. If one of these divides evenly into n, then n is not prime; if none of these divides n, then n has no proper divisors, so n must be prime. This is known as "the method of trial divisions".

We'll assume that the input integer is greater than 2, and we'll use t as the variable for the trial divisors.

Algorithm 1.2.1: Prime Tester #1

```
Begin
   t ← 1;

   Repeat
      t ← t + 1;
   Until (t|n) or (t = n - 1);

   If (t|n) Then
      Output(t, "is a proper divisor of",n);
   Else
      Output(n, "is prime");
   End ;
End.
```

// As this is the first example of the pseudo-code we will use to present algorithms
// in this book, let us expand, for a moment, on the syntax (form) and semantics
// (meaning) of that code.

// There are three "steps" here; each is terminated by a semicolon and all of them
// lie between the tokens, "**Begin**" and "**End.**" which mark the beginning and the
// end of the list of steps.

// The first is an *assignment*: The value of the expression to the right of the symbol
// "←" is assigned to the variable on the left of the symbol. (*t* is initialized to the
// value 1.)

// The second is a *repeat-loop*: The steps between "**Repeat**" and "**Until**" (the *body*
// of the loop) are done again and again until the "condition" following the "**Until**"
// occurs.

// The third is a *conditional statement*: When the condition between the "**If**" and
// "**Then**" is True the steps between "**Then**" and "**Else**" are done, but when the
// condition is False the steps between the "**Else**" and "**End**" are done.

Walkthrough Algorithm 1.2.1 with input $n = 35$:

// A "walkthrough" is the construction of a table to show the changes in the values
// of the variables and expressions in the algorithm. As time passes we move
// down the table. This table is commonly called a "trace" of the operation of the
// algorithm.

t	t|n	t = n − 1	output
1	---	---	---
2	F	F	---
3	F	F	---
4	F	F	---
5	T	F	5 is a proper divisor of 35

Walkthrough with input $n = 11$:

t	t|n	t = n − 1	output
1	---	---	---
2	F	F	---
3	F	F	---
4	F	F	---
5	F	F	---
6	F	F	---
7	F	F	---
8	F	F	---
9	F	F	---
10	F	T	11 is prime

This algorithm simply reflects the definition of a prime, so it correctly determines whether or not the input integer n is a prime. The repeat-loop is sure to terminate after at most $n - 2$ iterations. A **worst case** occurs when n is prime. A **best case** occurs when n is even, and we do only one iteration of the loop.

Suppose now that your supervisor asks you to test an integer, N, that's 25 digits long using a machine that can do 10^9 iterations of the loop per second. How long might this algorithm take?

If N has 25 digits,

$$10^{24} = 1\underbrace{000\ldots000}_{24-zeros} <= N <= \underbrace{9\,999\ldots999}_{23-nines} = 10^{25} - 1.$$

If N were prime, the number of iterations to be done would be

$$N - 2 >= 10^{24} - 2 > 10^{24} - 10^{23} = (10)10^{23} - 10^{23} = (9)10^{23}.$$

These would take more than

$$(9)10^{23}/10^9 \text{ seconds} \quad \begin{aligned} &= (9)10^{14} \text{ seconds} \\ &= (15)10^{12} \text{ minutes} \\ &= (25)10^{10} \text{ hours} \\ &> 10^{10} \text{ days} \\ &> (2)10^7 \text{ years} \end{aligned} \quad \begin{aligned} &= (900)10^{12}/60 \text{ minutes} \\ &= (1500)10^{10}/60 \text{ hours} \\ &= (25)10^{10}/24 \text{ days} \\ &= (1000)10^7/365.2 \text{ years} \\ &= 20\,000\,000 \text{ years} \end{aligned}$$

Knowing your supervisor is not that patient, let's look again at the problem and try to construct a faster algorithm.

How large can a proper divisor of n be? If $n = a \times b$ and $1 < a < n$, then

$$2 <= a \text{ and so } 2 \times b <= a \times b = n. \text{ Thus, } b <= n/2.$$

Therefore, we can stop our search for a proper divisor of n immediately after we try dividing by $t = n/2$. But if n is odd, $n/2$ is not an integer, and we could stop when t is the next integer below $n/2$. We will find it useful to have a notation for that integer (and for the next integer above $n/2$). Let's digress for a moment to discuss

1.2.2 Real Numbers

The following three strings of symbols all represent the same number:

$$17 \qquad XVII \qquad \text{seventeen.}$$

The first uses Arabic numerals, the second uses Roman numerals, and the last uses an English word written in lower case Latin letters. Numbers themselves are entities that are independent of their representation.

Imagine a line extending infinitely in both directions but having two points marked on it, one labeled 0 and, slightly to the right of that, one labeled 1.

If we use the distance between the points labeled 0 and 1 as a standard unit of length, any point to the right of 0 can be labeled with a number, x, equal to the distance from the point labeled 0 to that point in standard units. Points between 0 and 1 are labeled with **fractions** (of a standard unit). Points to the left of 0 are given a negative sign. In this way, **real numbers** are realized as lengths. Points on this line are geometric representations of real numbers, and the line itself is a geometric representation of **R**, the set of all real numbers. When a and b are real numbers, $a < b$ <u>means</u> a is left of b on this line.

Every real number is either an integer or lies between two consecutive integers. For any real number x, the **floor** function of x, $\lfloor x \rfloor$, is defined to be the largest integer $<= x$. Then

$$\lfloor 23.45 \rfloor = 23, \quad \lfloor 6 \rfloor = 6, \quad \text{and} \quad \lfloor -9.11 \rfloor = -10.$$

Also, the **ceiling** function of x, $\lceil x \rceil$, is defined to be the smallest integer $>= x$. Then

$$\lceil 23.45 \rceil = 24, \quad \lceil 6 \rceil = 6, \quad \text{and} \quad \lceil -9.11 \rceil = -9.$$

Either x is an integer and $\lfloor x \rfloor = x = \lceil x \rceil$ or, $\lfloor x \rfloor < x < \lceil x \rceil$ and $\lceil x \rceil = \lfloor x \rfloor + 1$.

// For all real numbers, $x - 1 < \lfloor x \rfloor <= x <= \lceil x \rceil < x + 1$,
// the largest integer $< x$ is $\lceil x \rceil - 1$, and the smallest integer $> x$ is $\lfloor x \rfloor + 1$.

1.2.3 More Prime Testing

We can now give an improved, more efficient method.

Algorithm 1.2.2: Prime Tester #2

```
Begin
   t ← 1;

   Repeat
      t ← t + 1;
   Until (t|n) or (t = ⌊n/2⌋);

   If (t|n) Then
      Output(t, "is a proper divisor of",n);
   Else
      Output(n, "is prime");
   End;
End.
```

This version will run about twice as fast as the first version in a worst case, but testing a 25-digit integer might still take a long time, more than 10 000 000 years. Your supervisor will probably not be too impressed with this improvement. So let's try again.

Algorithm 1.2.2 actually searches for the <u>smallest</u> proper divisor of n. How large can the smallest proper divisor of n be? If $n = a \times b$ and $1 < a <= b < n$, then

$$a <= b \text{ and so } a \times a <= a \times b = n. \quad \text{Thus, } a <= \sqrt{n}.$$

In fact, $a <= \lfloor \sqrt{n} \rfloor$, and we have the following:

Algorithm 1.2.3: Prime Tester #3

Begin
    t ← 1;

  Repeat
     t ← t + 1;
  Until (t|n) or (t = $\lfloor \sqrt{n} \rfloor$);

  If (t|n) **Then**
    Output(t, "is a proper divisor of",n);
  Else
    Output(n, "is prime");
  End;
End.

Walkthrough with input $n = 107$.   // $\lfloor \sqrt{n} \rfloor = 10$

t	t|n	t = $\lfloor \sqrt{n} \rfloor$	output
1	---	---	---
2	F	F	---
3	F	F	---
4	F	F	---
5	F	F	---
6	F	F	---
7	F	F	---
8	F	F	---
9	F	F	---
10	F	T	107 is prime

// This version is faster when n is prime, but how much of an improvement has
// been made? Can this test a 25-digit integer in less than a million years?

If N has 25 digits, the number of iterations of the body of the repeat-loop in Algorithm 1.2.3 will be less than

$$\sqrt{N} \text{ which is less than } \sqrt{10^{25}} = \sqrt{10^{24} \times 10} = 10^{12} \times \sqrt{10}.$$

These will take less than

$$(10^{12} \times \sqrt{10})/10^9 \text{ s} = 10^3 \times \sqrt{10} \text{ s}$$
$$= 1000 \times (3.162\ 277\ldots) \text{ s}$$
$$< 3163 \text{ s}$$
$$= 52.716\ 6\ldots \text{ min}$$
$$< 53 \text{ min}.$$

Now your supervisor will be pleased! The first moral of this story about prime testing is that *efficiency might matter* a lot – an intractable problem instance might become doable after some thought is invested in the design of your algorithm.

Can we speed up our prime tester even more? We've seen that the algorithm searches for the smallest proper divisor of n. *That smallest proper divisor must be a prime.*

// If d is a proper divisor of n but d is not prime, then d itself has a proper divisor f
// which must be a proper divisor of n that is smaller than d.

If 2 does not divide evenly into n, then no even number will. So no even numbers larger than 2 need be tested. We could make the algorithm almost twice as fast again by trying 2 separately and then only trying odd numbers up to $\lfloor \sqrt{n} \rfloor$. When working by hand, we can restrict ourselves to trying only *prime* values of t in the range from 2 up to $\lfloor \sqrt{n} \rfloor$.

However, there is a second moral to this story: *Better algorithms need better conceptual (mathematical?) foundations.* Our algorithm was based only on the definition of a prime and a search for a proper divisor. Primes have many amazing properties, which are studied in "number theory". In 2002, an extremely efficient (but completely different) algorithm for prime testing was published where the number of steps required to test an input integer N depends directly on the number of digits needed to represent N.

1.2.4 Prime Factorization

The so-called *Fundamental Theorem of Arithmetic* states that any integer n greater than one can be factored (uniquely) as a product of primes:

$$n = p_1 \times p_2 \times p_3 \times \ldots \times p_k \quad \text{where} \quad p_1 <= p_2 <= p_3 <= \ldots <= p_k.$$

// How do we find p_1, p_2, p_3, and so on?

$$26\ 040 = 2 \times (13\ 020)$$ // The smallest prime factor of 26 040 is 2.
$$= 2 \times 2 \times (6\ 510)$$ // The smallest prime factor of 13 020 is 2.
$$= 2 \times 2 \times 2 \times (3\ 255)$$ // The smallest prime factor of 6 510 is 2.
$$= 2 \times 2 \times 2 \times 3 \times (1\ 085)$$ // The smallest prime factor of 3 255 is 3.
$$= 2 \times 2 \times 2 \times 3 \times 5 \times (217)$$ // The smallest prime factor of 1 085 is 5.
$$= 2 \times 2 \times 2 \times 3 \times 5 \times 7 \times (31)$$ // The smallest prime factor of   217 is 7.
// We're done because 31 is also prime.

The prime testing algorithm will find the smallest proper divisor of n which we know must be prime (and therefore must be p_1) or will tell us that n itself is prime. Therefore, we can adopt the following strategy:

Find the smallest prime factor p of n.   // Either $p = n$ or $p <= \lfloor \sqrt{n} \rfloor$.
If $(p = n)$ then we're done.   // We have all the prime factors of n.
Else, let $Q \leftarrow n$ DIV p
        now the prime factorization of n equals
        $p \times$ (the prime factorization of Q).

// We also know that the smallest prime factor of Q is at least as big as p, and
// therefore, if $p > \lfloor \sqrt{Q} \rfloor$, then Q is also prime.

We can now give a pseudo-code version of this algorithm to find the prime factorization of any input integer n <u>greater than one</u>. It incorporates the prime tester, Algorithm 1.2.3, and uses a variable Q for the integer that still has to be factored.

Algorithm 1.2.4: Prime Factorization

```
Begin
  Q ← n;
  t ← 2;

  While (t <= ⌊√Q⌋) Do
    If (t|Q) Then
      Output(t, "×");
      Q ← Q DIV t;
    Else
      t ← t + 1;
    End ;
  End ;

  If (Q = n) Then
    Output(n, "is prime");
  Else
    Output(Q, " = ", n);
  End;
End.
```

// This pseudo-code contains a slightly different structure, a *while-loop*:
// As long as the condition between "**While**" and "**Do**" is true,
// execute the steps between "**Do**" and "**End**" (the *body* of the loop).

// The body of a repeat-loop is always done (at least) once, but if the condition of
// a while-loop is not satisfied the first time it's checked, the body of the loop is
// never done. That happens here if the input value, n, is 2 or 3.

Walkthrough with input $n = 74382$:

Q	$\lfloor\sqrt{Q}\rfloor$	t	$t <= \lfloor\sqrt{Q}\rfloor$	t|Q	Q=n	output-so-far
74382	272	2	T	T	–	$2 \times$
37191	192	"	T	F	–	$2 \times$
"	"	3	T	T	–	$2 \times 3 \times$
12397	111	"	T	F	–	$2 \times 3 \times$
"	"	4	T	F	–	$2 \times 3 \times$
"	"	5	T	F	–	$2 \times 3 \times$
"	"	6	T	F	–	$2 \times 3 \times$
"	"	7	T	T	–	$2 \times 3 \times 7 \times$
1771	42	"	T	T	–	$2 \times 3 \times 7 \times 7 \times$
253	15	"	T	F	–	$2 \times 3 \times 7 \times 7 \times$
"	"	8	T	F	–	$2 \times 3 \times 7 \times 7 \times$
"	' '	9	T	F	–	$2 \times 3 \times 7 \times 7 \times$
"	"	10	T	F	–	$2 \times 3 \times 7 \times 7 \times$
"	"	11	T	T	–	$2 \times 3 \times 7 \times 7 \times 11 \times$
23	4	"	F	–	F	$2 \times 3 \times 7 \times 7 \times 11 \times 23 = 74382$

// On each iteration of the while-loop, either Q decreases or t increases (but not
// both), so eventually, $t > \lfloor\sqrt{Q}\rfloor$ and the condition controlling the while-loop
// becomes false.

//X For positive integers t and Q, $(t <= \lfloor\sqrt{Q}\rfloor)$ is equivalent to $(t \times t <= Q)$ and
//   this second form is much more suitable for a computer program.

   Algorithm 1.2.4 terminates and is correct. The number of trial divisions is at
most $\lfloor\sqrt{n}\rfloor - 1$; a worst case occurs when no proper factors are found because n is
prime. On the other hand, how many prime factors might there be? If $n = p_1 \times p_2$
$\times p_3 \times \ldots \times p_k$, how large can k be? To answer this question, let's digress for a
moment to discuss

1.2.5 Logarithms

We know how to calculate integer powers of 2:

n	2^n	2^{-n}		
0	1	1/1	=	1
1	2	1/2	=	0.5
2	4	1/4	=	0.25
3	8	1/8	=	0.125
4	16	1/16	=	0.062 5
5	32	1/32	=	0.031 25
$\ldots$	$\ldots$	$\ldots$	=	$\ldots$
10	1024	1/1024	=	0.000 976 562 5

In every case, 2^n is a positive real number.

> **We will assume the following facts about real numbers.**
> Whenever b is some particular real number greater than one:
> 1. For any real number x, b^x is a certain, positive real number.
> 2. For any real numbers x and y, if $x < y$ then $b^x < b^y$.
> 3. For any *positive* real number z, there is a real number x such that $z = b^x$.

When $z = b^x$, we say that x is the **logarithm of z to base b** and write $\log_b(z) = x$. Thus, if z is any positive real number,

$$z = b^{\log_b(z)}.$$

For instance,
$$\log_2(32) = 5 \quad \text{because} \quad 32 = 2^5$$
$$\log_2(1024) = 10 \quad \text{because} \quad 1024 = 2^{10}$$
$$\log_{10}(10000) = 4 \quad \text{because} \quad 10000 = 10^4$$
$$\text{and,} \quad \log_{10}(0.001) = -3 \quad \text{because} \quad 0.001 = 10^{-3}.$$

Furthermore, for any real number y, $\log_b(b^y) = y$.

Logarithm tables (and anti-log tables) were created to simplify arithmetic in the 1600s. These tables were used to convert multiplication problems to addition and convert exponentiation to multiplication. If x and y are positive, then

$$xy = b^{\log_b(x)} \times b^{\log_b(y)} = b^{\log_b(x) + \log_b(y)} \quad \text{so} \quad \log_b(xy) = \log_b(x) + \log_b(y)$$

If z is any real number, $x^z = \left(b^{\log_b(x)}\right)^z = b^{z \times \log_b(x)}$ so $\log_b(x^z) = z \times \log_b(x)$.

Calculators have a **log**-button for "common" logarithms where the base is 10, and they have a **ln**-button for "natural" logarithms where the base is (a very unnatural number perhaps named for Euler) $e = 2.718\ 281\ 828\ 44\ldots$.

Throughout this book, we will frequently use logarithms to base 2 for which there is no calculator button. We will use $\lg(x)$ to denote $\log_2(x)$ and give you an algorithm (actually, a formula) to calculate values of $\lg(x)$. Suppose that x is a given positive number:

$$x = 2^{\lg(x)} \quad \text{so} \quad \log_b(x) = \log_b\left(2^{\lg(x)}\right) = \lg(x) \times \log_b(2) \quad \text{and therefore}$$

So
or
$$\lg(x) = \log_b(x)/\log_b(2). \qquad \text{// for any base } b$$
$$\lg(x) = \log(x)/\log(2) \qquad \text{// when } b = 10$$
$$\lg(x) = \ln(x)/\ln(2). \qquad \text{// when } b = e$$

Returning to the question that prompted this digression into logarithms:
If $n = p_1 \times p_2 \times p_3 \times \ldots \times p_k$, how large can k be? Since each prime factor is at least as big as 2, $n >= 2 \times 2 \times \ldots \times 2 = 2^k$. If $n >= 2^k$ then $\lg(n) >= \lg(2^k) = k$ so $k <= \lfloor \lg(n) \rfloor$.

Logarithm functions are increasing but grow very slowly.

n	$\sqrt{n}$	$\lg(n)$
4	2	2
16	4	4
64	8	6
256	16	8
1024	32	10
4096	64	12
16384	128	14
65536	256	16
262144	512	18
1048576	1024	20

If the **complexity function of an algorithm** (the number of steps it takes for completion when the input is of size n) were logarithmic, it would be <u>very efficient</u>. We will later show that RPM is such an algorithm, as is the bisection algorithm for solving equations (done at the end of the chapter). And so is the next algorithm, Euclid's Algorithm for the greatest common divisor of two integers (used since ~ 300 BC).

1.2.6   Greatest Common Divisor

A *common divisor* of two integers x and y is any integer d that divides evenly into both x and y. My introduction to common divisors was in about grade IV where they were used to "reduce" fractions. Because 4 is a common divisor of both 40 and 60, 40/60 $=$ 10/15. But this fraction can be reduced further. The <u>greatest</u> common divisor of 40 and 60 is 20, and when 20 is divided into the numerator and the denominator, 40/60 $=$ 2/3, and this fraction is in what my teacher called "lowest terms."

// What would 3568/10035 be when reduced to lowest terms?
// Do we need to try all possible divisors up to a certain point?
// Or is there another strategy for solving this problem?

We want an efficient algorithm to find the greatest common divisor of two positive integers x and y, denoted **GCD(x,y)**.   // Will $GCD(x,y) = GCD(y,x)$?
Suppose that $x >= y >= 1$. Since $1|x$ and $1|y$, and since y is the largest integer that divides y, we know $1 <= GCD(x,y) <= y$. If $y|x$, then (because $y|y$) y is a common divisor and is as large as possible so $GCD(x,y) = y$.

// That case is easy and didn't involve a search for divisors. What about the other
// case?

Otherwise (y does not divide evenly into x so) $x > y$ and

$$x = y(q) + r \quad \text{where} \quad 0 < r < y. \qquad\qquad \text{// } q >= 1 \text{ and } r = x \text{ MOD } y$$

Euclid proved that in this case

$$GCD(x, y) = GCD(y, r). \qquad\qquad \text{//And we'll prove this in Chap. 3.}$$

// This is an example of *recursion*; the *GCD* of x and y will (sometimes) be found
// by finding the *GCD* of two other (but smaller) integers, y and r.

An iterative algorithm for this problem can be constructed by observing that *GCD* is a function with two "parameters", say A and B. These are initialized to be the input integers, x and y, respectively; and are revised whenever we calculate a positive remainder.

Algorithm 1.2.5: Euclid's Algorithm for GCD(x,y)

```
Begin
    A ← x;
    B ← y;
    R ← A MOD B;
    While (R > 0) Do
        A ← B;
        B ← R;
        R ← A MOD B;
    End;
    Output ("GCD(", x, ",", y, ")=", B);          // or Return(B)
End.
```

Walkthrough with input $x = 10035$ and $y = 3568$:

A	B	R	R > 0	
10035	3568	2899	T	// $10035 = 3568(2) + 2899$
3568	2899	669	T	// $3568 = 2899(1) + 669$
2899	669	223	T	// $2899 = 669(4) + 223$
669	223	0	F	// $669 = 223(3) + 0$

```
output: GCD(10035 , 3568) = 223.
```

// Is this correct? $10035 = 223 \times 45 = 223(3 \times 3 \times 5)$ and
// $3568 = 223 \times 16 = 223 (2 \times 2 \times 2 \times 2)$.

Walkthrough with $x = 2108$ and $y = 969$:

A	B	R	R > 0	//	GCD(2108,969)
2108	969	170	T	//	$= $ GCD(969,170)
969	170	119	T	//	$= $ GCD(170,119)
170	119	51	T	//	$= $ GCD(119, 51)
119	51	17	T	//	$= $ GCD(51, 17)
51	17	0	F	//	$= 17$

output: GCD(2108 , 969) $= 17$.

// Is this correct? $2108 = 17 \times 124 = 17(2 \times 2 \times 31)$ and $969 = 17 \times 57 = 17(3 \times 19)$.

// Is this algorithm guaranteed to terminate?

This algorithm generates a sequence of integer values for A, B, and R, where

$$
\begin{aligned}
& A_1 = x, & B_1 = y & \quad \text{and } 0 <= R_1 < y = B_1; \\
\text{if } 0 < R_1, \ & A_2 = B_1, & B_2 = R_1 & \quad \text{and } 0 <= R_2 < R_1 = B_2; \text{ and} \\
\text{if } 0 < R_2, \ & A_3 = B_2 = R_1, & B_3 = R_2 & \quad \text{and } 0 <- R_3 < R_2 = B_3.
\end{aligned}
$$

The R-values decrease but are never negative, so eventually some $R_k = 0$.

// How large can that k be?

We will see in Chap. 8 that if k iterations are required then

$$
y >= \left(\frac{1 + \sqrt{5}}{2} \right)^k \text{ and from this, } k <= \lfloor (3/2) \lg(y) \rfloor
$$

// Is that believable? That its complexity function is related to such a strange
// number?

Euclid's Algorithm is effective and very efficient.

// It works even when $x < y$, but then takes 1 extra iteration. How? Why?

The Most Important Ideas in This Section.
This section gave more examples of algorithms on positive integers. Something we all learned in elementary school is *that if n is any integer and d is any
positive integer then there are (unique) integers q and r where $n = d(q) + r$
and $0 <= r < d$*. We use two operators to describe integer-division: n DIV d
produces q and n MOD d produces the (nonnegative) remainder r.

When n MOD $d = 0$, d is a *factor* or a *divisor* of n. Furthermore, $d|n$
denotes the *statement* "d divides evenly into n" (so $d|n$ is sometimes True and
sometimes False). When $d|n$ is True and $1 < d < n$, d is a *proper divisor* of n.

(continued)

(continued)

A prime is an integer greater than one that has no proper divisors. <u>One is not a prime; the smallest prime is two.</u>
   Prime tester #1 implements this definition of prime and introduces the pseudo-code we use to present algorithms in this book. Before presenting another (faster) version, we described the "real" line, a geometric representation of **R**, the set of all real numbers. For any real number x, the *floor* of x, $\lfloor x \rfloor$, is the largest integer $<= x$, and the *ceiling* of x, $\lceil x \rceil$, is the smallest integer $>= x$.
   Prime tester #2 runs twice as fast as the first version (in a worst case), but still may be too slow. Prime tester #3 runs much, much faster.
   The first moral of this story about prime testing is that *efficiency might matter* – an intractable problem instance might become doable after some thought is invested in the design of your algorithm. However, there is a second moral to this story: *Better algorithms need better conceptual foundations*. Our algorithms were based only on the definition of a prime and a search for a proper divisor. In 2002, an extremely efficient (but completely different) algorithm for prime testing was published.
   The *Fundamental Theorem of Arithmetic* states that any integer n greater than one can be factored (uniquely) as a product of primes

$$n = p_1 \times p_2 \times p_3 \times \ldots \times p_k \quad \text{where} \quad p_1 <= p_2 <= p_3 <= \ldots <= p_k$$

Algorithm 1.2.4 finds the factors in nondecreasing order.
We assume that whenever b is some particular real number greater than one:
1. For any real number x, b^x is a certain, positive real number.
2. For any real numbers x and y, if $x < y$, then $b^x < b^y$.
3. For any *positive* real number z, there is a real number x such that $z = b^x$.
   When $z = b^x$, x is the *logarithm of z to base b* (written $\log_b(z) = x$). We use $\lg(x)$ to denote $\log_2(x)$.
   If the *complexity function of an algorithm* (the number of steps it takes for completion when the input is of size n) were logarithmic, it would be <u>very</u> efficient. We will later show that RPM is such an algorithm, as is the bisection algorithm for solving equations (done at the end of this chapter). And so is *Euclid's Algorithm for the greatest common divisor* of two positive integers (used since ~ 300 BC).

1.3   Machine Representation of Numbers

The usual representation of numbers by means of Arabic numerals uses the ten *digits* 0, 1, 2, 3, 4, 5, 6, 7, 8, and 9 as basic symbols.
                                         // Digitus is the Latin word for finger.

Ten symbols are enough because the notation is ***positional***; that is, the position of a digit in a numeral carries information about the meaning of that digit.

563.6 *means* $5 \times 10^2 + 6 \times 10^1 + 3 \times 10^0 + 6 \times 10^{-1} = 500 + 60 + 3 + 6/10.$

This is called the ***decimal*** system because the ***base*** is 10.

// "Decem" is Latin for "ten."

We'll consider positional notation in other bases later, and in every case, we'll refer to the "." in the numeral as the ***base point***, not the decimal point.

// It is only a decimal point in the decimal number system.

Multiplying by the base simply ***shifts*** the base point one space to the right; dividing by the base shifts it one space left. If k is a positive integer, the base raised to the k-th power is a one followed by k zeros ($10^4 = 10\,000$), and the $(-k)$-th power is the base point followed by $(k-1)$ zeros followed by a one ($10^{-4} = .0001$).

Before considering how computers might represent numbers internally, let's examine how calculators display numbers. Mine displays integers as

> *a minus sign if the number is negative,*
> *then up to* 10 *digits*
> *(right-justified in the display window)*   // placed at the right side
> *and then the base point.*

The largest integer it can display is 9 999 999 999. If this number were squared, the answer would be 99 999 999 980 000 000 001 which is too long to be displayed as an integer. My machine displays (an approximation of this number)

> 9.999 999 998 E 19   where   "E 19"   *means*   "times 10^{19}."

This is known as *floating point* notation. The part to the left of the E is the *mantissa*, and the part to the right is the *exponent*. Either or both of these parts may have a negative sign. When my calculator uses floating point notation, the mantissa always has one nonzero digit to the left of the base point and is right-justified if it has fewer than 10 digits, and the exponent always has 2 digits.

// When the exponent is between −9 and +9, the base point is moved to the correct
// position in the display window.
// 3.217 E 6 appears as 3217000 and 3.217 E −6 appears as 0.000003217.

The largest floating point number that can be displayed is 9.999 999 999 E 99. If that number is squared, the answer is too big to be displayed (even approximately), and my calculator stops and displays "error 2". This type of error, where the exponent of the number is just too big to fit, is called an "overflow error."

The smallest, positive floating point number that can be displayed is "1. E−99". If that number were divided by 2, the answer (in my calculator's floating point format) would be "5. E−100" which is too small to be displayed. But my calculator doesn't stop this time; instead, it displays 0 as an approximation of this very small number.

1.3.1 Approximation Errors

This problem, caused by numbers with positional representations that are too big to fit into my calculator's display window, is unavoidable when using a physical object to do arithmetic as in any (digital) computer. My calculator "says"

$$625/26 - 24.038\ 461\ 54. \qquad // \text{ in the display}$$

If I subtract 24 and multiply by 1000, it displays "38.461 538 4," and then, if I subtract 38 and multiply by 1000, it displays "461.538 4." My calculator seems to "think"

$$625/26 = 24.038\ 461\ 538\ 4. \qquad // \text{ in its memory}$$

Using "short division", 26 divides into 625 exactly 24 times with a remainder of 1. However, when we do this calculation by *"long division"*, we find the digits of the quotient one at a time by short division which creates a <u>remainder</u> at each iteration. Part of this algorithm "extends" the dividend (the 625) by adding zeros after the base point; let's look at the remainders constructed during this stage.

```
        24.038 461 538 461 538 …
       ─────────────────────────────
  26  ) 625.000 000 000 000 000 …
        52
        ───
        105
        104
        ───
          1 0 .   .   .   .   .   .   R1 =  1
            0
          ───
          1 00 .   .   .   .   .   .   R2 = 10 ←
            78
          ────
            220     .   .   .   .   .   R3 = 22
            208
          ─────
             12 0   .   .   .   .   .   R4 = 12
             10 4
          ──────
              1 60 .   .   .   .   .   R5 = 16
              1 56
          ──────
                40  .   .   .   .   .   R6 =  4
                26
          ──────
                14 0 .   .   .   .   R7 = 14
                13 0
          ──────
                1 00    .   .   .   R8 = 10 ←
```

At this point, when R8 is the same as R2, the algorithm will repeat the calculation of the six different remainders and will repeat the calculation of the block **384 615** in the quotient. And this will continue forever. When dividing by 26, there are only 26 possible remainders so there <u>must</u> be a repetition of a remainder, and there <u>must</u> be a repeated block of digits in the quotient, like we saw here.

// Must this sort of thing happen in every long division?
// What happens when some remainder equals zero? Is it repeated?

The value of 625/26 can be represented exactly (by people) using **bar notation** to indicate a repeating block of digits. Then

$$625/26 = 24.0\overline{384615} \qquad\qquad \text{// and } 1/3 = 0.\overline{3}.$$

A **rational number** is any number that can be expressed as a *ratio* (or a *quotient*) of integers. Every rational number represented in positional notation must be repeating. However, repetitions of a block of zeros are never written and such rationals are called *terminating*, like

$$1/625 = 0.0016 \quad \text{not} \quad 0.0016\overline{0}.$$

//X When does $1/n$ have a terminating positional representation in base 10?
//   Is that frequent or rare? Do most have an infinite repeating decimal
//   representation?

No physical object could contain an infinite repeating decimal, and they don't use bar notation to shorten the representation, as people do. They must do something else to shorten the representation of the mantissa to a certain number of digits.

The simplest thing to do is just *cut off* all the digits after the <u>most significant</u> ones; this is called **truncation.**

// The most significant digits are the leftmost; they contribute most to the
// magnitude of the number.

If $A = 625/26$, we can describe several approximations of A.

$A1 = 24.03$ which is A truncated to 4 significant figures, and
$A1 = 24.0384$ which is A truncated to 4 places after the base point.

A is between 24.03 and 24.04,

but nearer to 24.04 so a better approximation of A would be

$A3 = 24.04$ which is A **rounded** to 4 significant figures.

The (usual) rounding rule for base 10 is

> *if the next digit is 5 or bigger then add 1 to the least significant digit*
> *in the truncated approximation.*
>
>   // In our example, the next digit is 8.

We might also approximate A by

$A4 = 24.0385$ which is A **rounded** to 4 places after the base point.

                                                    // In our example, the next digit is 6.

// My calculator appeared to truncate 625/26 to 12 significant figures in memory
// and then display that value rounded to 10 significant figures.

If B is an approximation of A, we would like to know how good an approximation it is. The **error** in B is the difference $B - A$. Usually we're interested in the size
of that error; the **absolute error** in B is the absolute value of the difference, $|B - A|$
where for x in **R**

$$|x| = \begin{cases} x \text{ if } x >= 0 \\ -x \text{ if } x < 0 \end{cases}.$$

But an error of ± 1 in an approximation of 5.285 is much more serious than an
error of ± 1 in an approximation of 528.5. So sometimes we're interested in the size
of the error compared to the size of A; the **relative error** in B is

$$\frac{|B - A|}{|A|},$$ // assuming $A \neq 0$

and this quotient is frequently expressed as a percentage. For the four approximations of $A = 625/26$:

	Error	**Relative error**
For $A1 = 24.03$	$24.03 - A$	$|\text{Error}| \div A$
	$= \dfrac{2403}{100} - \dfrac{625}{26}$	$= \dfrac{22}{2600} \times \dfrac{26}{625}$
	$= \dfrac{62478 - 62500}{100 \times 26}$	$= 0.000352$
	$= \dfrac{-22}{2600}$	$= 0.0352\%$
	$= -0.00\overline{846153}$	
		// similarly, you can show
For $A2 = 24.0384$	$\dfrac{-16}{260000} = -0.0000\overline{615384}$	$0.000\ 256\%$
For $A3 = 24.04$	$\dfrac{+4}{2600} = 0.00\overline{153846}$	$0.006\ 4\%$
For $A4 = 24.0385$	$\dfrac{+10}{260000} = 0.0000\overline{384615}$	$0.000\ 16\%$

For large numbers, truncating to k significant figures means keeping the k most significant digits *and replacing the others by zeros*; rounding to k significant figures works similarly. So

$$345678 \quad \text{truncated to 3 significant figures is } 345000$$

and $\qquad 345678 \quad$ rounded to 3 significant figures is 346000.

Working to k significant figures limits the relative error; working to k places after the base point limits the absolute error. ***Truncating is easier but rounding is better.*** // Actually, rounding is only better half the time; // the other half of the time it is truncation.

Sometimes numbers are presented that are clearly approximations, but the precision of the approximation is not made explicit. For instance, the sign outside Hamilton used to say that the population is 306,000. The most natural reading of this is that when the sign was erected, the population rounded to the nearest thousand was 306 thousand. That is, one would assume that there are three significant figures in the sign (306) and three zeros that do not convey precise information but are there to display the size of the population – several thousands. This ambiguity could be removed by using floating point notation where the mantissa contains all the significant figures. // even zeros!

Not all numbers can be represented in my calculator nor displayed. All "representable numbers" are rational, and as we'll show in the next section, the representable numbers are "denser" near zero. Furthermore, most calculators and computers represent numbers in base 2 using a fixed length mantissa and exponent, and this is another source of (unavoidable) approximation error.

1.3.2 Base 2, 8, and 16

Imagine someone buys and sells gold dust by weight, and he has an accurate pan balance and very precisely manufactured standard weights of 1 unit and 2 units. He can use this equipment to weigh out 3 units of dust by putting both standard weights in one pan. The next standard weight he should buy is a 4-unit weight. Then he can weigh 4 units of dust, $5 = 4 + 1$ units, $6 = 4 + 2$, and $7 = 4 + 2 + 1$. The next standard weight he needs is an 8-unit weight. Then he can weigh 8 units of dust, $9 = 8 + 1$ units, $10 = 8 + 2$, and so on up to $15 = 8 + 7 = 8 + (4 + 2 + 1)$. The next standard weight he needs is a 16-unit weight. Then he can weigh 16 units of dust, 17 units, 18 units, and so on up to $31 = 16 + 15$. If we line up the standard weights largest to smallest and write underneath them a 1 if that standard weight is used and a 0 if it's not used to balance some dust, we obtain this table.

16	8	4	2	1	Total
0	0	0	0	0	0
0	0	0	0	1	1
0	0	0	1	0	2
0	0	0	1	1	3
0	0	1	0	0	4
0	0	1	0	1	5
0	0	1	1	0	6
0	0	1	1	1	7
0	1	0	0	0	8
0	1	0	0	1	9
0	1	0	1	0	10
0	1	0	1	1	11
0	1	1	0	0	12
0	1	1	0	1	13
0	1	1	1	0	14
0	1	1	1	1	15
1	0	0	0	0	16
1	0	0	0	1	17
1	0	0	1	0	18
1	0	0	1	1	19
1	0	1	0	0	20

We've seen earlier that every positive integer can be expressed as a sum of distinct powers of 2. That is, every positive integer n can be described as a sequence of 0s and 1s indicating which powers of 2 are used to balance n units of dust.

The **binary system** represents numbers in positional notation using the two digits, 0 and 1. The numeral

$$101\ 111.011\ \underline{means}\ 1 \times 2^5 + 0 \times 2^4 + 1 \times 2^3 + 1 \times 2^2 + 1 \times 2^1 + 1 \times 2^0 + 0 \times 2^{-1}$$
$$+ 1 \times 2^{-2} + 1 \times 2^{-3}.$$

The RHS is known as the *literal expansion* of the LHS. The LHS could be interpreted as a decimal numeral. So, to avoid this ambiguity, we will append the base in braces; that is,

$$101\ 111.011\{2\} = 1 \times 2^5 + 0 \times 2^4 + 1 \times 2^3 + 1 \times 2^2 + 1 \times 2^1 + 1 \times 2^0 + 0 \times 2^{-1}$$
$$+ 1 \times 2^{-2} + 1 \times 2^{-3}\{10\}$$
$$= 32 + 8 + 4 + 2 + 1 + 1/4 + 1/8\{10\}$$
$$= 47 + 0.25 + 0.125\{10\}$$
$$= 47.375\{10\}.$$

The binary system is used in machines because it only requires a physical representation of the two **binary digits** (*bits*): on or off, voltage above or below a certain threshold. (A decimal machine would have to be able to distinguish between ten different levels of some physical phenomenon.) As well, the addition and multiplication tables for binary digits are small (and so simple compared to the base 10 tables we had to memorize). The tables for these operations are these:

$\times$	**0**	**1**
0	0	0
1	0	1

$+$	**0**	**1**
0	0	1
1	1	10

The only complication is that $1 + 1 = 10$ in base 2. Digit by digit addition and multiplication can be done in base 2 the same way it is done in base 10.

There are other similarities with base 10: The base itself is represented by "10" in both cases; multiplying by the base simply *shifts* the base point one space to the right and dividing by the base shifts it one space left. If k is a positive integer, the base raised to the k-th power is a one followed by k zeros ($10^4 = 10\,000$), and the $(-k)$-th power is the base point followed by $(k - 1)$ zeros followed by a one, ($10^{-4} = 0.0001$).

As an example of binary arithmetic, let's look again at the instance of Russian Peasant Multiplication that began this chapter; we wanted to find the product $M \times N$ when $M = 73$ and $N = 41$.

A	B
73	41
~~146~~	20
~~292~~	10
584	5
~~1168~~	2
2336	1
2993	

Each new B-value was obtained by "dividing by 2 and reducing the result by a half if necessary to obtain an integer." When the B-values are written in binary, we divide by the base and truncate to an integer. The effect of this is to chop off the last bit off the binary representation of B.

$B\{10\}$	$B\{2\}$	
41	101001	$// \ 41 = 32 + 8 + 1 = 2^5 + 2^3 + 2^0$
20	10100	$// \ 20 = 16 + 4 \quad\ \ = 2^4 + 2^2$
10	1010	$// \ 10 = \ \ 8 + 2 \quad\ \ = 2^3 + 2^1$
5	101	$// \ \ 5 = \ \ 4 + 1 \quad\ \ = 2^2 + 2^0$
2	10	
1	1	

// <u>Now</u>, do you think the B-value <u>must</u> eventually equal one exactly?
// The number of rows in the table = the number of bits in $N\{2\}$.
// The B-value is even when $B\{2\}$ ends in a zero and is odd when $B\{2\}$ ends in a
// one.

// What about the A-values?

Multiplying by two adds a zero to the binary representation of A.

$A\{10\}$	$A\{2\}$	
73	100 100 1	// $73 - 64 + 8 + 1 = 2^6 + 2^3 + 2^0$
146	100 100 10	
292	100 100 100	
584	100 100 100 0	
1168	100 100 100 00	
2336	100 100 100 000	

If we were to do the multiplication (by the usual algorithm), we would get

73{10}=	1001001		
× 41{10}=	× 1001001		
	1001001	73	// multiply by the digit 1
	~~1001001x~~	~~146~~	// shift and multiply by 0
	~~1001001xx~~	~~292~~	// shift and multiply by 0
	1001001xxx	584	// shift and multiply by 1
	~~1001001xxxx~~	~~1168~~	// shift and multiply by 0
	1001001xxxxx	2336	// shift and multiply by 1
	101110110001	2993	

// 101 110 110 001$\{2\}$ $= 2^{11} + 2^9 + 2^8 + 2^7 + 2^5 + 2^4 + 2^0\{10\}$
// $= 2048 + 512 + 256 + 128 + 32 + 16 + 1\{10\}$
// $= 2993\{10\}$

RPM appears to be just a disguised version of multiplication in base 2.

The disadvantage of using base 2 is that numerals are about 3 times longer and with just two digits are hard for <u>people</u> to read. Is 10010110101010 = 10010110101010?

If b is <u>any</u> integer greater than 1, numbers can be written in positional notation in base b where the digits are single symbols for 0, 1, $\ldots$, $(b - 1)$. Then

$$d_p d_{p-1} \ldots d_0 \cdot d_{-1} \ldots d_{-q} \ \{b\}\underline{means}$$
$$d_p \times b^p + d_{p-1} \times b^{p-1} + \ldots + d_0 \times b^0 + d_{-1} \times b^{-1} + \ldots + d_{-q} \times b^{-q}.$$

Any number written in base b can be converted to base 10 by converting b to base 10, the digits to base 10, and then doing the arithmetic in the literal expansion.

In the **octal** system, the base is **8** and the digits are (single symbols representing) the integers from 0 to 7. Then

$$
\begin{aligned}
502.71\{8\} &= 5 \times 8^2 \ + \ \ 0 \times 8^1 \ \ + \ \ 2 \times 8^0 \ + \ 7 \times 8^{-1} + \ \ 1 \times 8^{-2} \quad\quad \{10\} \\
&= 5 \times 64 \ + \quad\quad\quad\quad\quad 2 \times 1 \ \ + \quad 7/8 \ \ + \quad 1/64 \quad\quad\quad\ \{10\} \\
&= \ \ \ 320 \ \ + \quad\quad\quad\quad\quad\ \ 2 \ \ \ + \ \ 0.875 \ + \ \ 0.015\ 625 \quad \{10\} \\
&= 322.890\ 625 \ \{10\}.
\end{aligned}
$$

In the **hexadecimal** system, the base is **16** and the digits are single symbols representing the integers from 0 to 15. The standard convention is to use 0–9 for the first ten and A, B, C, D, E, and F for 10–15.

Decimal	Binary	Octal	Hexadecimal
0	0	0	0
1	1	1	1
2	10 =b	2	2
3	11	3	3
4	100 $=b^2$	4	4
5	101	5	5
6	110	6	6
7	111	7	7
8	1000 $=b^3$	10 =b	8
9	1001	11	9
10 =b	1010	12	A
11	1011	13	B
12	1100	14	C
13	1101	15	D
14	1110	16	E
15	1111	17	F
16	10000 $=b^4$	20	10 =b
17	10001	21	11
18	10010	22	12
19	10011	23	13
20	10100	24	14

Then

$$
\text{A02.D4}\{16\} = \text{A} \times 16^2 \ \ + 0 \times 16^1 + 2 \times 16^0 + \text{D} \times 16^{-1} + 4 \times 16^{-2}
$$

//using both bases

$$
\begin{aligned}
&= (10) \times 256 + \quad\quad\quad\ 2 \quad\ + (13)/16 \ + 4/256 \quad\quad \{10\} \\
&= 2\ 560 \quad\ + \quad\quad\quad\quad\ 2 \quad\ + 0.812\ 5 \ \ + 0.015\ 625 \quad \{10\} \\
&= 2\ 562.828\ 125 \ \{10\}.
\end{aligned}
$$

Because 8 and 16 are integer powers of 2, conversion from binary to octal and from binary to hexadecimal is easy. The octal digits correspond to the 3-bit binary numbers and the hexadecimal digits correspond to the 4-bit binary numbers.

Group the bits from the base point to the left and then from the base point to the right. The binary number's value is not changed if we add extra zeros at the two ends.

$$\underbrace{101}_{5}\underbrace{000010}_{0}.\underbrace{111}_{7}\underbrace{001}_{1} \ \{2\} = 502.71 \ \{8\}$$

$$\underbrace{0001}_{1}\underbrace{0100}_{4}\underbrace{0010}_{2}.\underbrace{1110}_{E}\underbrace{0100}_{4} \ \{2\} = 142.\text{E}4 \ \{16\}$$

// Are these all equal to 322.890 625 [10]?

To convert from octal to binary, just expand each octal digit as 3 bits; to convert from hexadecimal to binary, just expand each hex digit as 4 bits.

Conversion from base 10 to base b is somewhat more complicated. This may be done using two different algorithms: one to convert integers and a second to convert fractions.

Suppose n is some positive integer. We want to determine the digits of its expansion in base b; that is, we want to determine the d_j's where

$$n = d_p d_{p-1} \ldots d_1 d_0 \{b\} \text{ and each } d_j \text{ is one of } 0, 1, \ldots, (b-1).$$

If

$$n = d_p \times b^p + d_{p-1} \times b^{p-1} + \ldots + d_1 \times b + d_0$$
$$= \left[d_p \times b^{p-1} + d_{p-1} \times b^{p-2} + \ldots + d_1 \right] \times b + d_0$$

then, dividing n by the base b, we get $n = (q) \times b + r$ where $0 <= r < b$. The remainder r must equal d_0 and the integer quotient q must equal

$$d_p \times b^{p-1} + d_{p-1} \times b^{p-2} + \ldots + d_2 \times b + d_1 = d_p d_{p-1} \ldots d_1 \{b\}.$$

Thus, the remainder is the rightmost digit, d_0, and the integer quotient q has the rest of the digits as its expansion in base b. Therefore, **to convert an integer n to base b**, generate the digits one at a time in order (from the base point to the left) by finding the integer quotients and remainders ***until*** the quotient is zero.

// Must $q = 0$ eventually?

Example 1.3.1: Convert 322 from base 10 to base 2, to base 8, and to base 16

$$\begin{array}{ll}
322 = 2(161) + 0 & \text{so } d_0 = 0 \\
161 = 2(\ 80) + 1 & \text{so } d_1 = 1 \\
80 = 2(\ 40) + 0 & \text{so } d_2 = 0 \\
40 = 2(\ 20) + 0 & \text{so } d_3 = 0 \\
20 = 2(\ 10) + 0 & \text{so } d_4 = 0 \\
10 = 2(\ \ 5) + 0 & \text{so } d_5 = 0 \\
5 = 2(\ \ 2) + 1 & \text{so } d_6 = 1 \\
2 = 2(\ \ 1) + 0 & \text{so } d_7 = 0 \\
1 = 2(\ \ 0) + 1 & \text{so } d_8 = 1
\end{array}$$

Therefore, $322\{10\} = 101\,000\,010\,\{2\}$. $// = 2^8 + 2^6 + 2^1 = 256 + 64 + 2$

$$322 = 8(40) + 2 \qquad \text{so } d_0 = 2$$
$$40 = 8(\ 5) + 0 \qquad \text{so } d_1 = 0$$
$$5 = 8(\ 0) + 5 \qquad \text{so } d_2 = 5$$

Therefore, $322\{10\} = 502\,\{8\}$. $// = (5)8^2 + (2)8^0 = 320 + 2$

$$322 = 16(20) + 2 \qquad \text{so } d_0 = 2$$
$$20 = 16(\ 1) + 4 \qquad \text{so } d_1 = 4$$
$$1 = 16(\ 0) + 1 \qquad \text{so } d_2 = 1$$

Therefore, $322\{10\} = 142\,\{16\}$. $// = (1)16^2 + (4)16^1 + (2)16^0 = 256 + 64 + 2$

// How long is $n\{2\}$? How many bits are there in the binary representation of n?
// More generally, how many digits are used when n is written in base b?

If n written in base 2 uses exactly k bits, then

$$2^{k-1} = 1\underbrace{00\ldots0}_{k-1}\{2\} <= n <= \underbrace{111\ldots1}_{k}\{2\} = 2^k - 1.$$

Hence, $2^{k-1} <= n < 2^k$, and so $k - 1 = \lg(2^{k-1}) <= \lg(n) < \lg(2^k) = k$.
Therefore, $k - 1 = \lfloor \lg(n) \rfloor$ and so

$$k = \lfloor \lg(n) \rfloor + 1. \qquad // \text{ and similarly for any base } b?$$

This k is the smallest integer strictly larger than $\lg(n)$.

// What's the complexity of RPM?

Suppose f is some positive *fraction*, so $0 < f < 1$. We want to determine the digits of its expansion in base b; that is, we want to determine the d_{-j}'s where

$$f = .d_{-1}d_{-2}d_{-3}d_{-4}\ldots\{b\} \text{ and each } d_{-j} \text{ is one of } 0, 1, \ldots, (b-1).$$

Multiplying f by the base b, we get   // The base point shifts 1 space right.

$$f \times b = d_{-1}.d_{-2}d_{-3}d_{-4}\ldots\{b\} \text{ where } 0 <= d_{-1} < b. \qquad // d_{-1} <= f \times b < 1 \times b$$
$$= d_{-1} + .d_{-2}d_{-3}d_{-4}\ldots\{b\}.$$

The integer part of $f \times b$ is the leftmost digit, d_{-1}, and the new fraction part has the rest of the digits as its expansion in base b. Thus, **to convert a fraction f to base b**, generate the digits one at a time in order (from the base point to the right) by multiplying by b and finding the integer parts and fraction parts. We can stop if the new fraction is ever equal to zero.   // Will that always happen?

Example 1.3.2: Convert 0.890625 from base 10 to base 2, to base 8, and to base 16

$$
\begin{aligned}
2(.890\ 625) &= 1.781\ 25 &\quad \text{so } d_{-1} &= 1\\
2(.781\ 25\ ) &= 1.562\ 5 &\quad \text{so } d_{-2} &= 1\\
2(.562\ 5\ \ ) &= 1.125 &\quad \text{so } d_{-3} &= 1\\
2(.125\ \ \ \ ) &= 0.25 &\quad \text{so } d_{-4} &= 0\\
2(.25\ \ \ \ \ ) &= 0.5 &\quad \text{so } d_{-5} &= 0\\
2(.5\ \ \ \ \ \ ) &= 1. &\quad \text{so } d_{-6} &= 1
\end{aligned}
$$

Therefore, $0.890\ 625\ \{10\} = 0.111\ 001\ \{2\}$.

$$
\begin{aligned}
8(.890\ 625) &= 7.125 &\quad \text{so } d_{-1} &= 7\\
8(.125\ \ \ \ ) &= 1. &\quad \text{so } d_{-2} &= 1
\end{aligned}
$$

Therefore, $0.890\ 625\ \{10\} = 0.71\ \{8\}$.

$$
\begin{aligned}
16(.890\ 625) &= 14.25 &\quad \text{so } d_{-1} &= \text{E} &\quad \text{// } 14\{10\} = \text{E}\{16\}\\
16(.25\ \ \ \ ) &= 4. &\quad \text{so } d_{-2} &= 4
\end{aligned}
$$

Therefore, $0.890\ 625\ \{10\} = 0.\text{E}4\ \{16\}$.

Example 1.3.3: Convert .7 from base 10 to base 8, to base 16, and to base 2
For base 8:

$$
\begin{aligned}
8(.7) &= 5.6 &\quad \text{so } d_{-1} &= 5\\
\rightarrow\quad 8(.6) &= 4.8 &\quad \text{so } d_{-2} &= 4\\
8(.8) &= 6.4 &\quad \text{so } d_{-3} &= 6\\
8(.4) &= 3.2 &\quad \text{so } d_{-4} &= 3\\
8(.2) &= 1.6 &\quad \text{so } d_{-5} &= 1
\end{aligned}
$$

But now the *new fraction*, .6, has appeared before, and hence the four digits, **4631**, will repeat in a block forever.
Therefore, $0.7\ \{10\} = 0.5\overline{4631}\ \{8\}$.

For base 16:

$$
\begin{aligned}
16(.7) &= 11.2 &\quad \text{so } d_{-1} &= \text{B} &\quad \text{// } 11\{10\} = \text{B}\{16\}\\
\rightarrow\quad 16(.2) &= 3.2 &\quad \text{so } d_{-2} &= 3
\end{aligned}
$$

But the *new fraction*, .2, appeared before as the *old fraction* and hence the hex digit, **3**, will repeat forever.
Therefore, $0.7\ \{10\} = 0.\text{B}\overline{3}\ \{16\}$.

For base 2:

$$2(.7) = 1.4 \qquad \text{so } d_{-1} = 1$$
$$\rightarrow \quad 2(.4) = 0.8 \qquad \text{so } d_{-2} = 0$$
$$2(.8) = 1.6 \qquad \text{so } d_{-3} = 1$$
$$2(.6) = 1.2 \qquad \text{so } d_{-4} = 1$$
$$2(.2) = 0.4 \qquad \text{so } d_{-5} = 0$$

But the *new fraction*, .4, has appeared before, and hence the four digits, **0110**, will repeat in a block forever.

Therefore, $0.7 \{10\} = 0.1\overline{0110} \{2\}$.

Because this conversion process does not stop in some natural way (unlike all the other, previous conversion methods), an artificial termination condition must be imposed. In Example 1.3.3, we stopped when we could express the result exactly using bar notation. Most often the process is terminated after a certain, fixed number of base b digits are found. Sometimes the next base b digit is found and the value is rounded. Conversion (of fractions) to base 2 for internal representation in computers is another source of approximation error.

// When does $1/n$ have a terminating positional representation in base 2?
// Is it only when n has no prime factor other than 2?
// Is that frequent or rare?
// Do almost all rational numbers have an infinite repeating binary representation?

// We asked earlier "Is $10010110101010 = 10010110101010$?" Converting these
// numbers to base 16, that question becomes "Is $25AA = 25AA$?" and is very easy
// to answer.

The Most Important Ideas in This Section.
Computers and calculators represent numbers using positional notation, as k-sequences of digits in some base, where k is a fixed integer. But most numbers cannot be written as such a sequence; most numbers must be approximated. This section shows the difference between *truncation* and *rounding*, between *absolute error* and *relative error*, between rounding to *s significant figures* and to *p places after the base point*, and between representing integers and representing real numbers (in *floating point* notation).

Computers represent numbers using base 2. Four conversion algorithms are given: convert an integer from base b to base 10, convert an integer from base 10 to base b, convert a fraction from base b to base 10, and convert a fraction from base 10 to base b. Conversion of fractions to base 2 for internal representation is another source of approximation error.

1.4 Numerical Solutions

The last section emphasized that arithmetic (especially division) on machines entails approximation errors, and (sometimes) it simply cannot be done (absolutely) accurately. **The *best* we can do is to calculate *approximate* answers.** In this section, we will try to make that a *virtue*.

1.4.1 Newton's Method for Square Roots

To find a "good" approximation of $\sqrt{A}$, when A is a given <u>positive</u> number:
1. Make a *guess* at the square root, x_0.   // some <u>positive</u> number
2. Revise the last *guess*, x_i, to obtain a new *guess*:

$$x_{i+1} = \frac{x_i + A/x_i}{2}. \qquad\qquad \text{// until ?}$$

Let's see what happens when $A = 144$ (so we know the answer) and $x_0 = 10$.

$$x_1 = \frac{10 + 144/10}{2} = \frac{10 + 14.4}{2} = \frac{24.4}{2}$$

$$= 12.2$$

$$x_2 = \frac{12.2 + 144/12.2}{2} = \frac{12.2 + 11.80327869\ldots}{2} = \frac{24.00327869\ldots}{2}$$

$$= 12.001\,639\,344\,2\ldots \qquad \text{// my calculator truncates to 12 digits}$$

$$x_3 = 12.000\,000\,111\,9\ldots$$

$$x_4 = 12$$

then x_5 will also be 12, and *all* other x_i's will be 12. Newton's method found the square root exactly after only 4 revisions of our initial guess of 10.

// Isn't that magical? And worth looking at a bit more closely?
// Would you believe that we only got the right answer <u>because of</u> roundoff error?
// (and that if we did exact arithmetic, we would never reach 12.)

Let E_i denote the error in x_i as an approximation of $\sqrt{A}$; that is,

$$E_i = x_i - \sqrt{A} \text{ and so } x_i = \sqrt{A} + E_i.$$

Then the next error

$$E_{i+1} = x_{i+1} - \sqrt{A} = \frac{x_i + A/x_i}{2} - \sqrt{A} = \frac{x_i + A/x_i - 2\sqrt{A}}{2}$$

$$= \frac{x_i^2 - (2\sqrt{A})x_i + A}{2x_i} = \frac{(x_i - \sqrt{A})^2}{2x_i} = \frac{E_i^2}{2x_i}.$$

After the first guess, the errors are always positive (>0).   // So $x_i > \sqrt{A} > 0$.

The positive errors get smaller and smaller; since $E_i = x_i - \sqrt{A} < x_i$,

$$0 < \frac{E_i}{x_i} < 1 \quad \text{and} \quad E_{i+1} = \frac{E_i^2}{2x_i} = \frac{E_i}{x_i} \times \frac{E_i}{2} < \frac{E_i}{2}$$

The errors that occur when $A = 144$ and $x_0 = 10$ are

$E_0 = x_0 - \sqrt{A} = 10 - 12 = -2$ // we know the exact value of $\sqrt{A}$

$E_1 = \dfrac{E_0^2}{2x_0} = \dfrac{(-2)^2}{2(10)} = \dfrac{4}{20} = 0.2$

$E_2 = \dfrac{E_1^2}{2x_1} = \dfrac{(0.2)^2}{2(12.2)} = \dfrac{0.04}{24.4} = 0.001\ 639\ 344\ 262\ 29 \ldots$

$E_3 = 0.000\ 000\ 111\ 961\ 771\ 811 \ldots = 1.119\ 617\ 718 \times 10^{-7}$

$E_4 = 5.223\ 099\ 263 \times 10^{-16}$ // This shows that there are
                                                // many more "representable"

$E_5 = 1.136\ 698\ 580 \times 10^{-32}$ // numbers near zero than
                                                // near 12 on my calculator.

$E_6 = 5.383\ 681\ 922 \times 10^{-66}$

$E_7 = 0$

// The last five equal signs here should be the symbol "$\cong$" and must be read as
// "appears to be very close to", not "is exactly identical to".
// When $A > 1$, the errors are (more or less) squared at each iteration,
// and the number of correct digits in the x_i (more or less) doubles.

When we don't know the exact value of $\sqrt{A}$ (which is the objective of the algorithm), when should we stop generating the x_i? When is x_i a good enough approximation?

Suppose our supervisor will be satisfied by an approximation z of $\sqrt{A}$ with an absolute error $< 0.000\ 000\ 1$. Let's use δ to denote this bound. We want to generate the x_{i+1} until we're <u>certain</u> that $|E_{i+1}|$ is $< \delta = 0.000\ 000\ 1$. Without knowing the exact value of $\sqrt{A}$, we cannot calculate the exact value of E_{i+1}. We'll have to approximate it and show that whatever its true value is, that value is $< \delta$. We know for $i > 0$

$$0 < E_{i+1} < E_i/2 < E_i \text{ and so } 0 < x_{i+1} - \sqrt{A} < x_i - \sqrt{A}.$$

Therefore, $\sqrt{A} < x_{i+1} < x_i.$

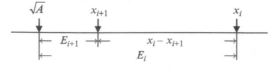

Then

$$E_i = x_i - \sqrt{A} = x_i - x_{i+1} + x_{i+1} - \sqrt{A} = (x_i - x_{i+1}) + E_{i+1} < (x_i - x_{i+1}) + E_i/2.$$

Thus, $\qquad E_{i+1} < E_i/2 < (x_i - x_{i+1}).$

Furthermore, if we generate new guesses, x_{i+1}, until $(x_i - x_{i+1}) < \delta$, that is, until the change between successive guesses is $< \delta$, we will be able to guarantee to our supervisor that this last guess, $z = x_{i+1}$, is an approximation of $\sqrt{A}$ with an absolute error less than $\delta = 0.000\ 000\ 1.$ $\qquad$ // or whatever value of δ she specifies

Newton's method for square roots produces very good approximate answers very quickly. Finding $\sqrt{A}$ is equivalent to solving $x^2 - A = 0$ for a positive "root." The method can be generalized to quickly find a good approximate solution to an equation of the form $f(x) = 0$, but calculus and the derivative of f are required.

On the other hand, we can give a simple, intuitive and efficient algorithm for finding approximate solutions to equations.

1.4.2 The Bisection Algorithm

It would be nice to have a formula to solve equations like

$$x^3 + 2^x = 200.$$

But probably no such formula (for x) exists. However, we can design a simple and efficient algorithm which will find a good, numerical approximation z of an exact solution x^*; a number z that we can guarantee has an absolute error less than any limit δ our supervisor specifies.

Our *strategy* will be to find *an interval that must contain* x^*. Each iteration will determine whether x^* is in the lower half of that interval or the upper half. The original interval is bisected (cut in half) again and again until its length is $< 2\delta$. At that point, the algorithm will return the midpoint of that last interval.

Let $f(x)$ denote the function $x^3 + 2^x$ and let T denote the **target value** 200.

If $x = 5$ then $\qquad x^3 + 2^x = 5^3 + 2^5 = 125 + 32 = 157 < 200,$
and if $x = 10$ then $\qquad x^3 + 2^x = 10^3 + 2^{10} = 1000 + 1024 = 2024 > 200.$

At 5, the function value is too small; at 10, the function value is too big, so for some x-value between 5 and 10, the function value is just right.

$\qquad\qquad\qquad\qquad\qquad\qquad\qquad\qquad\qquad\qquad$ // Goldilock's Theorem?

Let's try halfway between.

If $x = 7.5$ then $\qquad x^3 + 2^x = 7.5^3 + 2^{7.5}$
$$= 421.875 + 181.019\ 336\ldots > 200.$$

So x^* lies between 5 and 7.5. Let's try halfway between again.

If $x = 6.25$ then
$$x^3 + 2^x = 6.25^3 + 2^{6.25}$$
$$= 244.140\ 625 + 76.109\ 255\ldots > 200.$$

So x^* lies between 5 and 6.25. If we make our next guess halfway between these values

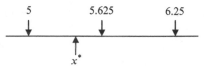

then $|5.625 - x^*| < |5.625 - 5| = 0.625$.

In general, if the function value at A is too small and the function value at B is too big then for some x^* between A and B, the function value is just right. If we let z be the value that is halfway between A and B, then the point z <u>bisects</u> the interval from A to B. If $f(z)$ is too small, then x^* lies between z and B, and on the next iteration we can use z for A. If $f(z)$ is too big, then x^* lies between A and z, and on the next iteration we can use z for B. Also, if we know that x^* lies somewhere between A and B, and z is halfway between A and B, then

$$|z - x^*| < |z - A| = |B - A|/2.$$

On the next iteration, this error bound is halved, and so with enough iterations, it can be made as small as our supervisor requires.

Algorithm 1.4.1: The Bisection Algorithm for Solving $f(x) = T$

```
Begin
    z ← (A + B) / 2;
    While (|z − A| >= δ) Do
        If (f(z)<= T) Then
            A ← z;
        End;
        If (f(z)>= T) Then
            B ← z;
        End;
        z ← (A + B) / 2;
    End;
    Return(z);
End.
```

// This pseudo-code contains 2 *conditional statements* with no **Else** part:
// When the condition between the "**If**" and "**Then**" is true, the steps between
// "**Then**" and "**End**" are done, but when it is false nothing at all is done.
// What would happen if at some point, $f(z) = T$?

Walkthrough with input $f(x) = x^3 + 2^x$, $T = 200$, $A = 5$, $B = 10$, and $\delta = 0.005$:
// Each line in the table corresponds to fixed values of A and B (and z between them).

A	z	B	$\mid z - A \mid$	f(z)
5	7.5	10	2.5	602.894 ...
"	6.25	7.5	1.25	320.249 ...
"	5.625	6.25	.625	227.329 ...
"	5.3125	5.625	.3125	189.672 ...
5.3125	5.46875	"	.15625	207.840 ...
"	5.390625	5.46875	.078125	198.596 ...
5.390625	5.4296875	"	.0390625	203.177 ...
"	5.41015625	5.4296875	.01953125	200.876 ...
"	5.400390625	5.41015625	.009765625	199.733 ...
5.400390625	5.4052734375	"	.0048828125	(200.304 ...)

The algorithm returns $z = 5.405\ 273\ 437\ 5$ as an approximation of the solution of the equation, x^*, <u>and</u> this number has an absolute error that is $< \delta = 0.005$.

// If we continue, we find $5.402\ 668\ 655 < x^* < 5.402\ 668\ 656$,
// but no one can ever know the exact (numerical) value of x^*.

Before this algorithm can be executed, certain "**_preconditions_**" must be met:
1. $f(x)$ must be a *continuous* and *computable* function.
2. A target value T for the function $f(x)$ must be specified
3. An x-value A where $f(A) < T$ must be specified.   // "guessed" somehow
4. An x-value B where $f(B) > T$ must be specified.   // "guessed" too
5. A bound δ on the absolute error in the approximation must be given.

// *Continuity* is a concept from calculus; it ensures f has no sudden "jumps" in
// value so that the Intermediate Value Theorem applies.
// And f must be in a form that can be evaluated fairly accurately despite roundoff
// errors. In practical applications, this precondition is almost always met.

// Very often the target value T for the function is taken to be zero.

// Preconditions 3 and 4 imply that $A \neq B$ and (at least one) *exact* solution x^* is
// between A and B.
// Therefore, either $A < x^* < B$ <u>or</u> $B < x^* < A$.

// δ provides "quality control"; it specifies how "good" an approximation we will
// get and gives a termination criterion for the algorithm.

// But how many iterations will be done?

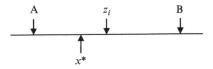

Let A_1 and B_1 denote the input values of A and B and let z_i denote the i-th midpoint calculated. Then

$$|z_1 - x^*| < |B_1 - A_1|/2 = |z_1 - A|$$
$$|z_2 - x^*| < |B_1 - A_1|/4 = |z_2 - A|$$
$$|z_3 - x^*| < |B_1 - A_1|/8 = |z_3 - A|$$

and in general,

$$|z_k - x^*| < |B_1 - A_1|/2^k = |z_k - A|.$$

The algorithm terminates when $|z_k - A| < \delta$. This is sure to happen when

$$|B_1 - A_1|/2^k < \delta \quad \text{or} \quad |B_1 - A_1| < \delta \times 2^k \quad \text{or} \quad |B_1 - A_1|/\delta < 2^k.$$

Taking logs to base 2, we get $\lg(|B_1 - A_1|/\delta) < \lg(2^k) = k$. Therefore, the number of midpoints z_i that the algorithm calculates is (at most)

$$\lfloor \lg(|B_1 - A_1|/\delta) \rfloor + 1 \qquad\qquad \text{// the smallest integer} > \lg(|B_1 - A_1|/\delta)$$

and this will guarantee that

> if the preconditions are met and the algorithm is run, it terminates and then, the following "**post-condition**" holds
>
> the value z returned has $|z - x^*| < \delta$
> where x^* is an (exact) solution of the equation $f(x) = T$.

// This algorithm finds a good, numerical approximation z of an exact solution x^*;
// a number z that we can be certain has an absolute error less than any limit δ
// our supervisor might specify.

When $A_1 = 5$, $B_1 = 10$, and $\delta = 0.005$

$$\lfloor \lg(|B_1 - A_1|/\delta) \rfloor + 1 = \lfloor \lg(|10 - 5|/0.005) \rfloor + 1$$
$$= \lfloor \lg(1000) \rfloor + 1$$
$$= 10.$$

The bisection algorithm is short, simple, effective, and efficient and perhaps the most practical algorithm in this book.

The Most Important Ideas in This Section.
The previous section showed that arithmetic (especially division) on machines causes unavoidable truncation or roundoff errors. The *best* we can do is to calculate *approximate* answers. In this section, we made approximate answers our goal.

(continued)

(continued)

> Newton's method for square roots produces very good approximate answers very quickly. (Finding $\sqrt{A}$ is equivalent to solving $x^2 - A = 0$ for a positive "root." The method can be generalized to quickly find a good approximate solution to an equation of the form $f(x) = 0$ where calculus and the derivative of f are used.)
>
> The (very simple) bisection algorithm for solving $f(x) = T$ was examined, shown to *terminate* after at most $\lfloor \lg(|B_1 - A_1|/\delta) \rfloor$ iterations of the while-loop and shown to *correctly* determine an approximate solution with an absolute error $< \delta$.

Exercises

1. Express 2015 as a sum of distinct powers of 2.
2. Is 83 prime?
3. Find the smallest prime larger than 800.
4. A positive integer, n, has property P if n is equal to the sum of its positive divisors (not including itself). Demonstrate that the following numbers have property P:
   (a) 6
   (b) 28
   (c) 496
   (d) 8128
5. A positive integer, n, has property Q if n is less than the sum of its positive divisors (not including itself). Demonstrate that the following numbers have property Q:
   (a) 12
   (b) 20
   (c) 30
   (d) 36
6. Show that if $n = 3^4 \times 5^4 = 2475$ then n has property Q (defined in the previous question).
7. Demonstrate that for positive integers t and Q,
   $\left(t <= \lfloor \sqrt{Q} \rfloor\right)$ is equivalent to $(t \times t <= Q)$.
   Show that if $\left(t <= \lfloor \sqrt{Q} \rfloor\right)$, then $t \times t <= Q$ and also show that if t is an integer and $t \times t <= Q$, then $\left(t <= \lfloor \sqrt{Q} \rfloor\right)$.
8. Is $n^2 + n + 17$ always prime (when n is a nonnegative integer)?
9. Show that if n is a positive integer and d is the smallest proper divisor of n, then d must be a prime; that is, explain why it is true that:
   If d is a proper divisor of n but d is not prime, then d itself has a proper divisor f which must be a proper divisor of n that is smaller than d.

10. Let K be your birth year. Use Algorithm 1.2.4 to factor K, $K + 1$, and $K + 2$ into primes.

11. (a) Compute $\lg(128)$, $\lg(8,192)$ and $\lg(1,048,576)$.
    (b) Compute $\lfloor \lg(1,000) \rfloor$, $\lfloor \lg(10,000) \rfloor$ and $\lfloor \lg(10,000,000) \rfloor$.
    (c) Compute $\lg(100)$ correct to 4 significant figures.
    (d) Compute $\lg(1,000,000)$ correct to 4 decimal places.

12. If you use your calculator to divide 4678352 by 1974 and it displays 2369.985816:
    (a) Is 4678352 DIV 1974 equal 2369?
    (b) Is 4678352 MOD 1974 equal 1974×0.985816?
    (c) Should 1974×0.985816 equal 4678352 MOD 1974?
    (d) Why is 1974×0.985816 not exactly equal 4678352 MOD 1974?

13. For all positive values of x, $\log_a(x)$ is a constant times $\log_b(x)$, though that constant is often irrational. Find the value of the constant C where $\log_2(x) = C \times \log_{10}(x)$ but round your answer to 6 significant figures.

14. Find the relative error in each of the following approximations expressed as a percentage and rounded to 2 decimal places.
    (a) $A = 2.3456$ is approximated by $A1 = 2.35$.
    (b) $B = 2.3541$ is approximated by $B1 = 2.3$.
    (c) $(A - B)$ is approximated by $(A1 - B1)$.
    (d) Why is the relative error in (c) so large, even though the absolute error is small?

15. Let K be your birth year and let N be your 7-digit phone number. Use Euclid's Algorithm to find GCD(N, K), GCD(N, $K + 1$), and GCD(N, $K + 2$).

16. Use Euclid's Algorithm to find GCD($N + 1$, N) for <u>any</u> positive integer N. Does this show that N and $N + 1$ never have a prime factor in common?

17. Use Euclid's Algorithm to find GCD($2N+1$, $3N+1$) for <u>any</u> positive integer N.

18. Use "long division" to show that $\dfrac{1}{81} = 0.\overline{012345679}$.   // 8 is missing

19. Convert the following numbers from base 2 to base 10:
    (a) 1101{2}
    (b) 10 0110{2}
    (c) 1111 1111{2}

20. Convert the following numbers from base 16 to base 10:
    (a) 1D{16}
    (b) 88{16}
    (c) ABC{16}

21. Convert the following base 10 numbers to base 2, base 8 and base 16:
    (a) 22
    (b) 77
    (c) 105

22. Convert 111001101.1011 from base 2 to base 10.

23. What's the complexity of RPM? (How many times will B be divided by 2?)

24. Convert 1203.201 from base 10 to base 2, but round your answer:

(a) To 6 significant figures and to 12 significant figures.

(b) To 3 places after the base point.

(c) What is the rounding rule for base 2?

25. (a) Use the Fundamental Theorem of Arithmetic to explain why $1/n$ has a terminating positional representation in base b only when every prime factor of n is also a factor of b.

(b) When does $1/n$ have a terminating positional representation in base 2?

(c) When does $1/n$ have a terminating positional representation in base 10?

(d) Do almost all rational numbers have an infinite repeating binary representation?

(e) Do almost all rational numbers have an infinite repeating decimal representation?

26. Use Newton's Method for Square Roots to compute the square root of 157, starting from an initial guess of $x_0 = 78.5$, accurate to 4 decimal places. How many iterations are needed to reach this level of precision?

27. Use the bisection algorithm to find an approximate solution z to the equation $x^{5.3} + (3.5)^x = N$ where N is your 7-digit phone number, and:

(a) z is correct to 2 significant figures.

(b) z is correct to 2 decimal places.

28. Let X^* denote the approximation obtained by rounding X to 2 places after the base point. If $X^* = 45.67$, then $45.665 <= X < 45.675$.

In general, $X^* - 0.005 <= X < X^* + 0.005$.

If $A^* = B^* = z$, then $|B - A| < 0.010$.

Is it true that if $|B - A| <= 0.010$, then $A^* = B^*$?

Or could we have $|B - A| <= 0.000\ 000\ 1$ but $A^* \neq B^*$?

Sets, Sequences, and Counting

2

Sets and sequences are the fundamental objects of study in discrete mathematics, and constructions and enumeration of these are the main elements of combinatorics. Our objectives in this chapter are to give you the basic vocabulary and formulas to describe and count these so we can apply them to analyze the complexity of algorithms.

2.1 Naïve Set Theory

Mathematicians are sometimes very sensitive about the precision of the language used to describe this area of the foundations of mathematics, but we will simply give set theoretic terms their commonly used meanings. Some fuss is often made about a "set" being a "primitive notion" that's not defined itself, but it is used to define other "derived notions". Let's agree that

*A **set** is a well-defined collection of objects called its **elements***

// By "well-defined", we mean that for any object that might possibly be in the set,
// there is a way of deciding whether it is in the set or not.
// We've left "collection" and "object" as primitive (that is, undefined) terms
// (but we hope "you know what I mean").

When S is a set and x is an object,

$$x \in S \quad \underline{means} \quad x \text{ is an element of } S;$$
$$x \notin S \quad \underline{means} \quad x \text{ is not an element of } S.$$

A set then is completely determined by its elements. When a set has only a few elements, it may be possible to list them all. Such an explicit listing of elements places them between a matching pair of braces; for example,

© Springer International Publishing AG, part of Springer Nature 2018
T. Jenkyns and B. Stephenson, *Fundamentals of Discrete Math for Computer Science: A Problem-Solving Primer*, Undergraduate Topics in Computer Science,
https://doi.org/10.1007/978-3-319-70151-6_2

$B = \{0,1\}$ or $H = \{0,1,2,3,4,5,6,7,8,9,A,B,C,D,E,F\}$ or $A = \{a,b,c,\ldots,z\}$.

The three dots in the list of A form an "ellipsis"; the reader is supposed to understand that the list goes "and so on up to z" and that the set A is "the lower case Roman alphabet from a to z". Using ellipses, we can define three important sets of numbers:

The *positive* integers $\mathbf{P} = \{1,2,3,\ldots\}$ // All "whole" numbers > 0

The *nonnegative* integers $\mathbf{N} = \{0,1,2,3,\ldots\}$ // Zero is not negative

and   all *integers*, $\mathbf{Z} = \{\ldots,-3,-2,-1,0,1,2,3,\ldots\}$.

// $\mathbf{Z}$ is commonly used to denote the integers − from Zahl − the German word for
// number.

If listing all the elements, even with ellipses, is not convenient or not possible, a set may be specified by giving a list of properties that determine if an object is in the set or not. For instance,

$$W = \{x: x \in \mathbf{Z}, 0 < x <= 99 \text{ and } 3|x\}$$

means   "W is the set of all objects x **such that** x is an integer **and**
            $0 < x$ **and** $x <= 99$ **and also** 3 divides evenly into x".

// The colon is read "such that" and the comma is read "and".

Thus, W is the set of positive multiples of 3 up to $99 = 3 \times 33$.

Recall that a ***rational number*** is any real number that can be expressed as a *ratio* or a *quotient* of integers. The set of rational numbers,

$$\mathbf{Q} = \{x: x \in \mathbf{R}, x = p/q \text{ where } p, q \in \mathbf{Z}, \text{but } q \neq 0\}   // Q \text{ is for quotients.}$$

// But remember that $2/3 = 6/9 = 400/600 = -8/(-12) = \ldots$. This single (real)
// number can be expressed as a quotient of integers in many, many ways.

We might also define the set

$$D = \{x: x \text{ is a number that can be displayed on my calculator}\}.$$

Then D contains some but not all the elements of $\mathbf{Q}$, but $\mathbf{Q}$ contains every element of D.

Two sets are ***equal*** (or perhaps, two descriptions of a set are equivalent) when they have the same elements.

$\{0,1,4,9,4,1,0,0,9,1\} = \{0,1,4,9\}$ // Repeating elements in the list does not
                                          // change <u>which</u> elements are in the set;
                                          // repetitions are redundant and can be removed.

$\{4,0,9,1\} = \{0,1,4,9\}$ // Changing the order of the elements in the list
                                           // does not change <u>which</u> elements are in the set.
                                           // The <u>order</u> of the elements <u>doesn't matter</u>, so
                                           // <u>we can pick</u> any convenient order for the list.

We could also say $\{0,1,4,9\} = \{x^2: x \in \mathbf{N}, x < 4\}$ and $W = \{3k: k = 1,2,\ldots,33\}$.
If A and B are sets,

> A is a **subset** of B [written $A \subseteq B$] <u>means</u>
>
> every element of A is also an element of B.

So $D \subseteq \mathbf{Q}$ and also $\mathbf{P} \subseteq \mathbf{N} \subseteq \mathbf{Z} \subseteq \mathbf{Q} \subseteq \mathbf{R}$. If every element of A is an element of B
<u>and</u> every element of B is an element of A, then A and B have the same elements;
that is,

> if $A \subseteq B$ and $B \subseteq A$,   then $A = B$.

Every set is a subset of itself. On the other hand,

> A is a **proper subset** of B [written $A \subset B$] <u>means</u>
>
> A is a subset of B but $A \neq B$.

Then, $D \subset \mathbf{Q}$ and also $\mathbf{P} \subset \mathbf{N} \subset \mathbf{Z} \subset \mathbf{Q} \subset \mathbf{R}$.
    It will be useful to be able to speak of the set that has no elements at all, the
empty set, which we'll denote with the symbol $\varnothing$. Now, if B is some specific given
set, is $\varnothing$ a subset of B? Is <u>every</u> element of $\varnothing$ also an element of B? If that were not
True, then some element of $\varnothing$ would not be in B, but that cannot happen because
there are no elements in $\varnothing$. Therefore,

> $\varnothing \subseteq B$ for any set B.

// Aristotle (384–322 BC) seems to have thought that it is nonsense to talk of
// things that don't exist (like elements of the empty set) and attribute properties to
// them (like membership in set B). But the modern use of the empty set does
// make sense, and it is very useful; Aristotle has been superseded.

    If A is any set, the **power set** of A,

$$\mathscr{P}(A) = \{S: S \subseteq A\}. \qquad \text{// the set of all subset of } A$$

For example, if $A = \{a,b,c\}$ then

$$\mathscr{P}(A) = \{\varnothing, \{a\},\{b\},\{c\}, \{a,b\},\{a,c\},\{b,c\}, \{a,b,c\}\}.$$

That is, A contains 8 subsets: 1 of size 0, 3 of size 1, 3 of size 2, and 1 of size 3. So sets are objects and may be elements of other sets.

// Could a set be an element of itself?

There definitely are sets that are not elements of themselves. For instance,

$$\varnothing \notin \varnothing \quad \text{// because } \varnothing \text{ has no elements}$$

and $\quad \mathbf{Z} \notin \mathbf{Z}. \quad$ // because the elements of $\mathbf{Z}$ are finite numbers

// and $\mathbf{Z}$ itself is an infinite set.

It is said that when Bertrand Russell (1872–1970) was a student, he read a description of set theory as a basis for the foundations of mathematics by Gottlob Frege (1848–1925) much like the last few pages here and wondered about the set

$$K = \{x: x \notin x\},$$

the set of all sets that are not elements of themselves. This set is paradoxical because

if $K \in K$, then (K must satisfy the condition for membership in K so) $K \notin K$,

but   if $K \notin K$, then (K does satisfy the condition for membership in K so) $K \in K$.

Russell's Paradox may be removed by constructing a much more "sophisticated" set theory (Zermelo-Fraenkel set theory) that assigns types to the objects, classes, and sets or that restricts objects to a fixed "universe of discourse". But let's just not worry about the paradox hoping it will never cause us problems like

2.1.1 The Diabolical Librarian

Suppose a certain library contains a large but fixed collection of books. Each book has a "title" that's written on its cover, and each book has a "text" that's written on its pages. Often the title occurs in the text. Sometimes it is many pages into the text that you find its title, as in "Silence of the Lambs" or "Catcher in the Rye".

This library also has one particular book whose title is "The Special Catalogue" (written on its cover) but whose pages are blank. The librarian offers you a substantial amount of money to write the text for The Special Catalogue. He wants to know which books in the collection do <u>not</u> have their title somewhere in their text. He wants you to list, on the pages of The Special Catalogue, the titles of <u>all</u> the books in the library that do not have their title somewhere in their text, but you must <u>only</u> include the titles of books in the library that do not have their title in their text.

You work diligently in the library over the summer, and with some help from your friends, you complete the task. Then you go to present the librarian with the finished Special Catalogue and to collect your payment.

Before giving you the money, he asks you about "The Special Catalogue" itself. It is a book in the library, it has a title and a text; does its title occur in its text? If you answer "no", he'll say you haven't finished the job and refuse to pay you; if you say "yes", he'll say you haven't done the job correctly and refuse to pay you.

2.1.2 Operations on Sets and Cardinality

If A and B are sets, several other sets can be constructed from them:

the ***intersection*** of A and B, $A \cap B = \{x: x \in A \text{ and } x \in B\}$;
       // $B \cap A = A \cap B$.

the ***union*** of A and B, $A \cup B = \{x: x \in A \text{ or } x \in B\}$;
       // $B \cup A = A \cup B$.

and   the ***set difference***, A but not B, $A \setminus B = \{x: x \in A \text{ and } x \notin B\}$.
       // Is $B \setminus A = A \setminus B$?

// The set $A \setminus B$ is sometimes called the "relative complement" of B in A.

When $A \cap B = \emptyset$, sets A and B are said to be ***disjoint***.
       // A and B have no common element.

The number of elements in a set S is called the ***cardinality*** of S and denoted by $|S|$. When this is a finite number, then $|S| \in \mathbf{N}$, and when $|S| = n$, we'll say that S is an ***n-set***. For any pair of sets,

$$|A \cup B| = |A| + |B| - |A \cap B|,$$

and when A and B are disjoint,

$$|A \cup B| = |A| + |B|. \qquad\qquad \text{// since } A \cap B = \emptyset$$

Furthermore, we always have

$$|A \cup B| = |A \setminus B| + |B \setminus A| + |A \cap B|.$$

Example 2.1.1: Operations, Sizes, and Subsets

Suppose A is the set of odd integers less than 10 and B is the set of primes less than 10. Then

$$A = \{1,3,5,7,9\} \quad \text{and} \quad B = \{2,3,5,7\}$$
$$A \cap B = \{3,5,7\} \quad \text{and} \quad A \cup B = \{1,2,3,5,7,9\}$$
$$A \setminus B = \{1,9\} \quad \text{and} \quad B \setminus A = \{2\};$$

$$6 = |A \cup B| = |A| \quad + |B| \quad - |A \cap B| = 5 + 4 - 3$$
$$= |A \setminus B| + |B \setminus A| + |A \cap B| = 2 + 1 + 3.$$

Each element of $A \cup B$ is in exactly one of the sets $A \setminus B$, $B \setminus A$, and $A \cap B$. More generally,

> subsets S_1, S_2, S_3, … S_k of T form a ***partition*** of T *means*
> every element of T belongs to exactly one of the sets S_j.

// The sets $S_1 = A \setminus B$, $S_2 = B \setminus A$ and $S_3 = A \cap B$ form a partition of $T = A \cup B$.
// In general, $S_1 \cup S_2 \cup S_3 \cup \ldots \cup S_k \subseteq T$ because each S_j is a subset of T.
// $T \subseteq S_1 \cup S_2 \cup S_3 \cup \ldots \cup S_k$ because each element of T is in some subset S_j.
// Therefore, $T = S_1 \cup S_2 \cup S_3 \cup \ldots \cup S_k$.

The subsets in a partition are *mutually disjoint*; that is, any two are disjoint sets.

// If $p \neq q$, $S_p \cap S_q = \emptyset$ because no element (of T) belongs to more than one S_j.

When S_1, S_2, S_3, … S_k forms a partition of T, then

$$|T| = |S_1| + |S_2| + |S_3| + \ldots + |S_k|.$$

The ***Cartesian product*** of sets A and B, named for René Descartes (1596–1650), is

$$A \times B = \{(a,b): a \in A \text{ and } b \in B\},$$

where (a,b) denotes an *ordered pair* of objects; there is a *first* entry and a *second* entry in each ordered pair. *Parentheses indicate that order matters.*
                      // Braces indicate it doesn't.

// $\{0, 1\} = \{1, 0\}$, but $(0, 1) \neq (1, 0)$ – in sets, order doesn't matter; in ordered pairs
// it does.

// $\{1, 1\} = \{1\}$, but $(1, 1) \neq (1)$ – in sets, repetitions don't matter; in ordered pairs
// they do.

Example 2.1.2: Two Cartesian Products
   If $A = \{1,3,5,7\}$ and $B = \{2,3,5\}$, then

$$A \times B = \{(1,2),(1,3),(1,5),(3,2),(3,3),(3,5),(5,2),(5,3),(5,5),(7,2),(7,3),(7,5)\}$$

and

$$B \times A = \{(2,1),(2,3),(2,5),(2,7),(3,1),(3,3),(3,5),(3,7),(5,1),(5,3),(5,5),(5,7)\}.$$

// $(A \times B) \cap (B \times A) = \{(3,3),(3,5),(5,3),(5,5)\}$, so $A \times B \neq B \times A$.

In this example, both $|A \times B| = 12$ and $|B \times A| = 12$. For any object z,

$$\{z\} \times B = \{(z,2),(z,3),(z,5)\} \text{ and so } |\{z\} \times B| = |B|.$$

Since $S_1 = \{1\} \times B$, $S_2 = \{3\} \times B$, $S_3 = \{5\} \times B$, and $S_4 = \{7\} \times B$ form a partition of $A \times B$,

$$|A \times B| = |\{1\} \times B| + |\{3\} \times B| + |\{5\} \times B| + |\{7\} \times B| = |B| + |B| + |B| + |B|$$
$$= |A| \times |B|$$

This formula applies in general, for all sets A and B // even when $A = B$?

$$|A \times B| = |A| \times |B|$$ // which equals $|B| \times |A| = |B \times A|$

This ***product rule for counting*** is often given without reference to a Cartesian product of sets as

> *if a first thing can be done in p different ways, and (no matter how it was done)*
> *a second thing can be done in q different ways, then*
> *the two things can be done (together) in p $\times$ q different ways.*

2.1.3   The Pigeonhole Principle

Suppose that 27 pigeons fly into a coop (an apartment house for pigeons) with 5 pigeonholes (apartments). What's the maximum number M that could fly into any single pigeonhole? It's clear that $M <= 27$. But is there a lower bound on M? If the holes were numbered from 1 to 5, and X_i is the number that fly into hole number i, then

$$27 = X_1 + X_2 + X_3 + X_4 + X_5 <= M + M + M + M + M = 5M.$$

Therefore, $M >= 27/5 = 5.4.$ // which is the average per pigeonhole

In fact, $M >= \lceil 27/5 \rceil = 6.$ // because M is an integer

The ***pigeonhole principle*** asserts that

> *If P pigeons fly into H pigeonholes, then (at least) one pigeonhole*
> *contains (at least) $\lceil P/H \rceil$ pigeons.*

Example 2.1.3: How Many Socks Are Enough?

Suppose that while you're in the bath you ask your color-blind roommate to open your sock drawer (where there are 10 red socks, 12 blue socks, and 9 black socks) and take out a bunch of socks so you can find a matching pair. How many socks must he select to guarantee that there is a matching pair in his selection? All 31? How *few* socks can he select and still guarantee that there is a matching pair in his selection?

// Let the colors be the pigeonholes; now find the smallest P so $\lceil P/3 \rceil >= 2$.

The pigeonhole principle is really a theorem about partitions.

// The pigeonholes divide the set of pigeons into mutually disjoint subsets.

If S_1, S_2, S_3, ... S_k forms a **partition** of an n-set T, then

$$n = |T| = |S_1| + |S_2| + |S_3| + \ldots + |S_k|.$$

Hence,   the average size of a subset S_j is n/k;   // not necessarily an integer

the largest size of a subset S_j is at least $\lceil n/k \rceil$,   // and at most n

and   the smallest size of a subset S_j is at most $\lfloor n/k \rfloor$.   // and at least 0.

// Not all the subsets can be smaller than average, and not all can be bigger than
// average. In fact, either all the subsets have the same size, or the largest is larger
// than average, <u>and</u> the smallest is smaller than average.

The Most Important Ideas in This Section.

A *set* is a well-defined collection of *elements*. Some important sets of numbers are $\mathbf{P} = \{1,2,3, \ldots \}$, $\mathbf{N} = \{0,1,2,3, \ldots \}$, $\mathbf{Z} = \{\ldots,-3,-2,-1,0,1, 2,3,\ldots \}$, the real numbers $\mathbf{R}$, and the *rational numbers* $\mathbf{Q} = \{x\colon x \in \mathbf{R}, x = p/q$ where p, $q \in \mathbf{Z}$, but $q \neq 0\}$.

A is a *subset* of B [written $A \subseteq B$] when every element of A is also an element of B. The *empty set* $\varnothing \subseteq B$ for any set B. The *power set* of B, $\mathscr{P}(B) = \{S\colon S \subseteq B\}$.

Russell's Paradox seems inherent in this description of sets and can produce impossible tasks like the diabolical librarian did.

If A and B are sets, other sets can be constructed from them: the *intersection* $A \cap B = \{x\colon x \in A$ and $x \in B\}$, the *union* $A \cup B = \{x\colon x \in A$ or $x \in B\}$, and the *set difference* $A \setminus B = \{x\colon x \in A$ and $x \notin B\}$. When $A \cap B = \varnothing$, sets A and B are *disjoint*. Sets S_1, S_2, S_3, ... S_k form a *partition* of set T if every element of T belongs to exactly one of the sets S_j. The sets in a partition are *mutually disjoint*.

The number of elements in a set S is called the *cardinality* of S and denoted by $|S|$. When S_1, S_2, S_3, ... S_k forms a partition of T, then

$$|T| = |S_1| + |S_2| + |S_3| + \ldots + |S_k|.$$

The *Cartesian product* of sets A and B, $A \times B = \{(a,b)\colon a \in A$ and $b \in B\}$ where (a,b) denotes an *ordered pair* of objects. The *product rule* for counting is *if a first thing can be done in p different ways, and (no matter how it was done) a second thing can be done in q different ways, then the two things can be done (together) in $p \times q$ different ways.*

The *pigeonhole principle* is *if P pigeons fly into H pigeonholes, then some pigeonhole contains (at least) $\lceil P/H \rceil$ pigeons.*

All these definitions will aid our analysis of algorithms, particularly our counts of operations.

2.2 Sequences

Sequences are special functions. A function f is a "rule" that *takes* an object x from a (nonempty) set called the "domain of f" and *makes* a (unique) object $f(x)$ in a set called the "codomain of f." More formally,

> f is a ***function from*** set $D \neq \varnothing$ ***into*** set C [written $f:D \to C$] *means*
>
> $f \subseteq D \times C$ where each $x \in D$ occurs in exactly one ordered pair of f.

As a set of ordered pairs, $f = \{(x, f(x)): x \in D\}$.

> A function $f:D \to C$ is **one-to-one** *means*
>
> if x and y are in D and $x \neq y$, then $f(x) \neq f(y)$,
>
> // One-to-one functions are also known as an "*injections*".

and   a function $f:D \to C$ is **onto** *means*
for every $z \in C$, there is (at least) one $w \in D$ such that $f(w) = z$.

> // Onto functions are also known as an "*surjections*".

Two sets, A and B, have the ***same cardinality*** (size) *means* there is a function $f:A \to B$ that is one-to-one and onto.

> // Such functions are also known as an "*bijections*".

In particular, if X is an n-set, there is a one-to-one *indexing function* from $\{1, 2, \ldots, n\}$ onto X, and using this function, the elements of X can be listed (in order) as $\{x_1, x_2, \ldots, x_n\}$.

Example 2.2.1: Bums on Seats

To determine whether or not the number of students in this classroom equals the number of seats, I could ask each student to sit on a seat. If no two (different) students sat on the same seat (at the same time) so each student got their own seat, then the number of students is not more than the number of seats. And, if no seats were left vacant, then the number of students is not less than the number of seats.

This description of "equal cardinality" is due to Georg Cantor (1845–1918), and while it agrees with our intuition about finite sets, it revolutionized (mathematicians') ideas about infinite sets in part because it implies there are several gradations of infinity.   // There's even an infinite number of bigger and bigger infinities!

If a and b are integers, let

> $\{a.. \}$ denote the set $\{x \in \mathbf{Z}: a <= x\}$

and $\{a..b\}$ denote the set $\{x \in \mathbf{Z}: a <= x \text{ and } x <= b\}$.

> // So $\{a..b\} = \varnothing$ if $b < a$.

An *interval of integers* is defined to be any subset of **Z** that is of one of those two types. The first type is an *infinite* interval; the second type is a *finite* interval where

$$|\{a,b\}| = b - a + 1 \text{ when } a <= b.$$

// $\{a..b\} = \{a+0, a+1, a+2, \ldots, a+(b-a)\}$ and $|\{0,1,2,\ldots,n\}| = n + 1.$

A *sequence* is defined to be a function S whose domain D is a nonempty interval of integers. S is an *infinite* sequence if D has the form $\{a..\}$.

// Usually a is 1 or 0.

S is a *finite* sequence if D has the form $\{a..b\}$ where $a <= b$. When $|D| = n$, we will say that S is an *n-sequence*. We will take the domain of an n-sequence to be the set $\{1..n\}$.   // But a could be 0, and then D is $\{0..(n-1)\}$.

The (natural) ordering of the domain of a sequence S gives a natural ordering to the ordered pairs in the set S. If S is a 5-sequence, then

$$S = \{(1, S(1)), (2, S(2)), (3, S(3)), (4, S(4)), (5, S(5))\}.$$

But this is just an awful, clumsy notation to represent S. However, it makes it clear that S has a first ordered pair, a second ordered pair, a third, and so on. Instead, we'll use the conventional notation and write

$$S = (S_1, S_2, S_3, S_4, S_5) \text{ where } S_i \text{ means } S(i).   \text{// the function value at } i$$

Parentheses again indicate that order matters; in sequences, the order of the entries is the fundamental characteristic.

Example 2.2.2: What's a Sequence?

Suppose $D = \{1..10\}$, and we define the function S on D by

$$S(i) = \text{the smallest prime factor of the integer } (1+i).$$

Then D is a finite interval of integers, and so S is the sequence denoted by

$$S = (2, 3, 2, 5, 2, 7, 2, 3, 2, 11).$$

// *In sequences, order matters and repetitions also matter.*

If $S = (S_1, S_2, S_3, \ldots, S_n)$ is a finite sequence of <u>numbers</u>, the corresponding *series* is the sum of the entries in S,

$$S_1 + S_2 + S_3 + \ldots + S_n.$$

There is a compact notation for series, called *sigma notation* where

$$\sum_{i=1}^{n} S_i \quad \underline{means} \quad \text{the sum of the values of the } S_i \text{ as } i \text{ goes from 1 up to } n.$$

// "Σ" is the Greek capital letter "sigma"; it became the letter S in the Latin
// alphabet and stands for "sum" – the output from addition.

It is sometimes useful to generalize sigma notation by allowing other lower limits
and other upper limits,

$$\sum_{i=a}^{b} S_i \quad \underline{means} \quad S_a + S_{a+1} + S_{a+2} + \ldots + S_b. \qquad \text{// when } a <= b$$

// This sum is given a "default" value of zero when $a > b$, and there are no terms
// added.

2.2.1 The Characteristic Sequence of a Subset

Suppose that U is some given n-set whose elements have been *indexed* (listed in a
certain order) so that $U = \{x_1, x_2, \ldots, x_n\}$. If A is a subset of U, the **characteristic
sequence** of A is the function whose domain is $\{1..n\}$ defined by

$$X_i^A = X^A(i) = \begin{cases} 1 & if \quad x_i \in A \\ 0 & if \quad x_i \notin A \end{cases}.$$

Example 2.2.3: Characteristic Sequences

If U is the set of the first 10 odd positive integers, A is the subset of primes in U,
and B is the set of multiples of 3 in U, then

```
U  ={1, 3, 5, 7, 9, 11, 13, 15, 17, 19}        // x_i = 2i − 1.
A  ={    3, 5, 7,    11, 13,     17, 19}
B  ={    3,       9,        15        }
X^A = (0, 1, 1, 1, 0,  1,  1,  0,  1,  1)
X^B = (0, 1, 0, 0, 1,  0,  0,  1,  0,  0).
```

Characteristic sequences may be used as an implementation model for subsets of
any given indexed set U. The set operations may be done on these sequences:

$$X^{A \cap B}(i) = X^A(i) \times X^B(i);$$

$$X^{A \cup B}(i) = X^A(i) + X^B(i) - X^A(i) \times X^B(i);$$

$$X^{A \setminus B}(i) = X^A(i) - X^A(i) \times X^B(i).$$

If $A \subseteq B$ then $\qquad X^A(i) <= X^B(i) \qquad$ for each index i,

and $\qquad\qquad\qquad |A| = \sum_{i=1}^{n} X_i^A.$

The Most Important Ideas in This Section.
f is a *function from* set $D \neq \emptyset$ *into* set C [written $f{:}D \rightarrow C$] when $f \subseteq D \times C$, and each $x \in D$ occurs in exactly one ordered pair of f; f is *one-to-one* means each element in the domain of the function maps to a different element in the codomain; f is *onto* means every element in the codomain can be generated by the function.

Two sets, A and B, have the *same cardinality* if and only if there is a function $f{:}A \rightarrow B$ that is one-to-one and onto.

If a and b are integers, $\{a.. \}$ denotes the set $\{x \in \mathbf{Z}: a <= x\}$, and $\{a..b\}$ denotes the set $\{x \in \mathbf{Z}: a <= x \text{ and } x <= b\}$. An *interval of integers* is a subset of $\mathbf{Z}$ that is of one of those two types. A *sequence* is a function S whose domain D is a nonempty interval of integers; S is an *infinite* sequence if D has the form $\{a.. \}$, and S is a *finite* sequence if D has the form $\{a..b\}$ where $a <= b$. When $|D| = n$, we will say that S is an *n-sequence*. Unlike sets, *in sequences, order matters and repetitions also matter.*

If $S = (S_1, S_2, S_3, \ldots, S_n)$ is a finite sequence of numbers, the corresponding *series* is the sum of the entries in S,

$$\sum_{i=1}^{n} S_i = S_1 + S_2 + S_3 + \ldots + S_n$$

When $U = \{x_1, x_2, \ldots, x_n\}$ and A is a subset of U, the *characteristic sequence* of A is the function whose domain is $\{1..n\}$ defined by

$$X_i^A = X^A(i) = \begin{cases} 1 & if \quad x_i \in A \\ 0 & if \quad x_i \notin A \end{cases}.$$

All these definitions will aid our analysis of algorithms, particularly our counts of operations.

2.3   Counting

Let's determine the number of telephone numbers, sequences that are composed of a 3-digit "area code", followed by a 3-digit "exchange", followed by 4 more digits. But we'll assume that these telephones are in some system where if you dial a 0, you get connected to an "operator", and if you dial a 1, you get connected to a "long-distance line", and there are a few other restrictions.

Suppose that an area code cannot begin with a 0 or a 1 and must have a 0 or a 1 as the middle digit. That is, an area code is a sequence (x,y,z) where

$$x \in \{2, 3, 4, 5, 6, 7, 8, 9\} \text{ and } y \in \{0, 1\} \text{ and } z \in \{0, 1, 2, 3, 4, 5, 6, 7, 8, 9\}.$$

There are 8 possible values for x, and after x is chosen, there are 2 possible values for y, so the number of possible choices for x and then y is $8 \times 2 = 16$. After both x and y have been chosen, there are 10 possible values for z, so the number of possible choices for x and y and then z is $16 \times 10 = 160$. The number of area codes in this system is 160.

Suppose also that an exchange cannot begin with a 0 or a 1 and cannot be mistaken for an area code because the middle digit must not be a 0 or a 1. That is, an exchange is a sequence (x,y,z) where

$$x \text{ and } y \in \{2,3,4,5,6,7,8,9\} \text{ and } z \in \{0,1,2,3,4,5,6,7,8,9\}.$$

There are 8 possible values for x, and after x is chosen, there are 8 possible values for y, so the number of possible choices for x and then y is $8 \times 8 = 64$. After both x and y have been chosen, there are 10 possible values for z, so the number of possible choices for x and y and then z is $64 \times 10 = 640$. The number of exchanges in this system is 640.

Finally, the number of 4-digit sequences that end the number is the number of sequences, $S = (w,x,y,z)$ where

$$w,x,y \text{ and } z \in \{0,1,2,3,4,5,6,7,8,9\}.$$

There are 10 possible values for w, and after w is chosen, there are 10 possible values for x, and after both w and x have been chosen, there are 10 possible values for y, and after all of w and x and y have been chosen, there are 10 possible values for z, so the number of possible choices for constructing the sequence S is $10 \times 10 \times 10 \times 10 = 10000$.

Hence, using the product rule, the number of telephone numbers is

$$\text{(the number of area codes)} \times \text{(the number of exchanges)} \times (10000)$$
$$= (160)(640)(10000) = 1\,024\,000\,000.$$

2.3.1 Number of k-Sequences on an n-Set

When the codomain of a sequence S is the set C, we say that S is a *sequence on* C. If both k and n are positive integers, then a *k-sequence on an n-set* is a function S from $\{1..k\}$ into some set $X = \{x_1, x_2, \ldots, x_n\}$ with exactly n elements, and we may write S as

$$S = (s_1, s_2, s_3, \ldots, s_k) \text{ where each } s_j \in X.$$

We can count these like we did the telephone numbers using the product rule. There are n possible choices for s_1.

After s_1 is chosen, there are n possible choices for s_2.

After s_1 and s_2 are chosen, there are n possible choices for s_3, and so on. Since there are n possible choices for each of the k entries in S,

the number of k-sequences on an n-set is $n \times n \times \ldots \times n = \mathbf{n^k}$.

In particular, the number of k-sequences on $\{0,1\}$ is 2^k.

2.3.2 Number of Subsets of an *n*-Set

If X is a set of size n, the elements of X may be indexed so that $X = \{x_1, x_2, \ldots, x_n\}$. Every subset of X has a unique characteristic sequence, and every characteristic sequence corresponds to a unique subset. The number of characteristic sequences is 2^n, so

the number of subsets of an n-set is $\mathbf{2^n}$.   // This formula even
                                                              // applies when $n = 0$.

2.3.3 Number of *k*-Permutations on an *n*-Set

Permutations may be defined in a number of ways, but in this book,

a ***permutation*** *is a sequence without repetitions.*

Then a *k-permutation on an n-set* is a function S from $\{1..k\}$ into some set $X = \{x_1, x_2, \ldots, x_n\}$ with exactly n elements that is ***one-to-one***; that is, if $i \neq j$, then $S(i) \neq S(j)$. We may write S as

$S = (s_1, s_2, s_3, \ldots, s_k)$ where the s_j's are distinct elements of X.

If $k > n$ (by the pigeonhole principle), at least two of the s_j's must be the same in any k-sequence S on an n-set. So there are no (zero) k-permutations on an n-set when $k > n$.

If $1 <= k <= n$, we can count these like we did sequences using the product rule. There are n possible choices for s_1.

After s_1 is chosen, there are $(n - 1)$ possible choices for s_2.

                                                              // s_2 cannot be the same as s_1.

After s_1 and s_2 are chosen, there are $(n - 2)$ possible choices for s_3.
$\qquad$ // s_3 cannot be s_1 or s_2, and so on.
After s_1 up to s_j have been chosen, there are $(n - j)$ possible choices for s_{j+1}.
$\qquad$ // and so on until
After s_1 up to s_{k-1} have been chosen and then there are $(n - [k - 1])$ possible
choices for s_k. $\qquad$ // since $(n - [k - 1]) = n - k + 1$
Thus, if $1 <= k <= n$,

$$\text{the number of k-permutations on an n-set is}$$
$$\boldsymbol{n \times (n - 1) \times (n - 2) \times \ldots \times (n - k + 1).}$$

The number of 4-permutations on an 8-set is $8 \times 7 \times 6 \times 5 = 1680$, but the number of 8-permutations on an 4-set is $4 \times 3 \times 2 \times 1 \times 0 \times (-1) \times (-2) \times (-3) = 0$. The formula correctly counts the number of k-permutations on an n-set for all positive integers, k and n.

2.3.4 *n*-Factorial

When $k = n$, each k-permutation uses *all* of the elements of the n-set X exactly once; we'll call these *full permutations* of X. The number of full permutations of an n-set is

$$n \times (n - 1) \times (n - 2) \times \ldots \times (2) \times (1).$$

It will be useful to have a notation for this decreasing product. The function ***n-factorial*** [written $n!$] is defined on **N** by $\qquad$ // So $n!$ is an infinite sequence.

$$n! = \begin{cases} n(n - 1)\ldots(2)(1) & \text{if } n > 0 \\ 1 & \text{if } n = 0 \end{cases}$$

// It may seem strange to define 0! at all, but it will be very convenient to have a
// value for 0! equal 1. (All definitions are supposed to facilitate communication.)
// Note also that $\qquad$ if $n >= 1$, then $n! = n \times (n - 1)!$
// $\qquad\qquad\qquad$ if $n >= 2$, then $n! = n \times (n - 1)(n - 2)!$
// $\qquad\qquad\qquad$ if $n >= 3$, then $n! = n \times (n - 1)(n - 2)(n - 3)!$

In general, if $n >= k > 0$, then $n! = n \times (n - 1) \times (n - 2) \times \ldots \times (n - k + 1) \times (n - k)!$. Therefore, the number of k-permutations on an n-set is also given by the expression

$$\frac{n!}{(n - k)!}. \qquad \text{// even when } n = k \text{ and we divide by 0!}$$

The function $n!$ grows very large very quickly.

n	$n!$
0	1
1	1
2	2
3	6
4	24
5	120
6	720
7	5 040
8	40 320
9	362 880
10	3 628 800
11	39 916 800
12	479 001 600

// The largest factorial value my calculator can display is
// $69! \cong 1.711\ 224\ 524$ E 98. That times 7 produces $1.197\ 857\ 167$ E 99, and so
// 70! causes an overflow error.

The number of full permutations of an n-set is $n!$, and each full permutation is an indexing function from $\{1..n\}$ onto X. That is, the elements of the set X may be listed in (exactly) $n!$ different orders.

2.3.5 Number of k-Subsets of an n-Set

The complex symbol $\begin{pmatrix} n \\ k \end{pmatrix}$ will denote the number of k-subsets of an n-set assuming $0 <= k <= n$. Suppose X is any n-set. The empty set is the only subset of X of size 0, and all of X is the only subset of size n, so

$$\begin{pmatrix} n \\ 0 \end{pmatrix} = 1 \text{ and } \begin{pmatrix} n \\ n \end{pmatrix} = 1. \qquad \text{// Even when } n = 0?$$

There are exactly n subsets of size 1, so

$$\begin{pmatrix} n \\ 1 \end{pmatrix} = n. \qquad \text{// But how many of size } k?$$

Suppose $1 < k <= n$. If we were to list all the k-permutations on X, each k-subset will appear several times in that list. The same k-subset occurs once for every possible reordering of its k elements. So each subset will occur $k!$ times in the list of all k-permutations on X. Each k-permutation on X is an ordering of k distinct elements of X and therefore

the total number of k-permutations on X
$=$ the total number of orderings of k-subsets of X
$=$ (the number of k-subsets of X)
$\times$ (the number of full permutations of a k-subset).

In algebra, $\dfrac{n!}{(n-k)!} = \dbinom{n}{k} \times k!$. Therefore, when $0 <= k <= n$

$$\binom{n}{k} = \frac{n!}{k! \times (n-k)!}.$$

// The argument justifying the formula wasn't applied when $k = 0$ or 1, but
// the formula produces the right number even then.

Example 2.3.1: Virginia's Boyfriends

Suppose Virginia has 5 boyfriends: Tom, Dick, Harry, George, and Alvah. She wants to choose two of them to take home to meet mother on the next long weekend. In how many ways can she do this?

Let $X = \{A, D, G, H, T\}$. Then we can list all the 2 subsets of X:

$\{A, D\}$,
$\{A, G\}, \{D, G\}$,
$\{A, H\}, \{D, H\}, \{G, H\}$,
$\{A, T\}, \{D, T\}, \{G, T\}$ and $\{H, T\}$.   // There are 10 ways.

So $\dbinom{5}{2} = 10$ and $\dfrac{5!}{2! \times (5-2)!} = \dfrac{5 \times 4 \times (3)!}{2 \times 1 \times (3)!} = 5 \times 2 = 10$.

Each time she chooses 2 boyfriends to take, she also chooses 3 to leave behind. The number of ways of choosing 3 out of 5 must be the same as the number of ways of choosing 2 out of 5.

So $\dbinom{5}{3} = 10$ and $\dfrac{5!}{3! \times (5-3)!} = \dfrac{5 \times 4 \times 3 \times (2)!}{3 \times 2 \times 1 \times (2)!} = 5 \times 2 = 10$.

This idea generalizes. For every k-subset A of an n-set X, the relative complement, $X \setminus A$ is an $(n-k)$-subset. Different k-subsets produce different $(n-k)$-subsets, and every $(n-k)$-subset is produced. Therefore, the number of $(n-k)$-subsets must equal the number of k-subsets. In terms of the formula, we have

$$\binom{n}{n-k} = \frac{n!}{(n-k)! \times [n-(n-k)]!} = \frac{n!}{(n-k)! \times [k]!} = \frac{n!}{k! \times (n-k)!} = \binom{n}{k}.$$

Example 2.3.2: Counting Subsets by Sizes

We know $\binom{5}{0} = 1$, // And the formula gives $\dfrac{5!}{0! \times (5-0)!} = \dfrac{5!}{1 \times (5)!} = 1$.

and we know $\binom{5}{1} = 5$. // And the formula gives $\dfrac{5!}{1! \times (5-1)!} = \dfrac{5 \times (4)!}{1 \times (4)!} = 5$.

k	$\binom{5}{k}$
0	1
1	5
2	10
3	10
4	5
5	$\dfrac{1}{32}$

// # subsets of a 5-set $= 2^5 = 32$

For any $n \in \mathbf{P}$, we could partition all the subsets according to their sizes and get

$$\sum_{k=0}^{n} \binom{n}{k} = 2^n.$$

This is a special case of the Binomial Theorem, which states that

$$(a+b)^n = \sum_{k=0}^{n} \binom{n}{k} a^k \times b^{n-k} \qquad \text{// take } a = 1 \text{ and } b = 1$$

// Prof. Moriarty, Sherlock Holmes' archenemy, is said to have written a treatise
// on the Binomial Theorem. We will prove the theorem in Chap. 3.

Because of this theorem, the numbers $\binom{n}{k}$ are often called *binomial coefficients*.

Example 2.3.3: The Bad Banana Theorem   // aka Pascal's Theorem

Suppose you are sent to buy 6 bananas. The store has 20 bananas altogether: 19 good ones and 1 bad one. Any selection of 6 either avoids the bad one or includes it. So the total number of selections equals the number containing only good bananas plus the number that contain the bad one and 5 good ones. That is,

$$\begin{aligned}
\binom{20}{6} &= \binom{19}{6} + \binom{19}{5} = \frac{19!}{6! \times 13!} + \frac{19!}{5! \times 14!} \\
&= \frac{19 \times 18 \times 17 \times 16 \times 15 \times 14}{6 \times 5 \times 4 \times 3 \times 2 \times 1} + \frac{19 \times 18 \times 17 \times 16 \times 15}{5 \times 4 \times 3 \times 2 \times 1} \\
&= 19 \times 17 \times 2 \times 3 \times 14 + 19 \times 18 \times 17 \times 2 \\
&= 27\,132 + 11\,628 = 38\,760.
\end{aligned}$$

This result applies in general, if $0 < k < n$, then

$$\binom{n-1}{k} + \binom{n-1}{k-1} = \frac{(n-1)!}{k! \times [(n-1)-k]!} + \frac{(n-1)!}{(k-1)! \times [(n-1)-(k-1)]!}$$

$$= \frac{(n-1)!}{k! \times [n-k-1]!} + \frac{(n-1)!}{(k-1)! \times [n-k]!} \qquad \textit{// we need a commom denominator}$$

$$= \frac{(n-1)!}{k! \times (n-k-1)!} \times \frac{(n-k)}{(n-k)} + \frac{k}{k} \times \frac{(n-1)!}{(k-1)! \times (n-k)!} \qquad \textit{// N! = N} \times \textit{(N} - 1)! \textit{ so}$$

$$= \frac{[(n-k)+k] \times (n-1)!}{k! \times (n-k)!} = \frac{n \times (n-1)!}{k! \times (n-k)!} = \frac{n!}{k! \times (n-k)!} = \binom{n}{k}.$$

2.3.6 Pascal's Triangle

The binomial coefficients may be displayed in a triangular array. For $n \in \mathbf{N}$ and $0 <= k <= n$, let $B[n,k] = \binom{n}{k}$. The rows are indexed by n and start with row 0; each row has $n + 1$ entries, one for each k. The first 7 rows of B are

		0	**1**	**2**	**3**	**4**	**5**	**6**
	0	1						
	1	1	1					
	2	1	2	1				
n	**3**	1	3	3	1			
	4	1	4	6	4	1		
	5	1	5	10	10	5	1	
	6	1	6	15	20	15	6	1

Each row begins and ends with a 1. $\textit{// } B[n, 0] = 1 = B[n, n].$

The Bad Banana Theorem allows us to fill in the table row by row since

$$\text{for } 0 < k < n \qquad B[n,k] = B[n-1,k-1] + B[n-1,k].$$

// $B[n, k]$ is the sum of two entries in the row above it.
// $B[n − 1, k]$ is the entry one step north, and $B[n − 1, k − 1]$ is the entry one step
// northwest.

Example 2.3.4: The Sailing Teams

Suppose that a sailing team is to be chosen from 30 people and it consists of a captain and a crew of 5. How many different teams can be chosen?

       // Different captains make different teams – the captain is BOSS.

There are (at least) 3 methods we might use to select a sailing team:
- The *dictator* method: a captain is chosen (somehow) and s/he chooses the crew.
- The *democratic* method: a team is chosen (somehow) and they elect the captain.
- The *completely random* method: a crew is chosen (at random) and then a captain is also chosen (at random).

The number of possible teams in each case is: // using the product rule

Case 1. $\binom{30}{1} \times \binom{29}{5} =$ _____ $\times$ _____ $=$ _____ .

Case 2. $\binom{30}{6} \times \binom{6}{1} =$ _____ $\times$ _____ $=$ _____ .

Case 3. $\binom{30}{5} \times \binom{25}{1} =$ _____ $\times$ _____ $=$ _____ .

// Do these all produce the same number? Is the number 3 562 650?
// Can this be generalized to *n* people and a crew of size *k*?   // assuming $k < n$

Case 1.
$$\binom{n}{1} \times \binom{n-1}{k} = n \times \binom{n-1}{k} = n \times \frac{(n-1)!}{k! \times ([n-1]-k)!} = \frac{n!}{k! \times (n-k-1)!}.$$

Case 2.
$$\binom{n}{k+1} \times \binom{k+1}{1} = \binom{n}{k+1}(k+1) = (k+1) \times \frac{n!}{(k+1)! \times (n-[k+1])!}$$
$$= \frac{n!}{k! \times (n-k-1)!}.$$

Case 3.
$$\binom{n}{k} \times \binom{n-k}{1} = \binom{n}{k} \times (n-k) = \frac{n!}{k! \times (n-k)!} \times (n-k) = \frac{n!}{k! \times (n-k-1)!}.$$

These results from counting sailing teams imply the following equations involving binomial coefficients: // again assuming $k < n$

From Cases 1 and 3, $n \times B[n-1, k] = (n-k) \times B[n, k]$, so

$$\binom{n}{k} = \frac{n}{n-k} \times \binom{n-1}{k}$$ // in column k, the values increase
                                                        // (unless $k = 0$).

From Cases 1 and 2, $n \times B[n-1, k] = (k+1) \times B[n, k+1]$ and putting j for $k+1$, we get $n \times B[n-1, j-1] = (j) \times B[n, j]$, so

$$\binom{n}{j} = \frac{n}{j} \times \binom{n-1}{j-1}.$$ // Stepping southeast, the values increase
                                                         // (unless $k = n-1$ and $j = n$).

From Cases 2 and 3, $(k + 1) \times B[n, k + 1] = (n - k) \times B[n, k]$, so

$$\binom{n}{k+1} = \frac{(n - k)}{(k + 1)} \times \binom{n}{k}.$$

// Going across row n, do the values increase? decrease? or stay the same?
//X   For which values of k, do they increase? decrease? stay the same?
//X   Is $B[n, k]$ is largest when $k = \lfloor n/2 \rfloor$?

Example 2.3.5: The Exploration Parties

Suppose there are 11 people aboard a yacht, 5 women and 6 men, and 4 of them take the dinghy to explore an island. How many exploration parties have at least two women?

A quick and dirty solution is this: choose 2 women from the 5 and then choose 2 others from the 9 remaining people. (This will certainly produce an exploration party with 2 or more women.) Applying the product rule, we know that this can be done in

$$\binom{5}{2} \times \binom{9}{2} = 10 \times 36 = 360 \text{ ways.} \qquad \text{// But is this correct?}$$

// Does each "way" produce a unique exploration party?

That number cannot be right because the total number of possible exploration parties is

$$\binom{11}{4} = \frac{11!}{4! \times 7!} = \frac{11 \times 10 \times 9 \times 8 \times 7!}{4 \times 3 \times 2 \times 1 \times 7!} = 11 \times 10 \times 3 = 330.$$

For $w = 0$ to 4, the number of parties that contain exactly w women can be counted by determining the number of ways w can be chosen from the 5 women, and then choosing the rest of the party from the men; that is, $(4 - w)$ men are chosen from the 6 men.

w	$4 - w$	$\binom{5}{w}$	$\times$	$\binom{6}{4-w}$				
0	4	1	$\times$	15	$=$	15		
1	3	5	$\times$	20	$=$	100		
2	2	10	$\times$	15	$=$	150	$// \times 1$	$= 150$
3	1	10	$\times$	6	$=$	60	$// \times 3$	$= 180$
4	0	5	$\times$	1	$=$	5	$// \times 6$	$= 30$
						330		360

The correct answer is $150 + 60 + 5 = 215$ parties contain at least 2 women.

// <u>not</u> 360

// What is wrong about the first "solution"? If the set of women were
// $\{A,B,C,D,E\}$, how often would the subset $\{A,B,D,E\}$ be counted in the first
// "solution"? Why 6?

2.3.7 Counting Algorithmically (Without a Formula)

In this subsection, we want to provide a strategy for counting sequences, using tree
diagrams, which can be applied when the set of possible values for s_j depends on
what the previous values in S actually are. We'll look at two examples.

Example 2.3.6: Sequences on P that sum to 5

How many sequences of positive integers have a sum that equals 5?

// Both $(1,4)$ and $(1,1,2,1)$ sum to 5, so we'll have to deal with sequences of
// varying lengths. But, since each entry is at least one, there can be at most 5
// entries.

Let's generate all such sequences, one entry at a time, by growing a "tree of
possibilities." Suppose $S = (s_1, s_2,\ldots, s_k)$ is such a sequence. The possible values for
s_1 are 1,2,3,4, and 5. This may be displayed by

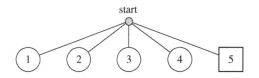

The vertex labeled *start*, at the top of the diagram, is called the "root" of the tree.
Attached to the root are five "branches," one for each possible value of s_1. These
possible values of s_1 are written inside the vertices (circles) at the ends of the 5
branches.   // Yes, the tree is upside down.

If $s_1 = 1$, then the possible values for s_2 are 1,2,3, and 4, and we may add 4 more
(downward) branches from the vertex corresponding to $s_1 = 1$. If $s_1 = 2$, then the
possible values for s_2 are 1,2, and 3, and we may add 3 more branches from the
vertex corresponding to $s_1 = 2$. But if $s_1 = 5$, then the current subsequence cannot
be extended; we cannot add more branches from the vertex corresponding to $s_1 = 5$.
These terminal vertices are called "leaves" and are indicated in the diagram as
squares (rather than circles).

From each vertex indicating a possible value of s_2 that's not a leaf, we may add
branches corresponding to values of s_3. And so on until we have a diagram of <u>all</u>
sequences on **P** that sum to 5.

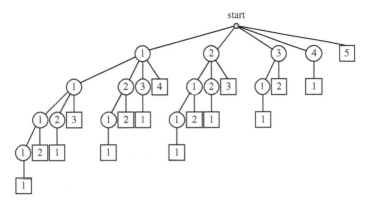

Now, by counting the leaves, we have determined that there are exactly 16 sequences on **P** which sum to 5.

We have consistently labeled the values below a vertex from smallest to largest across the diagram from left to right. If we take the leaves as they occur across the diagram from left to right, and write down the corresponding sequence, we get

1.	(1,1,1,1,1)	// aaaaa
2.	(1,1,1,2)	// aaab
3.	(1,1,2,1)	// aaba
4.	(1,1,3)	// aac
5.	(1,2,1,1)	// abaa
6.	(1,2,2)	// abb
7.	(1,3,1)	// aca
8.	(1,4)	// ad
9.	(2,1,1,1)	// baaa
10.	(2,1,2)	// bab
11.	(2,2,1)	// bba
12.	(2,3)	// bc
13.	(3,1,1)	// caa
14.	(3,2)	// cb
15.	(4,1)	// da
16.	(5)	// e

// There is something special about the order in which the sequences appear.
// If the first 5 digits were replaced by the first 5 letters, the sequences of digits
// produce "words" in alphabetic or *lexicographic* order.

Example 2.3.7: Subsets of P that Sum to 10

How many sets of positive integers have a sum that equals 10?

Let's again grow a tree of possibilities. Each set of positive integers may be listed (uniquely) in its natural order as a sequence (where the entries get larger and larger). We want to generate all such sequences $S = (s_1, s_2, \ldots, s_k)$ on **P** where

$$10 = s_1 + s_2 + \ldots + s_k \quad \text{and} \quad 1 <= s_1 < s_2 < \ldots < s_k <= 10.$$

// What are the possible values for s_1? 1,2,3,...,10?
// Will there be 10 branches from the root?

Either $k = 1$ and $s_1 = 10$,   or $k > 1$ and

$$10 = s_1 + s_2 + \ldots + s_k >= s_1 + s_2 > s_1 + s_1 = 2 \times s_1. \qquad \text{// So } s_1 < 5.$$

Therefore, the possible values for s_1 are 1,2,3,4, and 10. Furthermore, if $s_1 < 10$ then

either $k = 2$ and $s_2 = 10 - s_1$,   or $k > 2$ and

$$10 - s_1 = s_2 + s_3 + \ldots + s_k >= s_2 + s_3 > s_2 + s_2 = 2 \times s_2. \quad \text{// So } s_2 < (10 - s_1)/2.$$

We may "prune off" the useless branches and obtain a diagram of all subsets of **P** that sum to 10.

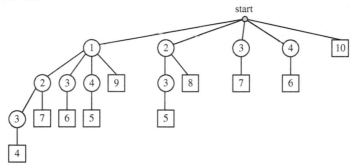

If we take the leaves as they occur across the diagram from left to right, and write down the corresponding sets, we get the 10 subsets of **P** which sum to 10.

// Subsets are written in braces.

    1. {1,2,3,4}
    2. {1,2,7}
    3. {1,3,6}
    4. {1,4,5}
    5. {1,9}
    6. {2,3,5}
    7. {2,8}
    8. {3,7}
    9. {4,6}
    10. {10}

// We could also convert sets to sequences by listing the elements in *decreasing*
// order, and then generating a tree, but this tree would look quite different.

// The generation of the tree diagrams by growing the tree across each "level" at a
// time is known as *breadth-first* generation. Reading the sequences from the root
// to each leaf is known as *depth-first* traversal.

The Most Important Ideas in This Section.

A *permutation* is a sequence without repetitions.

The function *n-factorial* [written $n!$] is defined on $\mathbf{N}$ by

$$n! = \begin{cases} n(n-1)\ldots(2)(1) & \text{if } n > 0 \\ 1 & \text{if } n = 0 \end{cases}.$$

This section derived the fundamental counting formulas:

1. *The number of k-sequences on an n-set is* $n \times n \times \ldots \times n = n^k$.
2. *The number of subsets of an n-set is* 2^n.
3. *The number of k-permutations on an n-set is*

$$n \times (n-1) \times (n-2) \times \ldots \times (n-k+1) = \frac{n!}{(n-k)!}.$$

4. *The number of k-subsets of an n-set is* $\displaystyle \binom{n}{k} = \frac{n!}{k! \times (n-k)!}.$

Some basic properties of these binomial coefficients were given:

$$\binom{n}{0} = 1 = \binom{n}{n} \qquad \text{for all } n >= 0.$$

$$\binom{n}{n-k} = \binom{n}{k} \qquad \text{whenever } 0 <= k <= n.$$

The Bad Banana Theorem: If $0 < k < n$, then

$$\binom{n}{k} = \binom{n-1}{k} + \binom{n-1}{k-1}.$$

When no standard paradigm applies, we can often generate a tree diagram of all the possibilities and then count them.

The next section looks at infinite sequences, especially complexity functions of algorithms.

2.4   Infinite Sequences and Complexity Functions

Infinite sequences are functions defined from an infinite interval of integers. Throughout this section, we'll suppose that S is a sequence with domain $\{a.. \}$ and codomain $\mathbf{R}$. Our main interest in infinite sequences will be studying *complexity functions* of algorithms, where $f(n)$ is the number of steps it takes for the algorithm to finish a problem instance of size n (in the worst case).

But there are other infinite sequences worth examining. For instance, in the walkthrough of the operation of the Bisection Algorithm in the example in Chap. 1, five sequences were generated as we attempted to find (a good approximation of) x^* where $f(x^*) = (x^*)^3 + 2^{(x^*)} = 200$.

// x^*, when rounded to 8 decimal places, is 5.402 668 66.

A	z	B	$e = \mid z - A \mid$	f(z)
5	7.5	10	2.5	602.894...
"	6.25	7.5	1.25	320.249...
"	5.625	6.25	.625	227.329...
"	5.3125	5.625	.3125	189.672...
5.3125	5.46875	"	.15625	207.840...
"	5.390625	5.468 75	.078125	198.596...
5.390625	5.4296875	"	.0390625	203.177...
"	5.41015625	5.4296875	.01953125	200.876...
"	5.400390625	5.41015625	.009765625	199.733...
5.400390625	5.4052734375	"	.0048828125	(200.304...)

Assuming the first row of the table corresponds to $i = 1$, the second row to $i = 2$, and so on and that the process continued indefinitely, we would expect:

The sequence of midpoints [the z_i] "converges to" x^*.

The sequence of function values [the $f(z_i)$] "converges to" 200. // $= f(x^*)$.

The sequence of error bounds [the e_i, which decrease by half on each iteration] "converges to" 0.

The sequence of lower bounds [the A_i, which never decrease and are always between $A_1 = 5$ and x^*] also "converges to" x^*.

The sequence of upper bounds [the B_i never increase and are always between x^* and $10 = B_1$] also "converges to" x^*.

// We will give a formal definition of *convergence* of a sequence momentarily.

// The algorithm itself determines these five functions on **P** – from the values in
// any row, it calculates the values in the next row. There is a "formula" for the
// error bounds, but there seems to be no formula for the other sequences.

When S is a sequence with domain $\{a.. \}$ and codomain **R**,

S is a **geometric sequence** <u>means</u>

there is a number r such that for all $n >= a, S_{n+1} = r \times S_n$.

// Each entry, after the first, is obtained by multiplying the previous entry by the
// constant value r. The equation "$S_{n+1} = r \times S_n$" is an example of a
// *recurrence equation.*

If $a = 0$ and S_0 equals some *initial* value, I, then

$$S_1 = r \times I$$
$$S_2 = r \times (r \times I) = r^2 \times I$$
$$S_3 = r \times (r^2 \times I) = r^3 \times I$$
$$S_4 = r \times (r^3 \times I) = r^4 \times I,$$

and in general,   // as we'll *prove* in Chap. 3

$$S_n = r \times \left(r^{n-1} \times I\right) = r^n \times I.$$ // Or $S_n = r^{n-1} \times r \times I = r^{n-1} \times S_1$.

The error bounds in the example are a geometric sequence with $r = \frac{1}{2}$. Here, $a = 1$ and $e_1 = 2.5 = 5/2$. We can (pretend the sequence began with $e_0 = 5$ and) use the formula to get

$$e_n = r^n \times I = \left(\frac{1}{2}\right)^n \times 5 = 5/(2^n).$$ // or use $S_n = r^{n-1} \times S_1$

If S is any geometric sequence and $I > 0$, then if $r > 1$, the entries in S get larger and larger, and if $0 < r < 1$, the entries get smaller and smaller.

//X   What happens if r or I (or both) are negative?

S is **increasing**   <u>means</u> $S_a < S_{a+1} < S_{a+2} < S_{a+3} < \ldots$

and S is **decreasing**   <u>means</u> $S_a > S_{a+1} > S_{a+2} > S_{a+3} > \ldots$

// e_n is decreasing.

Often we don't have strict inequality between consecutive entries.

S is **nondecreasing**   <u>means</u> $S_a <= S_{a+1} <= S_{a+2} <= S_{a+3} <= \ldots$

and S is **nonincreasing**   <u>means</u> $S_a >= S_{a+1} >= S_{a+2} >= S_{a+3} >= \ldots$

// A_n is nondecreasing and B_n is nonincreasing.

And, S is a **monotone** <u>means</u> either S is nonincreasing or nondecreasing.

// the midpoints z_i are <u>not</u> a monotone sequence.

It is clear that the error bounds, the e_i, get closer and closer to 0. But the midpoints, the z_i, have values that "bounce around" x^*: sometimes they're bigger, sometimes smaller, sometimes closer, sometimes further away. Nevertheless, there is an overall "trend" in the z_i toward x^*; in general, the larger the index i, the closer z_i is to x^*. The formal description of this is

A sequence S **converges** to L [written $S_n \to L$] <u>means</u>
for any number $\delta > 0$, there is an integer M such that
if $(n > M)$ then $|S_n - L| < \delta$.

// No matter how small δ is, from some point on, the S_n's are within δ of (the
// <u>number</u>) L.

// δ was the precision our supervisor specified for the absolute error of our
// approximation in the problem given in Chap. 1.

To show that the sequence of midpoints [the z_n] does converge to x^* under the definition given above, suppose a positive value for δ is given. Let $M = \lceil \lg(|B_1 - A_1|/\delta) \rceil$. Then as we saw in Chap. 1,

$$\text{if } (n > M), \text{ then} \qquad |z_n - x^*| < |B_1 - A_1|/2^n < \delta.$$

Since $e_n = |B_1 - A_1|/2^n = 5/2^n$, we also have

$$\text{if } (n > M) \text{ then} \qquad |e_n - 0| = e_n = |B_1 - A_1|/2^n < \delta,$$

and therefore, the sequence of error bounds, e_n, converges to 0.

// $| A_n - x^* | < |A_n - B_n| = 2e_n \rightarrow 0$ so $A_n \rightarrow x^*$ and also
// $| B_n - x^* | < |B_n - A_n| = 2e_n \rightarrow 0$ so $B_n \rightarrow x^*$.

A sequence S is **bounded** <u>means</u>
there are numbers A and B such that if $(n >= a)$ then $A <= S_n <= B$.

// The e_i are bounded by 0 and 2.5 (or -100 and $+100$),
// and the z_i are bounded by 5 and 10 (or -100 and $+100$).

// If a sequence converges, then it must be bounded.
// If a sequence is monotone and bounded, then it must converge.

Complexity functions are infinite sequences whose entries are nonnegative integers. They are bounded below by 0 and are usually increasing. Rarely are they bounded above, and bad complexity functions grow very large very rapidly.

2.4.1 The Towers of Hanoi

In the late 1800s, a toy was marketed called the Towers of Hanoi. The toy consisted of a base on which there were three posts, together with a set of discs. The discs were all different diameters, and each had a hole in the center so they could fit over any of the three posts as displayed in the illustration below.

The toy came with a blurb that said something like "In the Temple of Brahma in Hanoi there is a brass platform with three diamond needles and 64 golden discs all of different sizes. At the beginning of time the discs were placed on the first needle in a pile from largest up to smallest. The priests of the temple are transferring the discs to another needle one at a time so that no disc ever rests on a smaller disc. When they finish, time and the world will end."

// Is it possible to accomplish this divine task? Will the world ever end?
// Is there an algorithm for doing this? When will the world end?

Let's label the needles a, b, and c where the tower of discs starts on needle a. Also let's number the discs from smallest to largest, 1 to $n = 64$.

                                                                    // n is the problem size.

If there were only one disc, it could be transferred to another needle in one step; that is, one move of a single disc. Let $T(n)$ be the (minimum) number of single-disc-moves (steps) required to transfer a tower of n discs from needle a to some other needle. Then $T(1) = 1$.

If there were two discs,

The first step must move disc 1 from a to another needle, say needle b. Then disc 2 may be moved from a to c. Finally, disc 1 may be moved from b to c. So $T(2) = 3$.

// $T(2) >= 2$ because both discs must be moved, and it cannot be done with just 2
// moves.

Suppose there are k discs (in a tower) on needle a to be transferred. Before disc k can be moved, the $(k - 1)$ discs above disc k <u>must</u> be transferred off needle a, and there <u>must</u> be a needle with none of the small discs on it so that disc k may be moved to that needle. Thus, to transfer a tower of k discs from needle a, we <u>must</u>
1. Transfer the whole tower of the top $(k - 1)$ discs from a to b (or to c)
2. Move disc k from a to c (or to b)
3. Transfer the whole tower of the top $(k - 1)$ discs from b to c (or c to b)

If the transfer is to be as efficient as possible, each of these three parts must be done as efficiently as possible. Therefore,

$$T(n) = T(n - 1) + 1 + T(n - 1)$$
$$= 2T(n - 1) + 1.$$

Using this recurrence equation when $n >= 2$ and the starting value $T(1) = 1$, we get

$$T(1) = 1$$
$$T(2) = 2\,T(1) + 1 = 2 \times 1 + 1 \; = \; 3$$
$$T(3) = 2\,T(2) + 1 = 2 \times 3 + 1 \; = \; 7$$
$$T(4) = 2\,T(3) + 1 = 2 \times 7 + 1 \; = 15$$
$$T(5) = 2\,T(4) + 1 = 2 \times 15 + 1 = 31.$$

In every case,   // as we'll *prove* in Chap. 3

$$T(n) = 2^n - 1.$$

This recursive description of a solution indicates that it can be done and allows us to count how many steps must be done but doesn't explicitly tell us <u>how</u> to move whole towers of discs.

Any move consists of taking the top disc from the pile on one needle and then placing it on top of the (perhaps empty) pile of discs on some other needle.

The smallest disc is never covered by another disc, so it is always the top disc on some needle. The smallest disc can be moved to either of the other two needles. After the smallest disc is moved, the next step will not move the smallest disc again and will not place another disc on top of the smallest.

// The first step must move disc 1. What other moves are possible?
// We will show there is only one move possible that does not involve the
// smallest disc.

If all the discs are on one needle, we're either at the beginning or the end of the task. If we're in the midst of the transfer, there must be discs on at least two needles. Suppose disc 1 is now on needle x and disc i is at the top of the pile on needle y. Either the third needle, z, has no discs on it, or some discs are on needle z and disc j is at the top. In the first case, the <u>only</u> move is to take disc i from needle y to needle z. In the second case, the <u>only</u> move is to take the smaller of discs i and j and place it on top of the larger. After a move that didn't involve disc 1, we can only reverse that move <u>or</u> move disc 1.

<u>Any</u> solution algorithm (that uses as few moves as possible) must alternate moves of disc 1 with moves that don't involve disc 1. If we standardize the moves of disc 1 by always moving from a to b, from b to c, and from c to a, we obtain an (iterative) algorithm that solves the Towers of Hanoi problem.

// But how long will it take to transfer a tower of 64 discs?

The number of steps required is

$$2^{64} - 1 = 18\,446\,744\,073\,709\,551\,615 \cong 1.844\,674\,407 \times 10^{19}.$$

If the priests worked continuously in shifts day and night and move (on average) one disc per second, it will take at least 584.5 billion years to finish.

The Towers of Hanoi with 64 discs is an example of a problem that has an effective algorithmic solution in principle but not in practice. The number of steps involved is finite but just too large to be practical. [Certain modern methods of cryptography are based on this consideration; the codes can be broken but even with fast machines, it would take much too long to be of any value to the code-breaker.] The difficulty lies with the complexity functions which grow very large very quickly, which may be described as

2.4.2 Bad Complexity Functions

Let's say a problem instance is "doable" if a machine executing 10^9 operations per second could finish it in a *year*; that is,

$$\text{the \# steps } <= 10^9 \times 60 \times 60 \times 24 \times 365 = 3.1536 \times 10^{16}.$$

What is the maximum size of a doable instance for the following four bad complexity functions?

n	2^n	$\binom{2n}{n}$	$n!$	n^n
1	2	2	1	1
2	4	6	2	4
3	8	20	6	27
4	16	70	24	256
5	32	252	120	3125
6	64	924	720	46656
7	128	3432	5040	823543
8	256	12870	40320	16777216
9	512	48620	362880	387420489
10	1024	184756	3628800	1000000000
..	..	..	..	..
14	16384	40116600	8.718 E10	1.111 E16
15	32768	155117520	1.308 E12	4.379 E17
..	..	..	..	
18	262144	9075135300	6.402 E15	..
19	524288	3.535 E10	1.216 E17	..
..	..	..	..	..
29	536870912	3.007 E16	..	..
30	1073741824	1.183 E17	..	..
..	..	..	..	..
54	1.801 E16	..	..	..
55	3.603 E16	..	..	..

// The first is the complexity of the Towers of Hanoi algorithm, or of generating
// all subsets of an n-set; an instance is doable only when $n < 55$.
// The second is that of generating all n-subsets in a set twice as large;

// we need $n < 30$.
// The third is that of generating all full permutations of an n-set; n must be < 19.
// The last is that of generating all functions from one n-set into another; $n <= 14$.

Thus, if the complexity function grows very rapidly, only (relatively) small problem instances can be done (no matter how fast the machine is, and how much time is allowed for the task).

The Most Important Ideas in This Section.
A number of adjectives for a sequence $S:\{a..\} \to \mathbf{R}$ were defined:
S is a *geometric sequence* when there is a number r such that
for all $n >= a$, $S_{n+1} = r \times S_n$.
S is *increasing* when $S_a < S_{a+1} < S_{a+2} < S_{a+3} < \cdots$
S is *decreasing* when $S_a > S_{a+1} > S_{a+2} > S_{a+3} > \cdots$
S is *nondecreasing* when $S_a <= S_{a+1} <= S_{a+2} <= S_{a+3} <= \cdots$
S is *nonincreasing* when $S_a >= S_{a+1} >= S_{a+2} >= S_{a+3} >= \cdots$
S is a *monotone* when   either S is nonincreasing or nondecreasing.
S *converges* to L [written $S_n \to L$] when for any number $\delta > 0$ there is an integer M such that if $(n > M)$ then $|S_n - L| < \delta$.
S is *bounded* when there are numbers A and B such that
if $(n >= a)$ then $A <= S_n <= B$.
   Complexity functions are infinite sequences whose entries are nonnegative integers. They are bounded below by 0 and are usually increasing. Rarely are they bounded above, and bad complexity functions grow very large very rapidly. The section ends with several examples of bad complexity functions.

Exercises

1. Indicate whether each statement is true or false:
    (a) $\{4, 0, 3, 0\} = \{4, 4, 0, 3\}$
    (b) $\{4\} \subset \{0, 3, 4\}$
    (c) $\{0, 3, 4\} \subset \{4\}$
    (d) $\{0, 3, 4\} \subset \{0, 3, 4\}$
    (e) $\emptyset \subset \{0, 3, 4\}$

2. What is $\mathscr{P}(\{0, 3, 4, 7\})$?

3. Let $A = \{1, 2, 3, 4\}$ and $B = \{2, 3, 5, 8\}$. Evaluate each of the following expressions:
    (a) $A \cap B$
    (b) $A \cup B$
    (c) $A \setminus B$
    (d) $A \times B$

4. Let $C = \{a, c, e\}$ and $D = \{a, b, c, d, e\}$. Evaluate each of the following expressions:

  (a) $C \cap D$
  (b) $C \cup D$
  (c) $C \setminus D$
  (d) $C \times D$

5. Is $\{1, 3\}, \{2, 3\}, \{4\}$ a partition of $\{1, 2, 3, 4\}$? Justify your answer.

6. Consider the set $\{a, b, c, d, e\}$. Construct 3 different partitions of this set.

7. Consider a large family that has a mother, a father, and 11 children. Use the Pigeon Hole Principle to construct an argument that:

  (a) At least two family members were born on the same day of the week
  (b) At least two family members were born in the same month

8. The particular cultivar of raspberries in my garden has between 100 and 125 drupelets per berry. Each cane produces at least 55 berries. Does each cane produce at least two berries that have the same number of drupelets? Use the Pigeon Hole Principle to construct an argument that supports your answer.

9. Assume that a and b are integers where $a <= b$, and both S and T are sequences on **R**.

  (a) Explain why $\displaystyle\sum_{i=a}^{b} S_i + \sum_{i=a}^{b} T_i = \sum_{i=a}^{b} (S_i + T_i)$.

  (b) Explain why $\displaystyle\sum_{i=a}^{b} (c \times S_i) = c \times \left(\sum_{i=a}^{b} S_i \right)$ for any number c.

10. Let A denote all k-subsets of $\{1..n\}$ where $0 < k <= n$ and let B denote all increasing k-sequences on $\{1..n\}$. Show that the number of k-subsets in A equals the number of k-sequences in B.

11. (a) How many 4-sequences on $\{1..9\}$ are there?
  (b) How many 4-permutations on $\{1..9\}$ are there?
  (c) How many 4-permutations on $\{1..9\}$ begin with 3?
  (d) How many increasing 4-sequences on $\{1..9\}$ are there?
  (e) How many increasing 4-sequences on $\{1..9\}$ begin with 3?

12. Suppose that $a = 5$ and $b = 20$.
  (a) How many elements are there in $\{a..b\}$?
  (b) How many 4-sequences on $\{a..b\}$ are there?
  (c) How many 4-permutations on $\{a..b\}$ are there?
  (d) How many 4-permutations on $\{a..b\}$ begin with 8?
  (e) How many increasing 4-sequences on $\{a..b\}$ are there?
  (f) How many increasing 4-sequences on $\{a..b\}$ begin with 8?

13. Suppose that a and b are integers and $0 <= a < b - 4$.
  (a) How many elements are there in $\{a..b\}$?
  (b) How many 4-sequences on $\{a..b\}$ are there?
  (c) How many 4-permutations on $\{a..b\}$ are there?
  (d) How many increasing 4-sequences on $\{a..b\}$ are there?

14. (a) How many 4-sequences on $\{0..9\}$ are there?
    (b) How many 4-sequences on $\{0..9\}$ do not begin with 0?
    (c) How many 4-sequences on $\{0..9\}$ begin and end with 0?
    (d) How many 4-sequences on $\{0..9\}$ do not begin and end with 0?
    (e) How many 4-sequences on $\{0..9\}$ do not begin or end with 0?

15. Passwords on a certain system have exactly 5 letters that are either lowercase letters or uppercase letters.
    (a) How many possible passwords are there?
    (b) How many possible passwords are there that use only lowercase letters?
    (c) How many possible passwords are there that use only uppercase letters?
    (d) How many possible passwords are there that use at least one uppercase letter and at least one lowercase letter?

16. In LOTTO 6-49, a subset of six numbers is selected at random from $\{1..49\}$ as the "winning" numbers. How many different selections of winning numbers are there?

17. A personal identification number may be set to be any 4 digits.
    (a) How many possible PINs are there?
    (b) How many possible PINs are there that do not have a repeated digit?
    (c) How many possible PINs are there that do have a repeated digit?

18. Suppose that in a certain jurisdiction, license plates have 4 letters followed by 3 digits and all such character sequences are possible.
    (a) Show that the number of license plates that have exactly two T's and end in a 5 is 375,000.
    (b) Show that the number of license plates have the letter T and the digit 4 (someplace) in them is 17,981,121.

19. The number of <u>non-decreasing</u> k-sequences on $\{1..n\}$ is larger than the number of <u>increasing</u> k-sequences on $\{1..n\}$ because entries may be repeated. If $X = (x_1, x_2, x_3, \ldots, x_k)$ ia a non-decreasing k-sequences on $\{1..n\}$, define the sequence $Y = (y_1, y_2, y_3, \ldots, y_k)$ by

$$y_i = x_i + k - 1 \quad \text{for } i = 1, 2, \ldots, k.$$

    (a) Show that the sequence Y is an increasing k-sequence on $\{1..(n + k - 1)\}$.
    (b) Show that if Y is any increasing k-sequence on $\{1..(n + k - 1)\}$ then there is an X as above that would be transformed into Y.
    (c) Explain how this shows that the number of non-decreasing k-sequences on
$$\{1..n\} = \binom{n + k - 1}{k}.$$

    // If there are n different "kinds" of objects in a certain context, this allows
    // us to count the number of selections of k objects where several objects
    // of the same kind may be selected.

20. To find the number of k-sequences of positive integers that sum to n, one can use a caret-and-stick method. Suppose $k = 4$ and n $= 15$. Write down 15 ones in a row (sticks) then insert 3 carets (separators) into 3 different spaces between 2 ones. Viz.

$$111^\wedge 111111^\wedge 11^\wedge 1111 \text{ corresponds to the sequence } (3, 6, 2, 4)$$

(a) Does every such caret-and-stick configuration correspond to a k-sequence of positive integers that has a sum equal to n?

(b) Does every selection of $(k - 1)$ insertion points for the carets from $(n - 1)$ possible insertion points produce a k-sequence of positive integers that has a sum equal to n?

(c) How many 4-sequences of positive integers have a sum equal 15?

(d) How many k-sequences of positive integers have a sum equal n?

(e) Can this "method" be adapted to count how many k-sequences of non-negative integers have a sum equal to n?

21. Use a tree diagram to show that there are exactly 22 non-decreasing sequences of positive integers that add up to 8.

Boolean Expressions, Logic, and Proof

3

In the first two chapters, we made a number of arguments to try to convince you that certain algorithms were correct and that certain counting formulas applied. By an "argument", we don't mean a social disagreement; we mean a sequence of statements that leads to some conclusion. And by "leads to some conclusion", we mean increases your confidence in the truth of the conclusion to the point of certainty.

Logic is the art of reasoning and forms the basis of mathematics, which is a science of pure thought in which discoveries of new truths about the world are found (not by more careful observation but) by precise reasoning alone. Mathematics is not calculation; it's deduction. It's not formulas; it's proofs.

The objective of this chapter is to show you the structure of mathematical proofs. But before that, we want to try (once more) to illustrate the purpose of proofs with a look at

3.1 The Greedy Algorithm and Three Cookie Problems

Suppose you are a hungry 5-year-old and in front of you is a 6 × 6 sheet of just baked cookies, all of the same kind but each of a different size as indicated below – the larger the number, the bigger the cookie.

56	76	69	60	75	51
61	77	74	72	80	58
82	97	94	88	99	92
47	68	59	52	65	40
78	81	79	71	85	62
50	67	73	57	70	46

We'll consider three optimization problems associated with this array.

© Springer International Publishing AG, part of Springer Nature 2018
T. Jenkyns and B. Stephenson, *Fundamentals of Discrete Math for Computer Science: A Problem-Solving Primer*, Undergraduate Topics in Computer Science, https://doi.org/10.1007/978-3-319-70151-6_3

Cookie Problem #1

If you are allowed to take up to 6 cookies, what would be the best selection?

The *best* selection is the one with the *largest total* number of "cookie units". For this problem, the solution is obvious – take the 6 largest cookies.

56	76	69	60	75	51
61	77	74	72	80	58
82	97	94	88	99	92
47	68	59	52	65	40
78	81	79	71	85	62
50	67	73	57	70	46

An algorithm to construct this selection is

3.1.1   The Greedy Algorithm

```
Keep taking the best cookie you can
Until you can't take any more.
```

// The total for this selection is $99 + 97 + 94 + 92 + 88 + 85 = 555$.

// The number of selections of 6 cookies is $\binom{36}{6} = 1\,947\,792$.

// Could they all have different totals? Is the number of different totals < 555?

// Does the pigeonhole principle imply that some total occurs $> 3{,}500$ times?

// Does the worst selection have total, $40 + 46 + 47 + 50 + 51 + 52 = 286$?

Cookie Problem #2

If you are allowed to take at most one cookie from any (horizontal) row, what would be the best selection?

For this problem too, the solution is obvious – take the largest cookie in each row.

56	76	69	60	75	51
61	77	74	72	80	58
82	97	94	88	99	92
47	68	59	52	65	40
78	81	79	71	85	62
50	67	73	57	70	46

// The Greedy Algorithm will find this selection with total,

// $99 + 85 + 80 + 76 + 73 + 68 = 481$.

// The number of allowable selections of 6 cookies is $6^6 = 46\,656$;

// could they all have different totals? Does some total occur $>= 97$ times?

Cookie Problem #3

If you are allowed to take at most one cookie from any (horizontal) row and at most one cookie from any (vertical) column, what would be the best selection?

For this problem, the solution is not so obvious, but the Greedy Algorithm, which was effective for solving the previous two problems, may be applied here.

The best cookie of all is 99; after that is taken, the best we can take is 81;

56	76	69	60	75	51
61	77	74	72	80	58
82	97	94	88	99	92
47	68	59	52	65	40
78	81	79	71	85	62
50	67	73	57	70	46

after that is taken, the best we can take is 74;

56	76	69	60	75	51
61	77	74	72	80	58
82	97	94	88	99	92
47	68	59	52	65	40
78	81	79	71	85	62
50	67	73	57	70	46

after that is taken, the best we can take is 60; after that, the best we can take is 50; after that is taken, the best we can take is 40.   // It's the only cookie left.

56	76	69	60	75	51
61	77	74	72	80	58
82	97	94	88	99	92
47	68	59	52	65	40
78	81	79	71	85	62
50	67	73	57	70	46

The Greedy Algorithm will find this selection with total, $99 + 81 + 74 + 60 + 50 + 40 = 404$.   // Is this the best selection?

This selection is not the best. We can **prove** that by finding any allowable selection that's better. If we take 69 from the top row instead of 60, and we take 72 from the second row instead of 74, and keep the other four cookies, this selection has a <u>larger</u> total (411) than the greedy solution, so the greedy solution is not the best.

// Even if it's not the best selection, it's probably a fairly good selection.
// How bad could the greedy solution be? Would you believe the following
// assertion?

In this example of Cookie Problem #3,
the greedy solution is the *worst possible* selection of six cookies.

// Why would anyone believe that?
// Could someone provide enough evidence to make you believe that?   Let's try.

A *transversal* of an $n \times n$ array M is a selection of entries with exactly one in
each row and exactly one in each column. The *value* of a transversal is the sum of
its entries.

// We want to show that the transversal produced by the Greedy Algorithm in this
// example has the smallest value of all possible transversals (without generating
// all $n!$).

If 50 is subtracted from each entry in the top row, then (because each transversal
has exactly one entry in the top row) the value of each transversal is reduced by 50,
the transversal that was best before remains the best, and the transversal that was
worst before remains the worst.

More generally, if R_i is subtracted from each entry in row i, then (because each
transversal has exactly one entry in row i) the value of each transversal is reduced
by R_i, the transversal that was best before remains the best, and the transversal that
was worst before remains the worst.

Subtract

50 from row 1	56	76	69	60	75	51	and
56 from row 2	61	77	74	72	80	58	and
75 from row 3	82	97	94	88	99	92	and
40 from row 4	47	68	59	52	65	40	and
60 from row 5	78	81	79	71	85	62	and
45 from row 6	50	67	73	57	70	46	

$$// \sum R_i = 326.$$

The new matrix is

6	26	19	10	25	1
5	21	18	16	24	2
7	22	19	13	24	17
7	28	19	12	25	0
18	21	19	11	25	2
5	22	28	12	25	1

Now, if C_j is subtracted from each entry in column j, then (because each
transversal has exactly one entry in column j) the value of each transversal is
reduced by C_j, the transversal that was best before remains the best, and the
transversal that was worst before remains the worst. For each j, let C_j be the

minimum entry in column j and then subtract C_j from each entry in that column. This produces the following matrix: // $\sum C_j = 78$ and $326 + 78 = 404$.

1	5	1	0	1	1
0	0	0	6	0	2
2	1	1	3	0	17
2	7	1	2	1	0
13	0	1	1	1	2
0	1	10	2	1	1

The final element of the argument is this. If every entry in a matrix is $>= 0$, then the value of every transversal is $>= 0$. Therefore, if every entry in a matrix is $>= 0$ and T is a transversal with all-zero entries, then T is a worst transversal (in the final matrix and in the original matrix). This final all-zero transversal is the one produced by the Greedy Algorithm. Thus, for this example of Cookie Problem #3, the Greedy Algorithm produces the worst possible answer.

// What is the best transversal?

Using that last array, for each j let D_j be the maximum entry in column j and then subtract D_j from each entry in that column. This produces the following matrix:

−12	−2	−9	−6	0	−16
−13	−7	−10	0	−1	−15
−11	−6	−9	−3	−1	0
−11	0	−9	−4	0	−17
0	−7	−9	−5	0	−15
−13	−6	0	−4	0	−16

If every entry in a matrix is $<= 0$, then the value of every transversal is $<= 0$. Therefore, if every entry in a matrix is $<= 0$ and T is a transversal with all-zero entries, then T is a best transversal (in the final matrix and in the original matrix). Here the all-zero transversal is best, and therefore we know the best transversal in the original matrix. Its value is $75 + 72 + 92 + 68 + 78 + 73 \doteq 458$, and it does not contain the biggest cookie nor the second biggest nor the third biggest.

// $\sum D_j = 54$ and $404 + 54 = 458$.

56	76	69	60	75	51
61	77	74	72	80	58
82	97	94	88	99	92
47	68	59	52	65	40
78	81	79	71	85	62
50	67	73	57	70	46

The Most Important Ideas in This Section.
The moral of this story about cookies is: some plausible statements seem reasonable and probable but are not (always) true [statements like: "the best solution contains the biggest cookie" or "the Greedy Algorithm gives the best solution"], and some implausible statements seem unreasonable and unlikely but are true [statements like "the Greedy Algorithm produces the worst possible solution"]. We need more than plausibility (and our intuition) to determine which statements are really true. What we need is the subject of the remainder of this chapter.

3.2   Boolean Expressions and Truth Tables

A **Boolean variable**, p, is a symbol that takes a **Boolean value**; either p is True or p is False (never both at the same time, and never neither). They denote *assertions* from ordinary language and are named after George Boole (1815–1864) who was a pioneer in formal logic. Most high-level computer languages include such variables and allow evaluation of Boolean expressions. Boolean variables by themselves are the simplest Boolean expressions; more complicated Boolean expressions may be constructed using Boolean operators.

3.2.1   The Negation Operator

The **negation** of a Boolean expression P [written $\sim P$ and read "not-P"] is True when P is False and is False when P is True.   // " $\sim$" reverses the truth-value of P. The effect of the negation operator is summarized in the table below.

P	$\sim P$
T	F
F	T

This **truth table** gives the value of the expression " $\sim P$" for all possible values of P.

3.2.2   The Conjunction Operator

The **conjunction** of a Boolean expression P with a second Boolean expression Q [written $P \wedge Q$ and read "P and Q"] is True when both P and Q are True and is

False otherwise. The effect of the conjunction operator is summarized in the table below.

P	Q	P ∧ Q
T	T	T
T	F	F
F	T	F
F	F	F

This truth table gives the value of the expression "$P \wedge Q$" for all possible combinations of truth-values of P and Q.

The conjunction operator is meant to reflect "and" (or "but") in ordinary language; the compound statement "Today is Monday and it is raining" is true only when today is Monday and also it is raining.   // So, most of the time it's False.

3.2.3 The Disjunction Operator

The **disjunction** of a Boolean expression P with a second Boolean expression Q [written $P \vee Q$ and read "P or Q"] is False when both P and Q are False and is True otherwise.

// $P \vee Q$ is True when P is True or Q is True or both are True.
// $\vee$ is the "inclusive or".

The effect of the disjunction operator is summarized in the table below.

P	Q	P ∨ Q
T	T	T
T	F	T
F	T	T
F	F	F

This truth table gives the value of the expression "$P \vee Q$" for all possible combinations of truth-values of P and Q.

The disjunction operator is meant to reflect "either…or" in ordinary language; the compound statement "Either he is smart or he is very lucky" is true whenever he is smart or very lucky or both.

On the other hand, in ordinary language, "P or Q" is sometimes used to indicate that either P is True or Q is True but _not both_ as in "I'll either get an A or a B on this assignment" or "Either Brazil or Germany will win the World Cup". This "exclusive or" may be represented by the Boolean expression

$$(P \vee Q) \wedge \sim (P \wedge Q)$$

Let's construct the truth table for this expression.

These Boolean operators are common to web search engines as well as computer languages, and all follow the same conventions or rules for evaluating expressions (similar to the rules for evaluating arithmetic expressions):

work from left to right
but   *evaluate sub-expressions inside parentheses first*
and   *do negations before conjunctions*
and   *do conjunctions before disjunctions.*

P	Q	$(P \vee Q)$	$\wedge$	$\sim$	$(P \wedge Q)$
T	T	T	F	F	T
T	F	T	T	T	F
F	T	T	T	T	F
F	F	F	F	T	F
		↑	↑	↑	↑
		1	4	3	2

// The rows of the table contradict each other, so at any time only one row occurs.
// Each row corresponds to specific values for P and Q, and in each row
// $(P \vee Q)$ is evaluated first.
// $\wedge$ waits for $\sim$.
// $\sim$ waits for $(P \wedge Q)$, so evaluating $(P \wedge Q)$ is done second.
//   Now, $\sim$ is done (third).
//   Finally, $\wedge$ is done (fourth) and we obtain the value of the whole expression.

The shaded column of the truth table gives the value of the expression "$(P \vee Q) \wedge \sim (P \wedge Q)$" for all possible combinations of truth-values of P and Q.

Two Boolean expressions P and Q are **equivalent** [written $P \Leftrightarrow Q$] means that they have exactly the same truth tables. For instance,

P	Q	$\sim(P \wedge Q)$		$\sim P \vee \sim Q$			$\sim (P \vee Q)$		$\sim P \wedge \sim Q$		
T	T	F	T	F	F	F	F	T	F	F	F
T	F	T	F	F	T	T	F	T	F	F	T
F	T	T	F	T	T	F	F	T	T	F	F
F	F	T	F	T	T	T	T	F	T	T	T

shows that $\sim(P \wedge Q) \Leftrightarrow (\sim P) \vee (\sim Q)$ and that $\sim(P \vee Q) \Leftrightarrow (\sim P) \wedge (\sim Q)$.

// "Not both P and Q" carries the same information as "either $(\sim P)$ or $(\sim Q)$".
// "Neither P nor Q" carries the same information as "both $(\sim P)$ and $(\sim Q)$".
// These two logical equivalences are known as De Morgan's Laws.
// The most obvious equivalence is $P \Leftrightarrow P$.
// Another obvious equivalence is $\sim(\sim P) \Leftrightarrow P$.

3.2.4 The Conditional Operator

By far the most important Boolean operator in mathematics is the conditional operator which is denoted by "$\rightarrow$". The Boolean expression "$P \rightarrow Q$" is meant to reflect the conditional statement "if P then Q" in ordinary language. This conditional statement means that "whenever P is True, then Q must also be True," or "it cannot happen that P is True and Q is False". Therefore, "$P \rightarrow Q$" is (logically) equivalent to "$\sim(P \wedge \sim Q)$".

In formal terms, the **conditional operator** $\rightarrow$ is defined by the following truth table.

P	Q	$P \rightarrow Q$	$\sim$	$(P$	$\wedge$	$\sim Q)$
T	T	T	T	T	F	F
T	F	F	F	T	T	T
F	T	T	T	F	F	F
F	F	T	T	F	F	T
			↑		↑	↑
			3		2	1

// Then (by De Morgan's Laws), $P \rightarrow Q$ is (also) equivalent to $(\sim P) \vee Q$.

Because so many mathematical statements take the form of a conditional expression, names have been given to the two parts: P, the part before the operator, is the **antecedent**; Q, the part after the operator, is the **consequent**. Also names have been given to certain variations of it:

> The **converse** of "$P \rightarrow Q$"   is "$Q \rightarrow P$".
> The **contrapositive** of "$P \rightarrow Q$"   is "$\sim Q \rightarrow \sim P$".
> The **inverse** of "$P \rightarrow Q$"   is "$\sim P \rightarrow \sim Q$".

// Are these all equivalent?

P	Q	$(P \rightarrow Q)$	$\sim Q$	$\rightarrow$	$\sim P$		$Q \rightarrow P$	$\sim P$	$\rightarrow$	$\sim Q$
T	T	T	F	T	F		T	F	T	F
T	F	F	T	F	F		T	F	T	T
F	T	T	F	T	T		F	T	F	F
F	F	T	T	T	T		T	T	T	T
		↑_____↑				and	↑_____↑			

Any conditional expression is equivalent to its contrapositive.

// And so the converse is equivalent to its own contrapositive form, the inverse.

But a conditional expression and its converse are <u>not</u> equivalent.

// Are they that much different?

Example 3.2.1: Conditional Variations

Suppose p denotes the statement "you were 90 on your last birthday" and q denotes the statement "you are over 21":

$p \to q$ denotes the conditional statement
                  "***If*** you were 90 on your last birthday, ***then*** you are over 21."
                  which is true, no matter who you are.

$\sim q \to \sim p$ denotes the contrapositive
                  "***If*** you are not over 21, ***then*** you were not 90 on your last birthday."
                  which is true, no matter who you are.

$q \to p$ denotes the converse
                  "***If*** you are over 21, ***then*** you were 90 on your last birthday."
                  which is false for almost everyone over 21.

// If your age last birthday was $>= 21$ but $\neq 90$, the antecedent is true but the
// consequent is false.

$\sim p \to \sim q$ denotes the inverse   // the converse of the contrapositive
                  "***If*** you were not 90 on your last birthday, ***then*** you are not over 21".
                  which is false for almost everyone over 21.

// If your age last birthday was $\neq 90$ but was $>= 21$, the antecedent is true but the
// consequent is false.

A Boolean expression P ***implies*** Boolean expression Q [written $P \Rightarrow Q$] <u>means</u> that the conditional expression, $P \to Q$, is <u>always</u> True; that is, whenever P is True, Q must be True. We will sometimes use this symbol to indicate one statement implies another like

"You were 90 on your last birthday" $\Rightarrow$ "you are over 21."

Here, the relation of implication occurs because of the *meanings* of the statements. But it may also occur just because of the *forms* of the Boolean expressions. For instance,

$$P \wedge Q \Rightarrow P \text{ and } P \Rightarrow P \text{ and } P \Rightarrow P \vee Q \text{ and } Q \Rightarrow (P \to Q).$$

In general, the conditional expression $P \to Q$ may be False, <u>but</u> when P and Q are related by their meanings or by their forms so that $P \to Q$ is <u>always</u> True, P is said to imply Q.

3.2.5 The Biconditional Operator

The *biconditional operator* $\leftrightarrow$ is defined by the following truth table.

P	Q	$P \leftrightarrow Q$
T	T	T
T	F	F
F	T	F
F	F	T

We can interpret "*P only if Q*" to mean "if P is True, then (because P <u>only</u> occurs when Q occurs) Q must be True". Then "*P only if Q*" would be denoted by "$P \rightarrow Q$". Since "*P if Q*" is denoted by "$Q \rightarrow P$", "*P if and only if Q*" would be denoted by

$$(P \rightarrow Q) \wedge (Q \rightarrow P).$$

The operator $\leftrightarrow$ is called the *biconditional* because it's (logically) equivalent to the conjunction of the two conditional expressions $(P \rightarrow Q)$ and $(Q \rightarrow P)$.

P	Q	$(P \rightarrow Q)$	$\wedge$	$(Q \rightarrow P)$
T	T	T	T	T
T	F	F	F	T
F	T	T	F	F
F	F	T	T	T
		$\uparrow$	$\uparrow$	$\uparrow$
		1	3	2

// Since $(Q \rightarrow P) \Leftrightarrow (\sim P \rightarrow \sim Q)$, we also have
// $(P \leftrightarrow Q) \Leftrightarrow [(P \rightarrow Q) \wedge (\sim P \rightarrow \sim Q)]$.

$P \leftrightarrow Q$ is read "P if and only if Q", and $P \leftrightarrow Q$ is True when P and Q have the same truth-value, and is False when they disagree. Therefore, two Boolean expressions P and Q are *equivalent* means that $P \leftrightarrow Q$ is <u>always</u> True.

The Most Important Ideas in This Section.
Five standard Boolean operators are defined using truth tables: *negation* [written $\sim P$ and read "not-P"], *conjunction* [written $P \wedge Q$ and read "P and Q"], *disjunction* [written $P \vee Q$ and read "P or Q"], *conditional* [written $P \rightarrow Q$ and read "if P then Q"], and *biconditional* [written $P \leftrightarrow Q$ and read "P if and only if Q"].

P implies Q [written $P \Rightarrow Q$] <u>means</u> $P \rightarrow Q$ is always True; P and Q are *equivalent* [written $P \Leftrightarrow Q$] <u>means</u> $P \leftrightarrow Q$ is always True.

Because so many mathematical statements take the form of a conditional expression, names have been given to the parts: P is the *antecedent* and Q is

(continued)

(continued)

> the *consequent*. Names also have been given to certain variations of $P \rightarrow Q$:
> the *converse* is $Q \rightarrow P$, the *contrapositive* is $\sim Q \rightarrow \sim P$, and the *inverse* of
> is $\sim P \rightarrow \sim Q$.
>
> This formalism is very useful in describing the structure of valid arguments,
> which we will come to soon.

3.3 Predicates and Quantifiers

The truth-value of an assertion like "you were 90 on your last birthday" depends
on who "you" are, and the truth-value of "$x^2 > 25$" depends on what number x is.
This section concerns these sorts of statements.

A **predicate** is a function from a set D into the set $C = \{\text{True, False}\}$. For
instance, suppose that D is the set of positive integers, we might let

$P(k)$ denote the assertion "k is prime"// So $P(13)$ is True but $P(33)$ is False.
and $Q(n)$ denote the assertion "n is odd". // So $Q(23)$ is True but $Q(32)$ is False.

The truth-value of $P(x)$ and of $Q(x)$ depends on which value x takes in the set D.
Predicates using specific values from D are Boolean expressions and can be
connected by Boolean operators to make larger, more complicated Boolean
expressions.

More interesting Boolean expressions can be created using "quantifiers". The
universal quantifier $\forall$ is read "for every" or "for all" and

"$\forall x\ F(x)$" denotes the assertion
"For every x in the domain of the predicate F, $F(x)$ is True".

The **existential quantifier** $\exists$ is read "there exists a" and

"$\exists x\ F(x)$" denotes the assertion
"There exists a value x in the domain of the predicate F, where $F(x)$ is True".

// $\forall$ is an upside down A and stands for "All".
// $\exists$ is a backward E and stands for there "Exists".

For instance, taking D to be the set of positive integers and P and Q to be the
predicates defined above,

"$\forall x\ P(x)$" denotes "every positive integer is prime," so as a Boolean expression
$\forall x\ P(x)$ is False   // because 6 is a positive integer that's not prime

"$\exists x\ [P(x) \wedge \sim Q(x)]$" denotes "there exists a positive integer that is prime and not odd" so
$\exists x\ [P(x) \wedge \sim Q(x)]$ is True.   // because 2 is such an integer

Negations, in a sense, interchange the two quantifiers. Suppose that D is the
domain of the predicate F.

"$\sim[\forall x\ F(x)]$" denotes "it is not the case that for every x in D, $F(x)$ is True", that is, "for some x in D, $F(x)$ is False" and this is denoted by "$\exists x \sim F(x)$". Thus,

$$\sim[\forall x\, F(x)] \Leftrightarrow \exists x \sim F(x)$$

"$\sim[\exists x\, F(x)]$" denotes "it is not the case that there exists an x in D, where $F(x)$ is True,"
that is, "for every x in D, $F(x)$ is False" and this is denoted by "$\forall x \sim F(x)$". Thus,

$$\sim[\exists x\, F(x)] \Leftrightarrow \forall x \sim F(x).$$

In the example of Cookie Problem #3, let U denote the set of all transversals of the given 6×6 array. The assertion that the greedy solution (with total 404) is the *best* transversal is the statement

$$\forall T \in U, \text{the value of } T \text{ is} <= 404.$$

We saw that this is False, by constructing a transversal T^* whose value was not $<= 404$. That is,

$$\exists T^* \in U, \text{where the value of } T^* \text{ is} > 404.$$

One *example* proves an existential assertion is True. One *counterexample* to a universal assertion proves the universal assertion is False.

// How can universal assertions be proved to be True?

The Most Important Ideas in This Section.
Predicates are Boolean expressions whose truth-value depends on (one or more) parameters. The *universal quantifier* [written $\forall$ and read "for every" or "for all"] and the *existential quantifier* [written $\exists$ and read "there exists a" or "there is some"] are used to express statements asserting something occurs for every input of a certain kind or for some particular input of that kind. Such statements are essential for describing program correctness.

3.4 Valid Arguments

An *argument* is a sequence of statements called *premises*, followed by a statement, called the *conclusion*. The purpose of the argument is to convince the audience of the truth of the conclusion. The form of an argument is

$$P1$$
$$P2$$
$$P3$$
$$\ldots$$
$$\underline{Pk}$$
$$\therefore C \qquad // \therefore \text{read "therefore" signals the conclusion.}$$

where the first k statements are the premises and the last, C, is the conclusion.

An argument is **valid** <u>means</u> *if all the premises are True, then the conclusion must be True*. Certain arguments are valid just because of their *form*, independent of their content. We can test the validity of an argument form by constructing the conditional statement whose antecedent is the conjunction of all the premises and whose consequent is the conclusion,

$$[P_1 \wedge P_2 \wedge P_3 \wedge \ldots \wedge P_k] \to C.$$

The argument form is valid if and only if this conditional statement is always true (the conjunction of the premises <u>implies</u> the conclusion).

Example 3.4.1: Modus Ponens
This standard argument form is

$$
\begin{array}{ll}
P \to Q & \textit{If it's raining, then the streets are wet.} \\
\underline{P \qquad} & \underline{\textit{It is raining.}} \\
\therefore Q & \therefore \textit{The streets are wet.}
\end{array}
$$

That this form is valid (independent of the meanings of P and Q) is shown by the following truth table.

P	Q	$[(P$	$\to Q)$	$\wedge$	$P]$	$\to$	Q
T	T		T	T		T	T
T	F		F	F		T	F
F	T		T	F		T	T
F	F		T	F		T	F

$$
\begin{array}{ccc}
\uparrow & \uparrow & \uparrow \\
1 & 2 & 3
\end{array}
$$

In the case about the wet weather, the first premise is True, and if (at some time and place) the second premise is also True, then (at that time and place) the conclusion must also be True.

Example 3.4.2: Modus Tollens
This standard argument form is

$$
\begin{array}{ll}
P \to Q & \textit{If it's raining, then the streets are wet.} \\
\underline{\sim Q \qquad} & \underline{\textit{The streets are not wet.}} \\
\therefore \sim P & \therefore \textit{It's not raining.}
\end{array}
$$

That this form is valid (independent of the meaning of P and Q) is shown by the following truth table.

P	Q	$[(P \rightarrow Q)$	$\wedge$	$\sim Q]$	$\rightarrow$	$\sim P$
T	T	T	F	F	T	F
T	F	F	F	T	T	F
F	T	T	F	F	T	T
F	F	T	T	T	T	T
		↑	↑	↑	↑	↑
		1	3	2	5	4

In the case about the weather, the first premise is True, and if (at some time and place) the second premise is also True, then (at that time and place) the conclusion must also be True.

Example 3.4.3: Conditional Syllogism
This standard argument form is

$$P \rightarrow Q \qquad \textit{If it's raining, then the streets are wet.}$$
$$\underline{Q \rightarrow R} \qquad \underline{\textit{If the streets are wet then she'll wear galoshes.}}$$
$$\therefore P \rightarrow R \qquad \therefore \textit{If it's raining then she'll wear galoshes}$$

That this form is valid (independent of the meaning of P, Q, and R) is shown by the following truth table.   // Here there are 3 Boolean expression and $8 = 2^3$ rows

P	Q	R	$[(P \rightarrow Q)$	$\wedge$	$(Q \rightarrow R)]$	$\rightarrow$	$(P \rightarrow R)$
T	T	T	T	T	T	T	T
T	T	F	T	F	F	T	F
T	F	T	F	F	T	T	T
T	F	F	F	F	T	T	F
F	T	T	T	T	T	T	T
F	T	F	T	F	F	T	T
F	F	T	T	T	T	T	T
F	F	F	T	T	T	T	T
			↑	↑	↑	↑	↑
			1	3	2	5	4

In the case about the wet weather, the first premise is True, and if (for a certain lady) the second premise is also True, then (for that lady) the conclusion must also be True.

These examples of valid argument forms perhaps seem obvious, but we want to consider one more case.

Example 3.4.4: The Scientific Method?
A high school teacher described the "scientific method" as a means of discovering knowledge through experimentation and observation, a process of formulating and testing hypotheses. It seemed to be based on this (argument) paradigm:

If my theory is correct, then my experiment will produce such and such.
Look! Look! My experiment did produce such and such.
∴My theory is correct.

This argument has the form

$$P \rightarrow Q$$
$$\underline{Q}$$
$$\therefore P \qquad \qquad \text{// Is this a valid form of argument?}$$

That this form is not valid is shown by the following truth table.

P	Q	$[(P \rightarrow Q)$	$\wedge$	$Q]$	$\rightarrow$	P
T	T	T	T T		T	T
T	F	F	F F		T	T
F	T	T	T T		F	F
F	F	T	F F		T	F
↑		↑			↑	
1		2			3	

The third row shows that the two premises may both be True but the conclusion might be False. This invalid argument form is known as "*the fallacy of affirming the consequent*".

In the case about the wet weather, the argument would be:

> *If it's raining, then the streets are wet.*
> *The streets are wet*
> ———————————————————
> *∴It is raining.*

The first premise is True. But could it ever happen that (at some time and place) the streets are wet but it is not raining?

// after a thunderstorm is over
// as the snow is melting
// as the river floods the town
// after the street-washing truck passes
// ...

The conclusion would only be correct if Q only occurred when P occurs.

// That is, $Q \rightarrow P$

The methodology of testing hypotheses by experiment and observation is very important and, in general, very effective. It is the basis of much scientific progress. But when an experiment produces the results predicted by the theory, the scientist cannot be certain her/his theory is correct. The only thing she/he can be certain of is that her/his own experiment did not disprove her/his own theory.

// by modus tollens

Computer programmers too might be tempted to use the paradigm:

> *If my algorithm is correct, then my output will be such and such.*
> *Look! Look! My output is such and such.*
> ———————————————————
> *∴My algorithm is correct.*

The conclusion may be true, but this argument alone is not enough to guarantee its truth. (The first premise must be strengthened to say that the only way to get such and such output is by using a correct algorithm.) On the other hand, we will never risk making such a logical oversight; we'll prove our algorithms are correct by valid means.

> **The Most Important Ideas in This Section.**
> A *valid argument* is a (finite) sequence of *premises* followed by a *conclusion* where *if all the premises are True, then the conclusion must be True*. Several classical patterns of argument are shown to be valid and examples are given. The main point is that correct arguments have correct form (as well as content and meaning). This is applied and generalized in the next section on proofs.

3.5   Examples of Proofs

We've given a rather long-winded introduction to proofs emphasizing their form; now we want to add content to the arguments. We are particularly interested in proving universal statements about mathematical objects, especially algorithms.

> *A PROOF is a valid argument where all the premises are True.*
>
> // And therefore the conclusion must be True.

We'll look at several examples of proofs next. Remember that a proof is an attempt to strengthen your confidence that a certain statement is True to the point of certainty. Statements that can be proved to be True are known as ***theorems***.

Theorem 3.5.1:   For all integers n, $\lfloor n/2 \rfloor + \lceil n/2 \rceil = n$.

Proof. Either n is even or n is odd; that is, either $n = 2q$ or $n = 2q + 1$ where $q \in \mathbf{Z}$. If $n = 2q$ where $q \in \mathbf{Z}$, then $n/2 = q$ so $\lfloor n/2 \rfloor = q = \lceil n/2 \rceil$ and hence,

$$\lfloor n/2 \rfloor + \lceil n/2 \rceil = q + q = n.$$

If $n = 2q + 1$ where $q \in \mathbf{Z}$, then $n/2 = q + \frac{1}{2}$, so $\lfloor n/2 \rfloor = q$ and $\lceil n/2 \rceil = q + 1$ and hence,

$$\lfloor n/2 \rfloor + \lceil n/2 \rceil = q + (q + 1) = n.$$

Therefore, for all integers n, $\qquad \lfloor n/2 \rfloor + \lceil n/2 \rceil = n$. $\qquad\qquad$ ☐

// The universality of this argument is due to the "magic of algebra"; the *variable*
// n was used to denote an integer whose value is unspecified − so n represents any

// integer, and therefore all integers are represented at once.

// We will use ☐ as a symbol indicating the end of the proof, for QED.

The underlying form of this argument is

Example 3.5.1: Disjunctive Syllogism

$$P \vee Q \qquad \textit{Either } n \textit{ is even or } n \textit{ is odd.}$$
$$P \rightarrow R \qquad \textit{If } n \textit{ is even, then} \qquad \lfloor n/2 \rfloor + \lceil n/2 \rceil = n.$$
$$\underline{Q \rightarrow R \qquad \textit{If } n \textit{ is odd, then} \qquad \lfloor n/2 \rfloor + \lceil n/2 \rceil = n.}$$
$$\therefore R \qquad \qquad \therefore \lfloor n/2 \rfloor + \lceil n/2 \rceil = n.$$

That this form is valid (independent of the meaning of P, Q, and R) is shown by the following truth table.

P	Q	R	$[(P$	$\vee$	$Q)$	$\wedge$	$(P \rightarrow$	$R)$	$\wedge$	$(Q \rightarrow$	$R)]$	$\rightarrow$	R
T	T	T			T		T		T		T	T	T
T	T	F			T		F		F		F	T	F
T	F	T			T		T		T		T	T	T
T	F	F			T		F		F		T	T	F
F	T	T			T		T		T		T	T	T
F	T	F			T		T		F		F	T	F
F	F	T			F		T		F		T	T	T
F	F	F			F		T		F		T	T	F

$$\begin{array}{cccccc} \uparrow & \uparrow & \uparrow & \uparrow & \uparrow & \uparrow \\ 1 & 3 & 2 & 5 & 4 & 6 \end{array}$$

//X Theorem 3.5.1 may be generalized to

// If f is any real number and $g = (1 - f)$, then for all integers n, $\lfloor f \times n \rfloor + \lceil g \times n \rceil = n$.

// In Theorem 3.5.1, $f = \frac{1}{2}$ (and $g = \frac{1}{2}$).

// (A proof of the generalized theorem will be quite different.)

The Disjunctive Syllogism itself may be generalized to "Proof by Cases" as in

Theorem 3.5.2:   Among any three consecutive integers, there is a multiple of 3.

Proof. If the smallest of the three consecutive integers is denoted by n, then the three integers are n, $n + 1$, and $n + 2$. We must prove that

for all integers n, (at least) one of $n, n + 1$, and $n + 2$ is a multiple of 3.

Either $n = 3q$, or $n = 3q + 1$, or $n = 3q + 2$ where $q \in \mathbf{Z}$.   // n MOD 3 $\in \{0,1,2\}$.

Case 1. If $n = 3q$ where $q \in \mathbf{Z}$, then n itself is a multiple of 3.

Case 2. If $n = 3q + 1$ where $q \in \mathbf{Z}$, then $n + 2 = 3q + 3 = 3(q + 1)$, so $n + 2$ is a multiple of 3.

Case 3. If $n = 3q + 2$ where $q \in \mathbf{Z}$, then $n + 1 = 3q + 3 = 3(q + 1)$, so $n + 1$ is a
     multiple of 3.

Therefore, among any three consecutive integers, there is a multiple of 3.   ☐

//X Theorem 3.5.2 may be generalized to
// *Among any K consecutive integers, there is a multiple of K.* (A proof of the
// generalized theorem will be quite different and use more algebra.)
// You might also prove that
//   *Among any K consecutive integers, there is exactly one multiple of K,* and/or
//   *Among any K consecutive integers, there is a multiple of k for k = 2,3,...K.*
//   *The product of any K consecutive integers is a multiple of K!.*

The first two sentences of the proof of Theorem 3.5.2 are there to introduce
algebraic notation and then restate the theorem. The underlying form of the
argument in this proof is

Example 3.5.2: Proof by Cases

$$P1 \lor P2 \lor P3 \lor \ldots \lor Pk \qquad \text{// lists all cases}$$
$$P1 \to R \qquad\qquad\qquad\qquad \text{// case 1}$$
$$P2 \to R \qquad\qquad\qquad\qquad \text{// case 2}$$
$$P3 \to R \qquad\qquad\qquad\qquad \text{// case 3}$$
$$\ldots$$
$$\underline{Pk \to R \qquad\qquad\qquad\qquad} \text{// the last case}$$
$$\therefore R$$

To prove that "Proof by Cases" is a valid argument form for *all* (finite) positive
integers k, we cannot just construct truth tables. We'll have to give an argument that
shows that when all the premises are True, the conclusion R must be True.

Proof. (that "Proof by Cases" is a valid argument form)
    If all the premises are True, then (the first one which describes all possible cases)
$P1 \lor P2 \lor P3 \lor \ldots \lor Pk$ is True; because this is a disjunction, (at least) one of the
P's is True, and so $\exists j$ where $1 <= j <= k$ and Pj is True.
    If all the premises are True, then the premise $Pj \to R$ is True, and because we
know the antecedent of this conditional Pj is True, the consequent R must be True.
    Therefore, if all the premises are True, the conclusion R must be True; that is,
"Proof by Cases" is a valid argument form.   ☐

    That proof was constructed by expanding on the *meaning* of "a valid argument
form" and explaining what *must* result when all the premises are True in more and
more detail, until we knew that the conclusion of the argument must also be True.
This is a common form of proof and is often referred to as a

3.5.1   Direct Proof

Theorem 3.5.3:   If n is an odd integer, then n^2 is also an odd integer.

Proof. If n is an odd integer, then $n = 2q + 1$ where $q \in \mathbf{Z}$.
If $n = 2q + 1$ where $q \in \mathbf{Z}$, then

$$n^2 = (2q+1)^2 = (2q)^2 + 2(2q) + 1 \qquad\qquad // \ (a+b)^2 = a^2 + 2ab + b^2$$
$$= 4q^2 + 4q + 1$$
$$= 2(2q^2 + 2q) + 1 \ \ \text{where } (2q^2 + 2q) \in \mathbf{Z}$$

If $n^2 = 2(2q^2 + 2q) + 1$ where $(2q^2 + 2q) \in \mathbf{Z}$, then n^2 is an odd integer.
Therefore, if n is an odd integer, then n^2 is (also) an odd integer.   ☐

// Isn't the product of any two odd integers also an odd integer?
//X Construct a direct proof that for positive integers
//   if $a|b$ and $b|c$, then $a|c$.
//X Prove that the smallest proper divisor of an integer $n > 1$ must be prime.

The underlying form of that last argument (the proof of Theorem 3.5.3) is a sequence of conditional statements in the following pattern.

Example 3.5.3: Generalized Conditional Syllogism

$$P[1] \rightarrow P[2]$$
$$P[2] \rightarrow P[3]$$
$$P[3] \rightarrow P[4]$$
$$\dots$$
$$\frac{P[k] \rightarrow P[k+1]}{\therefore P[1] \rightarrow P[k+1]}$$

To prove that the Generalized Conditional Syllogism is a valid argument form for *all* (finite) positive integers k, we'll have to give an argument that shows that when <u>all</u> the premises are True, the conclusion $P[1] \rightarrow P[k+1]$ also must be True. However, it is somewhat easier to establish the contrapositive form:

If the conclusion "$P[1] \rightarrow P[k+1]$" is False,

then (at least) one of the premises is False.

// This may seem backward but remember that $(P \rightarrow Q) \Leftrightarrow (\sim Q \rightarrow \sim P)$;
// that is, a conditional and its contrapositive are logically equivalent.

Proof. (that the Generalized Conditional Syllogism is a valid argument form)
    If the conclusion $P[1] \rightarrow P[k+1]$ is False, then $P[1]$ is True and $P[k+1]$ is False. Let j be the smallest index where $P[j]$ is False, then $1 < j$ and $j <= k+1$.

// Must there be a smallest such index? A first case where $P[j]$ is False?

By this choice of j, we know that $P[j - 1]$ is not False. Since $1 <= j - 1 <= k$, "$P[j - 1] \to P[j]$" is one of the premises, where $P[j - 1]$ is True and $P[j]$ is False. Therefore the premise $P[j - 1] \to P[j]$ is False.   □

// Often this argument (a direct proof) takes a less formal, less repetitive, and more
// compact shape.
//
//   If $P[1]$ OR   "Suppose" or "Assume" $P[1]$.
//   then $P[2]$.   OR   "Hence" or "So" or "Therefore" $P[2]$.
//   Then $P[3]$.   "
//   ...   "
//   Then $\underline{P[k+1]}$.
// $\therefore P[1] \to P[k+1]$

Proving a conditional statement by showing its contrapositive is True seems strange but is correct. This "inverted" method of argument is more pronounced (and more counterintuitive) in an

3.5.2 Indirect Proof

This argument has the form

$$\frac{\begin{array}{l} \sim P \to Q \\ \sim P \to \sim Q \end{array}}{\therefore P}$$ // Can this be valid?

That this form is valid (independent of the meaning of P and Q) is shown by the following truth table.

P	Q	$[(\sim P \to$	$Q) \wedge$	$(\sim P \to \sim Q)]$	$\to$	P
T	T	T	T	T	T	T
T	F	T	T	T	T	T
F	T	T	F	F	T	F
F	F	F	F	T	T	F
		↑	↑	↑	↑	
		1	3	2	4	

// This truth table actually shows that $[(\sim P \to Q) \wedge (\sim P \to \sim Q)] \Leftrightarrow P$.
// Modus tollens shows that $[(\sim P \to Q) \wedge (\sim Q)] \Rightarrow P$.

Perhaps the most famous application of this method of proof is in the following story. Pythagoras, renowned for his theorem about right-angled triangles, established a school at Syracuse (in Italy) about 500 BC. Actually, it was more like a monastery. Mathematics was studied but numbers were revered. Numbers (for these ancient scholars) were the positive integers. One of their beliefs about numbers was that any two line segments were *commensurable* – that is, there is some unit of length such that both line segments were an integer number of units long. However, some unknown student showed that the diagonal and side of any square are <u>not</u> commensurable.

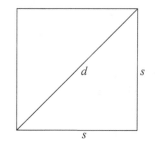

If the diagonal were *d* units long and the side were *s* units long, then by Pythagoras' famous theorem,

$$d = \sqrt{s^2 + s^2} = \sqrt{2s^2} = \sqrt{2} \times s \text{ and so } \sqrt{2} = d/s.$$

Thus, if the diagonal and side of a square were commensurable, then $\sqrt{2}$ would be a rational number. But the student proved

Theorem 3.5.4: $\sqrt{2}$ is an irrational number.

Proof. Suppose that $\sqrt{2}$ were a rational number.   // Suppose $\sim P$.
Then $\sqrt{2}$ may be written as a ratio of two integers; that is,

$$\sqrt{2} = A/B \text{ where both } A \text{ and } B \text{ are integers and } B \neq 0.$$

Because $\sqrt{2}$ is positive, both A and B may be taken to be positive. Then we could divide both A and B by GCD(A,B) and write the fraction in "lowest terms". Thus, we have

if $\sqrt{2}$ is a rational number
then $\exists a, b \in \mathbf{P}$ such that $\sqrt{2} = a/b$ and $GCD(a,b) = 1$.   // This is $\sim P \to Q$.

On the other hand, if $\sqrt{2} = a/b$ where $a,b \in \mathbf{P}$, then

$$\sqrt{2} \times b = a \qquad \text{// multiplying both sides by } b$$
so $$2 \times b^2 = a^2; \qquad \text{// squaring both sides}$$
that is, $$a^2 = 2 \times b^2.$$

Then, since a^2 is even, a itself must be even.

// the contrapositive of Theorem 3.5.3

Hence, $a = 2r$ for some integer r, and

$$2 \times b^2 = a^2 = (2r)^2 = 4 \times r^2,$$

so $b^2 = 2 \times r^2.$ // an even integer

Then, since b^2 is even, b (too) must be even.   // again by Theorem 3.5.3
We've shown that

$$\text{if } \sqrt{2} = a/b, \text{ then } GCD(a,b) \neq 1,$$

which is logically equivalent to $\sim\{(\sqrt{2} = a/b) \text{ and } (GCD(a,b) = 1)\}$. Thus, we
have

$$\textit{if } \sqrt{2} \textit{ is a rational number},$$
$$\textit{then } \forall a,b \in P \textit{ either } \sqrt{2} \neq a/b \textit{ or } GCD(a,b) \neq 1. \qquad \textit{// This is } \sim P \to \sim Q.$$

Therefore, $\sqrt{2}$ is <u>not</u> a rational number.   □

Because this was heresy at the academy in Syracuse, the clever student was "cast
out"; some say thrown off a cliff into the sea, but maybe "cast out" just means
expelled.

//   A similar indirect argument [and the prime factorization theorem]
//   can be used to prove two other theorems:
//X *if p is prime, then $\sqrt{p}$ is irrational*, and [the more general assertion]
//X *if n is a positive integer, then $\sqrt{n}$ is either an integer or is irrational.*
//   Almost all machine calculations involving square roots will entail round-off
//   error.

We will do a few more examples of indirect proofs. The pattern of these
arguments to prove some assertion P is:
1. Assume $\sim P$ [or suppose P is False].
2. Using that assumption, deduce Q.
3. Again, using that assumption (if necessary), deduce $\sim Q$.
4. Conclude P.
    The next example of an indirect proof is an ancient theorem concerning prime
numbers, but first let's look at an ancient algorithm for determining all primes $<= n$
known as **the sieve of Eratosthenes** (276–195 BC):
Step 1. Write down all the integers from 2 to n and set $p = 2$. // the smallest prime
Step 2. While $(p^2 <= n)$
                    – Strike out all multiples of p beginning at p^2 // They cannot be prime.
                    – Find the first number $q > p$ that's not struck out.   // q will be prime.
                    – Set $p = q$.

Now, all the integers that remain (the ones not struck out) are primes.
Walk through when $n = 25$:

2 3 4 5 6 7 8 9 10 11 12 13 14 15 16 17 18 19 20 21 22 23 24 25   // $p = 2$
2 3 4 5 6 7 8 9 ~~10~~ 11 ~~12~~ 13 ~~14~~ 15 ~~16~~ 17 ~~18~~ 19 ~~20~~ 21 ~~22~~ 23 ~~24~~ 25   // $p = 3$
2 3 4 5 6 7 **8** 9 ~~10~~ 11 ~~12~~ 13 ~~14~~ ~~15~~ ~~16~~ 17 ~~18~~ 19 ~~20~~ ~~21~~ ~~22~~ 23 24 25   // $p = 5$
2 3 4 5 **6** 7 **8** 9 ~~10~~ 11 ~~12~~ 13 ~~14~~ ~~15~~ ~~16~~ 17 ~~18~~ 19 ~~20~~ ~~21~~ ~~22~~ 23 ~~24~~ ~~25~~   // $p = 7$

The primes $<= 25$ are the nine integers that remain:

2 3 5 7   11   13   17   19   23

If we imagine this process being done on *all* the integers $>= 2$. Each time we find a new prime, an <u>infinite</u> number of multiples of it are struck out. Is it conceivable that after a certain (large) number of primes have been discovered, <u>all</u> larger numbers have been struck out? And there are no more primes? Is there a largest prime?

Theorem 3.5.5: There is an infinite number of primes.

Proof. Suppose that there were only a finite number of primes; let N be that number. Then the set of primes may be indexed so that

$$(p_1, p_2, p_3, \ldots, p_N) \text{ is a list of } \underline{\text{all}} \text{ primes.} \qquad // \text{This is } Q.$$

Now consider the integer $K = (p_1 \times p_2 \times p_3 \times \ldots \times p_N) + 1$, the product of all the primes plus one. No prime number in the list divides evenly into K because K MOD $p = 1$ for each p in the list. But K is a positive integer greater than one, so either K is prime or K has a smallest proper divisor q (which must be a prime). Thus, there is a prime p^* that divides evenly into K. Since p^* cannot be in the list,

$$(p_1, p_2, p_3, \ldots, p_N) \text{ is } \underline{\text{not}} \text{ a list of all primes.} \qquad // \text{ This is } {\sim}Q.$$

Therefore, the number of primes is infinite.   ☐

3.5.3   Cantor's Diagonalization Process

Our next example of an *indirect* proof illustrates Cantor's Diagonalization Process. Recall from Chap. 2 that two sets A and B have the same cardinality (size) if and only if there is a function $f: A \to B$ that is one to one and onto. We are going to show you two infinite sets that cannot have the same cardinality.

It would seem that the decimal expansion of every real number r between 0 and 1 corresponds to a sequence of digits; that is,

$$r = .d_{-1}d_{-2}d_{-3}\ldots$$

where each d_{-j} is in $D = \{0, 1, \ldots , 9\}$. Sometimes the sequence terminates and sometimes it continues forever.

Let $\mathfrak{S}$ denote all possible infinite sequences with domain $\mathbf{P}$ and co-domain D.

Theorem 3.5.6:   There cannot exist a function from P onto $\mathfrak{S}$.

// The (infinite) size of $\mathfrak{S}$ is not equal to the (infinite) size of $\mathbf{P}$; $\mathfrak{S}$ is bigger.

Proof. Suppose that there were a function f from $\mathbf{P}$ *onto* $\mathfrak{S}$. Then all the elements of $\mathfrak{S}$ can be indexed using f so that   // setting $S_j = f(j)$ for each $j \in \mathbf{P}$

$$\mathfrak{S} = \{S_j : j \in \mathbf{P}\}.$$ // This is Q.

// We will show that this "equation" is not correct by constructing a sequence T
// that is in $\mathfrak{S}$ but is different from every S_j.

If j is any (fixed) positive integer, S_j is an infinite sequence of digits, say

$$S_j = \left(d_{j_1}, d_{j_2}, d_{j_3}, d_{j_4}, \ldots, d_{jj}, \ldots \right).$$

Let $T = (t_1, t_2, t_3, t_4, \ldots)$ be the sequence of digits where for each $j \in \mathbf{P}$, $t_j = (9 - d_{jj})$. Then $T \in \mathfrak{S}$ because each entry $t_j \in D$. But for each $k \in \mathbf{P}$, the sequence $T \neq S_k$ because the kth entry in T is $\neq$ the kth entry in S_k; that is,

$\qquad\qquad t_k = (9 - d_{kk}) \neq d_{kk}.$ // If $9 - x = x$, then $x = 9/2 \notin \{0..9\}$.
Hence, $\qquad \mathfrak{S} \neq \{S_j : j \in \mathbf{P}\}.$ // This is $\sim Q$.

Therefore, there is no function from $\mathbf{P}$ *onto* $\mathfrak{S}$.   □

There is an elaboration of this argument that shows that there are more functions from $\mathbf{P}$ into $\{0..9\}$ (or into $\mathbf{P}$, or even into $\{0,1\}$) than there are computer programs (in any language, of any finite length). Therefore, there are (infinitely many) functions whose values cannot be computed (by computer programs)!

This theorem shows that there are infinite sets of different sizes (cardinalities).

// There are several "levels" of infinity?

Our last example of an indirect proof is

Theorem 3.5.7:   There is no smallest positive rational number.

Proof. Suppose there were a smallest positive rational number, let R_0 be that number. Then

$\qquad\qquad R_0$ *is the smallest positive rational number.*   // This is Q.

Then, since $0 < \frac{1}{2} < 1$ and since R_0 is positive,

$$0 = 0 \times R_0 < \frac{1}{2} \times R_0 < 1 \times R_0 = R_0.$$

If we let $S = \frac{1}{2} \times R_0$, then S is a rational number, S is positive, and S is less than R_0. So

R_0 is <u>not</u> the smallest positive rational number.   // This is $\sim Q$.

Therefore, there is no smallest positive rational number.   ▢

// There is no smallest positive real number either
//   − just replace the word "rational" by "real" throughout the above argument.
// But in every machine, there is a smallest positive representable number.

On the other hand, there is a smallest positive integer, namely 1. A stronger form of this statement is an *axiom* for the integers called

The Well-Ordering Principle:

   Every nonempty subset of positive integers has a smallest element.

The Most Important Ideas in This Section.
A *proof* is an attempt to strengthen confidence that a certain statement is True to the point of certainty. Statements that can be proved to be True are known as *theorems*.

   Several more patterns of argument were shown to be valid, and examples were given with content and meaning. *Disjunctive Syllogism* is extended to *Proof by Cases*. *Conditional Syllogism* is generalized and expanded to the idea of a *direct proof*. Proving a conditional statement by showing its contrapositive is True is generalized and expanded to the idea of an *indirect proof*, and several examples were given.

   The section ends with the *Well-Ordering Principle*, the basis for *Mathematical Induction* and *Strong Induction* given in the next section; these may be the most useful argument forms we'll present in this book.

3.6   Mathematical Induction

Mathematical Induction is used to prove universal assertions of the form

$$\forall \text{ integers } n >= a,\ P(n)$$

where P is some predicate whose domain is (or includes) the interval $\{a.. \}$. The form of the argument is

$$P(a)$$
$$\underline{P(k) \rightarrow P(k+1) \quad \text{for } \forall k \in \{a..\ \}}$$
$$\therefore \forall n \in \{a..\ \} \quad P(n)$$

The first premise asserts that

> *the predicate is True for the <u>one</u> value, a, the smallest element in $\{a..\ \}$.*

The second premise asserts that

> *if P is True for <u>any</u> particular value $k \in \{a..\ \}$,*
>
> *then P must also be True for the next value, $k+1$.*

We find that proofs by Mathematical Induction are most clearly presented as a three-step method where the conditional premise is divided into two parts.

Step 1. *Prove that P(a) is True.*   // Do the "basis" step.
Step 2. *Assume $\exists\ k \in \{a..\ \}$ where P(k) is True.* // State the "inductive hypothesis".
Step 3. *Use that assumption to prove that $P(k + 1)$ is also True.*

                                                                        // Do the "induction" step:

// Step 1 is done by substituting the value of a into the predicate.
//
// Step 2 is written with an <u>existential</u> quantifier to show we're not assuming
//   what we intend to prove (the universal assertion),
//   and is done by substituting a single but unspecified value
// k into the predicate.   // the magic of algebra again
//
// Step 3 is the hard part and usually involves some algebra and some argument.

Let's look at an instance where we can devise a universal conjecture and then construct its proof by Mathematical Induction.

Example 3.6.1: Comparing n^2 and 2^n

n	n^2		2^n
0	0		1
1	1		2
2	4	=	4
3	9	>	8
4	16	=	16
5	25		32
6	36		64

// In the table, $n^2 < 2^n$ except for three cases: $n = 2$, 3, and 4.
// Since 2^n grows larger faster than n^2, for large n it's likely that $n^2 < 2^n$. In fact,

Theorem 3.6.1: $\forall$ **integers** $n >= 5$, $n^2 < 2^n$.

// Here, $a = 5$ and $P(n)$ is the Boolean expression "$n^2 < 2^n$".
// $P(n)$ may be evaluated for any real number n, so $P(n)$ is defined for $\forall n \in \{a.. \}$.

Proof. // by Mathematical Induction
Step 1. If $n = 5$, then $n^2 = 25 < 32 = 2^n$.   // $P(5)$ is True.
Step 2. Assume $\exists$ $k \in \{5...\}$ where $k^2 < 2^k$.   // where $P(k)$ is True
Step 3. If $n = k + 1$, then

$$
\begin{aligned}
n^2 = (k+1)^2 &= k^2 + 2k + 1 \\
&< k^2 + 2k + k && // \ k >= 5 \text{ so } k > 1. \\
&= k^2 + 3k \\
&< k^2 + k \times k && // \ k >= 5 \text{ so } k > 3. \\
&= k^2 \times 2 \\
&< 2^k \times 2 && // \text{ by Step 2.} \\
&= 2^{k+1} = 2^n.
\end{aligned}
$$

// If $n = k + 1$, then $n^2 < 2^n$; that is, $P(k + 1)$ is True.

  Therefore, $\forall$ integers $n >= 5$, $n^2 < 2^n$.   ☐

//X Is $n < 2^n$ for $\forall$ integers $n \in \mathbf{N}$?
//X Is $n^2 < 3^n$ for $\forall$ integers $n \in \mathbf{N}$?

   The validity of Mathematical Induction as an argument form is based on the Well-Ordering Principle for integers. We can use it to prove the contrapositive of "if all the premises are True, then the conclusion must be True".

// We will show that if the conclusion is False,
// then one or other of the two premises must be False.

   If "$\forall n \in \{a.. \}$ $P(n)$" is not True, then $\exists n \in \{a.. \}$ where $P(n)$ is False. Let

$$Y = \{n \in \{a.. \}: P(n) \text{ is False}\}.$$

Then Y is a nonempty subset of the interval of integers $\{a.. \}$.

// The value of a might be negative and Y might include some negative numbers.

Let

$$X = \{1 + n - a: n \in Y\}.$$

When $n >= a$, $1 + n - a >= 1 + 0 > 0$. Therefore, X is a nonempty subset of positive integers, and by the Well-Ordering Principle, X has a smallest element x. This x must equal $1 + n^* - a$ for some $n^* \in Y$.

// If Y had an element $q < n^*$, then X would have an element
// $w = 1 + q - a < 1 + n^* - a = x$.
// If n^* is *not* the smallest element of Y, then x is *not* the smallest element of X;
// that is, if x is the smallest element of X, then n^* is the smallest element of Y.

Because n^* is the *smallest element* of Y,

$$P(n^*) \text{ is False,} \quad \text{but}$$
$$\text{if } a <= n < n^*, \text{ then } P(n) \text{ is } \underline{not} \text{ False.}$$

// In fact, $P(a) \wedge P(a+1) \wedge \ldots \wedge P(n^* - 1)$ would be True.

Either $n^* = a$ or $n^* >= a + 1$.
If $n^* = a$, then the first premise is False.   // $P(a)$ is False.
If $n^* >= a + 1$, then

$$n^* - 1 >= a, \quad P(n^* - 1) \text{ is True but } P(n^*) \text{ is False,}$$

and so the conditional statement "$P(n^* - 1) \to P(n^*)$" is False. But then the second premise,

$$\text{"}P(k) \to P(k+1) \quad \text{for } \forall k \in \{a.. \}\text{"}$$

is False because $\exists \ k \in \{a.. \}$ where $P(k) \to P(k + 1)$ is False, namely $k = n^* -1$.

Therefore, Mathematical Induction is a valid form of argument for any predicate P and any interval $\{a.. \}$.   □

Mathematical Induction [which we'll abbreviate to MI] is a technique for *proving* universal statements. Often it is of little help in discovering such theorems. However, when mathematical objects of size $(k + 1)$ are constructed (algorithmically) from objects of size k, many properties of *all* such objects can be easily proved using MI. *Mathematical Induction is particularly useful in analyzing algorithms.*

The rest of this section is a collection of proofs by Mathematical Induction. Section 3.7 gets back to algorithms and proving their correctness. The next few examples prove formulas for series.

Suppose we'd like to find a formula for the sum of the integers from 1 to n. Consider the array of naughts and crosses:

```
o   +   +   +   +   +
o   o   +   +   +   +
o   o   o   +   +   +
o   o   o   o   +   +
o   o   o   o   o   +
```

Counting from the top row down, the number of o's is $(1 + 2 + 3 + 4 + 5)$, and counting from the bottom row up, the number of +'s is $(1 + 2 + 3 + 4 + 5)$. The whole array is 5 rows where each row contains 6 symbols. Since the number of symbols equals the number of o's plus the number of +'s, we have

$$(1+2+3+4+5)+(1+2+3+4+5) = 5(6),$$
so $$(1+2+3+4+5) = 5(5+1)/2.$$

This pattern generalizes to $1 + 2 + 3 + \ldots + n = n(n + 1)/2$.

However, before we continue, we want to clarify the conventions that are usually followed when interpreting the *ellipses* in the formulas like the one on the left-hand side. To say that

"$1 + 2 + 3 + \ldots + n$" *means*

"add the integers, beginning with 1 and then 2 and then 3,

and so on, going up 1 unit at a time but stopping at n".

makes sense when n is bigger than 3.   // What about smaller values?

When $n = 3$ "$1 + 2 + 3 + \ldots + n$" *means* "$1 + 2 + 3$";   // start at 1 and stop at 3.
When $n = 2$ "$1 + 2 + 3 + \ldots + n$" *means* "$1 + 2$";   // start at 1 and stop at 2.
When $n = 1$ "$1 + 2 + 3 + \ldots + n$" *means* "1";   // start at 1 and stop at 1.
When $n = 0$ "$1 + 2 + 3 + \ldots + n$" would then mean

"add the integers starting at 1 going up 1 unit at a time but stopping at 0".

In this case, no integers are added. When $n = 0$, "$1 + 2 + 3 + \ldots + n$" is usually described as an "empty" sum. Empty sums are usually given the "default value" of zero.

Theorem 3.6.2: $\forall n \in \mathbf{P}$, $1 + 2 + 3 + \ldots + n = n(n + 1)/2$.

// Here $a = 1$ and $P(n)$ is an equation with a left-hand side (LHS)
//   and a right-hand side (RHS).

Proof. // by Mathematical Induction
Step 1. If $n = 1$, then LHS $= 1$ and RHS $= 1(1 + 1)/2 = 1$.   // $P(1)$ is True.
Step 2. Assume $\exists\, k \in \mathbf{P}$ where $1 + 2 + 3 + \ldots + k = k(k + 1)/2$.
                                                    // where $P(k)$ is True
Step 3. If $n = k + 1$, then

$$
\begin{aligned}
\text{LHS} &= 1 + 2 + 3 + \ldots + k & + (k + 1) \\
&= \{1 + 2 + 3 + \ldots + k\} & + (k + 1) \\
&= k(k + 1)/2 & + (k + 1) & \quad\text{// by Step 2} \\
&= k(k + 1)/2 & + (k + 1) \times 2/2 \\
&= \frac{k(k + 1) + (k + 1)2}{2} \\
&= \frac{(k + 1)[k + 2]}{2} \\
&= (k + 1)[(k + 1) + 1]/2 \\
&= \text{RHS.} & & \quad\text{// in the predicate } P
\end{aligned}
$$

// Therefore, $\forall n \in \mathbf{P}$, $1 + 2 + 3 + \ldots + n = n(n + 1)/2$.   □

// Is there a formula for the sum of the first n **even** positive integers?

// $2 + 4 + \ldots + (2n) = 2[1 + 2 + \ldots + n] = 2[n(n+1)/2] = n(n+1)$.

// Is there a formula for the sum of the first n **odd** positive integers?

Consider the 5×5 square array below and the "L"-shaped regions inside it – they each contain an odd number of small squares.

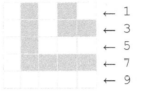

It looks as though the sum of the first 5 odd positive integers equals $5^2 = 25$.

// and the sum of the first 4 odd positive integers equals $4^2 = 16$,

// and the sum of the first 3 odd positive integers equals $3^2 = 9$.

// *Is the sum of the first n odd positive integers always equal n^2?*

The jth odd positive integer is $(2j - 1)$.   // $j >= 1$.

// This is obvious, but could be proved by Mathematical Induction.

Theorem 3.6.3: $\forall n \in \mathbf{P}, 1 + 3 + 5 + \ldots + (2n - 1) = n^2$.

// Here $a = 1$ and $P(n)$ is an equation with a LHS and a RHS.

Proof. // by Mathematical Induction

Step 1. If $n = 1$, then LHS $= 1$ and RHS $= 1^2 = 1$.   // $P(1)$ is True.

Step 2. Assume $\exists\, k \in \mathbf{P}$ where $1 + 3 + 5 + \ldots + (2k - 1) = k^2$.

                                                   // that is, $P(k)$ is True

Step 3. If $n = k + 1$, then

$$
\begin{aligned}
\text{LHS} &= 1 + 3 + 5 + \ldots + (2k - 1) &+ [2(k + 1) - 1] \\
&= \{1 + 3 + 5 + \ldots + (2k - 1)\} &+ [2k + 2 - 1] \\
&= k^2 &+ [2k + 2 - 1] &\quad \text{// by Step 2} \\
&= k^2 &+ 2k + 1 \\
&= (k + 1)^2 \\
&= \text{RHS}. &&\quad \text{// in the predicate } P
\end{aligned}
$$

// Therefore, $\forall n \in \mathbf{P}, 1 + 3 + 5 + \ldots + (2n - 1) = n^2$.   ⬚

When S is a sequence with domain $\{a.. \}$ and co-domain $\mathbf{R}$,

          S is an **arithmetic sequence**   means
    there is a number b such that for all $n >= a$, $S(n + 1) = S(n) + b$.

// $S = (6, 8, 10, 12, \ldots)$ is an arithmetic sequence with $b = 2$.

// Each entry, after the first, is obtained from the previous entry by adding the

// constant value b, which is called *"the common difference"*.
// The equation "$S(n + 1) = S(n) + b$" is an example of a *recurrence equation* [RE].

If $a = 0$ and S_0 equals some *initial* value, I, then

$$S_1 = I + b,$$
$$S_2 = (I + b) + b \; = I + 2b,$$
$$S_3 = (I + 2b) + b = I + 3b,$$
$$S_4 = (I + 3b) + b = I + 4b,$$

and in general, the *entries in an arithmetic sequence* are given by the formula $S_n = I + nb$.

Theorem 3.6.4: If S is an arithmetic sequence with common difference b, then $\forall n \in \mathbf{N}, S_n = I + nb$ where $I = S_0$.

// I is the initial value in S.

// Here $a = 0$ and $P(n)$ is the equation $S_n = I + nb$.

Proof. // by Mathematical Induction, using the RE: $S_{q+1} = S_q + b \; \forall q \in \mathbf{N}$.
Step 1. If $n = 0$, then $S_n = I$ and RHS $= I + 0 \times b = I$.   // $P(0)$ is True.
Step 2. Assume $\exists \, k \in \mathbf{N}$ where $S_k = I + kb$.   // $P(k)$ is True.
Step 3. If $n = k + 1$, then

$$
\begin{aligned}
S_n = S_{k+1} &= S_k + b & &\text{// using the RE} \\
&= \{I + kb\} + b & &\text{// by Step 2} \\
&= I + (kb + b) & & \\
&= I + (k + 1)b & & \\
&= \text{RHS.} & &\text{// in the predicate } P
\end{aligned}
$$

// Therefore, $\forall n \in \mathbf{N}, S_n = I + nb = S_0 + nb$.   □

We can also give a formula for the *sum of the first $(n + 1)$ entries in an arithmetic sequence*.

Theorem 3.6.5: If S is an arithmetic sequence, then $\forall n \in \mathbf{N}, S_0 + S_1 + S_2 + \ldots + S_n = (n + 1)[S_0 + S_n]/2.$

// Here $a = 0$ and $P(n)$ is the equation $S_0 + S_1 + S_2 + \ldots + S_n = (n + 1)[S_0 + S_n]/2$.

Proof. // by Mathematical Induction
   Assume that S is a arithmetic sequence with common difference b.
Step 1. If $n = 0$, then LHS $= S_0$ and RHS $= (0 + 1)[S_0 + S_0]/2 = S_0$.

// $P(0)$ is True.

Step 2. Assume $\exists \, k \in \mathbf{N}$ where $S_0 + S_1 + S_2 + \ldots + S_k = (k + 1)[S_0 + S_k]/2$.

Step 3. If $n = k + 1$, then

$$
\begin{aligned}
\text{LHS} &= S_0 + S_1 + S_2 + \ldots + S_k \quad + S_{k+1} \\
&= \{S_0 + S_1 + S_2 + \ldots + S_k\} \quad + S_{k+1} \times 2/2 \\
&= \{S_0 + S_1 + S_2 + \ldots + S_k\} \quad + \{S_{k+1} + S_{k+1}\}/2 \\
&= (k+1)[S_0 + S_k]/2 \quad + \{S_{k+1} + S_{k+1}\}/2 && \text{// by Step 2} \\
&= \frac{(k+1)[S_0 + S_k] + \{S_0 + (k+1)b\} + S_{k+1}}{2} && \text{// from Theorem 3.6.4} \\
&= \frac{[(k+1)+1]S_0 + (k+1)[S_k + b] + S_{k+1}}{2} \\
&= \frac{[(k+1)+1]S_0 + (k+1)S_{k+1} + S_{k+1}}{2} && \text{// from the RE} \\
&= [(k+1)+1][S_0 + S_{k+1}]/2 \\
&= \text{RHS.} && \text{// in the predicate } P
\end{aligned}
$$

// Therefore, $\forall n \in \mathbf{N}$, $S_0 + S_1 + S_2 + \ldots + S_n = (n+1)[S_0 + S_n]/2$. ☐

// The average of all the terms S_0, S_1, S_2, ... , S_n will be $(S_0 + S_n)/2$,
// which is the average of the first and last terms.

// $S = (0,1,2,3,\ldots,n)$ is an *arithmetic sequence* with $S_0 = I = 0$ and $b = 1$.
// Hence, $0 + 1 + 2 + \ldots + n = (n+1)[0 + n]/2 = n(n+1)/2$.
// as in Theorem 3.6.2
// $S = (-1,1,3,5,\ldots,[2n-1])$ is an *arithmetic* sequence with $S_0 = I = (-1)$
// and $b = 2$
// Hence,
$$
\begin{aligned}
-1 + 1 + 3 + 5 + \ldots + (2n-1) &= (n+1)[(-1) + (2n-1)]/2 \\
&= (n+1)[2n - 2]/2 \\
&= (n+1)(n-1) \\
&= n^2 - 1.
\end{aligned}
$$
// Therefore, $\quad 1 + 3 + 5 + \ldots + (2n-1) = n^2$. // as in Theorem 3.6.3

Theorem 3.6.6: $\forall n \in \mathbf{P}$, $(1)(2) + (2)(3) + (3)(4) + \ldots + (n)(n+1)$
$= n(n+1)(n+2)/3$.

Proof. // by Mathematical Induction $\qquad \{a = 1 \text{ and } P(n) \text{ is an equation}\}$
Step 1. If $n = 1$, then LHS $= (1)(2) = 2$ and RHS $= (1)(2)(3)/3 = 2$.
Step 2. Assume $\exists \ k \in \mathbf{P}$ where

$$(1)(2) + (2)(3) + (3)(4) + \ldots + (k)(k+1) = k(k+1)(k+2)/3.$$

Step 3.   If $n = k + 1$, then

$$
\begin{aligned}
\text{LHS} &= (1)(2) + (2)(3) + (3)(4) + \ldots + (k)(k+1) + (k+1)(k+2) \\
&= k(k+1)(k+2)/3 \qquad\qquad\qquad + (k+1)(k+2) \quad // \text{ by Step 2} \\
&= (k+1)(k+2)[k/3+1] \\
&= (k+1)(k+2)[(k+3)/3] \\
&= (k+1)[(k+1)+1][(k+1)+2]/3. \qquad = \text{RHS}.
\end{aligned}
$$

// Therefore, $\forall n \in \mathbf{P}$, $(1)(2) + (2)(3) + (3)(4) + \ldots + (n)(n+1) = n(n+1)(n+2)/3$.

$$\square$$

Is there a formula for the sum of the squares of the first n positive integers?

$$
\sum_{j=1}^{n} (j)(j+1) = \sum_{j=1}^{n} \{j^2 + j\} = 1^2 + 1
$$

$$
\begin{aligned}
&+ 2^2 + 2 \\
&+ 3^2 + 3 \\
&+ 4^2 + 4 \\
&\quad \ldots \\
&\underline{+ n^2 + n} \\
&= \sum_{j=1}^{n} j^2 + \sum_{j=1}^{n} j.
\end{aligned}
$$

Therefore, $n(n+1)(n+2)/3 = \sum_{j=1}^{n} j^2 + n(n+1)/2$, and hence,

$$
\begin{aligned}
\sum_{j=1}^{n} j^2 &= n(n+1)(n+2)/3 - n(n+1)/2 = n(n+1)\{(n+2)/3 - 1/2\} \\
&= n(n+1)\{(n+2) - 1/2 \times 3\}/3 \quad = n(n+1)\{n + 1/2\}/3 \\
&= n(n+1/2)(n+1)/3. \qquad\qquad\qquad\qquad // \text{ or } n(n+1)\{2n+1\}/6
\end{aligned}
$$

Recall from Chap. 2 that when S is a sequence with domain $\{a.. \}$ and co-domain $\mathbf{R}$,

S is a *geometric sequence* <u>means</u>

there is a number r such that for all $n >= a, S(n+1) = r \times S(n)$.

// $S = (4, 8, 16, 32, \ldots)$ is an geometric sequence with $r = 2$.
// Each entry, after the first, is obtained from the previous entry by multiplying by
// the constant value r, which is called "*the common ratio*."
// The equation "$S(n + 1) = r \times S(n)$" is another example of a
// *recurrence equation* [RE].

If $a = 0$ and S_0 equals some *initial* value, I, then we can prove

Theorem 3.6.7: **If S is a geometric sequence with common ratio r, then $\forall n \in \mathbf{N}$, $S_n = r^n \times I$ where $I = S_0$.**

Proof. // by Mathematical Induction, using the RE: $S_{q+1} = r \times S_q \ \forall q \in \mathbf{N}$.
    // Here $a = 0$ and $P(n)$ is the equation $S_n = r^n \times I$.
Step 1. If $n = 0$, then $S_n = I$ and RHS $= r^0 \times I = 1 \times I = I$.   // $P(0)$ is True.
Step 2. Assume $\exists \, k \in \mathbf{N}$ where $S_k = r^k \times I$.
Step 3. If $n = k + 1$, then

$$
\begin{aligned}
S_n = S_{k+1} &= r \times S_k & \text{// using the RE}\\
&= r \times \{r^k \times I\} & \text{// by Step 2}\\
&= \{r \times r^k\} \times I\\
&= r^{k+1} \times I\\
&= \text{RHS.}
\end{aligned}
$$

// Therefore, $\forall n \in \mathbf{N}$, $S_n = r^n \times I$.   □

//X   This can be generalized as follows:
//   If $S_a = I$ and $\forall q \in \{a.. \} \ S_{q+1} = r \times S_q$,
//   then $\forall n \in \{a.. \}$, $S_n = r^n \times K$ where $K = I \, / \, r^a$.

Is there a compact formula for the sum of the first $(n + 1)$ consecutive entries in a geometric sequence?

Theorem 3.6.8: **If $r \neq 1$, then $\forall n \in \mathbf{N}$, $I + rI + r^2 I + \ldots + r^n I = \dfrac{r^{n+1} - 1}{r - 1} \times I$.**

Proof. // by Mathematical Induction where $r - 1 \neq 0$, $a = 0$
    // and $P(n)$ is an equation
Step 1. If $n = 0$, then LHS $= I$ and RHS $= \dfrac{r^{0+1} - 1}{r - 1} \times I = I$.

Step 2. Assume $\exists \, k \in \mathbf{N}$ where $I + rI + r^2 I + \ldots + r^k I = \dfrac{r^{k+1} - 1}{r - 1} \times I$.

Step 3. If $n = k + 1$, then

$$
\begin{aligned}
\text{LHS} &= I + rI + r^2 I + \ldots + r^k I + r^{k+1} I\\
&= \frac{r^{k+1} - 1}{r - 1} \times I \quad + r^{k+1} I \times \frac{r - 1}{r - 1} & \text{// by Step 2}\\
&= \frac{\{r^{k+1} - 1 + r^{k+1} r - r^{k+1}\}}{r - 1} \times I\\
&= \frac{r^{[k+1]+1} - 1}{r - 1} \times I = \text{RHS.}
\end{aligned}
$$

// Therefore, $\forall n \in \mathbf{N}$, $I + rI + r^2 I + \ldots + r^n I = \dfrac{r^{n+1} - 1}{r - 1} \times I$.   □

// If $r = 1$, then $\forall n \in N$, $I + rI + r^2I + \ldots + r^nI = ?$ $\qquad$ $(n + 1)I?$

// If $I = 1$ and $r = 2$, $\quad 1 + 2 + 2^2 + 2^3 + \ldots + 2^n = 2^{n+1} - 1$.
// What would that last equation look like when written in positional notation in
// base 2?

// If $I = 1$ and $r = 10$, $1 + 10 + 10^2 + 10^3 + \ldots + 10^n = (10^{n+1} - 1)/9$.
// What would that last equation look like when written in positional notation in
// base 10?

//X   Is there a formula for the sum of the *any* $(n + 1)$ consecutive entries in a
//   geometric sequence?

Are there formulas for sequences defined by recurrence equations like the one
we saw for the number of disc transfers in the *Towers of Hanoi* problem?

**Theorem 3.6.9: If $S_0 = I$ and $\forall q \in N$ $S_{q+1} = 2 \times S_q + b$, then $\forall n \in N$,
$S_n = 2^n \times [I + b] - b$.**

Proof. // by Mathematical Induction
$\qquad$ // here $a = 0$ & $P(n)$ is the equation $S_n = 2^n \times [I + b] - b$.
Step 1. If $n = 0$, then $S_n = I$ and RHS $= 2^0 \times [I + b] - b = I$. $\qquad$ // $P(0)$ is True.
Step 2. Assume $\exists$ $k \in N$ where $S_k = 2^k \times [I + b] - b$.
Step 3. If $n = k + 1$, then

$$
\begin{aligned}
S_n = S_{k+1} &= 2 \times S_k + b & &\text{// using the RE} \\
&= 2 \times \left\{ 2^k \times [I + b] - b \right\} + b & &\text{// by Step 2} \\
&= 2 \times 2^k \times [I + b] - 2b + b & & \\
&= 2^{k+1} \times [I + b] - b & & \\
&= \text{RHS.} & &
\end{aligned}
$$

// Therefore, $\forall n \in N$, $S_n = 2^n \times [I + b] - b$. $\qquad\qquad\qquad$ ☐

This is the recurrence equation for counting the number of moves required in
the *Towers of Hanoi* problem when b is set to 1. S_1 (the number for moving a tower
with 1 disc) equals 1 so we may solve an equation for a value of I and thereby
prove the formula we "conjectured" in Chap. 2:

$$
\begin{aligned}
S_n &= 2^n \times [I + b] - b \\
S_1 &= 2^1 \times [I + b] - b = 2I + 2b - b = 2I + b = 2I + 1 \\
1 &= 2I + 1,
\end{aligned}
$$
so $I = 0$ and $\qquad S_n = 2^n \times [0 + 1] - 1 = 2^n - 1$.

This is the formula for $T(n)$ in Chap. 2.

// Is $S_0 = T(0)$, the number for moving a tower with 0 discs?

After all these examples, we're sure you recognize the pattern of a proof by Mathematical Induction. But we want to show you (an example of) a slight (but equivalent) variation called Strong Mathematical Induction, or just

3.6.1   Strong Induction

Strong Induction is used to prove assertions of the form

$$\forall \text{ integers } n >= a, \ P(n)$$

where P is some predicate whose domain is (or includes) the interval $\{a.. \ \}$. The form of the argument is

$$P(a)$$
$$[P(a) \wedge P(a+1) \wedge \ldots \wedge P(k)] \to P(k+1) \quad \text{for } \forall k \in \{a.. \ \}$$
$$\therefore \forall n \in \{a.. \ \} \quad P(n)$$

The first premise asserts that

the predicate is True for the <u>one</u> value a, the smallest element in $\{a.. \ \}$.

The second premise asserts that

if P is True for all integers from a up to <u>any</u> particular value $k \in \{a.. \ \}$,

then P must also be True for the next value, $k + 1$.

We can present proofs by Strong Induction as before, but with a revised inductive hypothesis.

Step 2. *Assume $\exists \ k \in \{a.. \ \}$ where $P(n)$ is True for all integers from a up to k.*
$$// \ n \in \{a..k\}.$$

// This is a much stronger assumption, so this form is called "**Strong**" Induction.

The validity of Strong Induction can be deduced from the Well-Ordering Principle for integers with only a slight modification to the proof we gave of the validity of ordinary MI.

// Why would we ever need this form?

Strong Induction is essential for proving the correctness of (many) recursive algorithms. But here, we give a simple illustration of this method.

Theorem 3.6.10: $\forall n \in \{2.. \}$ **either n is prime or n is a product of primes.**

Proof. // by Strong Induction on n
// $P(n)$ is the assertion "either n is prime or n is a product of primes".
Step 1. If $n = 2$, then n is prime.   // $P(2)$ is True.
Step 2. Assume $\exists\, k \in \{2.. \}$ where every integer in $\{2..k\}$ either is a prime or is the
        product of primes.
Step 3. If $n = k + 1$, then either $k + 1$ is prime or not.
If $k + 1$ is prime, then clearly $P(k + 1)$ is True.
Suppose that $k + 1$ is not prime. Then $k + 1$ has a proper divisor d; that is,

$$k + 1 = d \times m \text{ for some integer } m \quad \text{and} \quad 1 < d < k + 1.$$

Since $0 < d = d \times 1 < k + 1 = d \times m$, we have that $1 < m$. // dividing by d
Since $1 < d$, $k + 1 = d \times m > 1 \times m$, we have that $m < k + 1$.

Both d and m are in $\{2..k\}$ and so the inductive hypothesis applies to them both;
that is,

$$d = p_1 \times p_2 \times \ldots \times p_s, \qquad \text{where } s >= 1 \text{ and each } p_i \text{ is a prime,}$$
and $$m = q_1 \times q_2 \times \ldots \times q_t, \qquad \text{where } t >= 1 \text{ and each } q_j \text{ is a prime.}$$
Then $$k + 1 = d \times m = (p_1 \times p_2 \times \ldots \times p_s)(q_1 \times q_2 \times \ldots \times q_t)$$

which is a product of primes. Therefore, if $k + 1$ is not prime, then $P(k + 1)$ is True.

// Thus, $\forall n \in \{2.. \}$ either n is prime or n is a product of primes.   ☐

// Both d and m are $<= (k + 1)/2$ and are usually <u>much smaller</u> than k.
// We wanted to apply the inductive hypothesis to these two integers (not just to k),
// so this proof <u>required</u> the strong assumption of Strong Induction; in this
// argument, knowing $P(k)$ is True is of no use at all in showing $P(k + 1)$ is True.

The Most Important Ideas in This Section.
Mathematical Induction is used to prove universal assertions of the form

$$\forall \text{ integers } n >= a, \quad P(n)$$

and works in a very algorithmic fashion. The *validity* of *Mathematical
Induction* (and *Strong Induction*) is established from the *Well-Ordering
Principle*. This section gave many examples because the method is so important. Strong Induction is often essential for proving the correctness of recursive algorithms. Mathematical Induction itself is often used for proving the
correctness of algorithms as we show next.

3.7   Proofs Promised in Chap. 1

3.7.1   Russian Peasant Multiplication Is Correct

The purpose of the algorithm is to find the product of two input values M and N where N is a positive integer.   // The version in Chap. 1 assumed $N > 1$.

RPM can be given in pseudo-code as

Algorithm 3.7.1: Russian Peasant Multiplication #2

```
Begin
  Total ← 0;
  A ← M;
  B ← N;
  While (B > 1) Do
    If (B MOD 2 = 1) Then
      Total ← Total + A;
    End;
    A ← A × 2;
    B ← B DIV 2;
  End;
  Return(Total + A);
End.
```

// This version is slightly different; it doesn't keep all the intermediate values of A
// and B; it adds the current A-value to the variable Total when the B-value is odd.

Walk through with $M = 27$ and $N = 50$.

B > 1	B mod 2	Total	A	B
–	–	0	27	50
T	0	"	54	25
T	1	54	108	12
T	0	"	216	6
T	0	"	432	3
T	1	486	864	1
F	(1)			

// $50\{10\} = 110\ 010\{2\}$.

Return the value 1350   // $1350 = 486 + 864$.

We know that after $k = \lfloor \lg(N) \rfloor$ iterations, the value of B must be 1, and the while-loop will **terminate**.

// We want a formal mechanism to describe the action of a loop, iteration by
// iteration; something we can use to prove the correctness of algorithms
// containing loops.

A ***Loop invariant*** is *a statement describing the variables involved in the loop*
that is True <u>after each iteration</u> of the loop.

// And therefore, it must be True when the loop terminates.
// The truth-value of a loop invariant does not change (vary) with iterations of the
// loop. A loop invariant might be True before the loop is entered.

Loop invariants provide a very potent technique for designing and analyzing algorithms that contain loops. The *Bisection Algorithm* for finding an approximate solution to an equation revised the values of the two variables A and B so that "there is an exact solution between A and B" was a loop invariant. You will see many more loop invariants in this book.

For the loop in RPM, we can prove that the equation "AB + Total = MN" is a loop invariant.

Theorem 3.7.1: For any integer n, after n iterations of the loop in RPM, AB + Total = MN.   // This is $P(n)$.

Proof.   // by Mathematical Induction on n where $n \in \{0.. \}$

Step 1. After $n = 0$ iterations of the loop,   // that is, before the loop is done once
         AB + Total = MN + 0 = MN.   // $P(0)$ is True.

// Suppose that this equation has been True for the first few iterations.

Step 2. Assume $\exists \, k \in \mathbf{N}$ such that after k iterations of the loop
         AB + Total = MN.

// Now, what will happen on the next iteration, if there is one?

Step 3. Suppose there is another iteration, the $k + 1$-st.   // that is, B $>= 2$

The value of B at the beginning of this next iteration of the loop may be written as $2Q + R$ where R is either 0 or 1. Let A*, B*, and Total* denote the values of these variables after this next iteration. Then

$$\text{Total*} = \text{Total} + RA, \qquad\qquad // \; R = 1 \Leftrightarrow B \text{ is odd.}$$
$$A* = 2A,$$
$$B* = Q. \qquad\qquad // \; Q = \lfloor B/2 \rfloor = B \text{ DIV } 2.$$

Hence, $\text{A*B*} + \text{Total*} = (2A)(Q) + (\text{Total} + RA)$
$$= A\lfloor 2Q + R \rfloor + \text{Total}$$
$$= AB + \text{Total}$$
$$= MN. \qquad\qquad // \text{ from Step 2}$$

Therefore, $\forall n \in \mathbf{N}$, after n iterations of the loop in RPM, AB + Total = MN.   ▯

When the while-loop finally terminates, B = 1 and

$$MN = AB + \text{Total} \qquad\qquad // \text{ as we just saw}$$
$$= A \times 1 + \text{Total} = \text{Total} + A.$$

So the value returned by the algorithm *is* the product of the two input values *M* and *N*. **RPM *is correct*.** If N is any positive integer, then the algorithm correctly calculates the product MN.

                                                              // M may be any real number.

3.7.2 Resolving the Cake Cutting Conundrum

When N points are marked on the circumference of a circular cake and straight cuts are made joining all pairs of these points, how many pieces of cake, $P(N)$, are produced? In Chap. 1, we tabulated the number of points and the number of pieces:

N	$P(N)$
1	1
2	2
3	4
4	8
5	16
6	?

The number of pieces 6 points produce is 30 or 31; it's only 30 if the three "longest" cuts all pass through the center of the circle. The *maximum* number of pieces, $Q(N)$, will be obtained when *no three cuts go through a single point*. We will show you a formula for $Q(N)$.

// That equals 2^{N-1} for $N = 1,2,3,4$ and 5 but never again.

Theorem 3.7.2: $\forall N \in \mathbf{P}$, $\quad Q(N) = \dbinom{N}{4} + \dbinom{N}{2} + \dbinom{N}{0}$

Proof. // by Mathematical Induction on N where we follow the "convention" that

$$\dbinom{N}{k} = 0 \text{ unless } 0 <= k <= N.$$

// This notation was introduced in Chap. 2, to denote the number of k-subsets in
// an N-set

Step 1. If $N = 1$, there is only one point on the circumference of the circular cake. Then there are no pairs of distinct points, so no cuts are made. Hence the number of pieces, $Q(N)$, is 1. And the RHS

$$\dbinom{N}{4} + \dbinom{N}{2} + \dbinom{N}{0} = \dbinom{1}{4} + \dbinom{1}{2} + \dbinom{1}{0} = 0+0+1 = 1. \quad \text{// } P(1) \text{ is True.}$$

Step 2. Assume $\exists k \in \mathbf{P}$ such that the number of pieces produced by straight cuts joining all pairs of k points on the circumference of a circular cake and no 3 cuts "meet" is

$$Q(k) = \dbinom{k}{4} + \dbinom{k}{2} + \dbinom{k}{0} \qquad \text{// } P(k) \text{ is True.}$$

// Now, what will happen when one more point is added to k points on the
// circumference of a circular cake, and k more straight cuts are made from this
// $k + 1$-st point to the other points already there?

Step 3.   Suppose $N = k + 1$; that is, suppose that on some particular circular cake, there are $k + 1$ points marked on the circumference. Index these points from 1 to $k + 1$ clockwise around the cake. If straight cuts are made joining all pairs of points 1 up to k, then $Q(k)$ pieces are produced.

Now, k more straight cuts are made from point $k + 1$ to the other points numbered 1 to k.   // How many new pieces are produced?

Suppose $1 < j < k$ and consider the new cut from point $k + 1$ to point j. If this cut is made very slowly beginning at point $k + 1$, when it reaches the first of the old cuts, one old piece is divided in two, and one more piece is produced. As this new cut continues, when it reaches the second of the old cuts, another old piece is divided in two and one more piece is produced. As this new cut continues further, whenever it reaches an old cut, another old piece is divided in two, and one more piece is produced. And finally, when this new cut leaves the last old cut it crosses, it continues to point j, and another old piece is divided in two producing one more piece.

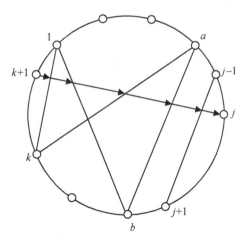

The number of old pieces divided into two new pieces by the new cuts from point $k + 1$ to the other points numbered 1 to k equals the number of old cuts crossed by new cuts *plus* one for each old point. That is,

$\qquad$ # additional pieces = # times a new cut crosses an old cut $+ k$.

// How many times do the new cuts cross old cuts?

If a new cut from point $k + 1$ to point j crosses an old cut, the old cut must be from a to b where $1 <= a < j < b <= k$. Since neither a nor b is equal j, $\{a,j,b\}$ is a 3-subset of $\{1..k\}$. Furthermore, if $\{x,y,z\}$ is any 3-subset of $\{1..k\}$ where $x < y < z$, then the new cut from point $k + 1$ to point y crosses the old cut from x to z. Therefore,

$$\text{\# times a new cut crosses an old cut} = \binom{k}{3}.$$

Hence,

$$Q(k+1) = Q(k) + \binom{k}{3} + k$$

$$= \binom{k}{4} + \binom{k}{2} + \binom{k}{0} + \binom{k}{3} + k \qquad \text{// from Step 2}$$

$$= \binom{k}{4} + \binom{k}{2} + \binom{k}{0} + \binom{k}{3} + \binom{k}{1} \qquad \text{// } \binom{n}{1} = n \ \forall n \in \mathbf{P}.$$

$$= \binom{k}{4} + \binom{k}{3} + \binom{k}{2} + \binom{k}{1} + \binom{k}{0} \qquad \text{// } 2^k \text{ iff } 0 <= k <= 1$$

$$= \binom{k+1}{4} + \binom{k+1}{2} + \binom{k}{0}$$

$$\qquad\qquad\qquad \text{// the Bad Banana Theorem}$$
$$\qquad\qquad\qquad \text{// in Example 2.3.3}$$

$$= \binom{k+1}{4} + \binom{k+1}{2} + \binom{k+1}{0} \qquad \text{// } \binom{n}{0} = 1 \ \forall n \in \mathbf{P}.$$

$$= \text{RHS}.$$

Therefore, $\forall N \in \mathbf{P}, \ Q(N) = \binom{N}{4} + \binom{N}{2} + \binom{N}{0}.$ ◻

// The N points can always be arranged, so no three cuts go through a single point.
// If more than 2 cuts do go through a single point,
// # additional pieces is < # times new cuts cross old cuts + k.

3.7.3 Casting Out Nines

For any *positive* integer K, let $R(K)$ denote the sum of the digits in K; that is,

if $K = d_a d_{a-1} \ldots d_1 d_0 \{10\}$ where all the $d_j \in \{0..9\}$ and $d_a > 0$,

then $R(K) = d_a + d_{a-1} + \ldots + d_1 + d_0.$ // $0 < d_a <= R(K) \in \mathbf{P}.$

// If $K < 10$, then K only has one digit, $a = 0$ and $R(K) = K$.

Suppose that K has more than one digit; that is, suppose $a >= 1$.

// We'll show that $\forall a \in \mathbf{P}, \ K - R(K) = 9Q$ where $Q \in \mathbf{P}$.

For each $n \in \mathbf{N}$, let T_n denote $1 + 10 + 10^2 + 10^3 + \ldots + 10^n$.   // each $T_n \in \mathbf{P}$
     // What is $T_n\{10\}$?

Then,

$$T_n = \left(10^{n+1} - 1\right)/9 \text{ and so } \left(10^{n+1} - 1\right) = 9T_n. \qquad \text{// from Theorem 3.6.8}$$

Hence,

$$K - R(K) = \sum_{j=0}^{a} d_j 10^j - \sum_{j=0}^{a} d_j = \sum_{j=0}^{a} d_j(10^j - 1) \qquad \text{// but } 10^0 - 1 = 0$$

$$= \sum_{j=1}^{a} d_j 9 T_{j-1} = 9 \times \sum_{j=1}^{a} d_j T_{j-1} = 9Q$$

where Q is the positive integer, $d_1 T_0 + d_2 T_1 + \ldots + d_a T_{a-1}$.   // $d_a T_{a-1} > 0$

The algorithm for casting out nines with an input $N \in \mathbf{P}$ may be given in pseudo-code as   // to discuss the algorithm not to implement it

Algorithm 3.7.2: Casting Out Nines #2

```
Begin
  j ← 0;
  K[j] ← N;
  While (K[j] >= 10) Do
    K[j+1] ← R(K[j]);                    // R(K) is evaluated here.
    j ← j+1;
  End;
  Return(K[j]);                          // This value is in {0..9}.
End.
```

Walkthrough with $N = 586\ 987\ 583$.

j	K[j]
0	586987583
1	59
2	14
3	5 return the value 5.

This algorithm generates a sequence of K-values in $\mathbf{P}$ where $K[0] = N$ and for $j >= 0$, $K[j+1] = R(K[j])$. The K-values decrease (by at least 9) with each iteration so eventually, some K-value must be < 10. The while-loop *must* **terminate**; let t denote the value j has when the loop terminates. When N itself is < 10, $t = 0$. If the output value is denoted by $<N>$, then $<N> = K[t]$ and $K[t]$ is a positive integer $<= 9$.

// Show that "$9|(N - K[j])$" is a loop invariant using MI on the number of
// iterations of the while-loop (starting from 0).

Theorem 3.7.3: $\forall N \in \mathrm{P}$, $<N> = 9$ **precisely when N is divisible by 9 and, in all** *other* **cases,** $<N>$ **is the remainder obtained when N is divided by 9.**

Proof. // We must prove: if $<N> = 9$ then $9|N$ and
      // if $<N> \neq 9$ then $<N> = N$ MOD 9 and is positive
      // so N is not divisible by 9.
      // We first show that $\exists\ Q^* \in \mathbf{N}$ such that $N = 9 \times Q^* + <N>$.

If $t > 0$, then for $j = 0, 1, 2, \ldots, (t-1)$, let $Q[j]$ be the positive integer such that

$$9 \times Q[j] = K[j] - R(K[j])$$
$$= K[j] - K[j+1].$$

Then, $\quad \displaystyle\sum_{j=0}^{t-1} \{K[j] - K[j+1]\} = \sum_{j=0}^{t-1} 9 \times Q[j] = 9 \times \sum_{j=0}^{t-1} Q[j]$

But $\displaystyle\sum_{j=0}^{t-1} \{K[j] - K[j+1]\} = \{K[0] - K[1]\} + \{K[1] - K[2]\} + \{K[2] - K[3]\} + \cdots$

$$\cdots + \{K[t-2] - K[t-1]\} + \{K[t-1] - K[t]\}$$
$$= K[0] - K[t] \qquad \text{// All the other terms “cancel out”.}$$
$$= N - <N>.$$

// A series of differences like this is sometimes called a "telescoping series"
// because the sum of all the differences collapses into one special difference.

Letting Q^* denote the positive integer $Q[0] + Q[1] + Q[2] + \ldots + Q[t-1]$, we have

$$N - <N> = 9 \times Q^* \quad \text{and so} \quad N = 9 \times Q^* + <N>.$$

If $t = 0$, then $N = <N>$; so letting $Q^* = 0$, we again have $N = 9 \times Q^* + <N>$.
   If $<N> = 9$, then

$$N = 9 \times Q^* + 9 = 9 \times (Q^* + 1) \quad \text{so} \quad 9|N. \qquad\qquad \text{// } (Q^* + 1) \in \mathbf{P}.$$

If $<N> \neq 9$, then

$$N = 9 \times Q^* + <N> \quad \text{where} \ <N> \text{ is positive and} < 9;$$

in these cases, $<N>$ *must be* the remainder obtained when N is divided by 9 (and because this remainder is positive, 9 cannot divide evenly into N).   □

3.7.4   Euclid's Algorithm for GCD Is Correct

The assertion underlying Euclid's Algorithm (1.2.5) for finding the greatest common divisor of two given positive integers x and y can be proven fairly easily. In the argument that follows, all variables are assumed to represent integer values. Let us begin by proving a *lemma* (a relatively small result to be used in proving other, larger, and more important results).

Lemma. If d is a common divisor of p and q then for <u>any</u> pair of integers a and b, d divides evenly into $pa + qb$.

Proof. If d is a common divisor of p and q, then there exist integers s and t such that $p = d \times s$ and $q = d \times t$. Then $as + bt$ is an integer, so d divides evenly into

$$d \times (as + bt) = das + dbt = ds \times a + dt \times b = pa + qb. \qquad \square$$

// If d is a common divisor of p and q, then d divides evenly into every "integer
// combination" of p and q (every number that can be expressed in the form
// $pa + qb$ where a and b are integers).

Theorem 3.7.4: (Euclid) For any two positive integers x and y, if $x = yq + r$ where $r > 0$ then $GCD(x,y) = GCD(y,r)$.

Proof. Suppose that $d1 = GCD(y,r)$ and $d2 = GCD(x,y)$.

// We will show that $d1 <= d2$ and $d2 <= d1$, and therefore $d1$ must equal $d2$.

    Because $d1$ is a common divisor of y and r, by our lemma, $d1$ also divides evenly into $y(q) + r(1) = x$. So $d1$ is a common divisor of x and of y, and therefore, $d1$ is $<=$ the greatest common divisor of x and y; that is, $d1 <= GCD(x,y) = d2$.

    Because $d2$ is a common divisor of x and y, (again by our lemma) $d2$ also divides evenly into $x(1) + y(-q) = r$. So $d2$ is a common divisor of y and r, and therefore, $d2 <= GCD(y,r) = d1$. $\square$

// What we actually have proved is that the set of all common divisors of x and y
// is the *same* as the set of all common divisors of y and r. Note that in the proof, r
// does not necessarily equal x MOD y and q may be <u>any</u> integer.

Theorem 3.7.5: For any two positive integers x and y,
(1) **$GCD(x,y) =$ the smallest element of the set $X = P \cap \{ax + by : a, b \in Z\}$;**
(2) **$\exists\ a_0, b_0 \in Z$ such that $GCD(x,y) = a_0 \times x + b_0 \times y$; and**
(3) **if f is a common divisor of x and y, then f also divides evenly into $GCD(x,y)$.**

Proof. Because x is positive and $x = (1)x + (0)y$, $x \in X$. Thus, X is a nonempty subset of the positive integers, so by the Well-Ordering Principle, X has a smallest element d. Since d is of the form $ax + by$ where $a,b \in Z$, $\exists\ a_0, b_0 \in Z$ such that $d = a_0 \times x + b_0 \times y$.

// But does this smallest element d really equal $GCD(x,y)$?
// We'll show that $d <= GCD(x,y)$ and $GCD(x,y) <= d$, so d must equal $GCD(x,y)$,
// and first we'll show that $d <= GCD(x,y)$ by showing that d is a common divisor of
// x and y.

Dividing x by d, we get $x = dq + r$ where $0 <= r < d$. Then,

$$r = x - dq = x - (a_0 \times x + b_0 \times y)q = x - qa_0 \times x - qb_0 \times y$$
$$= (1 - qa_0)x + (-qb_0)y.$$

If r were positive, then it would be an element of X that's smaller than d and that's not possible.   // That would contradict the fact that d is the smallest element of X.

Therefore r must be zero; that is, $d|x$.

// A completely "similar" argument, using y instead of x, gives the conclusion that
// $d|y$.

Dividing y by d, we get $y = dq + r$ where $0 <= r < d$. Then,

$$r = y - dq = y - (a_0 \times x + b_0 \times y)q = y - qa_0 \times x - qb_0 \times y$$
$$= (-qa_0)x + (1 - qb_0)y.$$

If r were positive, then it would be an element of X that's smaller than d and that's not possible. Therefore r must be zero; that is, $d|y$. Then, because d is a common divisor of x and y, d is $<= \text{GCD}(x,y)$.

// Now we show that $\text{GCD}(x,y) =< d$ by showing that $\text{GCD}(x,y)$ divides evenly
// into d.

Because $\text{GCD}(x,y)$ is a common divisor of x and y, by our lemma, $\text{GCD}(x,y)$ divides evenly into $a_0 \times x + b_0 \times y$ which equals d. Then,

$$d = \text{GCD}(x, y) \times t \text{ where } t \in \mathbf{Z}.$$

Since d is positive and $\text{GCD}(x,y) >= 1$, t must be positive, and therefore, $t >= 1$. But then $d = \text{GCD}(x,y) \times t >= \text{GCD}(x,y) \times 1 = \text{GCD}(x,y)$. Hence, $d = \text{GCD}(x,y)$, and we have proven (1).

Because $d = \text{GCD}(x,y)$, we have already shown (2) – that $\exists\ a_0, b_0 \in \mathbf{Z}$ such that $\text{GCD}(x,y) = a_0 \times x + b_0 \times y$.

Since $a_0, b_0 \in \mathbf{Z}$ and $\text{GCD}(x,y) = a_0 \times x + b_0 \times y$, the lemma implies (3) – if f is a common divisor of x and y, then f also divides evenly into $\text{GCD}(x,y)$.   □

// Euclid's algorithm to find $\text{GCD}(x,y)$ can be extended to also find integer values
// a and b such that $\text{GCD}(x,y) = a \times x + b \times y$.

The Most Important Ideas in This Section.
As we proved several assertions made earlier in the text, we gave a formal mechanism to describe the action of a loop, iteration by iteration, which is used to prove the correctness of algorithms containing loops. A *loop invariant* is a statement about the variables involved in the loop that is True after each iteration of the loop. Therefore, it must be True when the loop terminates.

3.8   The Proof Promised in Chap. 2

In Chap. 2, the Binomial Theorem was introduced. We prove it here using Mathematical Induction.   // and lots of algebra

Theorem 3.8.1: (The Binomial Theorem) For any two numbers a and b, and any nonnegative integer n

$$(a+b)^n = \sum_{k=0}^{n} \binom{n}{k} a^k \times b^{n-k}.$$

Proof. // by Mathematical Induction on n

Step. 1 If $n = 0$, then   LHS $= (a+b)^0 = 1$ and

$$\text{RHS} = \sum_{k=0}^{0} \binom{0}{k} a^k \times b^{n-k} = \binom{0}{0} a^0 \times b^{0-0} = (1)1 \times 1 = 1.$$

// $k = 0$ only.

// Even though it isn't necessary, let's check one more case.

//   If $n = 1$, then   LHS $= (a+b)^1 = a + b$ and

// $\text{RHS} = \sum_{k=0}^{1} \binom{1}{k} a^k \times b^{n-k} = \binom{1}{0} a^0 \times b^{1-0} + \binom{1}{1} a^1 \times b^{1-1}$

// $= (1)1 \times b \qquad\quad + (1)a \times 1$

// $= b \qquad\qquad\quad + a$

// $= a + b.$ // $k = 0$ and 1.

Step 2. Assume $\exists\, q \in \mathbf{N}$ where

$$(a+b)^q = \sum_{k=0}^{q} \binom{q}{k} a^k \times b^{q-k}.$$

Step 3. If $n = q + 1$, then

$$\text{LHS} = (a+b)^{q+1} = (a+b) \times (a+b)^q = a \times (a+b)^q + b \times (a+b)^q$$

$$= a \times \sum_{k=0}^{q} \binom{q}{k} a^k \times b^{q-k} + b \times \sum_{k=0}^{q} \binom{q}{k} a^k \times b^{q-k} \qquad \text{// by Step 2}$$

$$= \sum_{k=0}^{q} \binom{q}{k} a^{k+1} \times b^{q-k} \quad + \sum_{k=0}^{q} \binom{q}{k} a^k \times b^{q-k+1}.$$

Writing these sums out in detail as columns,

$$
\begin{aligned}
\text{LHS} = &\binom{q}{0}a^{0+1}b^{q-0} &&+ \binom{q}{0}a^{0}b^{q-0+1} \\
+ &\binom{q}{1}a^{1+1}b^{q-1} &&+ \binom{q}{1}a^{1}b^{q-1+1} \\
+ &\binom{q}{2}a^{2+1}b^{q-2} &&+ \binom{q}{2}a^{2}b^{q-2+1} \\
+ &\ \ldots\ldots \\
+ &\binom{q}{j}a^{j+1}b^{q-i} &&+ \ \ldots\ldots \\
+ &\ \ldots\ldots &&+ \binom{q}{j+1}a^{j+1}b^{q-(j+1)+1} \\
+ &\ \ldots\ldots &&+ \ \ldots\ldots \\
+ &\binom{q}{q}a^{q+1}b^{q-q} &&+ \binom{q}{q}a^{q}b^{q-q+1}.
\end{aligned}
$$

Writing the terms with equal powers of a (and b) on the same line,

$$
\begin{aligned}
\text{LHS} = &&&+ \binom{q}{0}a^{0}b^{q-0+1} \\
+ &\binom{q}{0}a^{0+1}b^{q-0} &&+ \binom{q}{1}a^{1}b^{q-1+1} \\
+ &\binom{q}{1}a^{1+1}b^{q-1} &&+ \binom{q}{2}a^{2}b^{q-2+1} \\
+ &\binom{q}{2}a^{2+1}b^{q-2} &&+ \binom{q}{3}a^{3}b^{q-3+1} \\
+ &\ \ldots\ldots &&+ \ \ldots\ldots \\
+ &\binom{q}{j}a^{j+1}b^{q-j} &&+ \binom{q}{j+1}a^{j+1}b^{q-(j+1)+1} \\
+ &\ \ldots\ldots &&+ \ \ldots\ldots \\
+ &\binom{q}{q-1}a^{(q-1)+1}b^{q-(q-1)} &&+ \binom{q}{q}a^{q}b^{q-q+1} \\
+ &\binom{q}{q}a^{q+1}b^{q-q}.
\end{aligned}
$$

By the Bad Banana Theorem from Example 2.3.3, we have for $j = 0, 1, \ldots, q - 1$

$$
\binom{q}{j} + \binom{q}{j+1} = \binom{q+1}{j+1}.
$$

Therefore,

$$\text{LHS} = \qquad\qquad\qquad + \binom{q}{0}a^0 b^{q+1}$$

$$+ \binom{q+1}{1}a^1 b^q$$

$$+ \binom{q+1}{2}a^2 b^{q-1}$$

$$+ \ldots\ldots$$

$$+ \binom{q+1}{j+1}a^{j+1} b^{q-j}$$

$$+ \ldots\ldots$$

$$+ \binom{q+1}{q}a^q b^1$$

$$+ \binom{q}{q}a^{q+1} b^{q-q}.$$

Since $\binom{q}{0} = \binom{q+1}{0} = 1$ and $\binom{q}{q} = \binom{q+1}{q+1} = 1,$

$$\text{LHS} = \sum_{k=0}^{q+1}\binom{q+1}{k}a^k \times b^{(q+1)-k} = \text{RHS}. \qquad\qquad \square$$

The Most Important Ideas in This Section.
Sometimes ugly algebra works (and is necessary) to prove a pretty theorem.

Exercises

1. Create truth tables for the following expressions:
   (a) $P \wedge \sim Q$
   (b) $\sim P \vee \sim Q$
   (c) $P \wedge Q \vee \sim(P \vee Q)$
   (d) $P \vee Q \sim Q \vee R$
   (e) $\sim P \wedge Q \wedge \sim R$
   (f) $P \wedge (\sim Q \vee R)$
2. Prove that the product of any two odd integers is also an odd integer.
3. Prove the following by cases: if $x,y \in \mathbf{R}$, then $|x \times y| = |x| \times |y|$.
4. Construct a direct proof that for positive integers

$$\text{if } a|b \text{ and } b|c \text{ then } a|c.$$

5. Prove that if $a-b$ is odd for integers a and b then a^3-b^3 is also odd. Use the same argument to prove that if $a+b$ is odd then a^2+b^2 is also odd.
6. Prove that the product of any two even integers is a multiple of 4.
7. Prove that for all integers n, if n is a multiple of 6 then n is also a multiple of 3.
8. Prove that for all integers, n, if n^2 is even then n is even.
9. Prove the smallest proper divisor of an integer $n > 1$ must be prime.
10. Generalize Theorem 3.5.1 by proving

$$\text{If } f \text{ is any real number and } g = (1-f), \text{ then}$$
$$\text{for all integers } n, \lfloor f \times n \rfloor + \lceil g \times n \rceil = n.$$
$$\text{// In Theorem 3.5.1, } f = \tfrac{1}{2} \text{ (and so } g = \tfrac{1}{2}\text{)}.$$

11. Disprove each of the following assertions:
    (a) $n^2 + n + 41$ is prime for every positive integer n.
    (b) the product of any two irrational numbers is irrational.
    (c) the product of any rational number and any irrational number is irrational.
12. Generalize Theorem 3.5.2 by proving
    (a) Among any K consecutive integers, there is a multiple of K.
    (b) Among any K consecutive integers, there is exactly one multiple of K.
    (c) Among any K consecutive integers, there is a multiple of k for $k = 2,3,...K$.
    (d) The product of any K consecutive integers is a multiple of $K!$.
13. Use the Fundamental Theorem of Arithmetic [Any integer n greater than one can be factored uniquely as a product of primes $n = p_1 \times p_2 \times p_3 \times ... \times p_k$ where $p_1 <= p_2 <= p_3 <= ... <= p_k$.] to prove two other theorems:
    (a) If p is prime, then $\sqrt{p}$ is irrational.
         Hint: Assume that $\sqrt{p}$ is rational. Then use the fact that if $k > 1$, then k^2 has an even number of prime factors.
    (b) If n is a positive integer, then $\sqrt{n}$ is either an integer or is irrational.
14. Find irrational numbers a and b such that a^b is rational.
15. Use truth tables to show the following are equivalent:
    (a) $(P \wedge Q) \to R$
    (b) $P \to (Q \to R)$
    (c) $[P \wedge (\sim R)] \to (\sim Q)$
16. Is $(P \vee Q) \to R$ equivalent to $(P \to R) \vee (Q \to R)$ or to $(P \to R) \wedge (Q \to R)$? Use truth tables to confirm your answer.
17. Use a truth table to determine the validity of the following argument form:

$$P \vee Q$$
$$\sim P \vee R$$
$$\underline{Q \to R}$$
$$\therefore \overline{Q \wedge R}$$

18. Use a truth table to determine the validity of the following argument form:

$$\sim P \vee Q$$
$$\sim[R \wedge (\sim Q)]$$
$$\therefore (P \vee R) \to Q$$

19. Use a Truth Table to show the validity of the following argument form:

$$(P \vee Q) \to \sim Q$$
$$(P \vee \sim Q) \to Q$$
$$\therefore \sim P$$

20. There is no function from X onto $\mathscr{P}(X)$. Hint: Consider the subset $B = \{x \in X : x \notin f(x)\}$ and an element $y \in X$ such that $f(y) = B$.

21. Prove, using mathematical induction, that:

$$\frac{1}{2} + \frac{1}{2^2} + \frac{1}{2^3} + \cdots + \frac{1}{2^n} = 1 - \frac{1}{2^n}$$

for all integers $n >= 1$.

22. Prove, using mathematical induction, that:

$$1^2 + 2^2 + 3^2 + \cdots + n^2 = \frac{n(n+1)(2n+1)}{6}$$

for all integers $n >= 1$.

23. Prove, using mathematical induction, that:

$$\frac{1}{1 \times 3} + \frac{1}{3 \times 5} + \frac{1}{5 \times 7} + \cdots + \frac{1}{(2n-1)(2n+1)} = \frac{n}{2n+1}$$

for all integers $n >= 1$.

24. Prove, using mathematical induction, that: // Hint: Use Theorem 3.6.2

$$\sum_{i=1}^{n} i^3 = \left(\sum_{i=1}^{n} i \right)^2$$

25. Is $n < 2^n$ for $\forall$ integers $n \in \mathbf{N}$?
    If your answer is "no," give a counterexample.
    If your answer is "yes," give a proof by MI.

26. Is $n^2 < 3^n$ for $\forall$ integers $n \in \mathbf{N}$?
    If your answer is "no," give a counterexample.
    If your answer is "yes," give a proof by MI.

27. Use MI to prove the following generalization of Theorem 3.6.4:

$$\text{If } S_a = I \text{ and } \forall q \in \{a.. \}, S_{q+1} = S_q + b,$$
$$\text{then } \forall n \in \{a.. \}, S_n = K + nb \text{ where } K = I - ab.$$

28. Use Mathematical Induction to prove that for all non-negative integers n, the number of subsets of an n-set is 2^n.

29. Use Mathematical Induction to prove that for all positive integers n:

$$1^2 + 3^2 + \ldots + (2n - 1)^2 = (2n - 1)(2n)(2n + 1)/6$$

30. Prove that for any arithmetic sequence S defined on $\{a.. \}$

$$\forall n \in \{a.. \} \; S_a + S_{a+1} + S_{a+2} + \ldots + S_n = (n - a + 1)[S_a + S_n]/2.$$

31. Use MI to prove the following generalization of Theorem 3.6.7:

$$\text{If } S_a = I \text{ and } \forall q \in \{a.. \} \quad S_{q+1} = r \times S_q \text{ where } r \text{ is not zero,}$$
$$\text{then } \forall n \in \{a.. \}, \quad S_n = r^n \times K \text{ where } K = I/r^a.$$

32. Find a formula for the sum of the *any* $(n + 1)$ consecutive entries in a geometric sequence. Use MI to prove your formula is correct.

33. Use Mathematical Induction to prove

$$\text{if } q \text{ is any (fixed) nonnegative integer, then } \forall n \in \mathbf{P},$$

$$\sum_{j=1}^{n} (j)(j+1)(j+2)\ldots(j+q) = n(n+1)(n+2)\ldots(n+q)(n+q+1)/(q+2)$$

// If $q = 0$, then $\sum_{j=1}^{n} (j)(j+1)(j+2)\ldots(j+q) = \sum_{j=1}^{n} j = 1 + 2 + \ldots + n$

// and RHS $= n(n + 1)(n + 2)\ldots(n + q + 1)/(q + 2) = n(n + 1)/2$

//   // as in Theorem 3.6.2.

// If $q = 1$, then $\sum_{j=1}^{n} (j)(j+1)(j+2)\ldots(j+q)$

// $= \sum_{j=1}^{n} (j)(j+1) = (1)(2) + \ldots + (n)(n+1)$

// and RHS $= n(n + 1)(n + 2)\ldots(n + q + 1)/(q + 2) = n(n + 1)(n + 2)/3$

//   // as in Theorem 3.6.6.

34. Find a formula for the sum

$$\frac{1}{(1)(2)} + \frac{1}{(2)(3)} + \ldots + \frac{1}{(n)(n+1)} \qquad \text{// try } n = 1,2,3,4, \text{ and } 5.$$

Use MI to prove your formula is correct for all $n \in \mathbf{P}$.

35. Prove by MI that $\forall n \in \mathbf{N}$,

(a) $\displaystyle\sum_{j=0}^{n}(j+1)2^j = n2^{n+1} + 1.$

(b) $\displaystyle\sum_{j=0}^{n}(j+1)3^j = \frac{[2n+1]3^{n+1} + 1}{4}.$

(c) $\displaystyle\sum_{j=0}^{n}(j+1)r^j = \frac{[(r-1)n + (r-2)]r^{n+1} + 1}{(r-1)^2}$ for all numbers $r \neq 1.$

36. Suppose that q is some fixed positive integer. Use MI to prove that $\forall$ integers $n >= q$,

$$\binom{q}{q} + \binom{q+1}{q} + \binom{q+2}{q} + \ldots + \binom{n}{q} = \binom{n+1}{q+1}.$$

37. Use the Binomial Theorem to prove that $\displaystyle\sum_{k=0}^{n}\binom{n}{k}^2 = \binom{2n}{n}$ for $\forall n \in \mathbf{N}.$

38. A non-recursive Square and Multiply Algorithm to calculate $(b)^n$.

Precondition: n is a positive integer and b is of any type that can be multiplied.
Postcondition: the value returned is equal $(b)^n$.

```
Begin
  product ← 1;
  square ← b;
  a ← n;
  While (a > 1) Do
    If (a is odd) Then
      product ← product*square;
    End;
    square ← square*square;
      a ← a DIV 2;
  End;
  Return(product*square);
End.
```

(a) Walk through the algorithm with $n = 53$ // $\lg(53) = 5.72\ldots$
     using the column headings

$a > 1$	a is odd	Product	Square	a	$(square)^a \times Product$
--	--	1	b	53	$(b)^{53} \times 1$
T	T	b	b^2	26	$(b^2)^{26} \times b$

...

Return (?)

(b) Walk through with $n = 710$.   // $\lg(710) = 9.47\ldots$
(c) Show that the algorithm terminates.
     Let a_k denote the value of a after the kth iteration of the while-loop, and let
     $s = \lfloor \lg(n) \rfloor$.
     Prove by Mathematical Induction on k that

     For any nonnegative integer k, after k iterations of the while-loop,
     $$2^{s-k} <= a_k < 2 \times 2^{s-k}.$$ // This is $P(k)$

     // Then for $k = 0,1,\ldots,(s-1)$, $a_k > 1$ and $a_s = 1$,
     // so the body of the while-loop is done exactly s times,
     // and the number of multiplication operations is $<= 2s + 1$.
(d) Proof of correctness.   // using a loop invariant as in Sect. 3.7 for RPM
     Use Mathematical Induction on k to prove

     For any nonnegative integer k, after k iterations of the while-loop,
     $$(square)^a \times product = (b)^n.$$

39. A Recursive Square and Multiply Algorithm to calculate $(b)^n$
    // based on the observation that when $n = 2q + r$ where $q >= 1$ and $r \in \{0,1\}$,
    // $b^n = b^{2q+r} = b^{2q} \times b^r = (b^q)^2 \times b^r$.
    Precondition: n is a positive integer and b is of any type that can be multiplied.
    Postcondition: the value returned, $y = (b)^n$ // every time the function is invoked

RECURSIVE FUNCTION Exponent(b, n)

                                             // returns a value of the type of b

```
Begin
  If (n = 1) Then
     Return(b);
  End;
  y ← Exponent(b, n Div 2);
  y ← y*y;
  If (n is odd) Then
     y ← y*b;
  End;
  Return(y);
End.
```

Walk through with $n = 53$:   // $\lg(53) = 5.72\ldots$

#	Recursive calls	(…Returned values…)
--	Exponent(b, 53)	$\rightarrow (b^{26})^2 \times b = b^{53}$
1	Exponent(b, 26)	$\rightarrow (b^{13})^2 = b^{26}$
2	Exponent(b, 13)	$\rightarrow (b^6)^2 \times b = b^{13}$
3	Exponent(b, 6)	$\rightarrow (b^3)^2 = b^6$
4	Exponent(b, 3)	$\rightarrow (b)^2 \times b = b^3$
5	Exponent(b, 1)	$\rightarrow b$

(a) Walk through the algorithm with $n = 710$.   // $\lg(710) = 9.47\ldots$
(b) Why is the number of subcalls of the function equal $\lfloor \lg(n) \rfloor$?
(c) Proof of correctness.
     Use Strong Induction on the value of parameter p to prove that

**For any positive integer p, if Exponent(b,p) is invoked
   then the value returned, $y = (b)^p$.**

// If $p = 1$, then the value returned, $y = b = (b)^1 = (b)^p$. $P(1)$ is True.

Searching and Sorting

4

After all that theory, this chapter returns to very practical problems, searching and sorting. Imagine maintaining the records of a bank, drugstore, or college. Most often, files of accounts are kept and updated periodically. Each account is identified by a "key": a bank account number or customer's name or a student's ID number. When a record is to be displayed and/or changed, the first step is to find it in the collection of records. Let's restrict our attention to the problem of determining whether or not a certain number occurs in a certain list of keys.

> // On the way, we'll admire a forest of Binary Trees.

4.1 Searching

The *search problem* is

> Given an array $A[1]$, $A[2]$, $A[3]$, ... , $A[n]$ and a target value T,
> find an index j where $T = A[j]$ or determine that no such index exists
> because T is not in the array A.

4.1.1 Searching an Arbitrary List

Names in telephone books are not kept in arbitrary order and probably neither are student records. But consider the problem of finding your car in a parking lot when you've forgotten where you left it. If parking spaces are selected at random by drivers as they arrive, *how would you find your car*?

```
// A solution algorithm sometimes applied by tipsy students is random walk:
//
//   (a)  Pick a direction N, S, E, or W at random and take a step in that direction;
//        if you bump into the edge of the parking lot, take a step back; but
//        if you bump into a car, try your car key and
//           if the key works, stop the search.          (Here is a key a key?)
```

© Springer International Publishing AG, part of Springer Nature 2018   137
T. Jenkyns and B. Stephenson, *Fundamentals of Discrete Math for Computer
Science: A Problem-Solving Primer*, Undergraduate Topics in Computer Science,
https://doi.org/10.1007/978-3-319-70151-6_4

// (b) Go back to (a) and pick again.
//
// Would you believe that this algorithm can be proved to be effective
// (in a certain "probabilistic" sense)?

A more practical algorithm would be as follows: Look at the parked cars one at a time in some order, say, from the beginning of the first row to the end of the first row, then back along the second row to its beginning, and so on until the last car in the last row is checked (if necessary). Stop searching when you've found your car or when you've checked all the positions in the lot (perhaps your car is not there because it's been stolen, seized by the bank, impounded by the police, borrowed by your sister, or left at home). The algorithm of comparing the target value T to the value in each position in turn is known as a sequential or a *Linear Search*. Returning to the case of numerical keys, we have

Algorithm 4.1.1: Linear Search

Begin
    j ← 0;
  Repeat
     j ← j+1;
  Until ((A[j] = T) **Or** (j = n));

  If (A[j] = T) **Then**
    Output("T is A[", j, "]");
  Else
    Output("T is not in A");
  End;
End.

This algorithm is clearly correct – it will surely find the target if it is present and will surely terminate with a correct report if the target is not present.

// It finds the first occurrence of T if T occurs in the array more than once.

The *cost* of a search algorithm is often taken as the number of *probes*, the number of array entries that must be retrieved and then compared to the target value. Linear Search makes at most n probes when T is in the array and makes exactly n probes when T is not present. A search typically takes longest (a worst case occurs) when it is unsuccessful.

Suppose that all the entries in A are distinct. If $T = A[1]$, the search stops after 1 probe; if $T = A[2]$, the search stops after 2 probes. For each index j, finding $A[j]$ as the target value requires exactly j probes. If we were to search for all the entries in A in turn, the average number of probes would be

$$\bar{p} = (1/n) \sum_{j=1}^{n} j = (1/n) \times \{n(n+1)/2\} = (n+1)/2. \qquad \text{// by Theorem 3.6.2}$$

// On average, Linear Search would go through about half the list.
// If n were 99, we'd average 50 probes; if n were 25,000 we'd average 12,500½
// probes.

4.1.2 Searching a Sorted List

Now consider adding some structure to the array A to facilitate searching, like sorting the entries into increasing order.   // Really just "nondecreasing" order. Words in a dictionary are kept in increasing alphabetical order, which does ease the task of finding a target word (when you know how to spell it). Suppose that we are given an array where

$$A[1] <= A[2] <= A[3] <= \ldots <= A[n].$$

// We use "$<=$" to denote that the elements are in nondecreasing order whether // they are numbers or not.

Of course, we could still apply Linear Search to this list. We could even change the Boolean expression controlling the repeat-loop to

$$\textbf{Until } ((A[j] >= T) \textbf{ Or } (j = n));$$

and this will stop the search sooner (when T is not present) because

if $T < A[j]$ then T cannot occur among $A[j+1], A[j+2], A[j+3], \ldots, A[n]$.

// since they are all at least as big as $A[j]$ which is bigger than T

Expanding on this idea, if we make **one** probe at $A[i]$, then:

1. If $A[i] = T$, then we're done.   // 1 entry
2. If $T < A[i]$, then T cannot occur among $A[i], A[i + 1], \ldots, A[n]$,
   and we need only search the list from $A[1]$ to $A[i - 1]$;   // $i - 1$ entries
                           // If $T = A[j]$, then j must be between 1 and $i - 1$.
3. If $A[i] < T$, then T cannot occur among $A[1], A[2], \ldots, A[i]$,
   and we need only search the list from $A[i + 1]$ to $A[n]$.   // $n - i$ entries
                           // If $T = A[j]$, then j must be between $i + 1$ and n.

// What is the best way to choose the position i for the first probe?
// If $n = 100$ and we probe at $A[5]$, then:
//   1. If $A[5] = T$, then we're done.
//   2. If $T < A[5]$, then we need to search the sublist $A[1], A[2], A[3], A[4]$.
//   3. If $A[5] < T$, then we need to search the sublist $A[6], A[7], \ldots, A[100]$.
// The worst and most likely of these three cases leaves us with a list of 95 entries.
// On the other hand, if we probe at $A[50]$, then the worst case leaves us with a list
// of 50 entries: $A[51], A[52], \ldots, A[100]$.

If we adopt a policy of making the worst case as favorable as we can, we should probe in the middle (or as near to the middle as possible) of the list (or sublist) we're searching. Then, if we don't find the target there, we will have at most half the list to search.   // Does this remind you of an earlier algorithm? Suppose the sublist we're searching is from $A[p]$ up to $A[q]$. The average of p and q is exactly halfway between them, but $(p + q)/2$ might not be an integer so let's probe at $A[i]$ where $i = \lfloor (p + q)/2 \rfloor$.

The resulting search algorithm where we probe in the middle of the current sublist is

Algorithm 4.1.2: Binary Search

Begin
  p ← 1;
  q ← n;
  Repeat
    j ← ⌊(p + q)/2⌋;
    If (A[j] < T) **Then**
      p ← j + 1;
    End;
    If (A[j] > T) **Then**
      q ← j − 1;
    End;
  Until ((A[j] = T) **Or** (p > q));
  If (A[j] = T) **Then** Output("T is A[", j, "]");
  Else Output("T is not in A");
  End;
End.

Walkthrough with $n = 12$ and $A = (3, 5, 8, 8, 9, 16, 29, 41, 50, 63, 64, 67)$

// $A[1] = 3$, $A[2] = 5$, $A[3] = 8$, $A[4] = 8$, $A[5] = 9$, $A[6] = 16$,
// $A[7] = 29$, $A[8] = 41$, $A[9] = 50$, $A[10] = 63$, $A[11] = 64$, & $A[12] = 67$

If $T = 9$, then

p	j	q	A[j]	relation	output
1	6	12	16	T < A[j]	–
1	3	5	8	A[j] < T	–
4	4	5	8	A[j] < T	–
5	5	5	9	A[j] = T	T is A[5]

If $T = 64$, then

p	j	q	A[j]	relation	output
1	6	12	16	A[j] < T	–
7	9	12	50	A[j] < T	–
10	11	12	64	A[j] = T	T is A[11]

If $T = 23.4$, then   // T and the entries in A might be real numbers.

p	j	q	A[j]	relation	output
1	6	12	16	A[j] < T	–
7	9	12	50	T < A[j]	–
7	7	8	29	T < A[j]	–
7	–	6	–	–	T is not in A

// T lies between A_6 and A_7.

If T = 99, then

p	j	q	A[j]	relation	output
1	6	12	16	A[j] < T	–
7	9	12	50	A[j] < T	–
10	11	12	64	A[j] < T	–
12	12	12	67	A[j] < T	–
13	–	12	–	–	T is not in A

// T lies beyond A_n.

// When $n = 12$, is the first probe always at $A[6]$ and the second, either at $A[3]$ or
// $A[9]$?
//
// In general:
//   Is this algorithm sure to terminate?
//   How many iterations of the loop could there be?
//   Is this algorithm correct? If T is in A is the algorithm guaranteed to find it?
//   If T is not in A, *must* p eventually become larger than q?

Suppose the sublist we're searching is from $A[p]$ up to $A[q]$ of length k and we
probe unsuccessfully at $A[j]$:   // $k = q - p + 1$.

$$\underbrace{A[p]\ldots A[j-1]}_{k1}\quad A[j]\quad \underbrace{A[j+1]\ldots A[q]}_{k2}.$$

If $A[j] > T$, then we will search from $A[p]$ up to $A[j - 1]$ of length $k1 = j - p$.
If $A[j] < T$, then we will search from $A[j + 1]$ up to $A[q]$ of length $k2 = q - j$.
We would like $k1$ and $k2$ to be equal, but if that's not possible (because $q - p$ is
odd), let's consistently make $k1$ the smaller value.

// How should we choose j so that $k1 <= k2$ and $k2$ is as small as possible?

To make $k2 = q - j$ as small as possible, we need to make j as large as possible.
We want

$$k1 + k2 <= k2 + k2,$$
that is, $$q - p <= 2(q - j) \qquad = 2q - 2j$$
or $$2j <= 2q - (q - p) = q + p.$$

The largest integer $j <= (q + p)/2$ is $\lfloor (q + p)/2 \rfloor$, and this is the j-value used in the
algorithm.

Theorem 4.1.1: On each iteration of Binary Search with $j = \lfloor (p + q)/2 \rfloor$, the lengths of the sublists will be $k1 = \lfloor (k - 1)/2 \rfloor <= k2 = \lceil (k - 1)/2 \rceil <= k/2$.

Proof. Since $\quad j <= (q+p)/2 < j+1$,
subtracting p from each of these three expressions gives

$$j - p <= (q+p)/2 - p < j - p + 1$$

or
$$k1 <= (q - p)/2 \quad < k1 + 1.$$
Thus,
$$k1 = \lfloor (q - p)/2 \rfloor = \lfloor (k - 1)/2 \rfloor.$$

Because $k1 + k2 = k - 1$, we know that

if k is odd, say $k = 2r + 1$, then $k - 1 = 2r$ and $k1 = r = k2 < k/2$, and
if k is even, say $k = 2r$, then $k - 1 = 2r - 1$ and $k1 = r - 1 < r = k2 = k/2$.

Thus, $\qquad k1 = \lfloor (k - 1)/2 \rfloor <= k2 = \lceil (k - 1)/2 \rceil <= k/2.$ ☐

Therefore, at each iteration of Binary Search, the length of the next sublist is at most half the length of the current sublist.   // The search-region is halved.

Theorem 4.1.2: Binary Search terminates after at most $\lfloor \lg(n) \rfloor + 1$ probes

Proof. Let $w = \lfloor \lg(n) \rfloor$. If, in some instance, Binary Search has <u>not</u> terminated after w (unsuccessful) probes, then the current value of p must be $<=$ the current value of q, and the length of the current sublist, k, must be $<= n/2^w$.

                                                //X Prove this by MI.

Since $w <= \lg(n) < w + 1$,   // $n = 2^{\lg(n)}$ so

$$2^w <= n < 2 \times 2^w \qquad \text{// then dividing each by } 2^w$$

so $1 <= n/2^w < 2.$ // which implies that $k = 1$

For the next iteration, $p = q$ and so $j = p$.

                      // And we probe the one remaining entry in A.

If $A[j] < T$, then $p \leftarrow j + 1 > q$, so Binary Search terminates after this probe;
if $A[j] > T$, then $q \leftarrow j - 1 < p$, so Binary Search terminates after this probe; and
if $A[j] = T$, then Binary Search terminates after this one last probe.

Therefore, Binary Search terminates after at most $\lfloor \lg(n) \rfloor + 1$ probes.   ☐

// Recall that on average, Linear Search would go through about half the list.
// If $n = 99$, it averages 50 probes; if $n = 25,000$, it averages $12,500\frac{1}{2}$ probes.
// For Binary Search, a worst possible case makes $\lfloor \lg(n) \rfloor + 1$ probes and
// $\lg(25000) \cong 14.609$. *If $n = 25,000$ Binary Search makes <u>at most</u> 15 probes.*
// If $n = 25,000$, Linear Search does >800 times as many probes on average.

In order to prove that Binary Search is correct, we will first establish a loop invariant. The algorithm proceeds by dividing the search region (the sublist from

$A[p]$ up to $A[q]$) in half at each iteration after probing in the middle and then changing either p or q. A ***loop invariant*** that is maintained is

"If T is in A,   then T must be in the sub-list from $A[p]$ to $A[q]$."
that is,   "If $T = A[i]$ then i must be between p and q."

Theorem 4.1.3: After k iterations of the loop,   // For any k,
if $T = A [i]$, then $p <= i <= q$ // this is $P(k)$.

Proof. // by Mathematical Induction on k where $k \in \{0.. \}$
Step 1. After $k = 0$ iterations of the loop,   // that is, before the loop is done once
   $p = 1$ and $q = n$, and therefore,

$$\text{if } T = A[i] \text{ then } p <= i <= q. \qquad // P(0) \text{ is True}$$

// Suppose that for the first few iterations of the loop (even though p and q may
// have been changed), this conditional statement has been True.

Step 2. Assume $\exists w \in \mathbf{N}$ such that
  after w iterations of the loop, if $T = A[i]$, then $p <= i <= q$.   // This is $P(w)$.

// Now, what will happen on the next iteration, if there is one?

Step 3. Suppose there is another iteration, the $w + 1^{st}$.

// That is, T has not been found in w unsuccessful probes, and now, $p <= q$.

On the next iteration, we calculate a new j-value,

$$j_{new} \leftarrow \lfloor (p+q)/2 \rfloor$$

Since $p <= q$, $p+p <= p+q$ $<= q+q$

and $p = (p+p)/2 <= (p+q)/2 <= (q+q)/2 = q$ and so $p <= j_{new} <= q$.

// In fact, if $p = q$, then $j_{new} = p$, and if $p < q$, then $p <= j_{new} < q$.

In the remainder of the proof, we will let p^* and q^* denote the values of p and q at the end of the iteration. There are three cases to consider:

Case 1. If $A[j_{new}] < T$, then T cannot occur at or before position j_{new} hence

  if $T = A[i]$, then $p < j_{new} + 1 <= i <= q$. // Here $p^* = j_{new} + 1$ and $q^* = q$.

Case 2. If $A[j_{new}] > T$, then T cannot occur at or after position j_{new} hence

  if $T = A[i]$,   then $p <= i <= j_{new} - 1 < q$. // Here $p^* = p$ and $q^* = j_{new} - 1$.

Case 3. If $A[j_{new}] = T$, then neither p nor q are changed hence, from Step 2

if $T = A[i]$, then $p <= i <= q$.   // In this case, $p^* = p$ and $q^* = q$.

Therefore, after this next iteration,   // in every case

if $T = A[i]$, then $p^* <= i <= q^*$.   ▯

When the repeat-loop in Binary Search terminates,

if $T = A[i]$, then $p <= i <= q$. // That is, this conditional statement is True.

But if $p > q$, there is no index i where $p <= i <= q$.

// The consequent must be False.
If $p > q$, there is no index i where $T = A[i]$.   // The antecedent must be False.
If $p > q$, then T cannot be in A. Therefore, when Binary Search terminates,

          the target has been found at position j
or   ($p > q$ and so) T is not in A.

Binary Search is correct and very efficient.   // compared to Linear Search

// Do other loop invariants hold for Binary Search? Perhaps
//   either $p = 1$ or $A[p - 1] < T$ and either $q = n$ or $T < A[q + 1]$.
//
// Do the p-values stay the same or grow larger? Do the q values never increase?
//
// If T is not found, where will T fit into the array? Where should it be inserted?
//   Is the final q-value always <u>one</u> less than the final p-value?
//   When is $A[q] < T < A[p]$?
//   If $p = 1$ (that is, p was never changed), is $T < A[1]$?
//   If $q = n$ (that is, q was never changed), is $A[n] < T$?

The Most Important Ideas in This Section.
The *search problem* is as follows: Given an array $A[1], A[2], \ldots, A[n]$ and a
target value T, find an index j where $T = A[j]$ if there is such an index j. *Linear
Search* compares T to the entries in turn until it finds T or exhausts the array.
Binary Search compares T to the middle entry, $A[j]$. If it finds T, it stops; if
$T < A[j]$, it searches to lower half of A, but if $T > A[j]$, it searches the upper
half of A. Binary Search is generally much, much faster. But Binary Search
requires a sorted input array (sorting is the subject of Sect. 4.3).

The next section gives a method to diagram the action of searches and then
gives a second version of Binary Search which makes fewer comparisons.

4.2 Branching Diagrams

A *General Branching Diagram* for an algorithm is a tree that shows all possible sequences of operations the algorithm might do. They can be huge structures even for simple algorithms, and (while they can be imagined) they are rarely constructed. However, sometimes constructing a portion of it is useful. When n is small, we can construct the *Branching Diagram of probes for Binary Search*, a tree diagram displaying all possible sequences of probes that Binary Search might make.

The first probe is *always* at $A[j]$ where $j = \lfloor (1 + n)/2 \rfloor$. This probe is placed at the top of the diagram.

// near the middle of the page like the "start" vertex in Chap. 2

After this probe, if $T \neq A[j]$ the algorithm continues along one of two "branches":

when $T < A[j]$, follow the tree *down* to the **left** to the next probe;

// $A[r]$ where $r = \lfloor j/2 \rfloor$

when $A[j] < T$, follow the tree *down* to the **right** to the next probe.

// $A[s]$ where $s = \lfloor (j + 1 + n)/2 \rfloor$

For $n = 12$, the diagram is

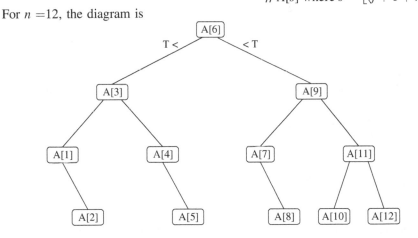

This kind of diagram is known as a *Binary Tree*: it's *"rooted"* at the vertex placed at the top of the diagram, and is binary in the sense that from any vertex, there are at most two edges downward in the diagram. The vertices with no downward edges are called *leaves*. The other vertices are called *internal* vertices. Every vertex, except the root, is the end of exactly one edge to it from a vertex above it in the diagram.

// Does this sound as though Binary Trees are drawn upside down?
// Does every entry in A appear exactly once in this tree? Why?

4.2.1 A Second Version of Binary Search

Jack and Jill are playing a game where Jill chooses a number between 1 and 1,000 and then Jack tries to discover Jill's number by asking "yes/no" questions about

Jill's number – Jack may <u>not</u> ask "What is your number?" but may ask "Is your number 6?" or "Is your number even and greater than 543?" *How should Jack formulate his sequence of questions to find Jill's number most quickly; that is, with fewest questions (in the worst case)?*

Binary Search as given in Algorithm 4.1.2 makes <u>three</u> comparisons with each probe (i.e., asks 3 questions about each number it investigates). If we wanted to reduce the number of comparisons, we could modify it to search for the target value *T* by making exactly one comparison per probe until the sublist we're searching has only one entry and then testing to see if that single entry equals *T*.

Algorithm 4.2.1: Binary Search #2   // version #2.0
Begin
   p ← 1;
   q ← n;
   While (p < q) **Do**
     j ← ⌊(p+q)/2⌋;
    If (A[j] < T) **Then**
      p ← j+1;
    Else
      q ← j;
    End;   // the if
   End;   // the while   // Now, p = q

   If (A[p] = T) **Then**
     Output("T is A[", p, "]");
   Else
     Output("T is not in A");
   End;   // the if
End.

Walkthrough with $n = 12$ and $A = (3, 5, 8, 8, 9, 16, 29, 41, 50, 63, 64, 67)$

// $A[1] = 3$, $A[2] = 5$, $A[3] = 8$, $A[4] = 8$, $A[5] = 9$, $A[6] = 16$,
// $A[7] = 29$, $A[8] = 41$, $A[9] = 50$, $A[10] = 63$, $A[11] = 64$,   & $A[12] = 67$

If $T = 9$, then (using **t** for True and **f** for False)   // and *T* for target

p	j	q	p<q	A[j]	A[j] < T	A[p] = T	output
1	6	12	t	16	f	–	–
1	3	6	t	8	t	–	–
4	5	6	t	9	f	–	–
4	4	5	t	8	t	–	–
5	–	5	f	–	–	t	T is A[5]

If T = 64, then

p	j	q	p < q	A[j]	A[j] < T	A[p] = T	output
1	6	12	t	16	t	–	–
7	9	12	t	50	t	–	–
10	11	12	t	64	f	–	–
10	10	11	t	63	t	–	–
11	–	11	f	–	–	t	T is A[11]

If T = 23.4, then   // T and the entries in A might be real numbers.

p	j	q	p < q	A[j]	A[j] < T	A[p] = T	output
1	6	12	t	16	t	–	–
7	9	12	t	50	f	–	–
7	8	9	t	41	f	–	–
7	7	8	t	29	f	–	–
7	–	7	f	–	–	f	T is not in A

// T lies between A_6 and A_7.

If T = 99, then

p	j	q	p < q	A[j]	A[j] < T	A[p] = T	output
1	6	12	t	16	t	–	–
7	9	12	t	50	t	–	–
10	11	12	t	64	t	–	–
12	–	12	f	–	–	f	T is not in A

// T lies beyond A_n.

//X Do the same loop invariants for the while loop hold for Version #2? Is it True
//   that
//   "after each iteration of the loop, if $T = A[i]$ then $p <= i <= q$"?
//
//X If T is not found, where will T fit in to the array? Where should it be inserted?
//X Suppose that at the last comparison, $A[p] \neq T$. Now,
//   if $1 < p < n$ is $A[p-1] < T < A[p]$?
//   if $1 = p < n$ is $T < A[1]$?
//   if $1 < p = n$ is $A[n-1] < T < A[n]$ or $A[n] < T$?

Suppose the sublist we're searching is from $A[p]$ up to $A[q]$ of length k and we
make the comparison "is $A[j] < T$?":   // $k = q - p + 1$.

$$\underbrace{A[p]\ldots A[j]}_{k1}\quad \underbrace{A[j+1]\ldots A[q]}_{k2}$$

If $A[j] < T$, then we will search the sublist from $A[j + 1]$ up to $A[q]$ of length $k2 = q - j$. If $A[j] >= T$, then we will search the sublist from $A[p]$ up to $A[j]$ of length $k1 = j - p + 1$. We would like $k1$ and $k2$ to be equal, but that's not possible when k is odd.

When j is given the value $\lfloor (q + p)/2 \rfloor$, as we saw in Theorem 4.1.1,

$$k2 = \lceil (k - 1)/2 \rceil.$$

But $\lceil (k - 1)/2 \rceil = \lfloor k/2 \rfloor.$ //X prove this $\forall k \in \mathbf{Z}$.

Since $k1 + k2 = k$, we have   // $k1 = k - k2$.

if k is even, say $k = 2r$, then $k1 = r = k2 = k/2$ and
if k is odd, say $k = 2r + 1$, then $k2 = r < k/2 < r+1 = k1$.

In general,

$$k2 = \lfloor k/2 \rfloor <= k/2 <= k1 = \lceil k/2 \rceil.$$ // Recall Theorem 3.5.1.

Therefore, at some iterations of binary search #2, the length of the next sublist is more than half the length of the previous sublist.   // but not much more

Let $L(w)$ denote the length of the sublist still to be searched after w iterations of the while loop.   // And $L(0) = n$.

// We've just shown that $\lfloor L(w)/2 \rfloor <= L(w+1) <= \lceil L(w)/2 \rceil$.

Theorem 4.2.1: After w iterations of the loop

$$\lfloor n/2^w \rfloor <= L(w) <= \lceil n/2^w \rceil.$$ // This is $P(w)$.

Proof. // by Mathematical Induction on w where $w \in \{0.. \}$
Step 1. After $w = 0$ iterations of the loop,   // before the loop is done once
         the length of the current sublist,

$$L(0) = n = \lfloor n/2^0 \rfloor = \lceil n/2^0 \rceil.$$ // $P(0)$ is True.

// Suppose that for the first few iterations of the loop (even though p and q may
// have been changed), these bounds on the length of the current sublist are
// maintained.

Step 2.   Assume $\exists m \in \mathbf{N}$ such that after m iterations of the loop

$$\lfloor n/2^m \rfloor <= L(m) <= \lceil n/2^m \rceil.$$ // This is $P(m)$.

// Now, what will happen on the next iteration, if there is one?

Step 3.   Suppose there is another iteration, the $m+1^{st}$.   // That is, now $p < q$.
We know that

$$\lfloor L(m)/2 \rfloor \;<=\; L(m+1) \;<=\; \lceil L(m)/2 \rceil.$$

// If we can to show that
//
// $\lceil L(m)/2 \rceil \;<=\; \lceil \lceil n/2^m \rceil /2 \rceil$ and $\lceil \lceil n/2^m \rceil /2 \rceil = \lceil n/2^{m+1} \rceil$,
//
// we will have the upper bound we want: $L(m+1) \;<=\; \lceil n/2^{m+1} \rceil$.
//
// We deal with these (and the corresponding inequalities with the floor function)
// in a more general setting in the next two lemmas.
//
// Remember that $\lceil r \rceil$ is (defined to be) the smallest integer $>=$ the (real) number r
// and $\lfloor r \rfloor$ is the largest integer $<= r$.
// Also, recall that whenever y is not an integer, $\lfloor y \rfloor < y < \lceil y \rceil = \lfloor y \rfloor + 1$.

Lemma A: If x and y are real numbers and $x < y$, then $\lfloor x \rfloor \;<=\; \lfloor y \rfloor$ and $\lceil x \rceil \;<=\; \lceil y \rceil$.

Proof. // of Lemma A, that the floor and ceiling functions are nondecreasing.

Since $x < y <= \lceil y \rceil \in \mathbf{Z}$ and $\lceil x \rceil$ is the smallest integer $>= x$, $\lceil x \rceil <= \lceil y \rceil$.
Since $\lfloor x \rfloor <= x < y$ and $\lfloor y \rfloor$ is the largest integer $<= y$, $\lfloor y \rfloor >= \lfloor x \rfloor$.   □

Lemma B: For any real number x, $\left\lfloor \dfrac{\lfloor x \rfloor}{2} \right\rfloor = \left\lfloor \dfrac{x}{2} \right\rfloor$ and $\left\lceil \dfrac{x}{2} \right\rceil = \left\lceil \dfrac{\lceil x \rceil}{2} \right\rceil$.

Proof. // of Lemma B

If x is an integer, then $\lfloor x \rfloor = x = \lceil x \rceil$, and therefore,

$$\left\lfloor \frac{\lfloor x \rfloor}{2} \right\rfloor = \left\lfloor \frac{x}{2} \right\rfloor \quad \text{and} \quad \left\lceil \frac{x}{2} \right\rceil = \left\lceil \frac{\lceil x \rceil}{2} \right\rceil.$$

Suppose now that x is <u>not</u> an integer, and let Q denote $\lfloor x/2 \rfloor$. Then, we have

$$Q \; < \; x/2 \; < \; Q+1 \qquad \text{// } x/2 \text{ cannot be an integer.}$$

$\Leftrightarrow$ $$2Q \; < \; x \; < \; 2Q+2.$$

In fact, $$2Q \; <= \lfloor x \rfloor \; < \; 2Q+2, \qquad \text{// because } 2Q \in \mathbf{Z}$$

and $$2Q \; < \; \lceil x \rceil \; <= \; 2Q+2.$$

Then, $$Q <= \frac{\lfloor x \rfloor}{2} < \frac{x}{2} < \frac{\lceil x \rceil}{2} <= Q + 1,$$

and so $\left\lfloor \dfrac{\lfloor x \rfloor}{2} \right\rfloor = \left\lfloor \dfrac{x}{2} \right\rfloor = Q$ and $\left\lceil \dfrac{x}{2} \right\rceil = \left\lceil \dfrac{\lceil x \rceil}{2} \right\rceil = Q + 1.$ ☐

Returning to the proof of Theorem 4.2.1, and taking $x = n/2^m$, we have // from Lemma B

$$\lfloor \lfloor n/2^m \rfloor /2 \rfloor = \lfloor n/2^{m+1} \rfloor \quad \text{and} \quad \lceil \lceil n/2^m \rceil /2 \rceil = \lceil n/2^{m+1} \rceil.$$

Since $\lfloor n/2^m \rfloor <= L(m) <= \lceil n/2^m \rceil,$ // from Step 2

$\lfloor n/2^m \rfloor /2 <= L(m)/2 <= \lceil n/2^m \rceil /2$ // and applying Lemma A

$$\lfloor \lfloor n/2^m \rfloor /2 \rfloor <= \lfloor L(m)/2 \rfloor \quad \text{and} \quad \lceil L(m)/2 \rceil <= \lceil \lceil n/2^m \rceil /2 \rceil.$$

Thus,

$$L(m+1) <= \lceil L(m)/2 \rceil <= \lceil \lceil n/2^m \rceil /2 \rceil = \lceil n/2^{m+1} \rceil$$

and $$L(m+1) >= \lfloor L(m)/2 \rfloor >= \lfloor \lfloor n/2^m \rfloor /2 \rfloor = \lfloor n/2^{m+1} \rfloor;$$

that is, $\lfloor n/2^{m+1} \rfloor <= L(m+1) <= \lceil n/2^{m+1} \rceil.$ // $P(m+1)$ is True.

The proof of Theorem 4.2.1 is complete. ☐

Theorem 4.2.2: After at most $\lceil \lg(n) \rceil$ comparisons of the form "is $A[j] < T$?" the length of the current sublist is 1. // And $p = q$.

Proof. // We will show that the while loop does at least $\lfloor \lg(n) \rfloor$ iterations and
// at most $\lceil \lg(n) \rceil$ iterations. So if $\lg(n) = Q \in \mathbf{N}$, it does exactly Q iterations.
//
// One more equality comparison is done, so Binary Search #2 stops after at most
// $\lceil \lg(n) \rceil + 1$ comparisons.

Let $Q = \lceil \lg(n) \rceil$. Then

$$Q - 1 < \lg(n) <= Q \text{ so } 2^{Q-1} < n <= 2^Q.$$

For any integer k

$$2^{Q-1-k} = \frac{2^{Q-1}}{2^k} < \frac{n}{2^k} <= \frac{2^Q}{2^k} = 2^{Q-k}.$$

Applying Theorem 4.2.1 and Lemma A, the length of the current sublist after k iterations is bounded:

$$L(k) <= \lceil n/2^k \rceil <= \lceil 2^Q/2^k \rceil \quad = 2^{Q-k} \qquad // \; Q - k \in \mathbf{P} \text{ so } 2^{Q-k} \in \mathbf{P}.$$

and $\quad L(k) >= \lfloor n/2^k \rfloor >= \lfloor 2^{Q-1}/2^k \rfloor = 2^{Q-1-k}. \qquad // \text{ And } 2^{Q-k-1} \in \mathbf{P}.$

In particular,

$$2 = 2^1 <= L(Q - 2) <= 2^2$$

and $\qquad\qquad 1 = 2^0 <= L(Q - 1) <= 2^1 = ?$

Thus, the while loop cannot stop before $Q - 1$ iterations are done, though it might stop after exactly $Q - 1$ iterations are done, but, if $L(Q - 1) = 2$, it must stop after (one more iteration) exactly Q iterations. $\qquad\qquad\qquad\qquad\qquad$ ☐

We can construct the **Branching Diagram of comparisons for Binary Search #2**. The first comparison is *always* "is $A[j] < T$?" where $j = \lfloor (1 + n)/2 \rfloor$, and this comparison is placed at the top of the diagram. $\qquad$ // in the middle of the page
When "$A[j] < T$" is **False**, follow the tree down to the **left** to the next comparison; when "$A[j] < T$" is **True**, follow the tree down to the **right** to the next comparison. The leaves are the (final) comparisons that take the form "is $A[p] = T$?"
When $n = 12$, the diagram we get is

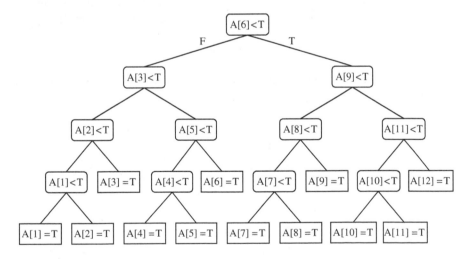

This diagram is also a **Binary Tree**. But in it, from <u>every</u> internal vertex, there are exactly two edges downward in the diagram (never just one). This kind of Binary Tree is said to be a *full* Binary Tree.

// The variable j is never equal to n so we never ask "is $A[n] < T$?"
// Does the tree have a unique internal vertex for all $(n - 1)$ other values of j?
// Are there n leaves each corresponding to one (possible) comparison
// "is $A[i] = T$?"

Theorem 4.2.3: If $\mathcal{T}$ is a full Binary Tree with m internal vertices, then $\mathcal{T}$ has $m + 1$ leaves

Proof. // by Strong Induction on m where $m \in \{0.. \}$

Step 1.   If $m = 0$, then $\mathcal{T}$ has a root vertex, r, but no internal vertices. Therefore, r must be a leaf and must be the only vertex in $\mathcal{T}$.   // $\mathcal{T}$ has $m + 1$ leaves.

Step 2.   Assume $\exists k \in \mathbf{N}$ such that if $0 <= m <= k$ and if $\mathcal{T}$ is a full Binary Tree with m internal vertices, then $\mathcal{T}$ has $m + 1$ leaves.

Step 3.   Suppose now that $\mathcal{T}^*$ is a full Binary Tree with $k + 1$ internal vertices.

// We must prove that $\mathcal{T}^*$ has $(k + 1) + 1 = k + 2$ leaves.

Let r^* denote the root of $\mathcal{T}^*$; r^* cannot be a leaf, so must be an internal vertex with exactly two vertices, u and v, below it in the diagram.

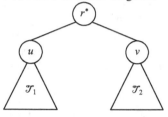

The part of the diagram from u down is full Binary Tree $\mathcal{T}_1$ with $m1$ internal vertices, and the part of the diagram from v down is full Binary Tree $\mathcal{T}_2$ with $m2$ internal vertices. Since every internal vertex of $\mathcal{T}$ is either in $\mathcal{T}_1$ or in $\mathcal{T}_2$ or r^*,

$$k + 1 = m1 + m2 + 1$$

where both $m1$ and $m2$ are nonnegative and $<= k$. Since every leaf of $\mathcal{T}$ is either a leaf in $\mathcal{T}_1$ or is a leaf in $\mathcal{T}_2$,

$$\# \text{ leaves in } \mathcal{T} = \# \text{ leaves in } \mathcal{T}_1 + \# \text{ leaves in } \mathcal{T}_2$$
$$= m1 + 1 \qquad + m2 + 1$$
$$= k + 2 \qquad\qquad\qquad\qquad \square$$

The Most Important Ideas in This Section.
A ***Branching Diagram*** for an algorithm (sometimes called a "decision tree") is a tree that shows all possible sequences of operations the algorithm might do. Sometimes, constructing a portion of it is useful. We constructed the ***Branching Diagram of probes for Binary Search*** when $n = 12$ displaying all possible sequences of probes that it might make. That diagram was a ***Binary Tree*** with 5 ***leaves*** and 7 ***internal*** vertices.

(continued)

(continued)

> A second version of Binary Search with fewer comparisons was given, and we constructed the **Branching Diagram of comparisons for Binary Search #2** for $n = 12$ displaying all possible sequences of comparisons it might make. That diagram was as a **full Binary Tree** with 12 **leaves** and 11 **internal** vertices.
>
> We proved that every full Binary Tree with m internal vertices has $m + 1$ leaves.   // a fact we'll use later when analyzing algorithms

---

4.3 Sorting

We showed that efficient searching requires sorted data. But even if sorting were useless, it is worth studying for other reasons. Sorting is an easily understood problem for which there is a truly surprising variety of solution strategies and algorithms.

> // And sorting methods are part of the common
> // knowledge of all computer scientists.

We'll look at just three such strategies in detail, though others are considered in the exercises.

The **sorting problem** is

Given an array $A[1], A[2], A[3], \ldots, A[n]$,   // of numbers in arbitrary order
rearrange the values in A so that   // maybe real numbers

$$A[1] \;<=\; A[2] \;<=\; A[3] \;<=\; \ldots \;<=\; A[n].$$

> // Is every list of length 1 already sorted?

4.3.1 Selection Sorts

Imagine your young nephew is visiting and has entertained himself by perusing your Illustrated Encyclopedia of Tantric Trigonometry. He's put the whole set back on its shelf but in a peculiar order. Could he use the following method to unscramble them?

Find volume 1.
If it's not in position 1, then
    set it on the floor.
    move whatever volume is in position 1 to the empty space
      that volume 1 left free
    pick up volume 1 and put it in position 1 that was just emptied.
Then do the same for volume 2, volume 3, and so on.

// Volume 1 "belongs" in the first position on the bookshelf,
// which value "belongs" in $A[1]$?

Selection sorts choose a particular element in the list (usually the smallest or the largest) and place it in the correct position; *MinSort* selects the minimum entry. So first let's construct an algorithm for finding the minimum entry in a list subject to the restriction that we can only compare two values at a time. Let's make a "pass" through the list keeping track of the minimum value we've found so far in the variable *min*.

Algorithm 4.3.1: Minimum

```
Begin
   min ← A[1];
   For j ← 2 To n Do
     If (A[j] < min) Then
       min ← A[j];
     End;   // the if
   End;     // the for-j loop
   Return(min);
End.
```

// This pseudo-code contains a new control structure, a *for-loop*, which looks like:
//
//   **For** variable ← expression1 **To** expression2 **Do**
//   (the *body* of the loop)
//   **End**;
//
// The execution of this construction is the same as the following:
//
//   variable ← expression1;
//   **While** (variable <= expression2) **Do**
//   (the *body* of the loop)
//   variable ← variable + 1;
//   **End**;
//
// For-loops are a natural way to manipulate arrays; the control variable starts at a
// particular value and goes up by ones until it reaches a second particular value.

To begin, the smallest so far is $A[1]$. Then, looking at $A[2], A[3] \ldots$ in turn, if one of these is smaller than the smallest so far we change *min* to be that A-value. That is,

$$\text{"}min \text{ is the smallest of } A[1], A[2] \ldots A[j]\text{"}$$

is a loop invariant of the for-j loop. Then, the final value of *min* must be the smallest value in the whole list, and it was obtained at the cost of $(n - 1)$ comparisons. *The cost of a sorting algorithm is usually taken to be the number of comparisons involving two array entries.*

                                         // comparisons involving values of the keys

We can sort the A-values by finding the minimum of all of them and interchanging it with $A[1]$, then finding the minimum of all the A-values from $A[2]$ to $A[n]$ of them and interchanging it with $A[2]$, and so on.

Algorithm 4.3.2: MinSort

```
Begin
  For k ← 1 To (n-1) Do        // the kᵗʰ pass selects a correct value for A[k]
    min ← A[k];                  // min will be the smallest in A[k]..A[n]
    index ← k;                   // & min occurs at entry A[index]
    For j ← k+1 To n Do
      If (A[j] < min) Then
        min ← A[j];
        index ← j;
      End;     // the if
    End;        // the for-j loop
    A[index] ← A[k];
    A[k] ← min;
  End;      // the for-k loop
End.
```

// If $A[1]$, $A[2]$, ... $A[n-1]$ have been selected correctly, what about $A[n]$?
// Can we remove the variable **min** and just use A[*index*]?

Walkthrough with $n = 5$ and $A = (3.1, 5.7, 4.3, 1.9, 3.1)$

// $A[1] = 3.1$, $A[2] = 5.7$, $A[3] = 4.3$, $A[4] = 1.9$, $A[5] = 3.1$

k	min	index	j	A[j]			A		
1	3.1	1	2	5.7	3.1	5.7	4.3	1.9	3.1
	"	"	3	4.3					
	"	"	4	1.9					
	1.9	4	5	3.1					
	"	"	–		**1.9**	5.7	4.3	**3.1**	3.1
2	5.7	2	3	4.3					
	4.3	3	4	3.1					
	3.1	4	5	3.1					
	"	"	–		1.9	**3.1**	4.3	**5.7**	3.1
3	4.3	3	4	5.7					
	"	"	5	3.1					
	3.1	5	–		1.9	3.1	**3.1**	5.7	**4.3**
4	5.7	4	5	4.3					
	4.3	5	–		1.9	3.1	3.1	**4.3**	**5.7**

We can prove that

$$\text{"}A[1] <= A[2] <= \ldots <= A[k] <= A[k+1], A[k+2], \ldots, A[n]\text{"}$$

is a loop invariant of the for-k loop.   // using MI on k
That will confirm that **MinSort** is correct.   // But is it efficient?

The number of comparisons in *MinSort* is

$(n - 1)$ for finding the minimum of $A[1], A[2], A[3], \ldots, A[n]$

$+ (n - 2)$ for finding the minimum of $\quad A[2], A[3], \ldots, A[n]$

$+ (n - 3)$ for finding the minimum of $\quad\quad A[3], \ldots, A[n]$

$\ldots$

$\ldots$

$+ (1)$ for finding the minimum of $\quad\quad\quad A[n - 1], A[n]$

$= n(n - 1)/2$ // Thm 3.6.2

$= (1/2) \times n^2 - (1/2) \times n$ // which is "of Order n^2"

// as we shall see in Chap. 7

Even if the array A is already sorted, *MinSort* will make this number of comparisons – it will compare each entry to every other entry. Every case is a worst case. To take advantage of some order already present in the array A (especially if A is already completely sorted), we need a different strategy.

4.3.2 Exchange Sorts

Even though exchanges occurred in *MinSort*, the main objective of the algorithm was selecting A-values to exchange. In *BubbleSort*, the primary element of the algorithm is exchanging A-values.

But let's first consider the problem of checking to see whether or not the A-values are already in sorted order, with

$$A[1] <= A[2] <= A[3] <= \ldots <= A[n].$$

The natural algorithm is as follows: Check each pair of consecutive entries $A[j]$ and $A[j + 1]$ to see that $A[j] <= A[j + 1]$. *BubbleSort* does exactly that, but when it finds a consecutive pair "out of order," it interchanges their values:

compare $A[1]$ and $A[2]$,

if $A[1] > A[2]$, then interchange them, so now $A[1] <= A[2]$;

compare $A[2]$ and $A[3]$,

if $A[2] > A[3]$, then interchange them, so now $A[2] <= A[3]$;

$\ldots$

compare $A[n - 1]$ and $A[n]$,

if $A[n - 1] > A[n]$, then interchange them, so $A[n - 1] <= A[n]$.

// Will this pass leave the list sorted?

Walkthrough a *BubbleSort* pass with $n = 11$ and $A = (5, 7, 6, 0, 9, 8, 2, 1, 5, 7, 8)$

// comparing and sometimes interchanging $A[j]$ and $A[j + 1]$ as j goes from 1 to $n - 1$.

j	A[j]	>	A[j+1]	A										
–	–	–	–	5	7	6	0	9	8	2	1	5	7	8
1	5	F	7	5	7	"	"	"	"	"	"	"	"	"
2	7	T	6	"	6	7	"	"	"	"	"	"	"	"
3	7	T	0	"	"	0	7	"	"	"	"	"	"	"
4	7	F	9	"	"	"	7	9	"	"	"	"	"	"
5	9	T	8	"	"	"	"	8	9	"	"	"	"	"
6	9	T	2	"	"	"	"	"	2	9	"	"	"	"
7	9	T	1	"	"	"	"	"	"	1	9	"	"	"
8	9	T	5	"	"	"	"	"	"	"	5	9	"	"
9	9	T	7	"	"	"	"	"	"	"	"	7	9	"
10	9	T	8	"	"	"	"	"	"	"	"	"	8	9

After the pass, $A = (5, 6, 0, 7, 8, 2, 1, 5, 7, 8, 9)$.

// This pass did not leave the list sorted. But is it closer to being sorted?
// What progress has been made? **7** "bubbled up" to A[4] and **9** "bubbled up"
// to A[11].
// What progress can we count on? Big values seem to bubble upward.

We can see that as j changes in the pass and some values are interchanged,

"$A[j+1]$ is always the largest entry in $A[1], A[2], \ldots A[j+1]$";

// that is,   "$A[1], A[2], \ldots, A[j] <= A[j+1]$"   is always True.
//   like a "loop invariant"

So at the end of the pass, the largest value in the list is now in the correct position, $A[n]$, and now, we just have to sort $A[1], A[2], \ldots, A[n-1]$.

A second (identical) **BubbleSort** pass will move the largest of $A[1], A[2], \ldots,$ $A[n-1]$ into $A[n-1]$. And then it will not move again because $A[n-1] <= A[n]$. The second pass can be shortened to terminate after comparing $A[n-2]$ with $A[n-1]$. At the end of the second pass, $A[n-1]$ and $A[n]$ are both correct, and the third pass can be shortened to terminate after comparing $A[n-3]$ with $A[n-2]$. At the end of the third pass, $A[n-2]$, $A[n-1]$ and $A[n]$ are all correct.

// How many passes are needed to guarantee the list is sorted?
// When $A[2], \ldots, A[n]$ are all correct, what about $A[1]$?

We'd like to make just one more observation before presenting **BubbleSort**, and that is, to interchange the values of $A[j]$ and $A[j+1]$, we need a third variable, x. If $A[j]$ has the value α and $A[j+1]$ has the value β (usually different from α), the assignment

$$A[j] \leftarrow A[j+1];$$

will copy the value of $A[j+1]$ into the storage position of the variable $A[j]$ overwriting any value that was there. So after that assignment statement is executed, both $A[j]$ and $A[j+1]$ have the value β, and α has been lost. The value α must be saved somewhere temporarily, like in an auxiliary variable, x.

Interchanging the values of $A[j]$ and $A[j + 1]$ requires three assignments.

$$x \leftarrow A[j] \quad \text{and} \quad A[j] \leftarrow A[j+1] \quad \text{and} \quad A[j+1] \leftarrow x.$$

// α β α

// In **MinSort**, when the values of $A[k]$ and $A[index]$ were interchanged,
// what was the auxiliary variable and where were the *three* assignments?

// Strictly speaking, you don't really need a third **variable** to interchange the
// values of variables y and z; you can do it with three assignments and some
// arithmetic as follows:
//
//
$$y \leftarrow y + z; \quad z \leftarrow y - z; \quad y \leftarrow y - z;$$
//
// If you begin when   the value of y is α and value of z is β, then
// after the first assignment,   the value of y is $\alpha + \beta$ and value of z is β;
// after the second assignment, the value of y is $\alpha + \beta$ and value of z is α;
// and after the third assignment, the value of y is β and value of z is α.
// But this is more work than just three assignments, and we can afford the extra
// variable x.

Algorithm 4.3.3: BubbleSort   // the standard version

```
Begin
  For k ← 1 To (n − 1) Do                // the kᵗʰ pass through A
    For j ← 1 To (n − k) Do
    If (A[j] > A[j + 1]) Then
      x ← A[j];
      A[j] ← A[j + 1];
      A[j + 1] ← x;
    End; // the if
    End;   // the for-j loop
  End;     // the for-k loop
End.
```

Walkthrough of **BubbleSort** with $n = 11$ and $A = (5, 7, 6, 0, 9, 8, 2, 1, 5, 7, 8)$

// showing the contents of A after each pass.

After pass k					A						
–	5	7	6	0	9	8	2	1	5	7	8
1	5	6	0	7	8	2	1	5	7	8	9
2	5	0	6	7	2	1	5	7	8	8	"
3	0	5	6	2	1	5	7	7	8	"	"
4	0	5	2	1	5	6	7	7	"	"	"
5	0	2	1	5	5	6	7	"	"	"	"
6	0	1	2	5	5	6	"	"	"	"	"
7	0	1	2	5	5	"	"	"	"	"	"
8	0	1	2	5	"	"	"	"	"	"	"
9	0	1	2	"	"	"	"	"	"	"	"
10	0	1	"	"	"	"	"	"	"	"	"

← // we're done! (at pass 6)

After 6 passes, $A = (0,1,2,5,5,6,7,7,8,8,9)$ and is sorted.

// The other passes make no (inter)changes.
// How can we stop sooner? How can we tell when the list has been sorted?

BubbleSort is correct. // But is it efficient?
The number of comparisons we want to count is the number of j-values where we ask "Is $A[j] > A[j + 1]$" in the inner for-loop.

When $k = 1$, j runs from 1 to $n - 1$,   so there are $n - 1$ comparisons.
When $k = 2$, j runs from 1 to $n - 2$,   so there are $n - 2$ comparisons.
When $k = 3$, j runs from 1 to $n - 3$,   so there are $n - 3$ comparisons.
. . .
When $k = n - 1$, j runs from 1 to $n - (n - 1)$, so there is only 1 comparison.

Thus, (just like **MinSort**), the number of comparisons done by **BubbleSort** in every case is

$$1 + 2 + \ldots + (n - 1) = n(n - 1)/2 = (\tfrac{1}{2}) \times n^2 - (\tfrac{1}{2}) \times n$$

// **BubbleSort** is of Order n^2.

To improve upon this and reduce the number of comparisons the algorithm does, we'd like to stop making passes as soon as we can determine that the list is sorted. The list must be sorted if we ever make an entire pass which produces no interchanges of consecutive entries. We'd also like to shorten the passes whenever we can.

Suppose that during the first pass, some exchanges did occur and the last one interchanged the values of $A[p]$ and $A[p + 1]$ in the list

$$A[1], A[2], A[3], \ldots, A[p], A[p+1], \ldots, A[n].$$

We know that now $A[p] < A[p + 1]$ and, since no interchanges occurred among $A[p + 1] \ldots A[n]$, these must be in sorted order. Thus,

$$A[p] < A[p+1] \ \mathrel{<=}\ A[p+2] \ \mathrel{<=}\ \ldots \ \mathrel{<=}\ A[n].$$

Furthermore, if we think of that pass just operating on $A[1]$ up to $A[p + 1]$, the largest value among these is now in the last position, so

$$A[1], A[2], A[3], \ldots, A[p] \ \mathrel{<=}\ A[p+1].$$

Therefore, the whole list will be sorted if we just sort $A[1], \ldots, A[p]$ where the *last entry* we need worry about now is $A[p]$.

To implement this observation, let's use a variable p, for the *position* (j-value) of the last interchange. Before each pass p is set to zero, and at each interchange of

$A[j]$ and $A[j + 1]$, p is updated to that value of j. After the pass is complete, p will give us the position of the *last* interchange.

// or will be 0 if there are no interchanges.

On the next pass j will run from 1 to $p - 1$, as we sort $A[1], \ldots, A[p]$. And we will avoid making another pass if the value of p is $<= 1$. We can implement this by using another new variable q, for the index of the last position we need to consider in the current pass. For the first pass, q is set to n, and after each pass, q is set to p.

Algorithm 4.3.4: BetterBubbleSort

```
Begin
  q ← n;
  Repeat          // passes through the array A
    p ← 0;
    For j ← 1 To (q − 1) Do
      If (A[j] > A[j + 1]) Then
        x ← A[j];
        A[j] ← A[j + 1];
        A[j + 1] ← x;
        p ← j;   // the "position" of the last exchange
      End;       // the if
    End;         // the for-j loop
    q ← p;
  Until (q <= 1);
End.
```

// Can the repeat-loop be rewritten as a while loop?

Walkthrough of ***BetterBubbleSort*** with $n = 11$ and $A = (5, 7, 6, 0, 9, 8, 2, 1, 5, 7, 8)$

// showing the contents of A after each pass and the final value of p
// using the same data as the previous walkthrough.

pass #	q				A							p
–	–	5	7	6	0	9	8	2	1	5	7	8	–
1	11	5	6	0	7	8	2	1	5	7	8	9	10
2	10	5	0	6	7	2	1	5	7	8	8	"	8
3	8	0	5	6	2	1	5	7	7	"	"	"	6
4	6	0	5	2	1	5	6	"	"	"	"	"	5
5	5	0	2	1	5	5	"	"	"	"	"	"	3
6	3	0	1	2	"	"	"	"	"	"	"	"	2
7	2	0	1	"	"	"	"	"	"	"	"	"	0

// After 6 passes, $A = (0,1,2,5,5,6,7,7,8,8,9)$ and is (completely) sorted.
// But we didn't discover that until we completed one more pass.

For ***BetterBubbleSort***, the number of passes, **#P**, and the number of comparisons, **#C**, depend on the input. In a best case, the first pass is enough.

A best case is when A is already sorted (or when the only interchange swaps $A[1]$ and $A[2]$). In a best case,

$$\#P = 1 \quad \text{and} \quad \#C = n - 1.$$

// What is a worst case for **BetterBubbleSort**?

   While large values move rapidly up the array to where they belong, small values move down the array one step at a time and only one step per pass. A worst case for **BetterBubbleSort** would occur if p is assigned the largest possible value on every pass. The largest possible value of p is the largest value j takes on the pass, $q - 1$. This happens when the smallest value in A is in the last position $A[n]$.

                                                                    // Are there other worst cases?

Then, **BetterBubbleSort** acts like standard **BubbleSort** and in a worst case

$$\#P = n - 1 \quad \text{and} \quad \#C = n(n - 1)/2.$$

The worst case complexity of **BetterBubbleSort** is of Order n^2.

// What about an average case? We look at average case complexity in Chap. 10.

// Is there a way to determine a minimum possible complexity,
// for **all possible** sorting algorithms that are based on comparing key values?

   Let's construct the **Branching Diagram of MinSort, when $n = 3$, with an internal vertex for each comparison, and a leaf for each possible rearrangement of the input values.** Suppose the input array $A = (x, y, z)$ where the entries are in some arbitrary order. At each comparison, take the left branch if the result is **False**, and the right branch if the result is **True**.

// On the first pass, $min = A[1] = x$; on the second pass, $min = A[2]$.
// The first pass ends at the dotted horizontal line.

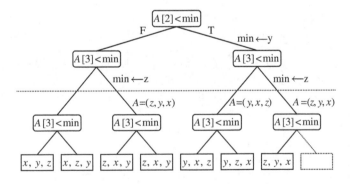

   The corresponding **Branching Diagram of BetterBubbleSort** is shown below for input $A = (x, y, z)$. At each comparison, take the left branch if the result is **False** and the right branch if the result is **True**.

// The first pass ends at the dotted horizontal line.

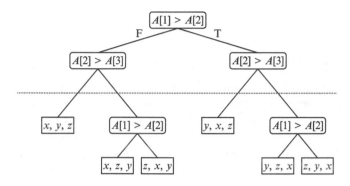

In both of these Branching Diagrams, we constructed a rooted Binary Tree with at least $6 = 3!$ leaves.

The Most Important Ideas in This Section.
The *sorting problem* is as follows: Given an array $A[1], A[2], A[3], \ldots , A[n]$, rearrange the values in A so that

$$A[1] \mathrel{<=} A[2] \mathrel{<=} A[3] \mathrel{<=} \ldots \mathrel{<=} A[n].$$

There is a surprising variety of solution strategies and sorting algorithms. Sorting methods are part of the common knowledge of all computer scientists. The *cost* of a sorting algorithm is usually taken to be the number of comparisons involving two array entries (key comparisons).

We described a "selection sort" called *MinSort* and an "exchange sort" called *BubbleSort*. And we showed that for both of these algorithms every case costs, $n(n - 1)/2 = (\frac{1}{2}) \times n^2 - (\frac{1}{2}) \times n$ comparisons. We then proposed consideration of a *BetterBubbleSort* where not all cases are worst cases, a best case costs only $(n - 1)$ comparisons, and in most cases, many comparisons are saved.

We constructed the *Branching Diagram of comparisons* for *MinSort* and for *BetterBubbleSort* when $n = 3$, displaying all possible sequences of comparisons they might make as the internal vertices, and the rearrangements of entries as the leaves (there must be at least $n!$ leaves.)

Next, we will look at such a Branching Diagram of comparisons for *any* possible sorting algorithm $\mathcal{A}$, and we'll use the diagram to prove that the average cost for $\mathcal{A}$ is at least $\lg(n!)$ comparisons (then we'll know what a best possible sorting algorithm must do – on average). Average-case complexity is the subject of Chap. 10.

4.4   Binary Trees with (at Least) *n*! Leaves

A **Binary Tree** is a tree $\mathcal{T}$ with a certain distinguished vertex, r, called the **root** of $\mathcal{T}$ and placed at the top of the diagram, and where each vertex, v, of $\mathcal{T}$ has at most two vertices directly below it in the diagram.

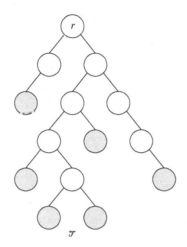

A **leaf** is a vertex with no vertices below it; the other (non-leaf) vertices are known as **internal** vertices. For any vertex, v, there is a unique path from it back to the root r.   // up the diagram and down the tree
The **length** of that path is the number of edges traversed and is called the **height** of vertex v and denoted by $h(v)$.   // so $h(r) = 0$
The **height of the tree** $\mathcal{T}$ equals the height of the highest vertex in $\mathcal{T}$.

// the highest leaf

Theorem 4.4.1: A Binary Tree with height h has at most 2^h leaves.

Proof. // by Strong Induction on h where $h \in \{0..\,\}$
Step 1.   If $\mathcal{T}$ has height $h = 0$, there can be no vertex in $\mathcal{T}$ other than the root, r. Then, r must be a leaf and so $\mathcal{T}$ has exactly $1 = 2^h$ leaves.
Step 2.   Assume $\exists q >= 0$ such that for any k where $0 <= k <= q$, any Binary Tree with height k has at most 2^k leaves.
Step 3.   Suppose $\mathcal{T}*$ is some particular Binary Tree with height $q + 1$ rooted at vertex r.   // We must prove that $\mathcal{T}*$ has at most 2^{q+1} leaves.

Case 1.   The root r has two vertices below it, u and w.

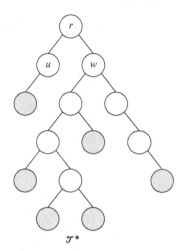

The portion of $\mathcal{T}^*$ below u is a Binary Tree $\mathcal{T}_u$ with height $i <= q$ rooted at u, and the portion of $\mathcal{T}^*$ below w is a Binary Tree $\mathcal{T}_w$ with height $j <= q$ rooted at w.
                              // In fact, either i or j (or both) must equal q.

From our assumption in Step 2,

$\quad\quad\quad\quad$ $\mathcal{T}_u$ has at most 2^i leaves and $\mathcal{T}_w$ has at most 2^j leaves.

Since every leaf in $\mathcal{T}$ occurs in $\mathcal{T}_u$ or in $\mathcal{T}_w$,   // but not in both

\# of leaves in $\mathcal{T}^* =$ \# of leaves in $\mathcal{T}_u +$ $\quad$ \# of leaves in $\mathcal{T}_w$

$\quad\quad\quad <=$ $\quad\quad$ 2^i $\quad + \quad$ 2^j

$\quad\quad\quad <=$ $\quad\quad$ 2^q $\quad + \quad$ 2^q $\quad\quad\quad$ // i and $j <= q$

$\quad\quad\quad = \quad\quad\quad$ 2^{q+1}.

Case 2.   The root r has only one vertex below it, u.
The portion of $\mathcal{T}^*$ below u is a Binary Tree $\mathcal{T}_u$ rooted at u, with height equal q exactly. Since every leaf in $\mathcal{T}$ occurs in $\mathcal{T}_u$,

$\quad\quad\quad$ \# of leaves in $\mathcal{T}^*$ $\quad = \quad$ \# of leaves in $\mathcal{T}_u$

$\quad\quad\quad\quad\quad\quad\quad <= \quad\quad\quad\quad$ 2^q

$\quad\quad\quad\quad\quad\quad\quad < \quad\quad\quad\quad$ 2^{q+1}. $\quad\quad\quad\quad\quad\quad\quad\quad\quad\quad$ □

From Theorem 4.4.1, we know that:
$\quad\quad$ If a Binary Tree has height $h < k$, then it has $<= 2^h < 2^k$ leaves.
$\quad\quad$ If a Binary Tree has height $h < \lg(n)$, then it has $<= 2^h < 2^{\lg(n)} = n$ leaves.

The contrapositive of this second statement is as follows:

If a Binary Tree has $>= n$ leaves, then it has a height $>= \lg(n)$.

And, in particular:

If a Binary Tree has $>= n!$ leaves, then it has a height $>= \lg(n!)$.

Since the Branching Diagram of comparisons for *any* Sorting Algorithm applied to an array of length n must have $>= n!$ leaves, the **Worst Case** for this algorithm (which corresponds to reaching a highest leaf) must do $>= \mathbf{lg(n!)}$ comparisons (of array entries).

// But what about the average case? Can the Branching Diagram give us a bound
// for it?

Theorem 4.4.2: The *average* height of a leaf in a Binary Tree with n leaves is $>= \lg(n)$.

Proof. Let $\mathscr{T}$ be some (fixed but arbitrary) Binary Tree with n leaves.

// We'll prove that the leaves in $\mathscr{T}$ have an average height $>= \lg(n)$ in four stages.

Let SHL($\mathscr{T}$) denote the **S**um of the **H**eights of the **L**eaves in $\mathscr{T}$, and let AHL($\mathscr{T}$) denote the **A**verage of the **H**eights of the **L**eaves in $\mathscr{T}$. Then,

$$\mathrm{AHL}(\mathscr{T}) = \mathrm{SHL}(\mathscr{T})/n.$$

// Stage 1.

If some vertex x, which is not the root, has only one vertex w below it, and we remove x but join the vertex u above x directly to w, we obtain a new tree $\mathscr{T}1$, with the same number of leaves but at least one leaf is now one step closer to the root.

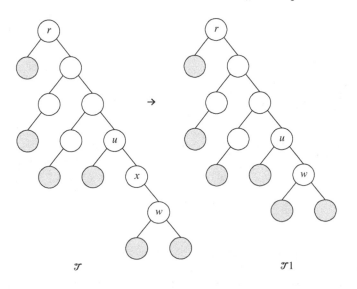

If the root r has only one vertex w below it, and we remove r and make w the root, we obtain a new Binary Tree $\mathcal{T}1$, with the same number of leaves but where every leaf is now one step closer to the root.

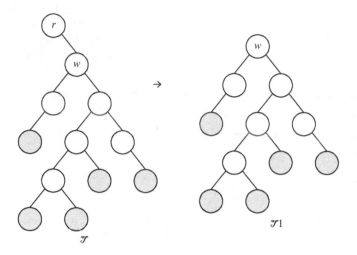

In either case, $\mathcal{T}1$ is a Binary Tree with n leaves. But since at least one leaf is now one step closer to the root,

$$\text{SHL}(\mathcal{T}) > \text{SHL}(\mathcal{T}1) \text{ and so AHL}(\mathcal{T}) > \text{AHL}(\mathcal{T}1).$$

If we remove all internal vertices with only one vertex below them in the same fashion, we obtain a **full** Binary Tree $\mathcal{T}_a$ with n leaves, where

$$\text{AHL}(\mathcal{T}) >= \text{AHL}(\mathcal{T}_a). \qquad\qquad // \ \mathcal{T} \text{ itself might be full.}$$

// Now we need to prove that $\text{AHL}(\mathcal{T}_a) >= \lg(n)$ $\qquad\qquad$ (in three more stages).

// Stage 2.

Let x be a "lowest" leaf with $p = h(x)$ as small as possible,
$\qquad\qquad\qquad\qquad\qquad\qquad$ // a leaf closest to the root
and let y be a "highest" leaf with $q = h(y)$ as large as possible.
$\qquad\qquad\qquad\qquad\qquad\qquad$ // farthest from the root
If $q > p + 1$, then we can construct a new tree $\mathcal{T}2$ that is still a full Binary Tree with n leaves, where

$$\text{SHL}(\mathcal{T}_a) > \text{SHL}(\mathcal{T}2).$$

Let s be the internal vertex above y; since s must have exactly two vertices directly below it, there is another vertex z directly below s along with y. Since $h(z) = h(y)$, z must also be a highest leaf. Construct $\mathcal{T}2$ by detaching them from s and reattaching them below x.

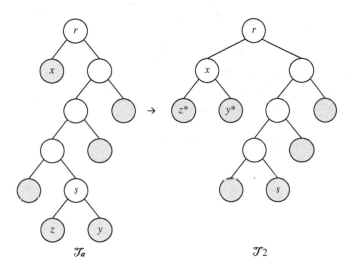

Then, $\mathcal{T}2$ is a full Binary Tree with n leaves, but

$$
\begin{aligned}
\mathrm{SHL}(\mathcal{T}2) &= \mathrm{SHL}(\mathcal{T}_a) - h(x) - h(y) - h(z) + h(z^*) \ \ + h(y^*) \ \ + h(s) \\
&= \mathrm{SHL}(\mathcal{T}_a) - p \quad\ \ - q \quad\ \ - q \ \ + (p+1) + (p+1) + (q-1) \\
&= \mathrm{SHL}(\mathcal{T}_a) + p \quad\ \ - q \ \ + 1 \\
&< \mathrm{SHL}(\mathcal{T}_a) \hspace{5.5cm} /\!/ \ q > p+1
\end{aligned}
$$

If we repeat this construction of detaching pairs of leaves at the highest level q and reattaching them below a vertex at the lowest level p whenever $q > p + 1$, we will obtain a full Binary Tree $\mathcal{T}_b$ with n leaves, where *all the leaves are at level p or (perhaps) $p + 1$.*

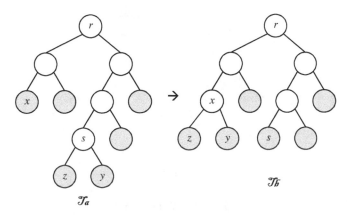

// Now we need to prove that $\mathrm{AHL}(\mathcal{T}_b) >= \lg(n)$ \hspace{2cm} (in two more stages).
// Stage 3.

We can show by Mathematical Induction that for $k = 0, 1, \ldots, p$ the number of vertices at level k in $\mathcal{T}_b$ is exactly 2^k.

\hspace{3cm} // because each internal vertex has 2 vertices below it

If all the vertices at level p are leaves, then all the leaves are at height p, $n = 2^p$, and

$$\text{AHL}(\mathcal{T}_b) = p = \lg(n).$$

If only t vertices at level p are leaves, then there are exactly $2^p - t$ internal vertices at level p. Since all other leaves are at level $p + 1$, there are exactly $2(2^p - t)$ leaves at level $p + 1$. Thus,

$$n = t + 2(2^p - t) = 2^{p+1} - t, \qquad\qquad // \text{ where } 0 < t < 2^p$$

$$\text{SHL}(\mathcal{T}_b) = t \times p + (n - t) \times (p + 1)$$
$$= tp + np - tp + n - t$$

and $\qquad\qquad \text{AHL}(\mathcal{T}_b) = \{(n)(p+1) - t\}/n$
$$= (p + 1) - t/n$$

// Stage 4. (finally)

The function $y = \lg(x)$ for $x > 0$ is concave down.

// Students of Calculus can prove this by showing that y'' is negative when $x > 0$.

This means that if any two points on the curve are joined by a straight line segment, that segment lies entirely below the curve.

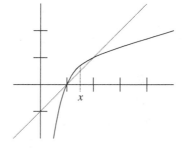

In particular, the line segment joining $(1,0)$ and $(2,1)$ given by $y = x - 1$ lies below the curve. $\qquad\qquad$ // That is, if $1 < x < 2$, then $x - 1 < \lg(x)$.

If n and t are integers where $0 < t < n$, then $1 < 1 + t/n < 2$, $\quad$ // $0 < t/n < 1$
so taking $x = 1 + t/n$, we get

$$x - 1 < \lg(x)$$

where $\qquad\qquad\qquad x - 1 = (1 + t/n) - 1 = t/n$

and $\quad \lg(x) = \lg(1 + t/n) \;\; = \lg(n/n + t/n)$
$$= \lg([n + t]/n) = \lg(n + t) - \lg(n). \qquad // \lg(a/b) = \lg(a) - \lg(b)$$

Hence, $\qquad t/n < \lg(n+t) - \lg(n)$
so $\qquad\qquad \lg(n) < \lg(n+t) - t/n.$

Because $n = 2^{p+1} - t, \quad n+t = 2^{p+1}$, so

$$\lg(n) < \lg(2^{p+1}) - t/n$$
$$= p + 1 - t/n$$
$$= \text{AHL}(\mathscr{T}_b).$$

Therefore, $\quad \text{AHL}(\mathscr{T}) >= \text{AHL}(\mathscr{T}_b) >= \lg(n).$ $\qquad\qquad\Box$

Since the Branching Diagram for **any** Sorting Algorithm applied to any array of length n must have $>= n!$ leaves, the **Average Case** for this algorithm (which corresponds to reaching a leaf at average height) must do $>= \mathbf{lg(n!)}$ comparisons (involving array entries).

Next, we find bounds on $\lg(n!)$ in terms of n and $\lg(n)$. Consider the table

n	$n!$		n^n		$(n!)^2$
1	1	=	1	=	1
2	2	<	4	=	4
3	6		27	<	36
4	24		256		576
5	120		3125		14400

Theorem 4.4.3: For all positive integers

(1) $n! <= n^n$ $\qquad$ but $\qquad\qquad n! < n^n$ $\qquad$ when $n > 1$;
(2) $n^n <= (n!)^2$ $\qquad$ but $\qquad\qquad n^n < (n!)^2$ $\qquad$ when $n > 2$;
(3) $\lg(n!) \qquad\qquad < n \times \lg(n) < 2 \times \lg(n!)$ $\qquad$ when $n > 2$;
and (4) $(1/2)n \times \lg(n) < \lg(n!) \qquad < n \times \lg(n)$ $\qquad$ when $n > 2$.

Proof. // We'll prove these results in order. $\qquad\qquad$ (using algebra, not MI)

$$n! = n \times (n-1) \times (n-2) \times \ldots \times (2) \times (1)$$
$$<= n \times (n) \times (n) \times \ldots \times (n) \times (n) = n^n.$$

Equality holds only when $n = 1$. $\qquad\qquad$ // We've proven (1).

// What about (2)?
// If $n = 5$, then $\qquad 5! = (5)(4)(3)(2)(1)$
// and $\qquad\qquad\qquad 5! = (1)(2)(3)(4)(5),$
// so
//
$$(5!)^2 = (5!)(5!) = (5 \times 1)(4 \times 2)(3 \times 3)(2 \times 4)(1 \times 5)$$
$$= \quad (5) \quad\;\; (8) \quad\;\; (9) \quad\;\; (8) \quad\;\; (5)$$
$$> \quad (5) \quad\;\; (5) \quad\;\; (5) \quad\;\; (5) \quad\;\; (5) = 5^5.$$
// Does this generalize?

In general,

$$n! = n \times (n-1) \times (n-2) \times \ldots \times (2) \times (1) = \prod_{r=1}^{n} (n+1-r)$$

and $n! = (1) \times (2) \times (3) \times \ldots \times (n-1) \times (n) = \prod_{r=1}^{n} r,$

where $\prod$ (the Greek capital letter π) is used in the way the Greek capital letter sigma is used in sigma-notation for sums, but it applies to **products**. Then,

$$(n!)^2 = \prod_{r=1}^{n} (n+1-r) \times \prod_{r=1}^{n} r$$
$$= \prod_{r=1}^{n} \{(n+1-r) \times r\}.$$

But $\quad (n+1-r)r = (n-r+1)r = (n-r)r \quad +r$
$\qquad\qquad\qquad = (n-r)([r-1]+1) \qquad +r$
$\qquad\qquad\qquad = (n-r)(r-1)+(n-r) \quad +r$
$\qquad\qquad\qquad = (n-r)(r-1)+n$
$\qquad\qquad\qquad >= n. \qquad\qquad$ // Both $(n-r) >= 0$ and $(r-1) >= 0$
$\qquad\qquad\qquad\qquad\qquad\qquad$ // when $1 <= r <= n$

Thus, $\quad (n!)^2 >= \prod_{r=1}^{n} n = n^n.$

Equality holds only when $n=1$ or 2. If $n >= 3$, there is at least one r-value in the product where $1 < r < n$ and then $n < (n+1-r)r$.

// Now we've proven (2). What about the logarithms in (3) and (4)?

Suppose that $n >= 3$. Since $n! < n^n < (n!)^2$, taking logarithms, we get (3),

$\lg(n!) < n \times \lg(n) < 2 \times \lg(n!).$ $\qquad\qquad$ // Recall $\log_b(x^y) = y \times \log_b(x)$

Dividing the last two terms by 2, we get

$$(1/2)n \times \lg(n) < \lg(n!)$$

and so we've proven result (4) as well. $\qquad\qquad\qquad\qquad\qquad\qquad\qquad\qquad$ ☐

The Most Important Ideas in This Section.
This section concerned three theorems about Binary trees which also apply to all possible sorting algorithms (based on comparisons of array entries).

A Binary Tree with height h has at most 2^h leaves. So if a rooted Binary tree has $>= n!$ leaves, then it has a height $>= \lg(n!)$. Therefore, in a **Worst Case**, **any** Sorting Algorithm must do $>= \lg(n!)$ comparisons.

(continued)

(continued)

> The **average** height of a leaf in a Binary Tree with n leaves is $>= \lg(n)$. Therefore, in an **Average Case**, **any** Sorting Algorithm must do $>= \lg(n!)$ comparisons.
>
> For all positive integers > 2, $n! < n^n < (n!)^2$, and so $(\frac{1}{2})n \times \lg(n) < \lg(n!)$ $< n \times \lg(n)$. Therefore, in an **Average Case**, **any** Sorting Algorithm must do $> (\frac{1}{2})n \times \lg(n)$ comparisons.
>
> We now know that a best possible sorting algorithm applied to an array of length n would cost about $n \times \lg(n)$ comparisons on average. How can we create such a sorting algorithm?
>
> If $\qquad\qquad A[i] < A[j] \quad$ and $\quad A[j] < A[k],$
>
> then (we know even without comparing them that)
>
> $$A[i] < A[k].$$
>
> If we can somehow remember the results of early comparisons, or somehow organize the sequence of comparisons so we remember the results, we can avoid many later (unnecessary) comparisons. We'll do that next.

4.5 Partition Sorts

We will examine one particular partition sort in detail, **QuickSort**. This is a very good sorting algorithm; it is very efficient on average, relative to the hoped for complexity of $n \times \lg(n)$ key comparisons and relative to the previous sorts. Also, it allows us to introduce and discuss an intrinsically **recursive algorithm**.

The strategy for **QuickSort** is this: Make a pass through the array $A[1]$ to $A[n]$ comparing each entry in turn to the value of one particular entry and interchanging certain entries so that after this pass, all entries smaller than $A[j]$ occur before position j in the array and all values larger than $A[j]$ occur after position j in the array.

$$\underbrace{A[1] \ldots A[j-1]}_{<= A[j]} \quad A[j] \quad \underbrace{A[j+1] \ldots A[n]}_{>= A[j]}$$

Now $A[j]$ is in the "correct" position, and we just have to sort the two sublists, $A[1]$ to $A[j-1]$ and $A[j+1]$ to $A[n]$, to obtain the whole list, $A[1]$ to $A[n]$, in sorted order.

The *optimum* partition divides the list as nearly in half as possible.   //X why?

In that case, the value of $A[j]$ would be the *Median* of all the values in A. However, we'll proceed by making a rather crude estimate M of the median, namely, we'll use $A[n]$. Actually, we will write **QuickSort** to work on a sublist of the array, $A[p]$ to $A[q]$ where $p < q$, and we will initialize M to be the value of $A[q]$.

// The median occurs somewhere; maybe it's at the end.

After we partition the whole list, as

$$\underbrace{A[1]\dots A[j-1]}_{<=M} \quad \underbrace{A[j]}_{=M} \quad \underbrace{A[j+1]\dots A[n]}_{>=M},$$

we'll **recursively** sort the sublists we've created (using **QuickSort** itself) until the input list is completely sorted. We'll assume that there is an "external" call of **QuickSort** that initiates a sort of the whole array from $A[1]$ to $A[n]$.

// And we'll assume that $1 <= n$.

There have been many versions of **QuickSort** devised since it was first created and published by C. A. R. Hoare as *Algorithms* 63 and 64 – *Partition* and *Quicksort* in the Communications of the ACM in 1961. The one included here is based on a version due to N. Lomuto which also appeared many years ago.

We will make one minor revision to the basic strategy: the values that are put before position j will be <u>strictly</u> smaller than M. The strategy will be implemented using a for-loop controlled by a variable k going from p up to $(q - 1)$. We'll alter the array A so that for each successive value of k, we have

$$\underbrace{A[p]\dots A[j-1]}_{<M} \quad \underbrace{A[j]\dots A[k]}_{>=M} \quad \underbrace{A[k+1]\dots\dots A[q-1]}_{?} \quad \underbrace{A[q]}_{=M}$$

// The part of A from $A[p]$ to $A[k]$ has been partitioned into those $< M$ and those
// not $< M$ the way we want, but
// the part from $A[k + 1]$ to $A[q - 1]$ is yet to be partitioned.

Now, if $A[k + 1] < M$, then interchange $A[k + 1]$ and $A[j]$ <u>and then increase j</u> by 1 (but, if $A[k + 1] >= M$, then do nothing), so we get

$$\underbrace{A[p]\dots\dots A[j-1]}_{<M} \quad \underbrace{A[j]\dots\dots A[k+1]}_{>=M} \quad \underbrace{A[k+2]\dots A[q-1]}_{?} \quad \underbrace{A[q]}_{=M}.$$

When k reaches $(q - 1)$, we'll have

$$\underbrace{A[p]\dots\dots A[j-1]}_{<M} \quad \underbrace{A[j]\dots\dots A[q-1]}_{>=M} \quad \underbrace{A[q]}_{=M}.$$

So if we now interchange $A[q]$ and $A[j]$, we'll have A **partitioned** the way we want at a cost of exactly $(q - p)$ key comparisons.   // And $(j - p) + 1$ exchanges.

Algorithm 4.5.1: QuickSort(p,q)

// Assumes an "external" invocation of the form QuickSort(1, n) and
// (recursively) sorts A[p], A[p + 1],..., A[q] into order:

```
Begin
  If (p < q) Then
    M ← A[q];
    j ← p;
    For k ← p to (q − 1) Do
      If (A[k] < M) Then
        x ← A[j];
        A[j] ← A[k];
        A[k] ← x;
        j ← j + 1;
      End;   // the if
    End;       // the for-k loop
    A[q] ← A[j];
    A[j] ← M; // this is the end of the "partitioning"

    QuickSort( p , j − 1);  // the first "recursive" sub-call
    QuickSort(j + 1, q);    // the second "recursive" sub-call

  End;       // the if-statement at the beginning
End.          // the recursive algorithm
```

Walkthrough assuming an **"external"** invocation of the form Q(1, 11); that is,

// QuickSort the array A from position, $p = 1$ to the last position, $q = 11$.

k	A[k]	A[k]<M													M / j
			5	7	6	0	9	8	2	1	5	7	3		M=3 1
−	−	−													
1	5	F	5	7	6	0	9	8	2	1	5	7	3		"
2	7	F	5	7	6	0	9	8	2	1	5	7	3		"
3	6	F	5	7	6	0	9	8	2	1	5	7	3		"
4	0	T	0	7	6	5	9	8	2	1	5	7	3		2
5	9	F	0	7	6	5	9	8	2	1	5	7	3		"
6	8	F	0	7	6	5	9	8	2	1	5	7	3		"
7	2	T	0	2	6	5	9	8	7	1	5	7	3		3
8	1	T	0	2	1	5	9	8	7	6	5	7	3		4
9	5	F	0	2	1	5	9	8	7	6	5	7	3		"
10	7	F	0	2	1	5	9	8	7	6	5	7	3		"
−	−	−	0	2	1	3	9	8	7	6	5	7	5		−

// the first Partition of A is: ⌈ <3 |3| >=3 ⌉ // and first **j** = 4

Now, $p = 1$ and $j − 1 = 3$, so the next "action" taken by QuickSort is the invocation of QuickSort itself with new parameter values, QuickSort(1,3).

Before continuing the walk through, we digress for a moment to describe a common mechanism for implementing recursion (in modern high-level computer languages). Each invocation of the algorithm, including the sub-calls inside the algorithm, causes the creation of a "call frame", a data structure specifically for that call. We can employ the analogy of using sheets of scrap paper for getting partial results throughout a long and complex calculation.

When the algorithm reaches a point where a sub-call is made, imagine starting the algorithm again but first marking where we are in the current execution/ calculation (this will be the "return address"). Then, getting a new page to work on; placing it on the top of the stack of scratch pages; making room on that new page for storing the values of the variables local to this instance of the algorithm (M, k, j, and x – but not the "global" variable A – all references to A must "point to" one master copy of the array); recording on this new page the values of the parameters in this invocation; and recording the return point on the page just below. Then, a run of the algorithm again from its beginning is started.

When the algorithm reaches another point where a new sub-call occurs, repeat that process. Start running the algorithm again from its beginning, with the new page to work on placed on the top of the stack of scratch pages.

If a run of a sub-call of the algorithm reaches completion, throw away the top page and return to the point on the previous page where that execution/calculation should continue.

Call #2 of QuickSort will be Q(1, 3).
                                    // Later, there will be a call Q(5, 11), but not
                                    // until we have finished sorting $A[1] \ldots A[3]$.

// QuickSort the array A from position $p = 1$ to position $q = 3$.

```
                               |              A              |
                             1 = p    q = 3
   k  A[k]  A[k] < M            ↓       ↓                                    j
   -    -       -            0  2  1  3  9  8  7  6  5  7  5  M=1          1
   1    0       T            0  2  1  "  "  "  "  "  "  "  "                2
   2    2       F            0  2  1  "  "  "  "  "  "  "  "                "
   -    -       -            0  1  2  "  "  "  "  "  "  "  "                -
```

// the second Partition of A is: | 0 | 1 | 2 | // and second $j = 2$

Now $p = j - 1 = 1$, and therefore, the next invocation,

Call #3 of QuickSort, will be Q(1, 1).   // And later, there will be Q(3, 3).
On this call, $p = q$, so nothing is done, and control returns to Call #2 of QuickSort where $j + 1 = q = 3$. The second sub-call is then done. Therefore, the next invocation,

Call #4 of QuickSort, will be Q(3, 3). On this call, $p = q$, so nothing is done, and control returns to Call #2 of QuickSort which is now complete.
                                    // Both sub-calls have been done.
At this point, we return to the first invocation of QuickSort and do the second sub-call;

Call #5 of QuickSort will be Q(5, 11).

// QuickSort the array A from position $p = 5$ to position $q = 11$.

```
                          |            A            |
                          5=p                  q=11
   k  A[k]  A[k]<M         ↓                    ↓           j
   -   -      -    0 1 2 3 9 8 7 6 5 7 5   M=5   5
   5   9      F    " " " " 9 8 7 6 5 7 5          "
   6   8      F    " " " " 9 8 7 6 5 7 5          "
   7   7      F    " " " " 9 8 7 6 5 7 5          "
   8   6      F    " " " " 9 8 7 6 5 7 5          "
   9   5      F    " " " " 9 8 7 6 5 7 5          "
  10   7      F    " " " " 9 8 7 6 5 7 5          "
   -   -      -    " " " " 5 8 7 6 5 7 9
```
// the new Partition of A is: | 5 | >=5 | // and current $j = 5$

Now, $p = 5$, but $j - 1 = 4$, and therefore, the next invocation,

Call #6 of QuickSort will be Q(5, 4).   // And later, there will be Q(6, 11).
But on this call, $p > q$, so nothing is done, and control returns to Call #5 of
QuickSort where $j = 5$ to do the second sub-call. Therefore, the next invocation,

Call #7 of QuickSort, will be Q(6, 11).

// QuickSort the array A from position $p = 6$ to position $q = 11$.

```
                          |            A            |
                          6=p                  q=11
   k  A[k]  A[k]<M         ↓                    ↓           j
   -   -      -    0 1 2 3 5 8 7 6 5 7 9   M=9   6
   6   8      T    " " " " " 8 7 6 5 7 9          7
   7   7      T    " " " " " 8 7 6 5 7 9          8
   8   6      T    " " " " " 8 7 6 5 7 9          9
   9   5      T    " " " " " 8 7 6 5 7 9          10
  10   7      T    " " " " " 8 7 6 5 7 9          11
   -   -      -    " " " " " 8 7 6 5 7 9          -
```
// the new Partition of A is: | <9 | 9 | // and current $j = 11$

Now, $p = 6$ and $j - 1 = 10$; therefore, the next invocation,

Call #8 of QuickSort, will be Q(6, 10).   // And later, there will be Q(12, 11)

// QuickSort the array A from position $p = 6$ to position $q = 10$.

```
                          |            A            |
                          6=p              q=10
   k  A[k]  A[k]<M         ↓                ↓             j
   -   -      -    0 1 2 3 5 8 7 6 5 7 9   M=7   6
   6   8      F    " " " " " 8 7 6 5 7 "          "
   7   7      F    " " " " " 8 7 6 5 7 "          "
   8   6      T    " " " " " 6 7 8 5 7 "          7
   9   5      T    " " " " " 6 5 8 7 7 "          8
   -   -      -    " " " " " 6 5 7 7 8 "          -
```
// the new Partition of A is: | <7 | 7 |>=7| // and current $j = 8$

Now, $p = 6$ and $j - 1 = 7$, so the next invocation,

Call #9 of QuickSort, will be Q(6, 7).   // And later, there will be Q(9, 10)

// QuickSort the array A from position $p = 6$ to position $q = 7$.

```
                    |               A                |
                      6 = p  q = 7
   k  A[k] A[k]<M            ↓ ↓                          j
   -   -       -      0 1 2 3 5 6 5 7 7 8 9    M=5     6
   6   6       F      " " " " " 6 5 " " " "            "
   -   -       -      " " " " " 5 6 " " " "            -
// the new Partition of A is:            5 6      // and current j = 6
```

Now $p = j = 6$, and therefore, the next invocation,

Call #10 of QuickSort, will be Q(6, 5).   // And later, there will be Q(7, 7)
On this call, $p > q$, so nothing is done, and control returns to Call #9 of QuickSort where $j + 1 = q = 7$, and therefore, the next invocation,

Call #11 of QuickSort, will be Q(7, 7). On this call, $p = q$, so nothing is done, and control returns to Call #9 of QuickSort which is now complete.
At this point, we return to Call #8 of QuickSort where $j = 8$ and do the second sub-call;

Call #12 of QuickSort will be Q(9, 10).

// QuickSort the array A from position $p = 9$ to position $q = 10$.

```
                    |                   A                |
                              9 = p  q = 10
   k  A[k] A[k]<M                ↓ ↓                          j
   -   -       -      0 1 2 3 5 5 6 7 7 8 9    M=8     9
   9   7       T      " " " " " " " " 7 8 "            10
   -   -       -      " " " " " " " " 7 8 "            -
// the new Partition of A is:                7 8     // and current j = 10
```

Now $p = j - 1 = 9$, and therefore, the next invocation,

Call #13 of QuickSort, will be Q(9, 9).   // And later, there will be Q(11,10).
On this call, $p = q$, so nothing is done, and control returns to Call #12 of QuickSort where $j + 1 = 11$ and $q = 10$ to do the second sub-call. Therefore, the next invocation,

Call #14 of QuickSort, will be Q(11, 10). On this call, $p > q$, so nothing is done, and control returns to Call #12 of QuickSort which is now complete.
At this point, we return to Call #8 of QuickSort which is now complete, so we return to Call #7 of QuickSort where $j = 11 = q$ and do the second sub-call.

Call #15 of QuickSort will be Q(12, 11). On this call, $p > q$, so nothing is done, and control returns to Call #7 of QuickSort which is now complete.
At this point, we return to Call #5 of QuickSort which is now complete.

Then, we return to the first invocation of QuickSort, and it is now complete, and we
can be certain that the input array is now completely sorted.

The diagram below is the Tree of Recursive Calls for this one instance of
QuickSort.   // It is not a Branching Diagram or Decision Tree for the algorithm.

The vertices are the invocations of QuickSort. It is a <u>full</u> Binary Tree rooted at
the "external" call, QuickSort(1,n), where $n = 11$ and where the vertices "below" an
internal vertex are the two sub-calls made. The leaves are sub-calls where $p >= q$,
and nothing is done (except to ascertain that $p >= q$).

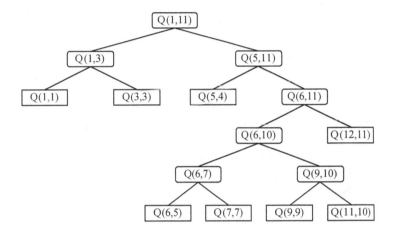

Theorem 4.5.1: QuickSort is a correct sorting algorithm

Proof. We will prove the following statement by Strong Induction on k where
$k \in \{1..\ \}$

If QuickSort(p, q) is invoked to sort the sublist $A[p] \ldots A[q]$ of length k,
it will terminate with the sublist correctly sorted.   // $k = q - p +1$.

Step 1. If $k = 1$, then p must equal q so QuickSort(p, q) terminates immediately
after ascertaining that $p = q$, and any sublist of length 1 is correctly sorted.

Step 2. Assume $\exists\ t >= 1$ such that for any k where $1 <= k <= t$
if QuickSort(p, q) is invoked to sort the sublist $A[p]\ldots A[q]$ of length k,
it will terminate with the sublist correctly sorted.

Step 3. Suppose QuickSort(p, q) is invoked to sort the sublist $A[p]\ldots A[q]$ of length
$t + 1$.

Since $t = q - p$ and $t >= 1$, q must be greater than p. In the partition portion of the algorithm, the value of j starts at p and is increased by 1 at most once for each value of k, so the value of j at the end of the for-k loop is at most $p + (q - p) = q$.

We will continue by considering three cases.   // of the value of j

Case 1. Suppose $j = p$. Because j was not increased above its initial value p, for every index k from p to $q - 1$, $A[k]$ was $>= M$. So the only exchange done is to interchange $A[q]$ and $A[j]$. Then

$$A[p] <= A[p+1], A[p+2], \ldots, A[q].$$

The first recursive sub-call is QuickSort(p, $p - 1$) which does nothing to array A. By step 2, the second recursive sub-call which is QuickSort($p + 1$, q) terminates and correctly sorts the sublist $A[p + 1]$ to $A[q]$ which has length t. But then, the algorithm terminates and the whole sublist $A[p]$ to $A[q]$ which has length $t+1$ has been correctly sorted.

Case 2. Suppose $j = q$. Because j must have been increased from its initial value p, for every index k from p to $q - 1$, each $A[k]$ was $< M$, each j-value equaled the current k-value, each $A[k]$ was interchanged with itself, and finally, $A[q]$ was interchanged with itself. Then

$$A[p], A[p+1], \ldots, A[q-1] < A[q]$$

By step 2, the first recursive sub-call which is QuickSort(p, $q - 1$) terminates and correctly sorts the sublist $A[p]$ to $A[q - 1]$ which has length t. The second recursive sub-call is QuickSort($q + 1$, q) which does nothing to array A. But then, the algorithm terminates and the whole sublist $A[p]$ to $A[q]$ which has length $t + 1$ has been correctly sorted.

Case 3. If $p < j < q$, the algorithm will partition the list $A[p] \ldots A[q]$ into three sections:

$A[p] \ldots A[j - 1]$ of length r where $1 <= r = j - p < q - p = t$;
$A[j]$ of length 1; and
$A[j + 1] \ldots A[q]$ of length s where $1 <= s = q - j < q - p = t$.

The first recursive sub-call is QuickSort(p, $j - 1$) which, by step 2, terminates and correctly sorts $A[p]$ to $A[j - 1]$ so that

$$A[p] <= \ldots <= A[j-1] < A[j] = M$$

The second recursive sub-call is QuickSort($j + 1$, q) which, by step 2, terminates and correctly sorts $A[j + 1]$ to $A[q]$ so that

$$A[j] = M <= A[j+1] <= A[j+2] <= \ldots <= A[q].$$

But then, the algorithm terminates and the whole sublist $A[p]$ to $A[q]$ which has length $t + 1$ has been correctly sorted.   □

The main virtues of this version of QuickSort are as follows: It is fairly simple despite the essential recursion, each partition makes one (left to right) traversal of the array, and it makes clear that there are exactly $q - p$ key comparisons done in each partition.

However, in our walk through of the example with $n = 12$, there were **8** "useless" sub-calls of QuickSort where $p >= q$ but only **7** "worthwhile" sub-calls (where $p < q$); there were also many "useless" interchanges of $A[j]$ with $A[k]$ when $k = j$ (in call #7) and two "useless" interchanges of $A[j]$ with $A[q]$ when $q = j$ (in call #7 and #12).

Can we remove these useless operations?	// Yes.
Can we make the algorithm more efficient?	// Maybe.
Can we still keep the algorithm simple?	// Probably not.

Any Tree of Recursive Calls of (any application of) QuickSort is a full Binary Tree rooted at the "external" call, QuickSort(1, n) where $n >= 1$. Each internal vertex has two vertices below it. The leaves are sub-calls where $p >= q$ and where nothing is done (so no sub-calls are made). Recall that Theorem 4.2.3 says that in a full Binary Tree with k internal vertices, there are $k + 1$ leaves – more than half the vertices are leaves. Thus, in the sub-calls of QuickSort, ***more than half of them do nothing*** (to alter array A), but each one has a cost in time and resources, as each induces the construction of a call frame, placing it on the call stack, checking to see that $p >= q$ then removing the frame from the stack, and then returning to the previous instance of QuickSort.

We can probably speed up the operation of QuickSort, if we check that a sub-call will actually do something *before* it is made. That is, make the sub-calls conditional.

If (p < j − 1) Then QuickSort(p, j − 1) End;

If (j + 1 < q) Then QuickSort(j + 1, q) End;

Then, we can remove the initial if-statement comparing the parameter values, that checks that $p < q$, provided we make this a precondition for any "external" call of QuickSort.

Furthermore, we can remove the useless interchanges of an entry with itself, if we search for an entry $A[j] >= M$ before the main partition loop. Implementing these two ideas give us

Algorithm 4.5.2: QuickSort2 (p, q) // version #2.0

// Assumes that any "external" invocation has $p < q$.   //X or $p = q$.
// (recursively) sorts $A[p]$, $A[p + 1]$, ..., $A[q]$ into order

```
Begin
    M ← A[q];
    j ← p;
    While (A[j] < M) Do
        j ← j + 1;
    End;
```
                                    // Since A[q] = M, this loop must terminate with j <= q.
                                    // At termination, A[j] >= M & perhaps j = q.
                                    // But j = q iff M is > all entries from A[p] to A[q − 1].

```
If (j=q) Then
  If (p < j − 1) Then
    QuickSort2(p, j − 1);
  End;
Else                              // when j < q & A[j] >= M, do the partitioning
  For k ← (j + 1) To (q − 1) Do
    If (A[k] < M) Then
      x ← A[j];
      A[j] ← A[k];                                              // j < k
      A[k] ← x;
      j ← j + 1;
    End;  // the if
  End;    // for k-loop

  A[q] ← A[j];
  A[j] ← M;                                                    // j < q

  If (p < j − 1) Then QuickSort2(p , j − 1) End;
  If (j + 1 < q) Then QuickSort2(j + 1, q) End;
End;        // the else part when j < q
End.        // the algorithm
```

Walkthrough assuming an "**external**" invocation of the form QuickSort2(1, 11).

// Sort the array A from position $p = 1$ to the last position $q = 11$.

			A		
			1= **p**	**q** = 11	j
j	**A[j]**	**A[j]<M**	↓	↓	
−	−	−	5 7 6 0 9 8 2 1 5 7 **3**	M=3	1
1	5	F	**5** 7 6 0 9 8 2 1 5 7 3		"

//At the end of the while-loop, $j<q$ so the statements following "Else" are done.

k	**A[k]**	**A[k]<M**	↓	↓	**j**
2	7	F	5 **7** 6 0 9 8 2 1 5 7 3		"
3	6	F	5 7 **6** 0 9 8 2 1 5 7 3		"
4	0	T	**0** 7 6 5 9 8 2 1 5 7 3		2
5	9	F	0 7 6 5 **9** 8 2 1 5 7 3		"
6	8	F	0 7 6 5 9 **8** 2 1 5 7 3		"
7	2	T	0 **2** 6 5 9 8 7 1 5 7 3		3
8	1	T	0 2 **1** 5 9 8 7 6 5 7 3		4
9	5	F	0 2 1 5 9 8 7 6 **5** 7 3		"
10	7	F	0 2 1 5 9 8 7 6 5 **7** 3		"
−	−	−	0 2 1 **3** 9 8 7 6 5 7 5		−

// the first Partition of A is: | <3 |3| >=3 | // and first $j = 4$

Now, $p = 1$ and $j − 1 = 3$, so the next "action" taken by QuickSort2 is the invocation of QuickSort2 itself with new parameter values.

Call #2 of QuickSort2 will be Q2(1, 3).   // Later, there will be Q2(5, 11)

// Sort the array A from position $p = 1$ to position $q = 3$.

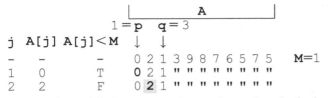

```
                              A
               1 = p    q = 3
 j  A[j] A[j]<M    ↓     ↓
 -   -      -    0 2 1 3 9 8 7 6 5 7 5    M=1
 1   0      T    0 2 1 " " " " " " " "
 2   2      F    0 2 1 " " " " " " " "
```

// At the end of the while-loop, $j < q$. But $j+1 > q-1$ so the for-loop is "empty".

```
 -   -      -    0 1 2 " " " " " " " "
```
// the second Partition is: |0|1|2| // and second $j = 2$

Now $p = j - 1$ and also $j + 1 = q$; therefore, neither sub-call is made, Call #2 is complete, and at this point, we return to the first invocation of QuickSort2 and do the second sub-call.

Call #3 of QuickSort2 will be Q2(5, 11).

// Sort the array A from position $p = 5$ to position $q = 11$.

```
                                   A
                       5 = p               q = 11
       j  A[j] A[j]<M     ↓                  ↓
       -   -      -    0 1 2 3 9 8 7 6 5 7 5        M=5
       5   9      F    " " " " 9 8 7 6 5 7 5
```
// At the end of the while-loop, $j<q$.
```
       k  A[k] A[k]<M                                            j
       6   8      F    " " " " " 9 8 7 6 5 7 5                   5
       7   7      F    " " " " " 9 8 7 6 5 7 5                   "
       8   6      F    " " " " " 9 8 7 6 5 7 5                   "
       9   5      F    " " " " " 9 8 7 6 5 7 5                   "
      10   7      F    " " " " " 9 8 7 6 5 7 5                   "
       -   -      -    " " " " " 5 8 7 6 5 7 9                   -
```
// the third Partition is: |5| >=5 | // and third $j = 5$

Now $p > j - 1$ but $j + 1 < q$; therefore, the next invocation,

Call #4 of QuickSort2, will be Q2(6, 11).

// Sort the array A from position $p = 6$ to position $q = 11$.

```
                                   A
                       6 = p             q = 11
 j  A[j] A[j]<M          ↓                ↓
 -   -      -    0 1 2 3 5 8 7 6 5 7 9        M=9
 6   8      T    " " " " " " 8 7 6 5 7 9
 7   7      T    " " " " " " 8 7 6 5 7 9
 8   6      T    " " " " " " 8 7 6 5 7 9
 9   5      T    " " " " " " 8 7 6 5 7 9
10   7      T    " " " " " " 8 7 6 5 7 9
11   9      F    " " " " " " 8 7 6 5 7 9
```
// the fourth Partition is: | <9 |9| // and fourth $j = 11$
// At the end of the while loop, $j = q$.
Since $p < j - 1$, the next invocation,

Call #5 of QuickSort2, will be Q2(6, 10).

// Sort the array A from position $p = 6$ to position $q = 10$.

```
                            |              A              |
                               6=p             q=10
   j   A[j]   A[j]<M            ↓               ↓
   -    -        -         0 1 2 3 5 8 7 6 5 7 9          M=7
   6    8        F         " " " " " 8 7 6 5 7 "
```

```
   k   A[k]   A[k]<M                                              j
   7    7        F        " " " " " 8 7 6 5 7 "                   6
   8    6        T        " " " " " 6 7 8 5 7 "                   7
   9    5        T        " " " " " 6 5 8 7 7 "                   8
   -    -        -        " " " " " 6 5 7 7 8 "                   -
```
// the fifth Partition is: $\boxed{<7\,|\,7\,|\,>=7}$ // and fifth **j = 8**

Now $p < j - 1$, so the next invocation,

Call #6 of QuickSort2, will be Q2(6, 7).   // And later, there will be Q2(9, 10).

// Sort the array A from position $p = 6$ to position $q = 7$.

```
                         |            A            |
                             6=p q=7
   j  A[j]  A[j]<M           ↓ ↓
   -    -       -        0 1 2 3 5 6 5 7 7 8 9        M=5
   6    6       F        " " " " " 6 5 " " " "
```
// At the end of the while-loop, $j<q$. But $j+1>q-1$ so the for-loop is "empty".
```
   -    -       -        " " " " " 5 6 " " " "
```
// the sixth Partition is: $\boxed{5\,|\,6}$ // and sixth **j = 6**

Now $p > j - 1$ and $j + 1 = q$, so Call #6 is complete. At this point, we return to Call #5 of QuickSort2 where $j = 8$ and do the second sub-call;

Call #7 of QuickSort2 will be Q2(9, 10).

// Sort the array A from position $p = 9$ to position $q = 10$.

```
                            |              A              |
                                          9=p q=10
   j   A[j]  A[j]<M                         ↓ ↓
   -    -       -         0 1 2 3 5 5 6 7 7 8 9          M=8
   9    7       T         " " " " " " " " 7 8 "
   10   8       F         " " " " " " " " 7 8 "
```
// the seventh Partition is: $\boxed{7\,|\,8}$ // and seventh **j = 10**
// At the end of the while-loop, $j = q$.

Now $p = j - 1$, so no sub-call is made, the else part is skipped over, and Call #7 is complete.

At this point, we return to Call #5 of QuickSort2 which is now complete.

Then, we return to Call #4 of QuickSort2, skip over the else part, and Call #4 is complete.
Then, we return to Call #3 of QuickSort2 which is now complete.
Then, we return to the first invocation of QuickSort2, and it is now complete, and
we can be certain that the input array is now completely sorted.

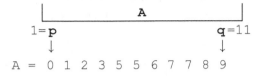

$$1 = p \qquad\qquad\qquad\qquad q = 11$$

$$A = 0 \quad 1 \quad 2 \quad 3 \quad 5 \quad 5 \quad 6 \quad 7 \quad 7 \quad 8 \quad 9$$

The diagram below is the Tree of Recursive Calls for this one instance of
QuickSort2.   // It is not a Branching Diagram or Decision Tree for the algorithm.
The vertices are the invocations of QuickSort2. It is no longer a <u>full</u> Binary Tree
rooted at the "external" call, QuickSort2(1, n), where $n = 11$. The vertices "below"
a vertex are the sub-calls made.

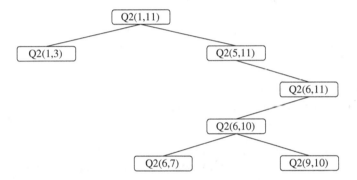

Compare this binary tree with the tree for the original QuickSort algorithm. Notice
that this is the same tree that we saw previously with all its leaves pruned off.
Or one might say that this is a sub-tree of the previous tree consisting of the internal
vertices only (together with those edges that join internal vertices to each other).

The Most Important Ideas in This Section.
We presented a "partition sort" called **QuickSort** and proved its correctness.
An exercise asks you to prove that it terminates after at most $n(n - 1)/2$
comparisons and to demonstrate that a worst case occurs when the input array
is already sorted.

We described a common mechanism for implementing recursion, a stack
of call frames. And we gave a way to diagram the action of recursion on a
specific input as a Tree of Recursive Calls. For QuickSort, that diagram is a
full Binary Tree. We used a general result about full Binary Trees to construct
a second version **QuickSort2** to try to speed up execution.

Recursive algorithms are often very easy to construct and to prove correct
(by Strong Induction).

The next section reports experimental results about the time taken for each
of the sorting algorithms – with some dramatic conclusions.

4.6 Comparison of Sorting Algorithms

To compare the running times of the sorting algorithms we've discussed in this chapter, each was implemented and run on a desktop computer using a sample of ten input arrays. The entries in these arrays were integers selected at random from 1 to 1,000,000. In each case, the average time in seconds was calculated.

// and rounded

	4,000 keys	8,000 keys	$T(2n)/T(n)$
BetterBubbleSort	5.467	21.992	4.023
BubbleSort	5.132	21.033	4.098
MinSort	2.445	9.782	4.002
QuickSort	0.0218	0.0465	2.131
QuickSort2	0.0217	0.0460	2.124

We also ran the two QuickSort algorithms on much longer arrays.

	200,000 keys	400,000 keys	$T(2n)/T(n)$
QuickSort	1.497	3.183	2.126
QuickSort2	1.454	3.165	2.177

From this data, we can see that:

1. BetterBubbleSort was just too clever – for random lists, the savings in comparisons was lost to the extra cost of maintaining the "flag" variable p.

   // p was updated at every exchange, and (usually) there are many
   // exchanges. BubbleSort is dumb and BetterBubbleSort is dumber.

2. Lowly MinSort spends much less time shuffling data and requires less than half the time of the BubbleSorts.   // Only one exchange is done on each pass.
3. QuickSort may seem complicated but is much, much quicker.
4. QuickSort2 is even quicker.   // but only slightly

4.6.1 Timings and Operation Counts

These data also show that counting comparisons is a reasonable way to determine "relative costs" of running these algorithms. If $T(n)$ is the time taken to sort a list of length n, and $f(n)$ is the number of key comparisons done, then we might expect that $T(n)$ is roughly proportional to $f(n)$. That is, $T(n) \cong c \times f(n)$ where c is a constant.

// The value of that constant would depend on the speed of the machine used, the
// optimization level of the object code, the computing environment, and other
// factors.

   If $f(n) \cong a \times n^2$,   // In a way, we'll make precise in Chap. 7.
as is the case for MinSort and the BubbleSorts, then doubling the size of the input more or less quadruples the running time.   // as shown in the data

$$\frac{T(2n)}{T(n)} \cong \frac{c \times a \times (2n)^2}{c \times a \times n^2} = 4$$

If $f(n) \cong b \times n \times \lg(n)$, as is the (average) case for the QuickSorts, then

$$\frac{T(2n)}{T(n)} \cong \frac{c \times b \times (2n) \lg(2n)}{c \times b \times n \lg(n)} = \frac{2[\lg(2) + \lg(n)]}{\lg(n)} = 2 + \frac{2}{\lg(n)}.$$

For $n = 4,000$, $\lg(n) = 11.96578\ldots$ and $2 + \frac{2}{\lg(n)} = 2.167143\ldots$

For $n = 200,000$, $\lg(n) = 17.60964\ldots$ and $2 + \frac{2}{\lg(n)} = 2.113574\ldots$

// The running time more than doubles (but not a lot more).

The Most Important Ideas in This Section.
BubbleSort is so bad it should never be shown to students, never used, and never remembered. MinSort works well for short lists. QuickSort is <u>very</u> quick.

Some improvement ideas work but some do not. Modern computers are fast and programming languages are complex, and it is often difficult to predict whether a change to an algorithm will actually result in a performance gain (or not). And even when a performance gain is realized, it may be very small.

But sometimes, a revision of an algorithm produces an enormous increase in efficiency! In Chap. 7, we will revisit these ideas about time-complexity.

Exercises

1. Consider the array A = (1, 2, 4, 8, 16, 32, 64, 128, 256, 512, 1024).
   (a) How many elements will be probed when performing a linear search for 32?
   (b) Can a binary search be performed on this array? If so, how many elements will be probed when performing a binary search for 32? If not, briefly explain why not.
2. Consider the array A = (0, 1, 127, 1023, 255, 3, 31, 511, 63, 7).
   (a) How many elements will be probed when performing a linear search for 32?
   (b) Can a binary search be performed on this array? If so, how many elements will be probed when performing a binary search for 32? If not, briefly explain why not.
3. (a) Draw the "Tree of Probes" for Linear Search when $n = 7$.
   (b) Draw the "Tree of Probes" for Binary Search when $n = 14$ and $n = 20$.

4. Prove by MI that for all $k \in \{0..~\}$, if Binary Search is applied to an array of length n and has not terminated after k (unsuccessful) probes, then the length of the current sublist must be $<= n/2^k$.

5. Why does every entry in A appear exactly once in the tree of Probes for Binary Search?

6. Do other loop invariants hold for Binary Search? Prove that:

   (a) Either $p = 1$ or $A[p - 1] < T$
   (b) Either $q = n$ or $T < A[q + 1]$

7. Prove that $\lceil (k - 1)/2 \rceil = \lfloor k/2 \rfloor$ for $\forall k \in \mathbf{Z}$.

8. Prove that $\lceil k/2 \rceil = \lfloor (k + 1)/2 \rfloor$ for $\forall k \in \mathbf{Z}$.

9. Suppose T is not found by Binary Search.
   // Where will T fit into the array? Where should it be inserted?

   (a) Is the final q-value always one less than the final p-value?
   (b) When is $A[q] < T < A[p]$?
   (c) If $p = 1$ (that is, p was never changed), is $T < A[1]$?
   (d) If $q = n$ (that is, q was never changed), is $A[n] < T$?

10. Do the same loop invariants for the while loop hold for Binary Search #2? Prove that "after each iteration of the loop, if $T = A[i]$ then $p <= i <= q$."

11. Suppose T is not found by Binary Search #2; then, in the last comparison, $A[p] \neq T$.

   (a) Show that if $1 < p < n$, then $A[p - 1] < T < A[p]$.
   (b) Show that if $1 = p < n$, then $\qquad\qquad T < A[1]$.
   (c) Show that if $1 < p = n$, then $A[n - 1] < T < A[n]$ or $A[n] < T$.

12. Suppose that $\mathcal{T}$ is a full Binary Tree where all leaves are at level p or higher. Prove by Mathematical Induction that for $k = 0, 1, \ldots, p$ the number of vertices at level k is exactly 2^k.
   // because each internal vertex has two vertices below it

13. (a) How would you sort $A[1], A[2], \ldots, A[400]$ if each entry is either 0 or 1?
    (b) How would you sort $A[1], A[2], \ldots, A[400]$ if each entry is one of ten possible values $X[1] < X[2] < \ldots < X[10]$?

14. The prototype for **InsertionSort** is a common method of "sorting" a hand of cards. Suppose $1 <= k < n$ and $A[1], A[2], \ldots, A[k]$ are in nondecreasing order. Now insert $A[k + 1]$ into its correct position among these previous entries. To make room for (the value of) $A[k + 1]$, those entries larger than $A[k + 1]$ can be shifted up one place, one entry at a time:

    (a) Write a program (or pseudo-code) for the Algorithm InsertionSort.
    (b) Would you say this is an exchange sort?
    (c) Is every case a worst case? How many comparisons would be done in a worst case?
    (d) What is a best case? How many comparisons would be done in a best case?

15. Prove that QuickSort terminates after at most $n(n - 1)/2$ key comparisons, where n is the length of the list being sorted.
   // Hint: Follow the pattern of the proof of Theorem 4.5.1.
   Does this argument show that a worst case for QuickSort is an already sorted list?

16. Does Quicksort2 work correctly if the "external" invocation has $p = q$?
17. In QuickSort, the list $A[p]\ldots A[q]$ of length k is partitioned into 3 sublists:
    $A[p]\ldots A[j-1]$ of length k_1, $A[j]$ of length 1, and $A[j+1]\ldots A[q]$ of length k_2.
    The total cost in key comparisons of sorting the list is described by

    $$\mathbf{C}(k) = (k-1) + \mathbf{C}(k_1) + \mathbf{C}(k_2).$$

    This exercise will show that $\mathbf{C}(k_1) + \mathbf{C}(k_2)$ is smallest when

    $$k_1 = \left\lfloor \frac{k-1}{2} \right\rfloor \quad \text{and} \quad k_2 = \left\lceil \frac{k-1}{2} \right\rceil. \qquad \text{//or visa versa}$$

    (a) Let f be any sequence on $\mathbf{N}$. The nth **increment** of f is
        $\Delta f(n) = f(n+1) - f(n)$.

        // So Δf is also a sequence on $\mathbf{N}$ (sometimes called the "first-differences").

        Prove that if $0 <= a < b$, then

        $$f(a) + \sum_{j=a}^{b-1} \Delta f(j) = f(b).$$

    (b) The function f has **increasing increments** means Δf is an increasing
        sequence. Assume that the cost of sorting an array in terms of the number
        of comparisons is such a function.
        Suppose that $0 <= a < r <= s < b$ and $N = a + b = r + s$.
        (i) Prove that if the function f has increasing increments, then

        $$f(r) + f(s) < f(a) + f(b).$$

        // Hint:
        // $f(a) + f(b) = f(a) + \{f(a) + [\Delta f(a) + \Delta f(a+1) + \ldots + \Delta f(b-1)]\}$
        //   // from (a)
        // $f(a) + [\Delta f(a) + \ldots + \Delta f(r-1) +$
        // $f(a) + [\Delta f(r) + \Delta f(r+1) + \ldots + \Delta f(b-1)]$
        // $= \ldots$
        // Remember $\Delta f(r+j) > \Delta f(a+j)$ whenever $j >= 0$
        // and that $b - 1 = r + (b-1-r) > a + (b-1-r) = N - r - 1 =$
        // $s - 1$.
        (ii) Use (i) to prove that of all choices of r and s where $r <= s$, $f(r) + f(s)$ is
              smallest when $r = \lfloor N/2 \rfloor$ and $s = \lceil N/2 \rceil$.
18. Walk through the operation of QuickSort when $n = 7$ and the input array is

    $$A = (11, 13, 12, 32, 31, 33, 20).$$

(a) Count the number of comparisons in the walk through.

(b) Evaluate 7!, lg(7!) and $7 \times$ lg(7).

(c) Construct a best-case example for QuickSort with $n = 15$.

19. The merging problem is as follows:

Given two sorted arrays

$$A[1] \; <= \; A[2] \; <= \; A[3] \; <= \; \ldots \; <= \; A[m]$$
$$\text{and} \quad B[1] \; <= \; B[2] \; <= \; B[3] \; <= \; \ldots \; <= \; B[n],$$

put these $m + n$ entries into an array C so that

$$C[1] \; <= \; C[2] \; <= \; C[3] \; <= \; \ldots\ldots\ldots \; <= \; C[m+n].$$

(a) Write a program (or pseudo-code) for the Algorithm Merge.

// with 3 array parameters

(b) How many comparisons of array entries would be done in a worst case?

(c) Can A and B be merged into C using $<= m+n$ key comparisons in every case?

(d) Can $A[p] <= \ldots <= A[j]$ and $A[j+1] <= \ldots <= A[q]$, where $p <= j <= q$, be merged directly into $C[p] <= C[p+1] <= \ldots <= C[q]$?

(e) Can $A[p] <= \ldots <= A[j]$ and $B[j+1] <= \ldots <= B[q]$, where $p <= j <= q$, be merged directly into $A[p] <= A[p+1] <= \ldots <= A[q]$?

(f) Can $A[p] <= \ldots <= A[j]$ and $A[j+1] <= \ldots <= A[q]$, where $p <= j <= q$, be merged directly into $A[p] <= A[p+1] <= \ldots <= A[q]$?

20. **MergeSort** is an easily described partition sort. To sort $A[p], A[p+1], \ldots, A[q]$ where $p < q$, divide the list in half, sort the first half (using this same method), sort the second half (also using this same method), and then Merge the two sorted sublists. The simplest possible "division into halves" can be used, namely, calculate $j = \lfloor (p + q)/2 \rfloor$ and let the "halves" be

$$A[p], A[p+1], \ldots, A[j] \quad \text{and} \quad A[j+1], A[j+1], \ldots, A[q].$$

// But if $p = q$, do nothing.

(a) Usually, one of these two halves is copied (entry by entry) into a different array B before the Merge operation can be done. Can B be a "global" variable like A?

(b) Write a program (or pseudo-code) for the **Recursive** Algorithm **MergeSort**.

(c) The number of comparisons of array entries can be shown to be less than $n \times \lceil \lg(n) \rceil$. But how much space will be used? How many computer words, if each array entry uses one word?

21. *HeapSort* is yet another efficient sorting algorithm. It uses a clever data structure known as a "heap" that has the properties of a Binary Tree <u>and</u> an array. Look up *HeapSort* on the web.

22. The Philosophy Prof's practical problem: Suppose Professor Plum has 600 marked midterm papers she wants to sort into alphabetical order. You know that an inefficient sorting algorithm might require $n(n-1)/2$ comparisons of names, and if it takes 1 second to do such a comparison, the whole job would take about 50 hours. How would <u>you</u> sort the papers? Would you partition the papers using the first letter of the last name? Would you merge small sets of papers?

Graphs and Trees

5

5.1 Introduction

We begin this chapter with three puzzle problems. Each is "set up" with a certain amount of "patter" in a story.

#1: Tartaglia's Pouring Problem ($\sim$ 1530)

Imagine three earthenware vessels (jars): one of unknown large capacity containing exactly 8 units (maybe pints) of some (maybe precious) liquid; a middle-sized one that's empty but can hold exactly 5 units when filled to the brim; and a small one that's empty but can hold exactly 3 units when filled to the brim. Using only these three vessels,

Can you divide the given liquid exactly in half?

// Hint 1. You may pour liquid from one vessel into another. But you only know
//   how much you've poured when you exactly fill a smaller vessel, or
//   (knowing how much the vessel holds) you completely empty a vessel.
// Hint 2. You can describe any "configuration" by a triple (x,y,z) where x is the
//   number of units in the large vessel, y the number in the middle-sized
//   one, and z the number in the small one.
// Now, you need a sequence of pourings that takes you from (8,0,0) to (4,4,0).

#2: The Missionaries and Cannibals Problem

Imagine three missionaries and three cannibals traveling together through the jungle. They come to a fairly wide river that they need to cross, in which there are threatening "monsters" of some kind (alligators, crocodiles, snakes, piranhas – who knows). Since they cannot safely swim or wade (or jump) across, they walk along the shore until they find a boat (with a paddle). But the boat is so small that only two people can fit into it.

© Springer International Publishing AG, part of Springer Nature 2018   191
T. Jenkyns and B. Stephenson, *Fundamentals of Discrete Math for Computer Science: A Problem-Solving Primer*, Undergraduate Topics in Computer Science,
https://doi.org/10.1007/978-3-319-70151-6_5

Using that boat, they can all get across. But if the cannibals ever outnumber the missionaries on either side, something dreadful will happen.

How do they all get across, so nothing dreadful happens?

#3: Konigsberg Bridge Problem (~ 1730)

The town of Konigsberg is situated at the place where two branches of a river join, and there is an island near that point. Seven bridges were built in the town as in the diagram below. The citizens entertained themselves (on Sunday afternoons when no one worked and everyone spent quality time with their families) by trying to find a route that crossed every bridge exactly once. It could start anywhere and end anywhere.

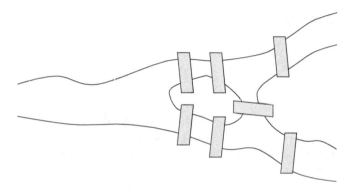

Can you find such a route?

Solution to #1: Tartaglia's Pouring Problem

One possible first move is to pour liquid from the large vessel into the smallest one and completely fill it; that is,

$$\text{go from } (8,0,0) \text{ to } (5,0,3).$$

Then a possible next move is to empty the smallest vessel into the middle-sized one; that is,

$$\text{go from } (5,0,3) \text{ to } (5,3,0).$$

From (5,3,0), we could pour *from* the middle vessel and go to (8,0,0) or to (5,0,3). But we've been in both of these configurations before. Let's adopt the policy of ***always going to a new configuration***, if we can. Then from (5,3,0), we could pour from the large vessel into the small one and go from (5,3,0) to (2,3,3).

Furthermore, a possible sequence of pourings would be:

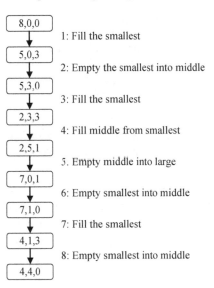

8,0,0

1: Fill the smallest

5,0,3

2: Empty the smallest into middle

5,3,0

3: Fill the smallest

2,3,3

4: Fill middle from smallest

2,5,1

5. Empty middle into large

7,0,1

6: Empty smallest into middle

7,1,0

7: Fill the smallest

4,1,3

8: Empty smallest into middle

4,4,0

So we have a solution that's not immediately obvious. It requires eight pourings but wasn't too hard to find. This is probably why this puzzle has been around for 500 years. But is this solution the "best" one? Can it be done with fewer pourings?

To see if it's the best, we can "grow a tree" of possible solutions: actually, all possible sequences of pourings that don't repeat a configuration. From any configuration, let's add "branches" down to *all new configurations that we can reach in one* more pouring, and let's organize this as a tree "rooted" at the initial configuration, (8,0,0).   // like the "start" vertex in the tree diagrams of Chap. 2

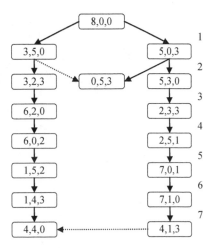

// Are you sure that this tree contains all possible branches to new configurations?

The first solution was not the best possible solution. This diagram proves that the (unique) best one uses only seven pourings.

Solution to #2: The Missionaries and Cannibals Problem

Let's begin with a representation of possible "configurations": say + will represent a missionary, o will represent a cannibal, | will represent the river, and * will represent the boat. Then the starting configuration is (+++ooo *|), and the configuration we want to reach is (|* +++ooo). A transition from one configuration to another will correspond to one or two people crossing the river in the boat. The five possibilities are:

    1. — + —   // a missionary paddles across
    2. — o —   // a cannibal paddles across
    3. — ++ —   // two missionaries paddle across
    4. — oo —   // two cannibals paddle across
    5. — +o —   // a missionary and a cannibal paddle across

But we want to restrict transitions so that we never reach a configuration where something dreadful happens. The first transition cannot be:

    1. — + —   // else we reach (++ooo |* +) and something dreadful
                              // happens on the left bank of the river

nor   3. — ++ —   // else we reach (+ooo |* ++) and ditto

If the first transition were

    2. — o —

we would reach (+++oo |* o) where nothing dreadful happens, but the cannibal must paddle back, and we must start again. To make progress, we must start by sending two people across, say a missionary and a cannibal.

    Let's again adopt the policy of ***always going to a new configuration***, if we can.

    // Restricted, of course, to the configurations where nothing dreadful happens.

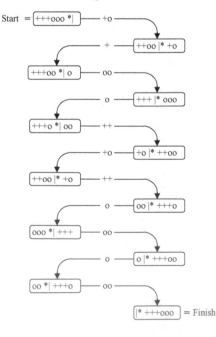

// Are these the only possible transitions? Are all these reversible?

// This solution is puzzling because there is an essential but "counterintuitive"
// step where two people are sent back.
// Perhaps Tartaglia's problem is puzzling because we took several steps that
// didn't appear to be taking us closer to the final target configuration.

Solution to #3: Konigsberg Bridge Problem (1736)

Leonard Euler (1707–1783) ruined this pastime for the citizens of Konigsberg by
proving that the problem has no solution; there is no route that crosses every bridge
exactly once, no matter where it starts or ends.

Tartaglia's Pouring Problem has survived for hundreds of years because it has a
solution. This problem has survived not because it has no solution but because
Euler's proof that it has no solution is said to mark the beginning of graph theory.

Euler abstracted the essential elements of the problem. He said (in effect) that:

(a) It doesn't matter where you are on the island; what matters is that you must
    leave or arrive on the island by crossing one of the 5 bridges to the island. Thus,
    the island may be thought of as a "point" (I) that may be "reached" by crossing
    the bridges.

(b) It doesn't matter where you are on the north shore; what matters is that you must
    leave or arrive on the north shore by crossing one of the 3 bridges to the north
    shore. Thus, the north shore may be thought of as a "point" (N) that may be
    "reached" by crossing the bridges.

(c) It doesn't matter where you are on the south shore; what matters is that you
    must leave or arrive on the south shore by crossing one of the 3 bridges to it.
    Thus, the south shore may be thought of as a "point" (S) that may be "reached"
    by crossing the bridges.

(d) It also doesn't matter where you are on the peninsula; what matters is that you
    must leave or arrive on the peninsula by crossing one of the 3 bridges to it.
    Thus, the peninsula shore may be thought of as a "point" (P) that may be
    "reached" by crossing the bridges.

When the bridges are represented as lines joining their two end points, the
problem may be diagrammed as follows.

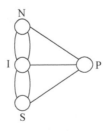

We'll come back to "routes" in the Konigsberg Bridge Problem after a short
digression introducing some of the basic definitions from graph theory.

// used in this book

A *graph G* consists of a (finite, nonempty) set *V* of *vertices* together with a (finite
but possibly empty) set *E* of *edges* where each edge "joins" two vertices. The

vertices are drawn as small circles, ovals, or squares, and the edges are drawn as line segments joining one vertex to another. An edge may join a vertex to itself; such an edge is called a *loop*. Vertices joined by an edge are said to be *neighbors*. Two different edges may join the same two vertices (like the two bridges from the north shore to the island in Konigsberg); such edges are said to be *parallel*. When there are no loops and no parallel edges, the graph is said to be *simple*.

A *path* (in a graph) is a sequence of the form

$$\pi = (v_0, e_1, v_1, e_2, v_2, e_3, \ldots, e_k, v_k)$$

where each v_j is a vertex and each e_i is an edge joining vertex v_{i-1} to vertex v_i.

// A path corresponds to a trip through (part of) the graph
// starting from vertex v_0 then following the edge e_1 to vertex v_1,
// then following the edge e_2 to vertex v_2, and so on until
// finally, the edge e_k is traversed and we end up at vertex v_k.

5.1.1   Degrees

Each edge has two ends and each edge-end occurs at a vertex. The number of edge-ends at any vertex, x, is called the *degree* of that vertex and will be denoted $d(x)$. Then, counting all the edge-ends in two ways, we have

$$2 \times |E| = \sum_{x \in V} d(x). \tag{5.1.1}$$

// In the Konigsberg bridge graph: $d(N) = 3 = d(S) = d(P)$, and $d(I) = 5$.

//X   Use Eq. 5.1.1 to prove that in any graph, the number of vertices with
//   odd degree must be even.

5.1.2   Eulerian Graphs

An *Euler Tour* of a graph is a path where each edge occurs exactly once; such a tour is called an *Euler Circuit* when it starts and ends at the same vertex.

// Puzzle problem #3 asks: Does the Konigsberg bridge graph contain an Euler
// Tour?

Lemma 5.1.1:   (Euler 1736) If $\pi = (v_0, e_1, v_1, e_2, v_2, e_3, \ldots, e_k, v_k)$ is a path in G containing each edge exactly once, then every "intermediate" vertex x in the path must have even degree in G, where an *intermediate* vertex occurs in the middle of the path in the sense that it's not an end so it's not equal v_0 nor v_k.

Proof. If x occurs as v_j in the path, then $0 < j < k$, and so one end of edge e_j and one end of edge e_{j+1} are at x. If x occurs q-times in the path, then $2q$ edge-ends are at x. Since every edge of G occurs exactly once in π, all the edge-ends at x have been counted exactly once; that is,

$$d(x) = 2q, \text{ an even number.} \qquad \square$$

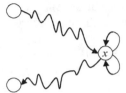

Thus, if G has an Euler Tour, then G has at most two vertices of odd degree.
// And if there are 2, they must occur as the ends of the tour.

The contrapositive form is: If G has more than two vertices of odd degree, then G cannot have an Euler Tour. Because the Konigsberg bridge graph contains 4 vertices of odd degree, it cannot have an Euler Tour.

// Finding an Euler Tour is easy. There are amusements for children that ask if
// they can "trace out a diagram (like the one below) without lifting their pencil,
// and never going over an edge more than once."

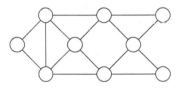

// More serious applications involve routing garbage trucks or mail carriers
// through a neighborhood so they go down every street but avoid going down
// any street more than once, or constructing a tour through an art gallery or
// museum so the most natural route takes patrons past every exhibit exactly once.
//
// You can look up Fleury's Algorithm to find an Euler Tour (1883).
// Two easily checked conditions characterizing Eulerian graphs are in an
// exercise.

5.1.3   Hamiltonian Graphs

A **Hamilton Tour** of a graph is a path where each <u>vertex</u> occurs exactly once; such a tour is called a **Hamilton Circuit** when there also is an edge joining the last vertex to the first vertex.   // And there are $>=$ 3 vertices.

// Unlike Eulerian graphs, characterizing Hamiltonian graphs is not easy. In fact,
// no one has (yet) given a list of easily checked conditions characterizing

// Hamiltonian graphs – that is, *G* is Hamiltonian if and only if all the conditions are
// present.
//
// Nor has anyone (yet) devised an efficient algorithm to find a Hamilton Tour.
// You could try all |V|-factorial permutations of the vertices in *G* and check to see
// if one of them is the vertex sequence of a path in *G*, but this "total-enumeration"
// method is too inefficient to be feasible in large graphs.
//
// The Clay Mathematics Institute has offered a $1,000,000 prize for solving a
// problem (known as the "P = NP Problem") that could be claimed by anybody
// who created an algorithm to find a Hamilton Tour (if there is one) in an arbitrary
// graph on *n* vertices, with a complexity function bounded by n^k for some constant
// *k*.

The Most Important Ideas in This Section.

A *graph G* consists of a set of *vertices* together with a set of *edges* where each
edge joins two vertices. A *simple* graph has no loops and no parallel edges.
A *path* is a sequence of the form $\pi = (v_0, e_1, v_1, e_2, v_2, e_3, \dots , e_k, v_k)$ where
each e_i is an edge joining vertex v_{i-1} to vertex v_i. The *degree* of a vertex *x*,
$d(x)$, is the number of edge-ends at *x*. An *Euler Tour* of a graph is a path
where each edge occurs exactly once; a *Hamilton Tour* is a path where each
vertex occurs exactly once. Eulerian graphs are easily characterized but
Hamiltonian graphs are not.

Many problems ask us to find a path (or tour) of a certain kind: such as an
Euler Tour or a Hamilton Tour. Some ask us to find a path from a particular
vertex *y* to a particular vertex *z*, and some problems ask us to find a shortest
path from *y* to *z*.

// Puzzle problem #2 asks: is there *any* path from (+++ooo *|) to
// (|* +++ooo)?

Designing solution algorithms for these kinds of problems is the underlying
purpose of this chapter.

The Clay Mathematics Institute has offered a $1,000,000 prize for solving
the "P = NP Problem," which some consider the most important problem in
theoretical computer science.

5.2   Paths, Circuits, and Polygons

Recall that a *path* (in a graph) is a sequence of the form

$$\pi = (v_0, e_1, v_1, e_2, v_2, e_3, \dots , e_k, v_k)$$

where each v_j is a vertex and each e_i is an edge joining vertex v_{i-1} to vertex v_i.

The path π is said to go *from* v_0 *to* v_k and to have **length** k, the number of edges
traversed.   // like pourings in puzzle problem #1

A graph is **connected** means for <u>any</u> two vertices y and z, there is a path from y to z.
A path is **simple** if all the vertices are distinct.   // If $i \neq j$, then $v_i \neq v_j$.
A path is **closed** if $v_0 = v_k$.   // It starts and ends at the same vertex.

A closed path is a *circuit* if $k >= 1$, and all the edges are distinct.

// The path $\pi = (v_0)$ is a simple, closed path of length zero from v_0 to v_0, often
// referred to as a "trivial" path, but it is <u>not</u> a circuit.
// If edge e is a loop at vertex v, then $\pi = (v, e, v)$ <u>is</u> a circuit (of length one).
// If e is an edge joining vertices v and w, then $\pi = (v, e, w, e, v)$ is a closed path
// of length two, which is <u>not</u> a circuit.
// If edges e and f are parallel edges joining vertices v and w, then $\pi = (v, e, w, f, v)$
// <u>is</u> a circuit (of length two).
// **G** is a simple graph $\Leftrightarrow$ **G** has no circuits of length 1 or 2.

Lemma 5.2.1: If $\pi = (v_0,\ e_1,\ v_1,\ e_2,\ v_2,\ e_3,\ \dots,\ e_k,\ v_k)$ **is a path from** v_0 **to**
a *different* **vertex** v_k**, then there is a simple path** π_s **from** v_0 **to** v_k **that is a**
subsequence of π**.**

Proof. If π is a simple path, we may use π itself as π_s. If π is not a simple path, then
there are indices i and j, where $0 <= i < j <= k$ and $v_i = v_j$. That is,

$$\pi = \left(v_0, e_1, v_1, e_2, \dots, e_i, v_i, e_{i+1}, v_{i+1}, \dots, e_j, v_j = v_i, e_{j+1}, v_{j+1}, \dots, e_k, v_k\right).$$

Let $\pi 1 = (v_0, e_1, v_1, e_2, \dots, e_i, v_i = v_j, e_{j+1}, v_{j+1}, \dots, e_k, v_k)$.

// Traverse the edges of π from v_0 to the *first occurrence* of v_i and then
// traverse the edges of π from the *second occurrence* of v_i to v_k.

Then $\pi 1$ is a path from v_0 to v_k that is a subsequence of π but has fewer edges
than π.   // e_{i+1} is not in $\pi 1$, and $\pi 1$ has fewer repeated vertices than π.
   If $\pi 1$ is not a simple path, we can find a repeated vertex and (as above) create a
new path $\pi 2$ from v_0 to v_k that is a subsequence of $\pi 1$ (and of π) that has fewer edges
than $\pi 1$.
   Because we remove at least one edge each time, this process must stop after at
most k iterations. And then we will have a **simple** path π_s from v_0 to v_k that is a
subsequence of π.   ▯

// Why must a <u>shortest</u> path (in the sense of having fewest edges) from one vertex
// to another vertex be a *simple* path?

5.2.1   Subgraphs Determined by Paths

Suppose that $\pi = (v_0, e_1, v_1, e_2, v_2, e_3, \ldots, e_k, v_k)$ is a path in graph G. To isolate the part of G occurring in π, let G_π denote the graph

whose vertex set is $V_\pi = \{v_0, v_1, v_2, \ldots, v_k\}$ // vertices of G occurring in π

and whose edge set is $E_\pi = \{e_1, e_2, \ldots, e_k\}$.   // edges of G occurring in π

// But these vertices and edges may not all be different.
// If the edges were all different, then π would be an Euler Tour of G_π.
// G_π is a connected graph.

If we denote by π^R the sequence we obtain by writing π in reverse order, then

$$\pi^R = (v_k, e_k, v_{k-1}, e_{k-1}, v_{k-2}, \ldots, v_2, e_2, v_1, e_1, v_0)$$

is a path, and this path *determines* the same subgraph as π.
If π is a *simple* path (having no repeated vertices), then π has no repeated edges and   // as we saw in Euler's Lemma 5.1.1

$$v_0 \text{ and } v_k \text{ have degree 1 in the graph } G_\pi,$$

$$v_1, v_2, \text{up to } v_{k-1} \text{ all have degree 2 in } G_\pi.$$

If π is a *simple* path from v to w and e is an edge joining v and w that is not an edge in G_π, then G_π together with e forms a graph where <u>every</u> vertex has degree 2.
    If π is a *circuit* (having no repeated edges but maybe having some repeated vertices) then

$$v_1 \text{ up to } v_k = v_0 \text{ all have } \textit{even} \text{ degree in } G_\pi.$$

A connected graph where all vertices have degree 2 is called a ***polygon***. Paths and circuits are algebraic objects; polygons are geometric. A polygon with $n \geq 2$ vertices is the subgraph determined by $2 \times n$ distinct circuits.

// We saw that every path joining different vertices contains a simple path as a
// subsequence. Do you think that any graph G_π where π is a closed path contains
// a polygon?

    If π is a *simple* path from v to w and e is an edge joining v and w that is not an edge in G_π, then G_π together with e forms a polygon.
    A subgraph containing all the vertices of G is called a ***spanning*** subgraph.

// When G has ≥ 3 vertices, G has a Hamilton circuit $\Leftrightarrow$ G has a spanning
// polygon.

Lemma 5.2.2: If $\pi = (v_0, e_1, v_1, \ldots, v_{j-1}, e_j, v_j, \ldots, e_k, v_k)$ is a closed path and edge e_j occurs *only once* in π, then the graph G_π contains a polygon that contains e_j.

Proof. If $v_j = v_{j-1}$, then e_j must be a loop and then (v_{j-1}, e_j, v_j) determines a polygon in the graph G_π.

If $v_j \neq v_{j-1}$,   then   // e_j is not a loop and

$$\pi 1 = \left(v_j, e_{j+1}, v_{j+1}, \ldots, e_k, v_k = v_0, e_1, v_1, \ldots, v_{j\,2}, e_{j-1}, v_{j-1}\right)$$

is a path from v_j to a *different* vertex v_{j-1}. Since e_j only occurs once in π, $\pi 1$ does not contain e_j. By Lemma 5.2.1, there is a simple path π_s from v_j to v_{j-1} (that is a subsequence of $\pi 1$). The path π_s has no repeated vertex so it can have no repeated edge, and in addition, π_s cannot contain e_j. This simple path together with e_j forms a polygon in the graph G_π.   ☐

// Construct an example of a closed path π, where G_π contains no polygons.
//X   If $\pi = (v_0, e_1, v_1, \ldots, e_k, v_k)$ is a closed path and k is <u>odd</u>,
//   then the graph G_π contains a polygon.

Lemma 5.2.3: If there are two distinct *simple* paths in G joining the same two vertices, then G contains a polygon.

Proof. Suppose

$$\pi 1 = \left(y, e_1, v_1, e_2, v_2, e_3, \ldots, e_k, z\right)$$
and
$$\pi 2 = \left(y, f_1, w_1, f_2, w_2, f_3, \ldots \ldots, f_q, z\right)$$

are two distinct *simple* paths in G joining the same two vertices, y and z.

Compare the sequences $\pi 1$ and $\pi 2$ entry by entry. They both begin with the same vertex, y. If $f_1 = e_1$, then w_1 must equal v_1; then if $f_2 = e_2$, then w_2 must equal v_2. But $\pi 1$ and $\pi 2$ are different sequences so there must be an entry where they differ, and the *first* entry where they differ must be an edge. Thus, there is an index j such that $0 <= j < k$ where (as sequences)   // Why is $j < k$?

$$\left(y, e_1, v_1, e_2, v_2, e_3, \ldots, e_j, v_j\right) = \left(y, f_1, w_1, f_2, w_2, f_3, \ldots, f_j, w_j\right)$$

but $e_{j+1} \neq f_{j+1}.$ // The next edges are not equal.

Let e denote the edge e_{j+1} and let x denote the vertex v_j.   // So x also equals w_j.

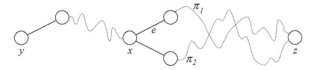

The edge e cannot occur twice in $\pi 1$ else $\pi 1$ would have a repeated vertex.

// namely, x

If e were in $\pi 2$ as f_r, then $r > j + 1$, and the end points of e_{j+1} would occur as w_r and w_{r+1}. But then the vertex $x = w_j$ would be repeated in $\pi 2$. Therefore, the edge e cannot occur at all in $\pi 2$.

Let π be the closed path constructed by following $\pi 1$ from y to z and then following $(\pi 2)^R$ from z back to y. The edge in e occurs exactly once in π. Therefore, by Lemma 5.2.2, G contains a polygon.   ⬜

If H is a polygon containing two distinct vertices y and z, then there are two different simple paths from y to z: one going "clockwise" around the polygon from y to z, and one going "anticlockwise" around the polygon from y to z.

The Most Important Ideas in This Section.

Suppose that $\pi = (v_0, e_1, v_1, e_2, v_2, e_3, \ldots, e_k, v_k)$ is a **path**. The path π is said to go **from v_0 to v_k** and to have **length** k. A graph is **connected** means for any two vertices y and z, there is a path from y to z. A path is **simple** if all the vertices are distinct, and **closed** if $v_0 = v_k$. A closed path is a **circuit** if $k >= 1$, and all the edges are distinct. G_π denotes the connected graph whose vertex set is the vertices of G occurring in π and whose edge set is the edges of G occurring in π. If π is a **circuit**, then all vertices in G_π have **even** degree. A **polygon** is a connected graph where all vertices have degree 2. If π is a **simple** path from v to w and e is an edge joining v and w that is not an edge in G_π, then G_π together with e forms a polygon. Polygons are used in the formal definition of trees, the subject of the next section.

5.3 Trees

Most of the graphs that we constructed earlier have been what were called "tree diagrams." They were all in one "connected" piece and had no polygons in them. We now formally define

a **tree** as a graph that is connected and has no polygons.

The diagram below is a "forest" of four trees.

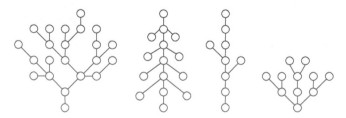

Trees are important data structures and useful conceptual models for a number of reasons: one of them being that *in a tree, there is a unique simple path from any vertex to any other vertex*.

                             // this follows from the contrapositive of Lemma 5.2.3.
In fact, a graph G is a tree if and only if there is a unique simple path between any two distinct vertices.

5.3.1   Traversals

In this subsection, we will describe two ways to travel through a (connected) graph so that every vertex is visited at least once, by *"growing" a spanning tree*. The first method will imitate the way we found our first solution to Tartaglia's Problem, and the second will imitate the way we found the best solution. By analyzing the algorithms for constructing the trees, we obtain a number of important theoretical results for free. Both methods may be used to search for a target vertex in a graph, and both methods may be modified to find a path joining any two target vertices.

We will describe the algorithms (throughout this chapter) in fairly informal terms and in a way that's independent of any "implementation" of graphs. Walkthroughs will be done using drawings, and we will focus on the "correctness" of the algorithms. Furthermore, we will use a "family tree" analogy and talk about the methods in terms of parents and children, ancestors, and descendants.

Depth-First Traversal   // a.k.a. Depth-First Search

The strategy is to pick a vertex y in G and construct a tree T rooted at y by always going to a new vertex (if possible). This method is called *depth-first* because we extend a *simple* path from y as "deeply" into the graph as we can, that is, until we reach a vertex x, where all neighbors of x are already in the path (and in T so far). Then we take a step back along the path and extend a second *simple* path from y as far as we can. Then we take a step back and extend another *simple* path from y as far as we can and so on.

The *input* is a graph G implemented in some manner, where the neighbors of a vertex can be quickly determined. The vertex objects might be given Boolean attributes like "InTree" and "Scanned".

                             // v has been "scanned" when all its neighbors are in T.
                             // And the *output* is also a graph, T.

Algorithm 5.3.1: Depth-First Traversal

Begin
    Mark all vertices "unscanned";
    Pick a vertex y and put it into the tree T as the root;   // y might be input.
    If (y has no neighbors) **Then**
        Mark vertex y as "scanned";
    End;
    Let $v = y$;   // v is the "current vertex".

```
While (y is still "unscanned") Do
    If (there is a neighbor w of v that is not yet in T) Then
        Call the edge joining w and v, the back-edge of w, BE(w);
        Add w and BE(w) to T;     // T stays connected & no polygon is created
        P(w) ← v;                              // call v the "Parent" of w
        v ← w;                                 // reset the "current vertex"

    Else                    // All the neighbors of v are already in the tree T.
        Mark vertex v as "scanned";
        If (v ≠ y) Then
            v ← P(v);
        End;                                   // reset the "current vertex"
                            // and "backtrack" 1 step along the path in T from y to v.
    End;     // the if-then-else statement
    End;     // the while-loop
End.
```

// Every vertex in *T*, except *y*, is put into *T* as the child of a (single) parent.
// (But please don't be confused by the mixed metaphor where neighbors get
// adopted as children.)

In the walkthrough below, we draw the input graph *G* using small circles for vertices, dotted lines for edges. We will use solid lines for edges that have been put into the tree *T*. And because there is a natural order of the vertices, we deal with the neighbors of a vertex in that order. After a vertex is marked "scanned," we draw it as a small square. In earlier tree diagrams the children of a vertex were drawn directly below that vertex; here, we'll draw the edge as an arrow from parent to child.

We also give a "table" indicating the progress in the traversal. It lists the unscanned vertices in the order they were added to *T*: v_j is the *j*-th vertex put into *T*, and P_j is the parent of v_j (provided that v_j is not the root of *T*).

Walkthrough with $V = \{0, 1, ..., 9\}$ and *y* set be 0.   // the "first" vertex in *V*

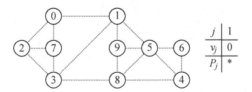

$v = 0 \quad w = 1$

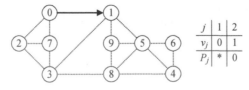

$v = 1 \quad w = 3$

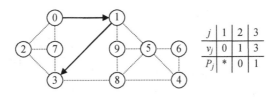

j	1	2	3
v_j	0	1	3
P_j	*	0	1

$v = 3 \quad w = 2$

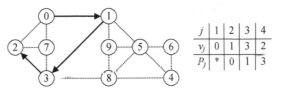

j	1	2	3	4
v_j	0	1	3	2
P_j	*	0	1	3

$v = 2 \quad w = 7$

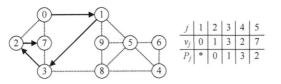

j	1	2	3	4	5
v_j	0	1	3	2	7
P_j	*	0	1	3	2

$v = 7 \quad$ 7 is marked "scanned" and removed from the list

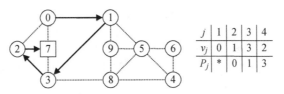

j	1	2	3	4
v_j	0	1	3	2
P_j	*	0	1	3

$v = 2 \quad$ 2 is marked "scanned" and removed from the list.

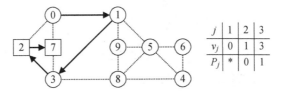

j	1	2	3
v_j	0	1	3
P_j	*	0	1

$v = 3 \quad w = 8$

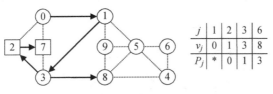

j	1	2	3	6
v_j	0	1	3	8
P_j	*	0	1	3

$v = 8 \quad w = 4$

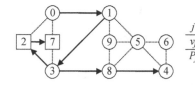
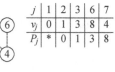

j	1	2	3	6	7
v_j	0	1	3	8	4
P_j	*	0	1	3	8

$v = 4$ $w = 5$

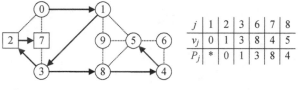

j	1	2	3	6	7	8
v_j	0	1	3	8	4	5
P_j	*	0	1	3	8	4

$v = 5$ $w = 6$

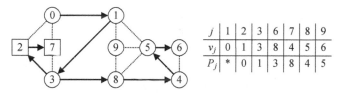

j	1	2	3	6	7	8	9
v_j	0	1	3	8	4	5	6
P_j	*	0	1	3	8	4	5

$v = 6$ 6 is marked "scanned" and removed from the list.

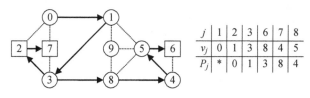

j	1	2	3	6	7	8
v_j	0	1	3	8	4	5
P_j	*	0	1	3	8	4

$v = 5$ $w = 9$

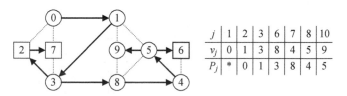

j	1	2	3	6	7	8	10
v_j	0	1	3	8	4	5	9
P_j	*	0	1	3	8	4	5

$v = 9$ 9 is marked "scanned" and removed from the list.

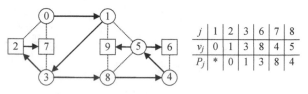

j	1	2	3	6	7	8
v_j	0	1	3	8	4	5
P_j	*	0	1	3	8	4

$v = 5$ 5 is marked "scanned" and removed from the list.

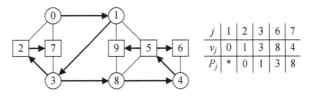

$v = 4$ 4 is marked "scanned" and removed from the list.

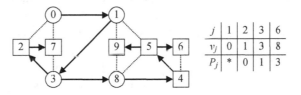

$v = 8$ 8 is marked "scanned" and removed from the list.

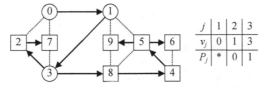

$v = 3$ 3 is marked "scanned" and removed from the list.

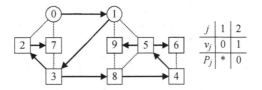

$v = 1$ 1 is marked "scanned" and removed from the list.

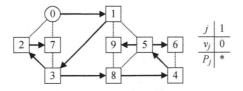

$v = 0$ 0 is marked "scanned" and removed from the list.

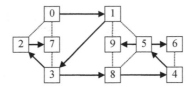

On every iteration of the body of the while-loop:

either   1. a (new) child w of v is added to T (as a leaf), and
         2. w is added to the "end" of the list L of unscanned vertices in T, and
         3. the current vertex v moves 1 step down the tree to w;

or   1. v is marked "scanned," and
         2. v is removed from the "end" of the list L, and
         3. the current vertex v moves 1 step up the tree to its parent (if v is not the root).

The list L of unscanned vertices in T is an example of a data structure called a *stack*: items are added to the "top" of the stack, and items are removed from the "top" of the stack – like plates or trays in a cafeteria.

                                // the right-hand end of the list is the "top" of the stack

Let n be the number of vertices in G. At most n vertices can be put into T and then later marked scanned, so there are at most $2n$ iterations of the while-loop performed. Thus, Algorithm 5.3.1 is certain to **terminate**.

Lemma 5.3.1: After k iterations of the while-loop, the current vertex v is (in the tree T and) at the top of the stack L, the root of T is at the bottom of the stack, and the vertex below any other vertex x is $P(x)$, the parent of x. However, after the last iteration the stack is empty.

Proof. // by Mathematical Induction on k.
Step 1. If $k = 0$, that is, before the loop is entered, y is the root of T and the only vertex in tree T and the list L, and y is the current vertex, v.
Step 2. Assume that $\exists$ an integer $q >= 0$ such that after q iterations of the while-loop, the current vertex v is at the top of L, the root of T is at the bottom of the stack, and the vertex below any other vertex x is $P(x)$, the parent of x.
Step 3. Suppose there is another iteration (because y is still unscanned), the $q+1^{st}$.

If there is a neighbor w of v that is not yet in T, then w is added to the tree. Because, only the current vertex is ever marked "scanned" and the current vertex is always in T, the new vertex w is still "unscanned." So w is put on the top of the stack, above v, the parent of w. Then the current vertex becomes w.

If all the neighbors of v are already in the tree T, then v is marked "scanned" and removed from the top of the stack L. If $v \neq y$, then the current vertex becomes $P(v)$, which is now the vertex at the top of the stack. If $v = y$, the stack is now empty and the while-loop terminates after this iteration.   ☐

From this Lemma it follows that:
1. When a parent is marked "scanned" <u>all its children</u> have already been marked "scanned".

2. When <u>any</u> vertex is marked "scanned" <u>all its descendants</u> (children of children, and children of children of children, and ...) have already been marked "scanned."
3. Since all the vertices in the tree are descendants of y, all the other vertices in T are marked "scanned" before y is marked "scanned" (and the algorithm ends).

We next prove the correctness of this algorithm: that every vertex in the graph is visited. We will show that if the **pre-condition** "G is connected" holds, then the **post-condition** "every vertex in G is visited at least once and put into the tree" also holds. This will follow from

Theorem 5.3.2: If there is a path from y to another vertex, z, then eventually z will be put into T by the Depth-First Traversal algorithm.

Proof. Suppose $\pi = (y = v_0, e_1, v_1, e_2, v_2, e_3, \ldots, e_k, v_k = z)$ is a path in G from y to z. We will prove each v_j in π will be put into T, by induction on j.

Step 1. If $j = 0$ then $v_j = y$, which is put into T as the root.

Step 2. Assume that $\exists$ an index q such that $0 <= q < k$ and v_q has been added to T.

Step 3. The algorithm does not terminate before v_q is marked "scanned," and when v_q is marked "scanned," all the neighbors of v_q including v_{q+1} must be in T. Thus, v_{q+1} will be put into T.

Therefore, eventually $z = v_k$ must be put into T.   ☐

If we are **searching for a path** in G from some **particular vertex y** to some other **particular vertex z**, we can use **Algorithm 5.3.1** to find such a path (or demonstrate that no such path exists). Start the algorithm by rooting the tree T at vertex y. Then run the algorithm until z enters the tree or until y is marked "scanned." In the first case, T contains a path π from y to z (and the list L is the vertex sequence of π). In the second case, z never enters T so (by the contrapositive of Theorem 5.3.2) there can be no path in G from y to z.

Backtracking is stepping up the tree toward the root, tracing the ancestry of a vertex back to y. Every vertex v except the root has a unique parent $P(v)$, grandparent $P(P(v))$, great-grandparent, and so on. Reversing this path gives the path in T from the root y to v.

Algorithm 5.3.1 provides a basis for proving some general results in graph theory.

Theorem 5.3.3: If G is connected then G contains a spanning tree.

Proof. Let T be the tree generated by **Algorithm 5.3.1** and suppose that T is rooted at vertex y. Let z be any other vertex in G. Since G is connected, there is a path from y to z. By **Theorem 5.3.2**, z will be put into T. Therefore T contains all the vertices of G so T is a spanning tree of G.   ☐

If T is the tree rooted at vertex y generated by **Algorithm 5.3.1**, then after y is put into T, each new vertex is added to T together with a new edge, its back-edge. Therefore, the number of edges in T equals the number of vertices minus one.

// This is true for all trees.

Lemma 5.3.4: **If H is the graph obtained from a tree T by adding one <u>new</u> edge e joining two vertices of T, then H contains a polygon, but only one polygon.**

Proof. Suppose that the new edge joins two vertices y and z. Any polygon in H must contain e.   // it can't be contained in T

If e is a loop at y, then y together with this loop is a polygon, and is the only polygon in H.

If y and z are distinct vertices, then there is a simple path π in T from y to z, and G_π together with e forms a polygon. If there were two polygons in H, they would determine two distinct simple paths from y to z in T. But Lemma 5.2.3 would imply that T contains a polygon. Since T contains no polygon, H contains exactly one polygon.   ⬚

Theorem 5.3.5: **In <u>any</u> tree, the number of edes equals the number of vertices minus 1.**

Proof. Suppose U is a given tree with n vertices. Let T be the tree generated by Algorithm 5.3.1 when $G = U$. Since U is connected, T is a subgraph of U that contains all n vertices and exactly $n - 1$ edges. Hence, U has <u>at least</u> $n - 1$ edges. If U had more edges, U would contain an edge e joining two vertices in T that is not contained in T. But then the polygon described in Lemma 5.3.4 would be contained in the tree U. Since U cannot contain a polygon, U must have exactly $n - 1$ edges.   ⬚

If G is not connected and consists of k connected "pieces" $G_1, G_2, \ldots, G_k$ but G does not contain a polygon, then each G_j is a tree.   // G is a forest of k trees.

So if G_j has n_j vertices, it must have $n_j - 1$ edges. Hence, if G has n vertices, it has $n - k$ edges.

// X If G is a connected graph with n vertices and has exactly $n - 1$ edges then G is
//   a tree.
// X If G has n vertices but no polygons and has exactly $n - 1$ edges then G is a tree.

Breadth-First Traversal   // a.k.a. Breadth-First Search

The strategy this time is to pick a vertex, y, in G and construct a tree T rooted at y by scanning the vertices v in T *in the order they were added* to T, where *scanning* a vertex x *means* adding to T <u>all</u> the neighbors of x that are not already in T.

// In Algorithm 5.3.1, a vertex was marked "scanned"
// after all its neighbors were in T.

This algorithm will grow a **broad**, bushy tree.

As before, vertices will be added to the tree one at a time as leaves. We include the height function in our description of the algorithm only to help in our analysis of its action; it is not necessary for constructing T.

The *input* is a graph G implemented in some manner, where the neighbors of a vertex can be quickly determined. The vertex objects might be given Boolean attributes like "InTree" and "Scanned." The *output* is also a graph, T.

Algorithm 5.3.2: Breadth-First Traversal

Begin
Mark all vertices "unscanned";
Pick a vertex y and put it into the tree, T, as the root;   // y might be input
$h(y) \leftarrow 0$;   // The "height" of the root is zero.

While (T has an unscanned vertex v) **Do**
                    // Find the "first" unscanned vertex in T and scan vertex v.

  While (there is a neighbor w of v that is not yet in T) **Do**
  Call the edge joining w and v, the back-edge of w, $BE(w)$;
  Add w and $BE(w)$ to T;   // T stays connected and no polygon is created.
  $P(w) \leftarrow v$;   // Call v the "Parent" of w.
  $h(w) \leftarrow h(v) + 1$;   // The height of a child is 1 more than
                                                  // the height of its parent.
  End;   // Now all the neighbors of v are in the tree T.

  Mark vertex v as "scanned";
  End;   // the outside while-loop
End.

In the walkthrough below, we draw the input graph G using small circles for vertices and dotted lines for edges. We will use solid lines for edges that have been put into the tree T. After a vertex is scanned, we draw it as a small square. In earlier tree diagrams, the children of a vertex were drawn directly below that vertex; here, we draw the edge as an arrow from parent to child.

We also give a "table" indicating the progress in the traversal. It lists the unscanned vertices in the order they were added to T: v_j is the j-th vertex put into T, and P_j is the parent of v_j (provided that v_j is not the root of T). The table also has a row for h_j, the "height" of vertex v_j which is equal the length of the (unique) simple path from the root vertex y to v_j. The height of the root is zero, and the height of any other vertex is 1 plus the height of its parent.

Walkthrough with $V = \{0, 1, \ldots, 9\}$ and y set be 0.   // the "first" vertex in V

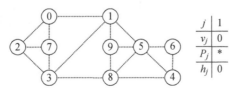

$v = 0$ and $h(v) = 0$

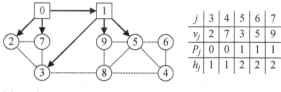

j	2	3	4
v_j	1	2	7
P_j	0	0	0
h_j	1	1	1

$v = 1$ and $h(v) = 1$

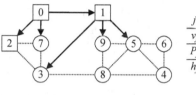

j	3	4	5	6	7
v_j	2	7	3	5	9
P_j	0	0	1	1	1
h_j	1	1	2	2	2

$v = 2$ and $h(v) = 1$.

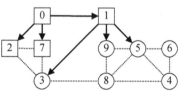

j	4	5	6	7
v_j	7	3	5	9
P_j	0	1	1	1
h_j	1	2	2	2

$v = 7$ and $h(v) = 1$

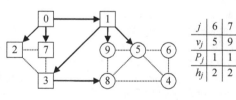

j	5	6	7
v_j	3	5	9
P_j	1	1	1
h_j	2	2	2

$v = 3$ and $h(v) = 2$

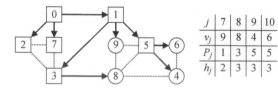

j	6	7	8
v_j	5	9	8
P_j	1	1	3
h_j	2	2	3

$v = 5$ and $h(v) = 2$

j	7	8	9	10
v_j	9	8	4	6
P_j	1	3	5	5
h_j	2	3	3	3

$v = 9$ and $h(v) = 2$

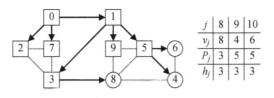

$v = 8$ and $h(v) = 3$

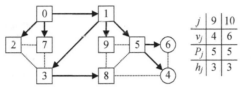

$v = 4$ and $h(v) = 3$

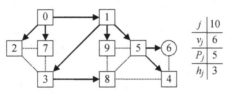

$v = 6$ and $h(v) = 3$

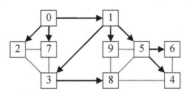

The list L of unscanned vertices in T is an example of a data structure called a *queue*: items are added to the "back" of the queue, and items leave from the "front" of the queue – like a line of customers at McDonald's or in a bank.

// The left-hand end of the list is the "front" of the queue.
// The right-hand end of the list is the "back" of the queue.

Let n be the number of vertices in G. At most n vertices can be put into T and then (later) scanned, so there are at most n iterations of the outer while-loop performed. Scanning a vertex v requires examining each of at most n neighbors of v. Thus, Algorithm 5.3.2 is certain to **terminate**.

Lemma 5.3.6: After k iterations of the outer while-loop, the next current vertex v is (in the tree T and) at the front of the queue L, and the children of v are at the back of the queue. However, after the last iteration, the queue is empty.

Proof. // by Mathematical Induction on k.

Step 1. If $k = 0$, that is, before the outer while-loop is entered, y is the root of T and the only vertex in tree T and the list L, and y is the current vertex, v. But v has no children in T yet.

Step 2. Assume that $\exists$ an integer $q >= 0$ such that after q iterations of the outer while-loop, the current vertex v is at the front of the queue L, and the children of v are at the back of the queue L.

Step 3. Suppose there is another iteration (because T still has an unscanned vertex v), the $q + 1^{st}$. The "first" unscanned vertex in T is at the front of the queue.

If there is a neighbor w of v that is not yet in T, then w is added to the tree as a child of v, and w is added to the back of the queue. When all the neighbors of v are in the tree T, then v has been scanned and is then removed from the front of the queue L. If the queue is now empty, the outer while-loop terminates after this iteration. $\square$

// Here, parents are scanned before their children are scanned –
// the opposite of the case for depth-first traversal.

Theorem 5.3.7: If there is a path from y to another vertex, z, then eventually z will be put into T by the Breadth-First Traversal algorithm.

Proof. Suppose $\pi = (y = v_0, e_1, v_1, e_2, v_2, e_3, \ldots, e_k, v_k = z)$ is a path in G from y to z. We prove that each v_j in π will be put into T by Mathematical Induction on j.

Step 1. If $j = 0$, then $v_j = y$, which is put into T as the root.

Step 2. Assume that $\exists$ an index q such that $0 <= q < k$, and v_q has been added to T as an unscanned vertex.

Step 3. The outer while-loop does not terminate before v_q is scanned, and after v_q is scanned, all the neighbors of v_q including v_{q+1} must be in T. Thus, v_{q+1} will be put into T.

Therefore, eventually $z = v_k$ will be put into T. $\square$

Lemma 5.3.8: After k iterations of the outer while-loop, the h-values of the vertices in the queue L are nondecreasing, and the h-value of the vertex at the back of the queue is $<= 1 +$ the h-value of the vertex at the front of the queue. However, after the last iteration, the queue is empty.

Proof. // by Mathematical Induction on k.

Step 1. If $k = 0$, that is, before the outer while-loop is entered, y is the only vertex in the list L and $h(y)$ has the value zero.

// The h-values of the vertices in the queue L are nondecreasing.
// The h-value of the vertex at the back of the queue is $<= 1 +$ the h-value of the
// vertex at the front of the queue.

Step 2. Assume that $\exists$ an integer $q \geq 0$ such that after q iterations of the while-loop, the h-values of the vertices in the queue L are nondecreasing, and the h-value of the vertex at the back of the queue is $\leq 1 +$ the h-value of the vertex at the front of queue.

Step 3. Suppose there is another iteration (because T still has an unscanned vertex v), the $q + 1^{st}$. The "first" unscanned vertex in T is at the front of the queue.

If there is a neighbor w of v that is not yet in T, then (w is added to the tree and) w is added to the back of the queue, and $h(w)$ is set equal to $h(v) + 1$. At this point, the h-values of the vertices in the queue L are still nondecreasing, and the h-value of the vertex at the back of the queue is equal to $1 +$ the h-value of the vertex at the front of the queue.

When all the neighbors of v are in the tree T, v is removed from the front of the queue L. If the queue is not empty, the h-values of the vertices in the queue L are still nondecreasing. Let v^* denote new vertex at the front of the queue and let w^* denote vertex at the back of the queue. Then

$$h(v) \;\; \leq h(v^*) \;\;\; \leq h(w^*), \qquad \text{// The } h\text{-values are nondecreasing.}$$
$$\text{and} \quad h(w^*) \leq h(v) + 1 \leq h(v^*) + 1.$$

Therefore, the h-value of the vertex at the back of the queue is $\leq 1 +$ the h-value of the (new) vertex at the front of the queue. If the queue is now empty then the outer while-loop terminates after this iteration. □

In fact, the whole sequence of h-values is nondecreasing, not just the current "active" portion that appears in the queue. Furthermore, for each vertex x in T, the unique path in T from y to x has length equal to $h(x)$.

$\qquad\qquad\qquad\qquad\qquad\qquad\qquad\qquad$ // This can be proved by induction too.

Lemma 5.3.9: If there is a path in G from y to vertex z of length k, then the (unique) path in the tree T from y to z has length $h(z) \leq k$.

Proof. // by Mathematical Induction on k

Step 1. If $k = 0$, then $y = z$, and the (trivial) path in T from y to z has length $h(z) = 0$.

Step 2. Assume that $\exists$ an index $q \geq 0$ such that if there is a path in G from y to z of length q, then the path in the tree T from y to z has length $h(z) \leq q$.

Step 3. Suppose that z is a vertex such that there is path π in G from y to z of length $q + 1$.

// We need to prove that the path in the tree T from y to z has length $h(z) \leq q + 1$.

If $\qquad\qquad\qquad\qquad \pi = \left(y = v_0, e_1, \ldots, e_q, v_q, e_{q+1}, v_{q+1} = z\right)$

then $\qquad\qquad\qquad \pi 1 = \left(y = v_0, e_1, \ldots, e_q, v_q\right)$

is a path in G from y to v_q of length q. By the inductive hypothesis in Step 2, the path in the tree T from y to v_q has length $h(v_q) <= q$. At some point, v_q is put into T and the queue L. Later, v_q is scanned. If the neighbor z of v_q is added to T when v_q is scanned, there will be a path in the tree T from y to z that has length $h(z) = h(v_q) + 1 <= q + 1$. Otherwise, the neighbor z of v_q must have been added to T before v_q is scanned and when some other vertex x was scanned. This other vertex x must have been added to T before vertex v_q was added. Therefore,

$$h(x) <= h(v_q).   \text{// All the h-values are nondecreasing.}$$

Then $h(z) = h(x) + 1 <= h(v_q) + 1 <= q + 1.$

Hence, there will be a path in the tree T from y to z with length $h(z) <= q + 1$. □

Theorem 5.3.10: If there is a path from y to another vertex z, then the path in the tree T (produced by Breadth-First Traversal) is a <u>shortest</u> path in G from y to z.

Proof.
If there is a path from y to another vertex z, then there is a <u>shortest</u> path π in G from y to z. If the length of π is k, by Lemma 5.3.9, the unique path π^* in the tree T from y to z has length $h(z) <= k$. But π^* is a path in G, so $h(z) >= k$. Thus, the path in the tree T is a shortest path.   □

The Most Important Ideas in This Section.
We described two ways to travel through a connected graph so that every vertex is visited, by *"growing" a spanning tree*. The first method was *Depth-First Traversal* and the second was *Breadth-First Traversal*. By analyzing the algorithms, we obtained a few important theoretical results about trees. Both methods may be used to search for a target vertex in a graph, and both methods may be modified to find a path joining any two target vertices.

We used a "family tree" analogy to talk about the methods in terms of parents and children, ancestors, and descendants. *Backtracking* is stepping up the tree toward the root, tracing the ancestry of a vertex back to y. Every vertex v except the root has a unique parent P(v), grandparent P(P(v)), great-grandparent, and so on. Reversing this path gives the path in T from the root to vertex v.

We also introduced two important data structures in our descriptions of the actions of these two algorithms: (1) a *stack* where items are added to the "top" of the stack and items are removed from the "top" of the stack – like plates or trays in a cafeteria, and (2) a *queue* where items are added to the "back" of the queue and items leave from the "front" of the queue – like a line of customers at McDonald's or in a bank.

5.4   Edge-Weighted Graphs

An "edge-weighted" graph is a graph $G = (V, E)$ together with a "weight" function on the edges

$$w: E \rightarrow R^+.$$

This function measures a "cost" of traversing an edge: the length in miles, the time in minutes, the price of a bus ticket, or whatever is relevant.

// All weights are *positive*.

A simple, though common, problem involving edge-weighted graphs is known as the **minimum connector problem**. Suppose a large company wants to rent or buy a secure communication network for its offices (or computers). Almost any pair of offices may be joined directly by a link in the system at a certain cost. But since communication may be made through intermediate offices, the company wants to select a subset of possible links so that there is a path from any office to any other office inside the company's network. As usual, the company wants the total cost of the system to be as small as possible.

// How do you decide which links to select?

Let's let G be the graph with a vertex for each office and an edge joining two vertices if a link between them can be constructed. The "weight" of an edge e is the cost of constructing the communication link e represents. We will have to determine a subgraph K of the graph G that contains all the vertices and where any two vertices are joined by a **path**; such a subgraph is a "connector". K must be a spanning subgraph which is connected. The cost of K is the sum of the costs of edges in K. A "minimum connector" is one that costs the least.

// There must be minimum connector if G is connected, even if it's all of G.

Since any connector K is a connected graph, we know it contains a spanning tree T. But then T itself would be a connector. Since all the edge weights are positive, a minimum connector must be a spanning tree. For this reason, a solution to this problem is often called a "minimum-cost spanning tree", an MST.

// There must be an MST if G is connected, even if it's all of G.

How do you find an MST? Here, the **Greedy Algorithm** works!

The *input* is a simple connected graph G implemented in some manner, together with a weight function $w: E \rightarrow \mathbf{R}^+$. And the *output* is also a graph, a minimum connector K.   // actually, a spanning tree with $n - 1$ edges

Algorithm 5.4.1: MST (Kruskal 1956)

Begin

Put all the vertices of G into K;   // K is spanning but has no
                                                          // edges yet and no polygons.

Sort the q edges by weight, and index them so that

$$w(e_1) <= w(e_2) <= w(e_3) <= \ldots <= w(e_q); \qquad // q = |E|.$$

```
For j = 1 To q Do
    If (K plus eⱼ does not contain a polygon) Then
        add eⱼ to K;                                 // K still has no polygons
    End;      // the if
  End;        // the for j-loop
End.
```

Input for Kruskal's Algorithm (an edge-weighted graph):

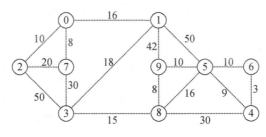

Walkthrough with $V = \{0, 1, \ldots, 9\}$.
The smallest edge-weight is 3. Add the edge joining 4 and 6.

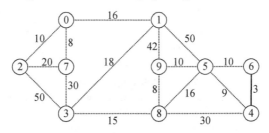

The next smallest edge-weight is 8. Both edges with weight 8 may be added.
                                       // one after the other, in any order

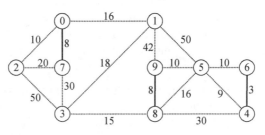

The next smallest edge-weight is 9. Add the edge joining 4 and 5.

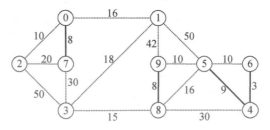

The next smallest edge-weight is 10. The edge joining 5 and 6 cannot be added. Both of the others with weight 10 may be added, one after the other, in any order.

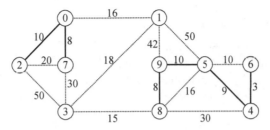

The next smallest edge-weight is 15. Add the edge joining 3 and 8.

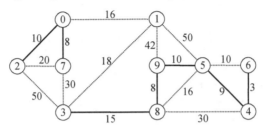

The next smallest edge-weight is 16. The edge joining 5 and 8 cannot be added. Add the edge joining 0 and 1.

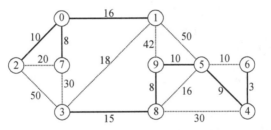

The next smallest edge-weight is 18. Add the edge joining 1 and 3.

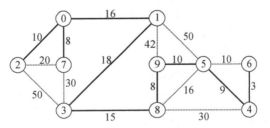

Now we have a spanning tree, and no other edges may be added (without creating a polygon).

There remain the questions: Is K always a tree? And is K always a minimum connector? The following theorems answer these questions. // "yes"

Theorem 5.4.1: K is a tree.

Proof. // We use an indirect argument to prove that K is connected.

Suppose that K is not connected. Then there must be distinct vertices x and y that are not joined by <u>any</u> path in K. Let W be all the vertices z in G such that there is a path from x to z in K. Since G is connected, there is a simple path in G from x to y, say

$$\pi = (x = v_0, e_1, v_1, e_2, v_2, e_3, \ldots, e_k, v_k = y).$$

Consider the vertices in π in turn. Vertex $x = v_0$ is in W, but vertex $y = v_k$ is not in W. Let j be the smallest index such that v_j is not in W. Then $0 < j <= k$, and e_j is an edge joining v_{j-1} in W and v_j which is not in W. Edge e_j is not in K.

Furthermore, there can be no (simple) path in K from v_{j-1} to v_j. Thus, K plus e_j cannot contain a polygon. But if K plus e_j doesn't contain a polygon, e_j would have been added to K by the algorithm. This contradiction implies that our supposition that K is not connected must be false. Therefore, K is connected but contains no polygons. K is a tree. ▯

If the edges put into K are $e[j_1]$, $e[j_2]$, $e[j_3]$, up to $e[j_p]$, then for $i = 1, 2, \ldots, p$, let E[i] denote the set $\{e[j_1], e[j_2], \ldots, e[j_i]\}$ and let E[0] denote the empty set. Also, let K_i denote the partially constructed subgraph of K with all the vertices of G but just the edges in E[i]. // K_0 has no edges.

Theorem 5.4.2: For $i = 0, 1, 2, \ldots, p$, there is an MST T_i that contains all the edges in E[i].

Proof. // by Mathematical Induction.
Step 1. If $i = 0$, then any MST contains all the edges in E[0], the empty set.
Step 2. Assume that $\exists$ an index k such that $0 <= k < p$ and $\exists$ an MST T_k that contains all the edges in E[k].
Step 3. // We need to find (or construct) an MST that contains all the edges in
// E[$k + 1$] $= \{e[j_1], \ldots, e[j_k], e[j_{k+1}]\}$.
Let e denote the next edge added to K, $e[j_{k+1}]$. If e is in T_k, then T_k is an MST that contains all the edges in E[$k + 1$] $= \{e[j_1], \ldots, e[j_k], e[j_{k+1}]\}$. Let T_{k+1} be the tree T_k.

If e is <u>not</u> in T_k, then T_k plus the edge e contains a unique polygon C by Lemma 5.3.4. The edges in C cannot all be edges in K else K would contain the polygon C. Therefore, C contains an edge f that is not in K but is in T_k. Let U be the graph obtained by adding e to T_k and removing edge f. Then U has the same number of edges as T_k does, $n - 1$, and U contains all the vertices that T_k does.

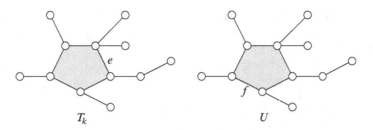

Since T_k plus the edge e contains one polygon C and f is in C, U contains no polygons. Because U contains no polygons and has $n - 1$ edges, U is a (forest consisting of a single) tree. In fact, U is a spanning tree of G.

// But is U of minimum cost?

The graph K_k was a subgraph of T_k, a tree that contains edge f. So K_k plus f does not contain a polygon. If $w(f) < w(e)$, then f would have been added to K before e was added. But f was never added to K. Therefore,

$$w(f) >= w(e).$$

But then the total weight of U
$$= \text{the total weight of } T_k + w(e) - w(f)$$
$$<= \text{the total weight of } T_k.$$

Since T_k is an MST, U cannot have a smaller total weight, so U is an MST that contains all the edges in $E[k + 1] = \{e[j_1], e[j_2], \ldots, e[j_k], e[j_{k+1}]\}$. Let T_{k+1} be the tree U.

Therefore, there is an MST T_{k+1} that contains all the edges in $E[k + 1]$. []

Since T_p is an MST that contains all the edges in K, and K itself is a spanning tree, K must equal T_p. Thus, the output of Kruskal's Algorithm is an MST.

// This implies that we can "exit" the for-loop if we detect that K (now) has $n - 1$
// edges.

// This gives a third algorithm for finding a spanning tree in a connected graph.

// Checking that "K plus e_j does <u>not</u> contain a polygon" may appear difficult to
// program, but the difficulty can be avoided. Look up Prim's Algorithm which
// finds an MST by growing a spanning tree.

5.4.1 Shortest Paths

If $\pi = (v_0, e_1, v_1, e_2, v_2, e_3, \ldots, e_k, v_k)$ is a path in G, then the **weight of** π is given by

$$w(\pi) = \sum_{j=1}^{k} w(e_j) \qquad \text{// the sum of the weights of edges in } \pi$$

A "lightest" (or fastest or cheapest) path from a vertex *x* to a vertex *y* is a path from *x* to *y* with smallest possible weight. It's much more common to refer to **shortest** paths than "lightest" paths, so we'll do that.

// If all edges "weigh" 1 unit, a shortest path is one with fewest edges.

The Most Important Ideas in This Section.
An edge-weighted graph is a graph $G = (V, E)$ together with a function on the edges $w: E \rightarrow \mathbf{R}^+$. In such a graph, the **minimum connector problem** is to find a spanning subgraph which is connected and has minimum total weight. There must be a minimum connector if G is connected. We prove that for this problem, the **Greedy Algorithm** produces the best possible solution, a "minimum-cost spanning tree."

In an edge-weighted graph, the weight of a path π is the sum of the weights of edges in π; a **shortest path** from a vertex *x* to a vertex *y* is a path from *x* to *y* with smallest possible weight.

But it may be that in a particular context, the cost of going from a vertex *x* to a neighbor *y* is not the same as going from *y* back to *x*. (If *y* were uphill or upstairs from *x*, the cost in time or energy would be different.) But we can fairly easily expand our model to accommodate this *asymmetry* due to the *direction* we're moving along certain edges. We continue the investigation of algorithms to find shortest paths in the more general model of **directed graphs** in the next chapter. That chapter begins with definitions parallel to those for "undirected" graphs.

5.5   Drawing and Coloring

The **Utilities Problem** is classic: Can the utilities Water, Gas and Electricity all be connected to each of three houses in such a way that no two utility lines cross each other?   // use pencil

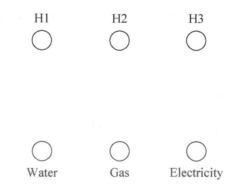

Real utility lines use 3 dimensions to avoid crossings. But can you make a blue-print
or schematic that does not have two utility connections that cross? This section
discusses drawings of graphs and colorings of graphs. It ends with some history of
the famous Four Color Problem and the critical role computers played in solving that
problem. But before that we will look at a few common classes of graphs.

The *complete graph* on n vertices, denoted K_n, is the (simple) graph where every
pair of distinct vertices is joined by an edge.

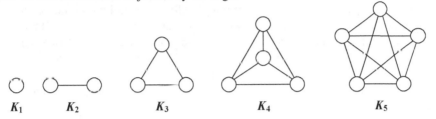

K_1 K_2 K_3 K_4 K_5

In any graph, a subset of vertices where every pair of distinct vertices is joined by an
edge is called a *clique*. The complementary idea is independence. In any graph, a
subset of vertices where no pair of distinct vertices is joined by an edge is called an
independent set.

5.5.1 Bipartite Graphs

A graph $G(V, E)$ is *bipartite* means that V may be partitioned into two independent sets.
So in a bipartite graph each vertex is in exactly one of two sets, R and B say, where
every edge joins a vertex in R to a vertex in B. Bipartite graphs model connections
between objects of different types: utilities and houses, tasks and workers, jobs and
machines, or TA's and courses. The *complete bipartite graph* with p vertices in R and
q vertices in B, denoted $K_{p,q}$, is the (simple) graph where every vertex in R is joined by
an edge to every vertex in B.   // The utilities problem is about "drawing" $K_{3,3}$.

A bipartite graph is sometimes said to be "2-colorable" in the sense that each vertex
is colored Red or Blue and every edge joins a Red vertex and a Blue one. A graph is
said to be *k-colorable* if V may be partitioned into k independent sets: $C_1, C_2, \ldots, C_k$.

Coloring a graph is assigning each vertex one of several colors so every edge
joins a vertex to one with a different color, and no edge joins two vertices of the
same color.   // Is every tree 2-colorable?

It is clear that if every component of a disconnected graph is bipartite, then the
whole graph is bipartite.

An edge e in a connected graph G is called a *bridge* (or an *isthmus*) if removing
e leaves a disconnected graph. Suppose that e is a bridge in G joining vertices a and b.

Let H be the disconnected graph obtained from G by removing edge e. Then H has exactly two connected "pieces", called *components*: H_a containing vertex a, and H_b containing vertex b. If both H_a and H_b are bipartite then so is all of G, because if a is colored Red in H_a and b is colored Blue in H_b then all vertices in G are colored so that no edge joins two vertices of the same color. On the other hand, if both a and b are given the same color, reverse the colors in H_b so a and b now have opposite colors and again, all vertices in G are colored so that no edge joins two vertices of the same color. Finally, if G has k bridges, removing all of the bridges would produce a bridgeless, disconnected graph with several components $C_1, C_2, \ldots, C_n$. If all these bridgeless components were bipartite, the bridges could be added back one at a time (and the colorings revised if necessary) so that all vertices in G are colored Red or Blue and no edge joins two vertices of the same color.

If G is bipartite, and $\pi = (v_0, e_1, v_1, e_2, v_2, e_3, \ldots, e_{2k+1}, v_{2k+1})$ is a path of odd length in G, then v_0 and v_{2k+1} must have different colors.

// This can be proved by induction on k.

Thus, if G is bipartite, then G cannot contain a closed path of   // $v_0 \neq v_{2k+1}$
odd length, and it also cannot contain an odd polygon. In fact

Theorem 5.5.1. A graph G is *bipartite* if and only if G does <u>not</u> contain an odd polygon.

Proof. We've just seen that if G is bipartite, then G cannot contain an odd polygon.

Suppose that G does not contain an odd polygon.   // How do we 2-color G? We will show how to 2-color each connected component of G, because then G itself will be 2-colored.

Suppose now that we want to 2-color a connected graph G that does not contain an odd polygon. Choose any vertex y of G. If Algorithm 5.3.2 (Breadth-First Traversal) were applied to G, it would produce a spanning tree T of G that contains a shortest path in G from y to every vertex z in G.

Color z <u>Red</u> if the path in T from y to z has <u>even</u> length, and
color z <u>Blue</u> if the path in T from y to z has <u>odd</u> length.

We can prove no pair of Red vertices is joined by an edge indirectly. Assume v and w are Red vertices that are joined by an edge e. // Must G contain an odd polygon? Then there is an even path π_1 from y to v and an even path π_2 from y to w. Consider the closed path

> $\pi = \pi_1$ from y to v
> > followed by the edge e from v to w
> > > followed by π_2^R from w in reverse back to y.

Since both π_1 and π_2 are even, π is a closed path of odd length in G. If π is not a polygon, it must contain a repeated vertex. Suppose a vertex is repeated in π (other than the last vertex repeating the first). That is, suppose $v_i = v_j$, where $0 <= i < j < k$ or $0 < i < j <= k$. Then

$$\pi = (v_0, e_1, v_1, \ldots, v_{i-1}, e_i, v_i, e_{i+1}, v_{i+1}, \ldots, v_{j-1}, e_j, v_j = v_i, e_{j+1}, v_{j+1}, \ldots, e_k, v_k = v_0).$$

Let $\qquad\qquad \pi_4 = (v_i, e_{i+1}, v_{i+1}, \ldots, v_{j-1}, e_j, v_j = v_i)$

and $\pi_3 = (v_0, e_1, v_1, \ldots, v_{i-1}, e_i, \qquad\qquad v_i = v_j, e_{j+1}, v_{j+1}, \ldots, e_k, v_k = v_0)$

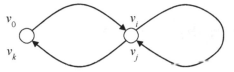

Both π_3 and π_4 are closed paths, and exactly one, π' say, is of odd length. This process may be repeated until we obtain $\pi*$, a closed odd path, which has no repeated vertices (except the last repeating the first). Therefore, $\pi*$ is an odd polygon in G. Since G has no odd polygon, no two Red vertices are joined by an edge.

Similarly, we can prove no two Blue vertices, v and w, are joined by an edge. If they were adjacent, then following the previous argument, both paths π_1 and π_2 are odd, so again the path constructed from them, π, is a closed path of odd length in G. And, as before, the existence of such a path implies that there is an odd polygon in G. □

Theorem 5.5.1 completely characterizes 2-colorable graphs. No one (yet) knows such a characterization for 3-colorable graphs. Deciding whether or not an arbitrary graph is 3-colorable is known to be NP-complete like the Hamilton Path problem.

// and worth $1M

5.5.2 Planar Graphs

Just as we did for set theory, we will take an intuitive approach to the topology of the plane and its geometry. A graph $G(V, E)$ is **planar** means that it can be "drawn" on a sheet of paper where the vertices are points (drawn as small circles, so they are visible) and the edges are drawn as "lines" joining the points that are the end-vertices of the edge, but where no two lines cross (at an interior point). We assume the lines representing edges are straight segments or nice, smooth curves that do not intersect (cross) themselves.

// Fary's Theorem from 1948 asserts that for a <u>simple</u> planar graph
// the vertices can be positioned so all edges can be drawn as
// straight line segments.

The complement of such a drawing partitions the <u>remainder of the plane</u> into connected regions called **faces**. For instance, K_4 can be drawn as

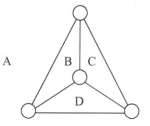

producing the 4 faces: A, B, C and D. A drawing of a path of length 3 is

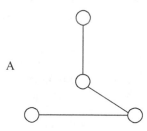

but only one face is produced. Every tree can be drawn in the plane producing
exactly one face.

      // This can be proved by induction on $|V|$ or giving an algorithm for
      // placing the vertices and joining adjacent vertices by straight lines.

The drawing of a polygon with 5 vertices (and 5 edges) has exactly 2 faces, one

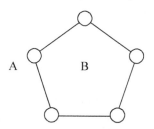

"interior" to the polygon and one "exterior" to it. Furthermore, each edge in the
polygon has the interior face on one side of it and the exterior face on the other side.

    In general, we will <u>assume</u> that any drawing of a polygon, P, produces a closed
curve in the plane that partitions the remainder of the plane into exactly two
regions, the "interior" of the curve and the "exterior" of it. Furthermore, if P is a
polygon in some planar graph G then in any drawing of G, each edge in the polygon
is on the "boundary" of a face in the interior of P on one side of it and is on the
"boundary" of a face in the exterior of P on the other side.

    If $G = (V, E)$ is **bridgeless** and **connected** then every edge is in some polygon,
and we <u>assume</u> that the boundary of every face f in any drawing of G is a circuit
(though not necessarily a polygon), and the "size" of the face f is the length of that
circuit. Then, since every edge is in the boundary of exactly 2 faces, we have

$$2|E| = \text{the sum of the sizes of the faces in any drawing of } G. \qquad (5.5.1)$$

// Draw a graph consisting of two disjoint triangles, one inside the other.
// Is the "boundary" of every face a circuit?
// Draw a graph consisting of two triangles one inside the other, that share one vertex.
// Is the "boundary" of the every face a polygon?

Given a drawing of a bridgeless graph G, $d(G)$, we can construct a drawing of another planar graph known as the **dual** of $d(G)$ by placing a vertex inside each face and for each edge e, joining the faces on either side of e by a new edge e'. Different drawings of the same graph may produce different dual graphs. For example,

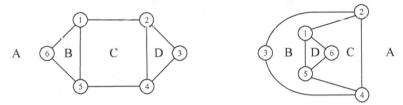

are both drawings of the same graph, but the dual drawings are quite different.

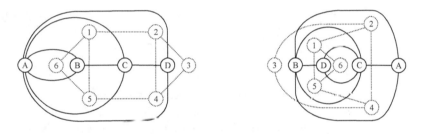

// Is the original drawing, the dual of the dual?
// Is every bridgeless planar graph the dual of a planar graph?
// Do all drawings of any particular graph have the same number of faces?

Given a drawing of graph $G(V, E)$ let F denote the set of faces in the drawing.

Theorem 5.5.2: Euler's Formula. If G is a connected, planar graph then in any drawing of G

$$|V| - |E| + |F| = 2. \qquad (5.5.3)$$

Proof. Let $G(V, E)$ be any connected, planar graph with $n \in \mathbf{P}$ vertices that has been drawn in the plane. We will establish the formula by Mathematical Induction on $|E|$.

Since G is connected, it contains a spanning tree (Theorem 5.3.3) and so it has at least $(n - 1)$ edges (Theorem 5.3.5).

Step 1.   If $|E| = n - 1$ then G itself is a tree and the drawing will have only one
          face. Hence,

$$|V| - |E| + |F| = n - (n - 1) + 1 = 2.$$

Step 2.   Assume that $\exists\ q >= n - 1$ such that any drawing of a connected, planar
          graph with n vertices and q edges satisfies Euler's formula.

Step 3.   Let G^* be any connected, planar graph with n vertices and $q + 1$ edges
          drawn in the plane, as $d(G^*)$. G^* has too many edges to be a tree, so G^*
          must contain a polygon, P. Let e be any edge in that polygon.

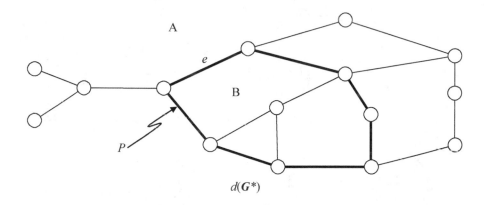

$d(G^*)$

The edge e is on the boundary of exactly 2 faces, A and B, in the drawing. If H is
the graph obtained from G^* by removing e, and $d(H)$ is obtained from $d(G^*)$ by
erasing the line representing e, then $d(H)$ is a drawing of a connected, planar graph
with n vertices and q edges, and so satisfies Euler's formula.   // by Step 2
    But the two faces, A and B, are now merged into one region.

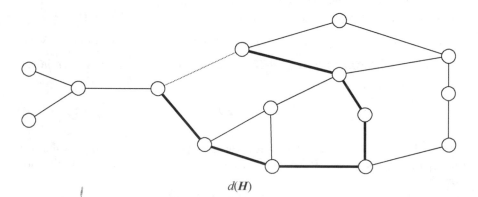

$d(H)$

Thus, $2 = |V(H)| \quad - \quad |E(H)| \qquad + \quad |F(d(H))|$
$= |V(G^*)| \quad - \{|E(G^*)| - 1\} \quad + \{|F(d(G^*))| - 1\}$
$= |V(G^*)| \quad - \quad |E(G^*)| \qquad + \quad |F(d(G^*))|$

Therefore, G^* satisfies Euler's Formula and the Theorem is proved.   ☐

One of our first examples in this book was the Cake Cutting Conundrum, where N points were marked on the circumference of a circular cake and then the cake was cut by making straight cuts joining all pairs of these points. How many pieces will N points produce?

The diagram below illustrates the case of 6 points.

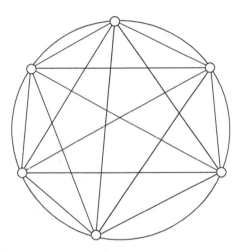

The number of pieces 6 points produce is 30 or 31; it's only 30 if the three "longest" cuts all pass through the center of the circle. The *maximum* number of pieces, $Q(N)$, will be obtained when *no three cuts pass through a single point*.

Chapter 3 contained a proof by Mathematical Induction that

$$\forall N \in \mathbf{P}, \; Q(N) = \binom{N}{4} + \binom{N}{2} + \binom{N}{0} \quad \text{where} \quad \binom{N}{k} = 0 \text{ unless } 0 <= k <= N.$$

This can also be proved using Euler's Formula.

Imagine a diagram like the one above for N points, but with a vertex at each of the given points on the circumference, and a vertex introduced wherever two cuts meet. This revised diagram is the drawing of a connected, planar graph, and therefore Euler's formula holds.

Recall the number of points where two cuts cross is $\binom{N}{4}$. In this revised diagram

$$|V| = N + \binom{N}{4}.$$

We can determine the number of edges using Eq. 5.1.1:

$$2|E| = \text{the sum of the degrees of all the vertices.}$$

Each "crossing" vertex will have degree 4, and each "point" vertex is the end of a cut to each of the other $N-1$ "point" vertices, and it is also the end of the 2 arcs on the circle next to it. Therefore

$$2|E| = \binom{N}{4} \times 4 + N \times (N+1) = \binom{N}{4} \times 4 + N \times (N-1) + 2N$$

and so

$$|E| = \binom{N}{4} \times 2 + \frac{N \times (N-1)}{2} + N = \binom{N}{4} \times 2 + \binom{N}{2} + N.$$

Hence, $|F| = 2 + |E| - |V| = \binom{N}{4} + \binom{N}{2} + 2.$

Since Euler's Formula counts the exterior face, the number of pieces of cake produced is

$$Q(N) = |F| - 1 = \binom{N}{4} + \binom{N}{2} + \binom{N}{0}.$$

Next we will use Euler's Formula to prove that $K_{3,3}$ and K_5 are not planar, by the indirect method of argument.

Suppose that $K_{3,3}$ were planar and that $d(K_{3,3})$ is a drawing of $K_{3,3}$. Then

$$2 = |V(K_{3,3})| - |E(K_{3,3})| + |F(d(K_{3,3}))| = \{3+3\} - \{3 \times 3\} + |F(d(K_{3,3}))|$$

and so $\qquad\qquad |F(d(K_{3,3}))| = 2 + 9 - 6 = 5.$

Since no edge of $K_{3,3}$ is a bridge, each line in the drawing is on the boundary of exactly two faces. Let f_j denote the number of faces in $d(K_{3,3})$ with exactly j edges on its boundary. Then

$$\left|F\left(d\left(K_{3,3}\right)\right)\right| = f_1 + f_2 + f_3 + f_4 + f_5 + f_6 + \ldots + f_9$$

since no circuit could have more than $|E(K_{3,3})| = 9$ edges. But $f_1 = 0$ because there are no loops; $f_2 = 0$ because there are no parallel edges; and because there are no odd circuits in $K_{3,3}$

$$\left|F\left(d\left(K_{3,3}\right)\right)\right| = f_4 + f_6 + f_8.$$

Counting the edges in the boundary circuits, each edge is counted exactly twice, so

$$2 \times \left|E\left(d\left(K_{3,3}\right)\right)\right| = 4 \times f_4 + 6 \times f_6 + 8 \times f_8$$
$$>= 4 \times f_4 + 4 \times f_6 + 4 \times f_8 = 4 \times \left|F\left(d\left(K_{3,3}\right)\right)\right| = 4 \times 5.$$

This asserts that $18 >= 20$ which is false, and therefore $K_{3,3}$ is not planar.

Suppose that K_5 were planar and that $d(K_5)$ is a drawing of K_5. Then

$$2 = |V(K_5)| - |E(K_5)| + |F(d(K_5))| = 5 - 10 + |F(d(K_5))|$$

and so $\qquad |F(d(K_5))| = 2 + 10 - 5 = 7.$

Since no edge of K_5 is a bridge, each line in the drawing is on the boundary of exactly two faces. Let f_j denote the number of faces in $d(K_{3,3})$ with exactly j edges on its boundary. Then

$$|F(d(K_5))| = f_1 + f_2 + f_3 + f_4 + f_5 + f_6 + \ldots + f_{10}$$

since no circuit could have more than $|E(K_5)| = 10$ edges. But $f_1 = 0$ because there are no loops, and $f_2 = 0$ because there are no parallel edges.

Counting the edges in the boundary circuits, each edge is counted exactly twice, so

$$2 \times |E(d(K_5))| = 3 \times f_3 + 4 \times f_4 + \ldots + 10 \times f_{10}$$
$$>= 3 \times f_3 + 3 \times f_4 + \ldots + 3 \times f_{10} = 3 \times |F(d(K_5))| = 3 \times 7.$$

This asserts that $20 >= 21$ which is false, and therefore K_5 is not planar.

A *subdivision* of a graph is formed by replacing its edges by paths of length one or more. It is fairly clear that if H is a subdivision of G, then H is planar if and only if G is planar. It's also clear that if H is a subgraph of a planar graph G, then H is planar. So if a graph G contains a subdivision of $K_{3,3}$ or K_5 then G is not planar.

We can use this to demonstrate that the Petersen Graph (1898) is non-planar.

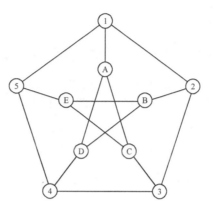

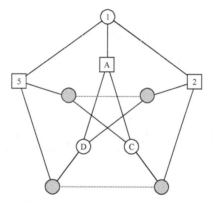

The two graphs, $K_{3,3}$ and K_5, are the fundamental non-planar graphs. In 1930 Kuratowski proved

A graph G is planar if and only if it does <u>not</u> contain a subdivision of $K_{3,3}$ or K_5.

5.5.3   Some History of the Four Color Theorem

In about 1850, Francis Guthrie asked his former mathematics professor (Augustus de Morgan) about coloring the map of the English counties with 4 colors so any two sharing a common boundary line had different colors. He asked "were 4 colors enough to color any map?" De Morgan wrote to Professor William Rowan Hamilton in Dublin in 1852 asking him about "Guthrie's Problem".

The assertion that 4 colors were enough to color any map became known as "the Four Color Conjecture". Much of the subsequent work on the conjecture focused on 4-coloring the vertices of the planar graph obtained as the dual of a map.

In 1879 Alfred Kempe announced he had a proof of the conjecture, and published a proof of the Four Color Theorem in the American Journal of Mathematics that same year. But the Theorem reverted to the Four Color Conjecture in 1890 when Percy Heawood showed that Kempe's proof was wrong and only demonstrated that 5 colors were enough.

The conjecture remained just that until 1976 when a complete proof of the Four Color Theorem was announced by Kenneth Appel and Wolfgang Haken. The details of their proof appeared in two articles in 1977. These were read very carefully, in part because of the disaster of Kempe's "proof", but mainly because their proof involved about 2000 cases, many of which were generated by computers and resolved by computer programs consuming an estimated 1200 hours of computing time. Their proof has survived 40 years of close scrutiny and some improvements have been made to the algorithms and the analysis of the cases, which now number fewer than 1500. Using a computer to assist with a mathematical proof in this manner was a breakthrough in the application of computing to theoretical mathematics.

> **The Most Important Ideas in This Section.**
> Both bipartite and planar graphs are characterized by forbidden subgraphs. A graph G is **bipartite** if and only if G does not contain an odd polygon. A graph G is **planar** if and only if it does not contain a subdivision of $K_{3,3}$ or K_5. Euler's Formula $|V| - |E| + |F| = 2$ was proved for connected, planar graphs. The section ends with some history of the Four Color Theorem and recognition of the role computers and computer programs play in modern Mathematics.

Exercises

1. Consider the following graph, G:

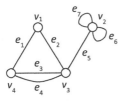

a. What is the vertex set of G?
b. What is it the edge set of G?
c. Which edges in G are loops?
d. What are the neighbours of v_1?
e. What is the degree of v_3?
f. Which edges in G are parallel?
g. Is G a simple graph? Why or why not?

2. Consider the following graph, **H**:

  a. What is the vertex set of **H**?
  b. What is it the edge set of **H**?
  c. Which edges in **H** are loops?
  d. What are the neighbours of v_1?
  e. What is the degree of v_3?
  f. Which edges in **H** are parallel?
  g. Is **H** a simple graph? Why or why not?
3. Draw a graph that has the stated properties:
  a. One vertex with degree 8
  b. Two vertices, each with degree 3
  c. Three vertices with degrees 1, 2, 3
  d. Four vertices with degrees 1, 2, 3 and 4
4. Try to draw a graph with 3 vertices of odd degree. Use Eq. 5.1.1 to prove that in any graph, the number of vertices with odd degree must be even.
5. Draw a graph that has both an Euler circuit and a Hamilton circuit.
6. Draw a graph that has an Euler circuit and a Hamilton circuit that are not the same.
7. Draw a graph that has an Euler circuit but does not have a Hamilton circuit.
8. Draw a graph that has a Hamilton circuit but does not have an Euler circuit.
9. Draw a graph that has a neither an Euler circuit nor a Hamilton circuit.
10. Which of the following graphs include an Euler Tour? If the graph includes an Euler Tour, list one of the paths that describes such a tour. If the graph does not include an Euler Tour, justify why no such path exists.

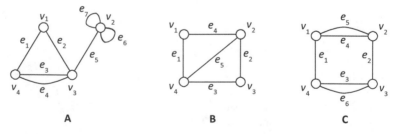

11. Which of the graphs from the previous question includes a Hamilton circuit? If such a circuit exists, list one of the paths that describes such a circuit.
12. Demonstrate that **G** has an Euler Tour if and only if **G** is connected and has $<= 2$ vertices of odd degree.
13. Look up Fleury's Algorithm to find an Euler Tour on the web.
14. Consider the path $\pi_1 = (v_1, e_1, v_2, e_2, v_3, e_3, v_2, e_4, v_4)$.
  a. Is π_1 a simple path? Why or why not?

    b. Is π_1 a closed path? Why or why not?

    c. Is π_1 a circuit? Why or why not?

15. Consider the path $\pi_2 = (v_1, e_1, v_2, e_2, v_1, e_3, v_3, e_4, v_1)$.

    a. Is π_2 a simple path? Why or why not?

    b. Is π_2 a closed path? Why or why not?

    c. Is π_2 a circuit? Why or why not?

16. Construct an example of a path π joining two distinct vertices v and w where there are two (or more) subsequences that are simple paths joining vertices v and w.

17. Why must a <u>shortest</u> path (in the sense of having fewest edges) from one vertex to another vertex be a *simple* path?

18. Why is a polygon with $n >= 2$ vertices the subgraph determined by $2 \times n$ distinct circuits?

19. Prove that there is no simple graph that has 4 vertices with degrees 1, 1, 3 and 3.

20. Prove that if π is a *simple* path from v to w, and e is an edge joining v and w that is not an edge not in G_π, then G_π together with e forms a polygon.

21. Prove that if $\pi = (v_0, e_1, v_1, \ldots, e_k, v_k)$ is a closed path and k is <u>odd</u>, then the graph G_π contains a polygon.

22. Draw a graph that contains n vertices and $n - 1$ edges that is not a tree.

23. Provide three different spanning trees for the following graph:

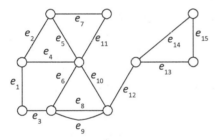

24. Prove that G is a tree if and only if there is a unique simple path between any two distinct vertices.

25. Prove that if all vertices have degree $>=2$, then G contains a polygon.

26. Prove that a tree with $>= 2$ vertices must have $>= 2$ *pendant* vertices, which are vertices of degree 1.

27. Prove that if T is a tree with a vertex of degree $k > 1$, then T must have $>= k$ pendant vertices.

28. A "disconnected" graph $G = (V, E)$ can be "divided" into connected "pieces" called ***components***, say $G_1, G_2, \ldots G_k$, where the vertex set V is partitioned into the vertex sets of the subgraphs, G_j. When all the components are trees, G is called a *forest*. Prove the following:

    (a) A graph G is a *forest* if and only if G has no polygons.

    (b) If G is a *forest* of q trees, then $|E| = |V| - q$.

29. Prove that if G has n vertices and $n - 1$ edges but no polygons, then G is a tree.
30. Prove that if a graph, G, includes an Euler circuit then G is connected.
31. Prove that if G is connected and has n vertices and exactly $n - 1$ edges, then G is a tree.
32. (a) Prove that if e is an edge in G then either e is a **bridge** or e is contained in a polygon.
    (b) Prove that G is a tree if and only if every edge is a bridge.
33. Prove that if there is a path from y to another vertex, z, then eventually z will be put into T, the tree produced by Algorithm 5.3.2 Breadth-First Traversal.
34. Prove that the paths in T (the tree produced by Algorithm 5.3.2 Breadth-First Traversal) are shortest paths from y (the root) to all other vertices in T.
35. Consider the following algorithm that "visits" every vertex in a connected graph.

   Algorithm: *Random Traversal*

```
Begin
     Mark all vertices "unscanned";
     Pick a vertex y and put it into the tree, T, as the root;      // y might be input
     While (T has an unscanned vertex) Do
         Find any unscanned vertex v in T;
         If (there is a neighbor of v that is not yet in T) Then
             Find any neighbor w of v that is not yet in T;
             Call the edge joining w and v, the back-edge of w, BE(w);
             Add w and BE(w) to T;  // T stays connected & no polygon is created.
             P(w) ← v;                                 // Call v the "Parent" of w.
         Else // now all the neighbors of v are in the tree T
             Mark vertex v as "scanned";
         End;  // the if-statement
     End;     // the while-loop    (now, all vertices in T have been "scanned")
End.
```

   (a) Prove that this algorithm terminates.
   (b) Prove that if there is a path $\pi = (y = v_0, e_1, v_1, e_2, v_2, e_3, \ldots, e_k, v_k = z)$ from y to another vertex, z, then eventually z will be put into T by this algorithm.
      // Hint: Prove each v_j in π will be put into T, by induction on j.
   (c) Explain how **Depth-First Traversal** uses a stack to store the unscanned vertices in T and takes v from the top of the stack when finding any unscanned vertex.
   (d) Explain how **Breadth-First Traversal** uses a queue to store the unscanned vertices in T and takes v from the front of the queue when finding any unscanned vertex.

36. Determine whether or not each of the following graphs is bipartite.

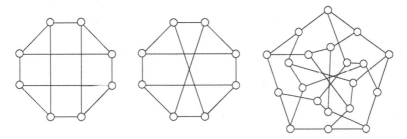

37. Determine whether or not each of the graphs in the previous question is planar.
38. Show that each of the graphs in question 36 has a Hamilton Circuit.
39. Suppose that G is a bridgeless, connected, planar graph.
    (a) Prove that if G is a *simple* graph then the average degree in G is < 6.
    (b) Construct an example of a bridgeless, connected, planar graph G^* (with loops and/or parallel edges) where the average degree in G^* is $>= 6$.
40. Suppose that G is a bridgeless, connected, planar graph.
    (a) Prove that if G has no vertex of degree < 3, the average size of a face in any drawing of G is < 6.
    (b) Construct a drawing of a bridgeless, connected, planar graph G^* (with some vertices of degree 2) where the average face size in G^* is $>= 6$.
41. A (3-dinensional) *polyhedron* is a (convex) crystal-like structure with flat faces, straight edges where two faces meet, and vertices at the corners where 3 or more faces meet. For example consider a cube: it has 6 square faces, 12 edges and 8 vertices. We can embed the surface of a polyhedron in the plane as follows: puncture the bottom face, and then stretch open the "hole" wider and wider until the surface of the polyhedron lies flat on a plane.
    // Assuming the surface is made of a pliable, flexible material.

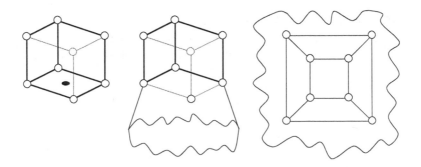

A *Platonic Solid* is a polyhedron where all the faces are (isomorphic and) the same size s and all the vertices have the same degree d. So a cube is a Platonic Solid. Five such solids have been studied since Plato:

(1) The *tetrahedron* with 4 triangular faces   ($s = 3$ & $d = 3$)
                                                      // a pyramid with triangular base
(2) The *cube* with 6 square faces   ($s = 4$ & $d = 3$)
(3) The *octahedron* with 8 triangular faces   ($s = 3$ & $d = 4$)
(4) The *dodecahedron* with 12 pentagonal faces ($s = 5$ & $d = 3$)
(5) The *icosahedron* with 20 triangular faces   ($s = 3$ & $d = 5$)

  (a) Draw the "planar" version of the surface of each of these 5 polyhedra.
  (b) Determine the dual of each "planar" version of the surface for each of
       these 5 polyhedra.
  (c) Use the results we have about planar graphs to prove that there are no
       other Platonic Solids.
42. Imagine that a vertex of an icosahedron is "sanded" down to create a plateau.
    That plateau would have 5 sides.

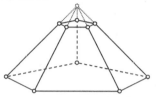

    If all the vertices were "sanded" down in that same way, we obtain a poly-
    hedron with 12 pentagonal sides and 20 hexagonal sides (resulting from the
    original triangles).
    (a) Draw the "planar" version of the surface of the resulting polyhedron.
         // This "rounded down" icosahedron is the pattern of the surface of most
         // modern soccer balls.
    (b) How many edges are there in the surface the resulting polyhedron?
43. Consider the Petersen graph.
    (a) Does it have a Hamilton Path?
    (b) Does it have a Hamilton Path between any pair of non-adjacent vertices?
    (c) Does it have a Hamilton Circuit?
44. Can the vertices of the Petersen graph be labelled with 2-subsets of $\{1, 2, 3, 4, 5\}$
    so that v is adjacent to w if and only if their subset labels are disjoint?
45. Imagine "shrinking" an edge e joining vertices a and b in some drawing of a
    planar graph $G = (V, E)$.

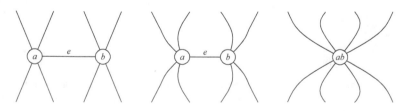

Shrinking edge e does not introduce any edge-crossing. If we denote the graph that results from shrinking edge e by $G \setminus e$, then we know that if G is planar then $G \setminus e$ is also planar.

(a) Explain why if H is obtained by shrinking a finite sequence of edges of G and H is not planar then G could not have been planar.

(b) Choose a sequence of edges in the Petersen Graph to shrink to obtain K_5, and thereby prove (again) that the Petersen Graph is not planar.

46. Suppose that G is a bridgeless, connected, planar graph. Prove that G is **bipartite** if in some drawing of G every face has an even size.

47. It is well known that the squares of a checker board can be colored with 2 colors so that no two squares that share an edge have the same color. Suppose that we draw a curve across the board from one edge of the board to another, and we draw a closed curve inside the border of the board.

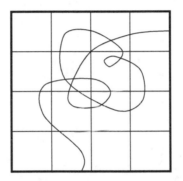

(a) Can the regions in this new drawing be colored with 2 colors so that no two regions that share an edge have the same color?

(b) Suppose that G is a bridgeless, connected planar graph where each vertex has even degree, and let $d(G)$ be any drawing of G. Use the dual graph and the result stated in question 46 to prove that the faces of $d(G)$ can be colored with 2 colors so that no two regions that share an edge have the same color.

(c) Explain how the result of part (b) relates to the drawing and coloring result in part (a).
// Think about "dropping" a square (or circular) frame onto the colored
// drawing and looking at the portion of it inside the frame.

Directed Graphs

6

This chapter gives a number of algorithms that apply to directed graphs, though many ideas and algorithms for undirected graphs apply to digraphs (and vice versa). The first is an algorithm to orient the edges of an undirected graph to produce a strongly-connected digraph. Then we consider acyclic digraphs, showing how they may be "topologically sorted", and how to use Dynamic Programming to count the number of dipaths from a to b, and find a shortest or a longest dipath from a to b.

Dijkstra's Algorithm grows a tree of shortest dipaths from a "single source", and the Floyd-Warshall Algorithm solves the "all-pairs" shortest dipath problem. Next we consider commodity Flow Networks. The Max-flow/Min-cut Theorem is proved by the action of the Ford-Fulkerson algorithm which finds a feasible flow, and gives it a "certificate of optimality" by producing a cut whose capacity equals the value of that flow. The final section shows that matching problems in bipartite graphs can be converted to flow problems and proves several classic results, including Hall's Marriage Theorem.

6.1 Introducing Directions

A *directed graph* (or *digraph*) D consists of a (finite, non-empty) set V of *vertices* together with a (finite but possibly empty) set A of *arcs* where each arc "goes" *from* one vertex *to* another vertex. The vertices are drawn as small circles, ovals or squares; and the arcs are drawn as line segments joining one vertex to another with an *arrowhead* indicating the "direction" of the arc.

                                       // like the arrows for one-way streets
An arc may join a vertex to itself; such an arc is (also) called a *loop*. All arcs have a "start"-end and a "finish"-end. The vertices at the finish-ends of arcs starting from vertex x are called *out-neighbours* of x; the vertices at the start-ends of arcs finishing at vertex y are called *in-neighbours* of y. Two different arcs may join the

© Springer International Publishing AG, part of Springer Nature 2018   241
T. Jenkyns and B. Stephenson, *Fundamentals of Discrete Math for Computer Science: A Problem-Solving Primer*, Undergraduate Topics in Computer Science, https://doi.org/10.1007/978-3-319-70151-6_6

same two vertices and go in the same direction; such arcs are said to be *strictly-parallel*. When there are no loops and no strictly-parallel arcs, the digraph is said to be *simple*.

A *directed path* (or *dipath*) is a sequence of the form:

$$\pi = (v_0, \alpha_1, v_1, \alpha_2, v_2, \alpha_3, \ldots, \alpha_k, v_k)$$

where each v_j is a vertex, and each α_i is an arc *from* vertex v_{i-1} *to* vertex v_i.

// A dipath corresponds to a trip through (part of) the directed graph
// starting from vertex v_0 then following the arc α_1 in the correct direction to vertex v_1,
// then following the arc α_2 (again in the correct direction) to vertex v_2, and so on until
// finally, the arc α_k is traversed (in its direction) and we end up at vertex v_k.
The dipath π is said to go *from* v_0 *to* v_k and have **length** k, where k is the number of arcs traversed.   // like paths in undirected graphs
We will say that v_k is **reachable** from v_0 when there is a dipath from v_0 to v_k.
A dipath is *simple* if all the vertices are distinct, and a dipath is *closed* if $v_0 = v_k$.
A closed dipath is *simple* if all vertices in the sequence are distinct except $v_0 = v_k$.
A closed dipath is a *cycle* if $k >= 1$, and all the arcs are distinct.

// The dipath $\pi = (v_0)$ is a simple, closed dipath of length zero from v_0 to v_0, often
//   referred to as a "trivial" dipath, but it is not a cycle.
// If arc α is a loop at vertex v, then $\pi = (v, \alpha, v)$ is a cycle (of length one).
// If α is an arc from v to w, and β is an arc from w to v then $\pi = (v, \alpha, w, \beta, v)$
//   is a cycle of length two.

Theorem 6.1.1: Suppose that $\pi = (v_0, \alpha_1, v_1, \alpha_2, v_2, \alpha_3, \ldots, \alpha_k, v_k)$ is a dipath in $D(V, A)$.

(a) If v_0 and v_k are different vertices, then there is a *simple* dipath π^* from v_0 to v_k in D.
(b) If π is a closed dipath, that is $v_0 = v_k$, then there is a *simple* cycle π^* in D.

Proof.

(a) If π is not a simple dipath, it must contain a repeated vertex. Suppose $v_i = v_j$, where $0 <= i < j <= k$. Then we can't have both $0 = i$ and $j = k$, and

$$\pi = (v_0, \alpha_1, v_1, \ldots, v_{i-1}, \alpha_i, v_i, \alpha_{i+1}, v_{i+1}, \ldots, v_{j-1}, \alpha_j, v_j = v_i, \alpha_{j+1}, v_{j+1}, \ldots, \alpha_k, v_k).$$

Then π may be replaced by

$$\pi_1 = (v_0, \alpha_1, v_1, \ldots, v_{i-1}, \alpha_i, \quad v_i = v_j, \alpha_{j+1}, v_{j+1}, \ldots, \alpha_k, v_k)$$

which is a shorter dipath from v_0 to v_k in D. This process may be repeated until we obtain π^*, a simple dipath from v_0 to v_k in D.

(b) If $v_0 = v_1$, then $\pi^* = (v_0, \alpha_1, v_1)$ is a *simple* cycle in D. If $v_0 \neq v_1$, then $\pi_1 = (v_1, \alpha_2, v_2, \alpha_3, \ldots, \alpha_k, v_k)$ is a dipath joining different vertices, so by part (a) there is a *simple* dipath π_2 from v_1 to v_k in D. This dipath π_2 together with arc α_1 forms a *simple* cycle π^* in D.

It could happen that $\pi_2 = (v_1, \alpha_k, v_k)$ but then α_1 and α_k are oppositely oriented arcs joining the same two vertices, and $(v_0, \alpha_1, v_1, \alpha_k, v_k)$ would be a *simple* cycle in D.   // What could happen in an undirected graph?

◻

Many problems ask us to find a dipath (or tour) of a certain kind. These occur in problems where "transitions" between configurations are not always reversible.
// What would be analogous to an Euler Tour or a Hamilton Tour in a directed graph?
// Is the corresponding Euler Tour problem still easy?
// Is the corresponding Hamilton Tour problem still hard, unsolved, and worth
// $1,000,000?

// Is there a dipath from a given vertex y to another specified vertex z?
// Can a Depth-First Traversal of all vertices reachable from a given vertex y be done?
// Can a Breadth-First Traversal of all vertices reachable from a given vertex y be done?

The Most Important Ideas in this Section.
A *directed graph* consists of a set of *vertices* together with a set of *arcs* where each arc goes *from* one vertex *to* another vertex. A *dipath* is a sequence of the form $\pi = (v_0, \alpha_1, v_1, \alpha_2, v_2, \alpha_3, \ldots, \alpha_k, v_k)$ where each v_j is a vertex, and each α_i is an arc *from* vertex v_{i-1} *to* vertex v_i. A closed dipath is a *cycle* if $k >= 1$, and all the arcs are distinct. Many ideas and algorithms for undirected graphs apply to digraphs (and vice versa).

6.2 Strong Connectivity

A digraph is *strongly-connected* means for any two vertices x and y, there is a dipath from x to y.   // Any vertex is reachable from any other.
// In a (completely) one-way street system, you must be able to get from any point
// to any other point following the one-way arrows.

An *orientation* of an undirected graph G, is an assignment of a direction to each edge of G, thereby producing a digraph, $O(G)$.

Now we consider two questions:
1. When can an undirected graph be oriented to produce a strongly-connected digraph?
2. Is there a simple algorithm to do this?

If $O(G)$ is strongly-connected then (ignoring the directions of the arcs) G itself must be (weakly) connected.   // But there's more.

An edge e in a connected graph is called a **bridge** if removing e leaves a dis-connected graph. If a bridge e joins vertices a and b, it provides the **only** route from a to b, and from b to a. Thus, if e is oriented from a to b, then a is not reachable from b; and if e is oriented from b to a, then b is not reachable from a. Hence, if G has a bridge, no orientation of G can be strongly-connected. On the other hand,

Theorem 6.2.1: If G is a bridgeless connected graph then G has an orientation that is strongly-connected.

Proof.

Choose any vertex y in G, and do a Depth-First Traversal of G (applying Algorithm 5.3.1), keeping track of the order in which all the vertices of G enter the tree T, then

$$V = (v_1 = y, v_2, v_3, \ldots, v_n).$$

As in the walk-through of Algorithm 5.3.1, orient the edges in the tree T from v_i to v_j when $i < j$. But also orient the edges that are <u>not</u> in the tree T from v_j back to v_i when $i < j$. Orienting the edges in the result of the walk-through of Algorithm 5.3.1 produces the digraph below.

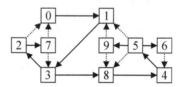

But to more clearly illustrate the relation between the Depth-First tree and the arcs that are not in that tree, consider this second drawing of this same digraph.

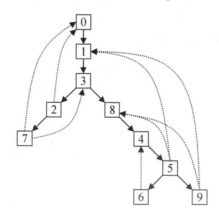

In general, with this orientation, there is a dipath in T from y to every vertex in G. We will prove by Mathematical Induction on k, that there is a dipath from vertex v_k back to vertex y.

<u>Step 1</u>. The trivial dipath (y) is a dipath from v_1 back to y.

<u>Step 2</u>. Assume $\exists$ an integer q where $1 <= q < n$ such that there is a dipath from vertex v_k back to y for $k = 1, 2, \ldots, q$.

<u>Step 3</u>. To show that there is a dipath from v_{q+1} back to y, imagine the Depth-First Traversal of G at the point when v_{q+1} is marked "scanned". Let A denote the set of vertices in T that are v_{q+1} and the "descendants" of v_{q+1}. Let B denote the set of other vertices in G.

The parent of v_{q+1}, $P(v_{q+1})$, is put in T before v_{q+1}. Let $x = P(v_{q+1})$ and let e be the edge in the tree joining x and v_{q+1} (which is oriented from x to v_{q+1}). But this edge e is <u>not</u> a bridge in G. Therefore there is a path (that avoids e) in G from v_{q+1} in set A to x in set B, say

$$\pi = \left(v_{q+1}, e_1, w_1, e_2, \ldots, w_{r-1}, e_r, w_r, \ldots, e_m, w_m = x\right)$$

where w_r is the first vertex along π that is in set B. Then w_{r-1} is in set A. In Depth-First Traversal all the descendants of a vertex are marked "scanned" before that vertex itself is marked "scanned", so w_{r-1} (a descendant of v_{q+1}) has been marked "scanned".

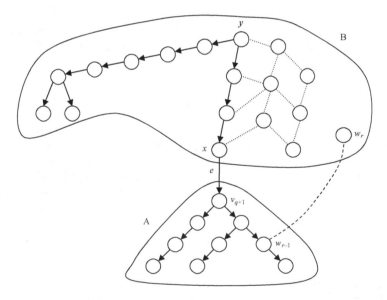

Because w_r has <u>not</u> been added to the tree as a child of w_{r-1}, it must be that w_r is in the tree already. Furthermore, w_r must have been added to the tree before any vertex in set A. Hence, $w_r = v_p$ for some $p <= q$. Then

there is a dipath in T from v_{q+1} to its descendant w_{r-1},
there is an arc from w_{r-1} to $w_r = v_p$ for some $p <= q$, and
there is a dipath from v_p to y.   // by Step 2.

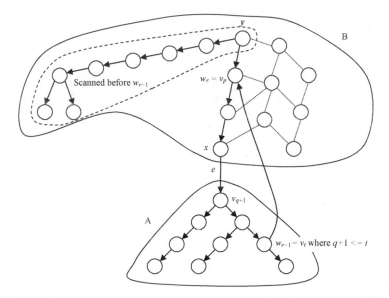

Thus, in this orientation of G, there is a dipath from every vertex of G back to y.

The final step of this proof is to show that if a and b are any vertices of G, there is a dipath from a to b. But we know there is a dipath from a to y, and a dipath (in the tree) from y to b, so there must be a dipath from a to b in this orientation of G.   ☐

One further observation is:

if a non-tree edge in G is oriented from v_β to v_α because $\alpha < \beta$ then v_α is an *ancestor* of v_β.

When v_α is put into the tree, it is marked "unscanned" and placed on top of the "stack" of unscanned vertices. This vertex, v_α, is not marked "scanned" and not removed from the stack until after its neighbor, v_β, is put into the tree, marked "unscanned" and placed on top of the "stack" of unscanned vertices. The stack itself (from top to bottom) is the vertex sequence of the parent, grand-parent, great-grand-parent, ... from v_β, down through v_α, back to $y = v_1$.

// Would Breadth-First Traversal work here?

The Most Important Ideas in this Section.
A digraph is *strongly-connected* means for any two vertices x and y, there is a dipath from x to y. An *orientation* of an undirected graph is an assignment of a direction to each edge. The Depth-First Traversal algorithm was used to prove that if G is connected and bridgeless, then G has an orientation that is strongly-connected.

6.3 Topological Sorting

Digraphs model processes where there is a priority involved: in college programs, certain courses are prerequisites for others; in project management, certain tasks must be completed before others can start. For example, think of building a house. Work must generally begin at the bottom and move up. A *topological sorting* of a digraph, in a sense, describes the precedence relations among all the vertices; it is a labelling, L, of all the vertices with the integers from 1 to $|V|$ in such a way that if an arc goes from *a* to *b* then $L(a) < L(b)$.

Again there are two questions to consider:
1. When can a digraph be topologically sorted?
2. Is there a simple algorithm to do this?

If a digraph, *D*, has been topologically sorted, and $\pi = (v_0, \alpha_1, v_1, \ldots, \alpha_j, v_j, \alpha_{j+1}, v_{j+1}, \ldots, \alpha_k, v_k)$ is a nontrivial dipath in *D*, then

$$L(v_0) < L(v_1) < \ldots < L(v_j) < L(v_{j+1}) < \ldots < L(v_k). \tag{6.3.1}$$

Thus $L(v_0) < L(v_k)$, so $v_0 \neq v_k$, and therefore *D* cannot contain a closed path, so it cannot contain a cycle. A digraph is said to be *acyclic* if it does not contain a cycle. If *D* can be topologically sorted then *D* **must** be acyclic.

A *source* is a vertex *s* in a digraph that is not the finish-end of any arc, so a source only has arcs leaving it. A *sink* is a vertex *t* that is not the start-end of any arc, so a sink only has arcs entering it.

If we grow a simple dipath as long as possible, say $\pi = (v_0, \alpha_1, v_1, \ldots, \alpha_k, v_k)$ then since no vertex is repeated, $k <= |V|$, and either v_k is a *sink*; or v_k has at least one out-neighbour *w*, but *w* has already occurred in π. Suppose that v_j is the last occurrence of *w* in π, then $\pi' = (v_j, \alpha_{j+1}, v_{j+1}, \ldots, \alpha_k, v_k, \alpha_{k+1}, w = v_j)$ is a cycle. Hence, *every digraph has a sink, or a cycle*, (or both).
                    // Does every digraph have a source, a cycle, or both?

If $D(V, A)$ has been topologically sorted, then the vertex labelled $n = |V|$ **must** be a sink. Furthermore, if a sink *t* and all arcs that end at *t* are removed, then any topological sorting of the remaining digraph on $n - 1$ vertices can be extended to a topological sorting of all of *D* by assigning $L(t)$ to be *n*.

Therefore, we can topologically sort any **acyclic** digraph, $D(V, A)$ by applying the following algorithm:

Algorithm 6.3.1: Topological Sort of a Digraph

Begin
   While $|V| > 0$ **Do**
      Find a sink, t, in $D(V, A)$;   // by beginning a Depth-First Traversal say
      $L(t) \leftarrow |V|$;
      Remove all arcs into t from A;
      Remove t from V;
   End;   // the while loop
End.

Walkthrough with $|V| = 14$ // And introducing an example that we
                                      // will use repeatedly in this chapter.

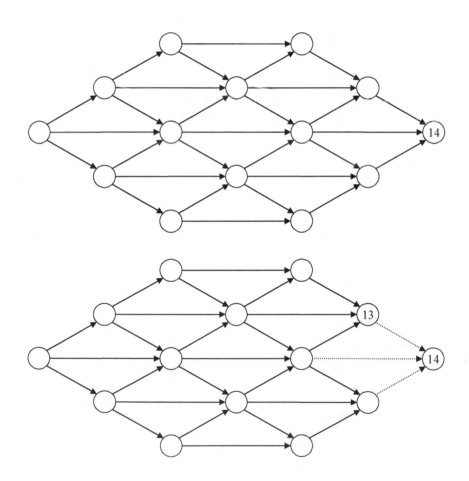

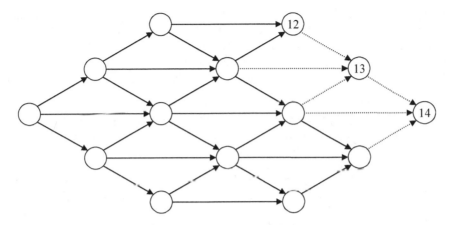

//X Walking through the remainder of Algorithm 6.3.1 to complete the labelling of the
//   graph is left as an exercise for the reader.

An "arc-weighted" digraph is a digraph, $D = (V, A)$ together with a "weight"
function on the arcs $w: A \to \mathbf{R}^+$.   // $\mathbf{R}^+$ is the set of all positive real numbers.

This function represents the "cost" of traversing an arc: the time in minutes, the
price of a one-way bus-ticket, the length in miles, or whatever is relevant to the
problem being considered. We usually use the "length" metaphor.

If $\pi = (v_0, \alpha_1, v_1, \alpha_2, \ldots, \alpha_k, v_k)$ is a dipath in D, then the *length* of π is given by

$$w(\pi) = \sum_{j=1}^{k} w(\alpha_j).\qquad\text{// the sum of the weights of the arcs in } \pi$$

A *shortest* dipath from v_0 to v_k is one of smallest possible length. Clearly a shortest
dipath has no repeated vertex; it must be a simple dipath. Furthermore, if
$0 <= i < j <= k$ then the portion of π from v_i to v_j must be a *shortest* dipath from
v_i to v_j. A *longest* dipath from v_0 to v_k is a <u>simple</u> dipath of largest possible length.

   Consider the following examples where edge-weights and arc-weights are all
equal to one unit.

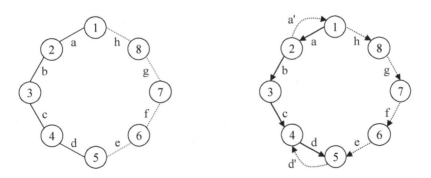

In the undirected graph, (1, a, 2, b, 3, c, 4, d, 5) is a longest <u>simple</u> path from vertex 1 to vertex 5. The portion (2, b, 3, c, 4) from 2 to 4 is *not* a longest such path; (2, a, 1, h, 8, g, 7, f, 6, e, 5, d, 4) is much longer. In the directed graph, (1, a, 2, b, 3, c, 4, d, 5) is a longest <u>simple</u> dipath from vertex 1 to vertex 5. The portion (2, b, 3, c, 4) from 2 to 4 is *not* a longest such dipath; (2, a', 1, h, 8, g, 7, f, 6, e, 5, d', 4) is much longer.

Theorem 6.3.1: If $D(V, A)$ is a simple, acyclic digraph that has been *topologically sorted*, and $P = (v_0, \alpha_1, v_1, \alpha_2, v_2, \alpha_3, \ldots, \alpha_k, v_k)$ is a dipath in D, then

1. P is a <u>simple</u> dipath;
2. if $0 <= i < j <= k$, and $Q = (v_i = y_0, \beta_1, y_1, \beta_2, \ldots, \beta_m, y_m = v_j)$ is a dipath from v_i to v_j then $R = (v_0, \alpha_1, v_1, \ldots, \alpha_i, v_i = y_0, \beta_1, y_1, \ldots, \beta_m, y_m = v_j, \alpha_{j+1}, v_{j+1}, \ldots, \alpha_k, v_k)$ is a simple dipath; and
3. if   P is a *longest* dipath in D from v_0 to v_k and $0 <= i < j <= k$,
    then the portion of P from v_i to v_j is a *longest* dipath from v_i to v_j.

Proof. From Eq. 6.3.1 we have

$$L(v_0) < L(v_1) < \ldots < L(v_i) < L(v_{i+1}) < \ldots < L(v_j) < L(v_{j+1}) < \ldots < L(v_k).$$

Thus the vertices in P are distinct, and part 1 holds. Applying Eq. 6.3.1. to the dipath Q we have

$$L(v_i) < L(y_1) < \ldots < L(y_r) < L(y_{r+1}) < \ldots < L(y_m) < L(v_{j+1}) < \ldots < L(v_k).$$

Thus the vertices in R are distinct, and part 2 holds.

To prove part 3 we will establish the contrapositive statement. Suppose that $0 <= i < j <= k$ but the portion of P from v_i to v_j is <u>not</u> a *longest* dipath from v_i to v_j. Then there is a longer dipath Q from v_i to v_j. Then, if R is constructed as in part 2 this dipath R would be longer than P. Hence, we have proved part 3.   ☐

In the remainder of this section, when we refer to a <u>topologically sorted digraph</u>, we will assume that $V = \{1, 2, \ldots, n\}$ and that if an arc goes from a to b then $a < b$.   // the labelling function is $L(j) = j$

The topological sorting of a digraph can be used to simplify answering certain common questions:

1. Is vertex z reachable from vertex y?
2. How many dipaths are there from vertex y to vertex z?
3. How can we find a <u>shortest</u> dipath from vertex y to vertex z?
4. How can we find a <u>longest</u> dipath from vertex y to vertex z?

Suppose that D is a simple, acyclic digraph that has been topologically sorted, and suppose a and b are distinct vertices in D. If

$$\pi = (x_0 = a, \alpha_1, x_1, \ldots, x_{k-1}, \alpha_k, x_k = b)$$

is *any* dipath from a to b, then x_{k-1} must be an in-neighbour of b. Furthermore, if

$$\pi' = (x_0 = a, \alpha_1, x_1, \ldots, x_{k-1}, \alpha_k, x_k = v)$$

is a dipath from a to an in-neighbour v of b, then there is a unique extension of π' to a dipath from a to b.   // D is simple, so has (at most) one arc from v to b.

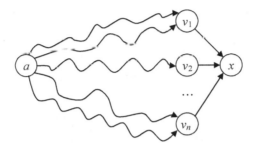

Define the function $f_a(b)$ to be the number of dipaths in D from a to b for each b in V. Then b is reachable from a if and only if $f_a(b) > 0$. Furthermore,

$$f_a(b) = \sum \{f_a(v) : v \text{ is an in-neighbour of } b\}$$

Therefore, we can evaluate $f_a(b)$ for all vertices b in a topologically sorted, simple, **acyclic** digraph, $D(V, A)$ by applying the following (iterative) algorithm:

Algorithm 6.3.2: Count the dipaths from a to b

Begin
   For $x \leftarrow 1$ **to** $(a - 1)$ **Do**
      $f_a(x) \leftarrow 0;$ // From Eq. 6.3.1, no dipath goes from a to x.
   End;
   $f_a(a) \leftarrow 1;$ // The trivial dipath is the only one from a to a.
   For $x \leftarrow a + 1$ **to** b **Do**
      $f_a(x) \leftarrow \sum \{f_a(v): v \text{ is an in-neighbour of } x\}$
                                 // $v < x$ so $f_a(v)$ has been evaluated.
   End;
End.

Three examples of applying Algorithm 6.3.2 to the topologically sorted digraph are shown below. The number of dipaths from a to each vertex is noted above each vertex.

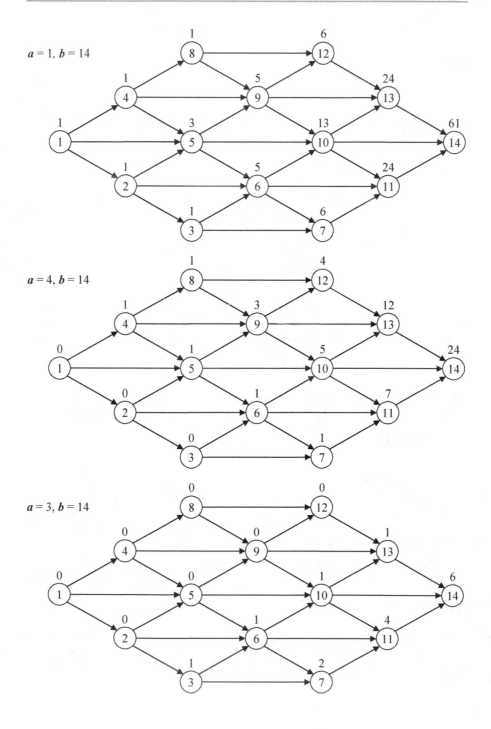

Algorithm 6.3.2 is a prototype of an algorithm design known as *"Dynamic Programming"*. The main problem, finding the value of $f_a(b)$, is computed iteratively by solving a sequence of similar sub-problems, tabulating their solutions, and combining their solutions to solve the next sub-problem in the sequence. The next two algorithms follow the same pattern of dynamic programming.

Define the function $g_a(b)$ to be the length of a *shortest* dipath in D from a to b for each b in V. We can evaluate $g_a(x)$ by growing a tree T of shortest paths from a similar to what we did for the traversals in Chap. 5.

Vertex a is reachable from a; any other vertex y is reachable from a if and only if some in-neighbour v of y is reachable from a.

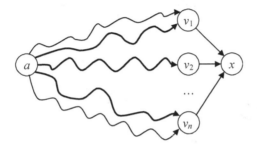

Algorithm 6.3.3: Find a shortest dipath from a to b

Begin
    Put a into T as the root;
    $g_a(a) \leftarrow 0$;   // the trivial dipath from a to a has total weight 0
    For $x \leftarrow a + 1$ **to** b **Do**
        If (x has an in-neighbour in T) **then**
            Find v, an in-neighbour of x in T that minimizes $g_a(v) + w(v, x)$;
                        // $v < x$ so $g_a(v)$ has been evaluated.
            $g_a(x) \leftarrow g_a(v) + w(v, x)$;
            Add x and the arc (v, x) to the tree T;
            $P(x) \leftarrow v$;   // call v the "Parent" of x
        End;
    End;
End.

We have updated our previous digraph so that it includes a weight on each arc. The diagrams below show the result of applying Algorithm 6.3.3 to this arc-weighted digraph. The length of the shortest dipath from a to each vertex is noted above the vertex.

                  // This arc-weighted digraph will be used in several
                  // additional examples later in this chapter.

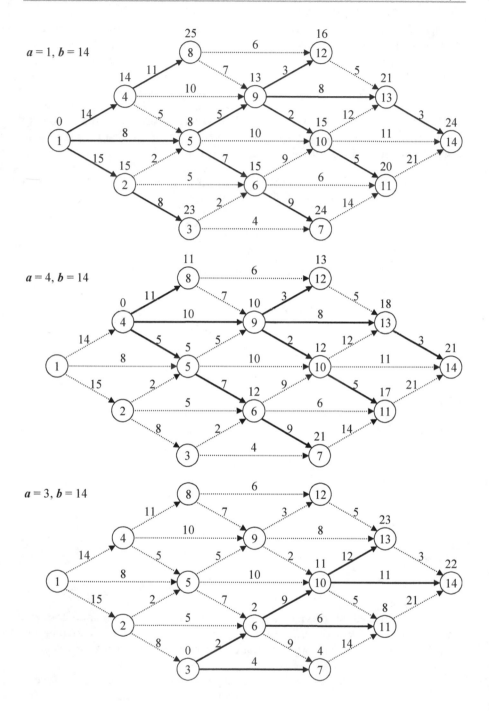

Theorem 6.3.2: Algorithm 6.3.3 is correct
When vertex x is added to the tree, T contains a shortest dipath from a to x and $g_a(x)$ equals the length of that dipath.

Proof. If $x < a$ then x is not reachable from a, and x is not put into T. We will proceed by mathematical induction on x, as x goes from a to b.

Step 1. When a is put into the tree, T contains a shortest dipath from a to a and $g_a(x)$ equals the length of the trivial dipath.

Step 2. Assume that $\exists$ an integer q such that $a <= q < b$, such that for each x in the range a up to q, when vertex x is added to the tree, T contains a shortest dipath from a to x and $g_a(x)$ equals the length of that dipath.

Step 3. On the next iteration of the body of the for-loop x equals $q + 1$. If x has no in-neighbour in T then x is not reachable from a, and x is not put into T. Otherwise, x is reachable from a through certain in-neighbours in T: $v_1, v_2, \ldots, v_k$ say. For each index j, $a <= v_j < x$ and when vertex v_j was added to the tree, T contained a shortest dipath from a to v_j and $g_a(v_j)$ equaled the length of that dipath.

If $\pi = (x_0 = a, \alpha_1, x_1, \ldots, x_{k-1}, \alpha_k, x_k = x)$ is <u>any</u> dipath from a to x then $x_{k-1} = v_j$, an in-neighbour of x in T.

$$\text{The length of } \pi = \text{the length of } \left(x_0 = a, \alpha_1, x_1, \ldots, x_{k-1} = v_j \right) + w(\alpha_k)$$
$$>= g_a\left(v_j\right) \qquad\qquad\qquad\qquad + w\left(v_j, x\right).$$

The length of a <u>shortest</u> dipath from a to x when $x_{k-1} = v_j$, is equal to $g_a(v_j) + w(v_j, x)$. Therefore, the length of a <u>shortest</u> dipath from a to x is equal to $g_a(v) + w(v, x)$ where v is the in-neighbor of x in T that minimizes $g_a(v_j) + w(v_j, x)$.

The algorithm finds <u>that very</u> in-neighbour v, sets $g_a(x)$ to be $g_a(v) + w(v, x)$, and adds x and the arc (v, x) to the tree. Now, as vertex x is added to the tree, T contains a shortest dipath from a to x and $g_a(x)$ equals the length of that dipath. □

This same technique can be used to find a ***longest*** dipath in a digraph D from vertex a to vertex b.   *// But why would a longest dipath be of any use?* Recall the example of "Project Management" we mentioned in the introduction of this section. The overall project is divided into smaller tasks or activities and these are the vertices of D. An arc goes from task a to task b when a must be completed before b can begin. Let the "weight" on arc (a, b) be the time it takes to do task a. A longest dipath in this digraph is known as a "Critical Path"; its length gives the total time needed to complete the whole project, and the tasks along the path are critical in the sense that if one of them is delayed, the whole project is delayed.

Define the function $h_a(b)$ to be the length of a ***longest*** dipath in D from a to b for each b in V. We can evaluate $h_a(x)$ by growing a tree T of longest paths from a just as we did in Algorithm 6.3.3 for shortest dipaths.

If $\pi = (x_0 = a, \alpha_1, x_1, \ldots, x_{k-1}, \alpha_k, x_k = x)$ is _any_ dipath from a to x then $x_{k-1} = v_j$, an in-neighbour of x in T.

$$\text{The length of } \pi = \text{the length of } \left(x_0 = a, \alpha_1, x_1, \ldots, x_{k-1} = v_j\right) + w(\alpha_k)$$
$$<= h_a(v_j) \qquad\qquad\qquad + w(v_j, x).$$

The length of a longest dipath from a to x when $x_{k-1} = v_j$, is equal to $h_a(v_j) + w(v_j, x)$. Therefore, the length of a longest dipath from a to x is equal to $h_a(v) + w(v, x)$ where v is the in-neighbor of x in T that maximizes $h_a(v_j) + w(v_j, x)$.

The algorithm finds that very in-neighbour v, sets $h_a(x)$ to be $h_a(v) + w(v, x)$, and adds x and the arc (v, x) to the tree. Now, as vertex x is added to the tree, T contains a longest dipath from a to x and $h_a(x)$ equals the length of that dipath.   ☐

The Most Important Ideas in this Section.
A *topological sorting* of a digraph is a labelling of the vertices with the integers from 1 to $|V|$ so that if an arc goes from a to b then $L(a) < L(b)$. A digraph is *acyclic* if it does not contain a cycle, and it can be topologically sorted if and only if it is acyclic. In topologically sorted digraphs the Dynamic Programming method can be used to count the number of dipaths from a to b, and to find a shortest or a longest dipath from a to b.

6.4   Shortest Paths in Digraphs (Acyclic or not)

From any simple, undirected, edge-weighted graph G we can construct a digraph D_G by replacing each (undirected) edge e of weight $w(e)$ joining vertices u and v, with *two arcs*: α_1 going from u to v with $w(\alpha_1) = w(e)$, and α_2 going from v to u with $w(\alpha_2) = w(e)$. Then it's obvious that:

(1) If G has a path from x to y with total weight K,
    then D_G has a dipath from x to y with total weight K.

and (2) If D_G has a dipath from x to y with total weight K,
    then G has a path from x to y with total weight K.

Therefore, an algorithm for finding shortest dipaths in (arbitrary) digraphs may be applied to digraphs like D_G, obtained from some undirected graph G, and it will find the shortest paths in G.

Many problems ask us to find a dipath from a particular vertex y to a particular vertex z; some problems ask us to find a shortest dipath from y to z.

6.4.1 Distance Function

Using the idea of dipath length, we can define a distance function on $V \times V$ by

$\delta(x, y) =$ the length of a *shortest* directed path <u>from</u> x <u>to</u> y.   // if there is one

It is tempting to say that "$\delta(x, y) = \infty$" when there is no dipath from x to y We will use the symbol "∞" in our descriptions of algorithms, but in implementations, we would "approximate" it by some **numerical value** M that is $>$ any shortest path length.

The distance function δ has the following properties. For all vertices x, y, and z

(1) $\delta(x, y) >= 0$.   // nonnegativity

(2) $\delta(x, y) = 0$ if and only if $x = y$.

(3) $\delta(x, z) <= \delta(x, y) + \delta(y, z)$.   // the triangle inequality

Property (3) holds because if there were a (shortest) dipath π_1 from x to y of length p and a (shortest) dipath π_2 from y to z of length q, then these could be **concatenated** to produce dipath π_3 from x to z of length $p + q$.

// Traverse π_1 and then traverse π_2 to form π_3.

Therefore, a shortest dipath from x to z will have length $<= \delta(x, y) + \delta(y, z)$.

6.4.2 Dijkstra's Algorithm

This algorithm finds a shortest dipath from a start vertex y to a finish vertex z in a simple, arc-weighted digraph.

The *input* is a simple arc-weighted digraph D (implemented in some manner where the out-neighbors of a vertex can be quickly determined) with a positive weight function w defined on the arc set A and two particular vertices, y and z.

The strategy is to grow a tree T rooted at y by adding one new vertex at a time (as a leaf) together with a "back-arc" <u>to</u> it <u>from</u> a vertex already in T. All the vertices in D are given labels where $L(v)$ is an "estimate" of $\delta(y, v)$.

// But always $>= \delta(y,v)$.

The *output* is also a digraph T which contains shortest dipaths from y, satisfying the *post-condition*:

1. If there is a dipath from y to z, T will contain a <u>shortest</u> dipath from y to z.
2. If there is no dipath from y to z in D, T will contain a shortest dipath from y to every vertex reachable from y.

Algorithm 6.4.1: Dijkstra's Algorithm (1959)

// for single-source shortest dipaths

Begin
  For every vertex v in D, set $L(v)$ to be ∞;
  Change $L(y)$ to be zero and put y in T as the root vertex;
  Set $v = y$;   // v is the "current vertex."

  While $((v \neq z)$ **and** $(L(v) < \infty))$ **Do**
    For (each out-neighbor x of v not yet in T) **Do**
      let α_x be the arc from v to x;
      If $(L(v) + w(\alpha_x) < L(x))$ **Then**
          $L(x) \leftarrow L(v) + w(\alpha_x)$;   // reduce $L(x)$ to be this new value
          $BA(x) \leftarrow \alpha_x$;   // call α_x the "Back-arc" of x
          $P(x) \leftarrow v$;   // call v the "Parent" of x
      End;   // the if
    End;   // the for-loop

    Find a vertex v that is not yet in T with the smallest label;

    If $(L(v) < \infty)$ **Then**
        add v and its back-arc $BA(v)$ to the tree T;
    End;   // T now contains a shortest dipath from y to v.
  End; // of the while-loop
End.   // of Dijkstra's Algorithm.

In the walkthrough below, we draw the input digraph D using small circles for vertices, dotted lines for each arc in D not assigned as the back-arc of any vertex (nor put into T), light lines for each back-arc of the vertex at its finish-end but not put into T, and heavy lines for those arcs that have been put into T. We display the configuration at the end of the for-loop before a new current vertex v is chosen. After vertex v is added to T, and the for-loop is executed, v is drawn as a square.

We also give a "table" indicating the progress in the algorithm. It lists the vertices w not yet put into T but having $L(w) < \infty$; P is the parent of w (provided that w is not the root of T), and L is the label on w, $L(w)$.
Input for Dijkstra's Algorithm (an arc-weighted digraph):

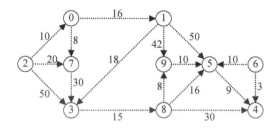

Walkthrough with $V = \{0, 1, \ldots, 9\}$ and y input as vertex 2 and z as vertex 4.

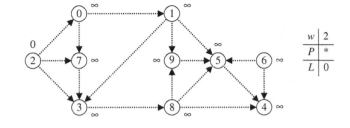

$v = 2$

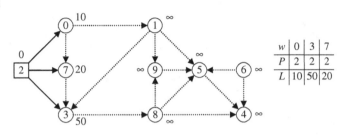

$v = 0$

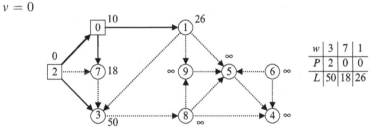

$v = 7$

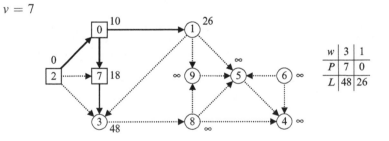

$v = 1$

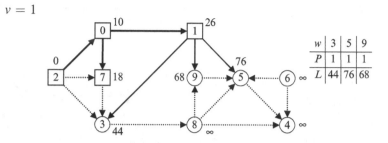

w	3	5	9
P	1	1	1
L	44	76	68

$v = 3$

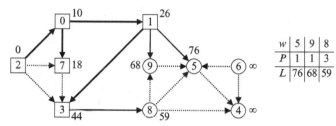

w	5	9	8
P	1	1	3
L	76	68	59

$v = 8$

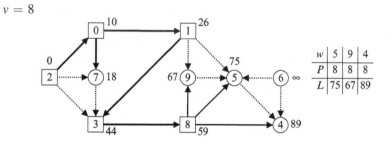

w	5	9	4
P	8	8	8
L	75	67	89

$v = 9$

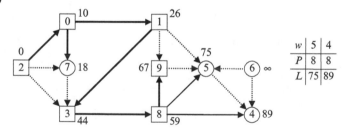

w	5	4
P	8	8
L	75	89

$v = 5$

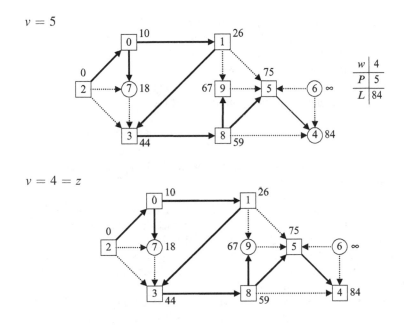

$v = 4 = z$

We note that the same final configuration would result if the input z was the vertex 6 where there is no dipath from $y = 2$ to $z = 6$.

// Every other vertex is reachable from 2.

Let n be the number of vertices in D. If z is selected as the current vertex, v, then the while-loop terminates. But if z is not selected, there is at least one vertex not yet in T, and there must be a vertex v not yet in T with the smallest label. If $L(v) = \infty$, the while-loop terminates. If $L(v) < \infty$, then a new vertex v is added to T. At most n vertices can be put into T so the while-loop must terminate after at most n iterations. Thus, Dijkstra's Algorithm is certain to **terminate**.

Furthermore, when the while-loop terminates, either z has been put into T or all vertices x of D not in T, including z, have $L(x) = \infty$.

We will prove Dijkstra's Algorithm is correct; that is, after Dijkstra's Algorithm is run, the following *post-conditions* are satisfied:

1. If there is a dipath from y to z, T will contain a <u>shortest</u> dipath from y to z.
2. If there is no dipath from y to z in D, T will contain a shortest dipath from y to every vertex reachable from y.

Theorem 6.4.1: When vertex v and its back-arc are added to the tree, the label on v equals $\delta(y,v)$, and the tree itself then contains a shortest dipath from y to v.

Proof. // by Strong Induction

Step 1.   When the first vertex y is added to the (empty) tree, $L(y) = 0 = \delta(y, y)$, and the tree contains $\pi = (y)$ which is a shortest dipath from y to y.

// When the second vertex v is added to the tree,
// the tree consists of the root vertex y and nothing else, and v is the "nearest"
// neighbor of y. If α_v denotes the arc from y to v, $L(v) = 0 + w(\alpha_v)$, and after
// v and α_v are added to the tree, it contains the dipath $\pi_0 = (y, \alpha_v, v)$ of length $L(v)$.
// Any different simple dipath π_1 from y to v must begin on some other arc
// and visit some other neighbor u of y before reaching v. But then
//
//   the length of π_1 >= $L(u)$ >= $L(v)$ = the length of π_0.
//
// Therefore, π_0 is a shortest dipath from y to v and is contained in the tree.
// Furthermore, $L(v)$ = the length of $\pi_0 = \delta(y, v)$.

Step 2. Assume $\exists$ a positive integer k less than the number of vertices in the final
         tree, such that when each of the first k vertices x (and their back-arcs) is
         added to the tree, the label on x was equal to $\delta(y, x)$, and the tree itself then
         contained a shortest dipath from y to x.
Step 3. Suppose now that a $k+1^{\text{st}}$ vertex v and its back-arc α_v are added to the tree
         T to produce a new (larger) tree $T1$.

Let u denote the start-end of α_v. The vertex u must already be in T, so it was
added earlier. By our induction hypothesis in Step 2, $L(u) = \delta(y, u)$, and the tree
contains a shortest dipath π_0 from y to u. Let π_1 be the dipath obtained by
appending arc α_v and vertex v to π_0. Then the tree $T1$ contains a dipath π_1, from y to
v extending π_0 along α_v to v. Because α_v is the back-arc of v,

$$L(v) = L(u) + w(\alpha_v) = \text{the length of } \pi_1 >= \delta(y, v).$$

// We'll be finished if we can prove that $L(v) = \delta(y, v)$,
// and that equality will follow if we can prove that $L(v) <= \delta(y, v)$.

Let π_2 denote some shortest dipath from y to v. This dipath starts at a vertex in T
but ends at a vertex outside of T. Let w be the first vertex along π_2 that is outside
of T; let x be the vertex before w along π_2 and let α denote the arc from x to w in π_2.
Then x is in T and so was added to the tree earlier. When x was added to the tree, the
labels on all the out-neighbors of x (outside the current tree at that time, including w)
were revised so we can be sure that now

$$L(w) <= L(x) + w(\alpha) = \delta(y, x) + w(\alpha).   \quad // L\text{-values only decrease.}$$

The portion of π_2 from y to x must be a shortest dipath from y to x, the portion of
π_2 from x to w must be a shortest dipath from x to w, and the portion of π_2 from w to
v must be a shortest dipath from w to v.

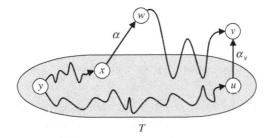

$$\delta(y, v) = \delta(y, w) + \delta(w, v) \ \ >= \delta(y, w)$$
$$= \delta(y, x) + \delta(x, w) \ \ = L(x) + w(\alpha)$$
$$>= L(w).$$

Because v is a vertex outside T with a smallest label, we have

$$L(v) <= L(w) <= \delta(y, v).$$

Therefore, when the $k + 1^{\text{st}}$ vertex v and its back-arc α_v are added to the tree T to produce a new (larger) tree $T1$, the label on v equals $\delta(y, v)$ and the new tree itself contains a shortest dipath π_1 from y to v.   ☐

If z is put into T, there may be many vertices w reachable from y that are not put into T because Dijkstra's Algorithm stops before w is considered for inclusion in T.

Theorem 6.4.2: Suppose z is <u>not</u> put into T. If there is a dipath π from y to vertex w, then eventually w will be put into T by Dijkstra's Algorithm.

Proof. Suppose $\pi = (y = v_0, \alpha_1, v_1, \alpha_2, v_2, \alpha_3, \ldots, \alpha_k, v_k = w)$. We will prove that each v_j in π will be put into T by Mathematical Induction on j.

Step 1. If $j = 0$, then $v_j = y$, which is put into T as the root.

Step 2. Assume that $\exists$ an index q such that $0 <= q < k$ and v_q has been added to T. When v_q is put into T, v_q was selected as the current vertex v and $L(v_q) < \infty$.

Step 3. On the next iteration of the body of the while-loop <u>after</u> v_q was put into T, all the out-neighbors w of v_q, including v_{q+1}, must be in T or now have a label $<= L(v_q) + w(\alpha)$ where α is the arc from v_q to w. Therefore

$$L(v_{q+1}) <= L(v_q) + w(\alpha_{j+1}) < \infty.$$

Eventually, this will be the smallest label on a vertex not yet put into T, and v_{q+1} will be selected as the current vertex v and v_{q+1} will be put into T.

Therefore, eventually $w = v_k$ will be put into T.   ☐

The argument in the proof of Theorem 6.4.2 applies even when $w = z$. It would then assert that

"If z is not put into T then,

if there is a dipath π from y to vertex z, then z will be put into T".

This statement has the form: $(\sim P) \rightarrow (Q \rightarrow P)$ which, using the methods in Chap. 3, is logically equivalent to $(Q \rightarrow P)$. Therefore, we have proven

"If there is a dipath π from y to vertex z, then z will be put into T".

Together, these two theorems prove Dijkstra's Algorithm is correct; that is, after Dijkstra's Algorithm is run, the following ***post-conditions*** are satisfied:
1. If there is a dipath from y to z, T will contain a <u>shortest</u> dipath from y to z.
2. If there is no dipath from y to z in D, T will contain a shortest dipath from y to every vertex reachable from y.

In addition, if there is no dipath from y to z in D, z will not be put into T and all vertices x of D not in T, including z, have $L(x) = \infty$.

If $\pi = (v_0 = x, \alpha_1, v_1, \ldots, \alpha_j, v_j = y, \alpha_{j+1}, v_{j+1}, \ldots, \alpha_k, v_k = z)$ is a shortest dipath from x to z, then the portion of π from x to y must be a <u>shortest</u> dipath from x to y, <u>and</u> the portion of π from y to z must be a <u>shortest</u> dipath from y to z. So

$$\text{the length of } \pi = \delta(x, z) = \delta(x, y) + \delta(y, z).$$

In particular, $\delta(x, z) = \delta(x, v_1) + \delta(v_1, z) = w(\alpha_1) + \delta(v_1, z)$.

Furthermore, if for some arc α from x to z

$$\delta(x, z) = w(\alpha) \tag{6.4.1}$$

then $\pi = (x, \alpha, z)$ is a shortest dipath from x to z. If for some vertex v and some arc α from x to v

$$\delta(x, z) = w(\alpha) + \delta(v, z) \tag{6.4.2}$$

then there is a shortest dipath from x to z that begins $(x, \alpha, v, \ldots)$ and continues with a shortest path from v to z. These observations can be used as the basis of an algorithm to construct a shortest dipath from x to z, for any two target vertices x and z, provided we know the functions w and δ.

// Search for v and α satisfying one of the equations.

In most applications, D is a simple digraph. // no loops or strictly parallel arcs

In this case, if $V = \{v_1, v_2, \ldots, v_n\}$, then w may be given in an $n \times n$ matrix W where

$$W[i, j] = \begin{cases} w(\alpha) & \text{if there is an arc } \alpha \text{ from } v_i \text{ to } v_j; \\ \infty & \text{if there is no such arc.} \quad \text{// } W \text{ may not be symmetric.} \end{cases}$$

In this case too, δ may be given in an $n \times n$ matrix δ where

$$\delta[i,j] = \begin{cases} \delta(v_i, v_j) & \text{if there is a dipath from } v_i \text{ to } v_j; \\ \infty & \text{otherwise.} \end{cases}$$

// The distance matrix δ is not necessarily symmetric, like **W**.

6.4.3 Floyd-Warshall Algorithm

This algorithm evaluates the entries in the distance matrix δ for a digraph without strictly parallel arcs and having vertex set $V = \{1, 2, \ldots, n\}$.

// Our description of the algorithm uses the symbol "∞" to represent
// some **numerical value** that is > any shortest dipath total weight.

The *input* is an $n \times n$ Matrix W where

$$W[a,b] = \begin{cases} \text{the weight of the arc from } a \text{ to } b & \text{if there is such an arc;} \\ \infty & \text{if there is no such arc.} \end{cases}$$

// It is amazing that we can determine δ(x, y), the length of a shortest dipath from
// x to y, without first finding a dipath π* from x to y and then showing π* is
// shortest.

The strategy is to start with the matrix W and then to use the triangle inequality, δ(x, z) <= δ(x, y) + δ(y, z), to reduce the entries (when that's possible) for all triples of vertices.

The *output* is an $n \times n$ matrix D where for all vertices p and q, if $p \neq q$, then $D[p, q] = \delta[p, q]$.

Algorithm 6.4.2: Floyd-Warshall Algorithm (1959)
                    for all-pairs shortest dipaths

```
Begin
    D ← W;                        // Copy the values from W into matrix D.
    For B ← 1 To n Do             // B is the intermediate vertex y.
      For A ← 1 To n Do           // A is the start vertex x.
        For C ← 1 To n Do         // C is the final vertex z.
                                  // B must control the outside loop.
          If (D[A, C] > D[A, B] + D[B, C]) Then
              D[A, C] ← D[A, B] + D[B, C];
          End;    // the if-statement
        End;      // the inner for-loop
      End;        // the middle for-loop
    End;          // the outer for-loop
End.
```

Walkthrough with $n = 5$ and W from the arc-weighted digraph below.

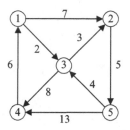

// $D_{(k)}$ denotes the matrix D after the iteration of the outer for-loop with $B = k$.

$$W = \begin{array}{c|ccccc} & 1 & 2 & 3 & 4 & 5 \\ \hline 1 & \infty & 7 & 2 & \infty & \infty \\ 2 & \infty & \infty & \infty & \infty & 5 \\ 3 & \infty & 3 & \infty & 8 & \infty \\ 4 & 6 & \infty & \infty & \infty & \infty \\ 5 & \infty & \infty & 4 & 13 & \infty \end{array}$$ // the input matrix

$$D_{(1)} = \begin{array}{c|ccccc} & 1 & 2 & 3 & 4 & 5 \\ \hline 1 & \infty & 7 & 2 & \infty & \infty \\ 2 & \infty & \infty & \infty & \infty & 5 \\ 3 & \infty & 3 & \infty & 8 & \infty \\ 4 & 6 & 13 & 8 & \infty & \infty \\ 5 & \infty & \infty & 4 & 13 & \infty \end{array}$$

$$D_{(2)} = \begin{array}{c|ccccc} & 1 & 2 & 3 & 4 & 5 \\ \hline 1 & \infty & 7 & 2 & \infty & 12 \\ 2 & \infty & \infty & \infty & \infty & 5 \\ 3 & \infty & 3 & \infty & 8 & 8 \\ 4 & 6 & 13 & 8 & \infty & 18 \\ 5 & \infty & \infty & 4 & 13 & \infty \end{array}$$

$$D_{(3)} = \begin{array}{c|ccccc} & 1 & 2 & 3 & 4 & 5 \\ \hline 1 & \infty & 5 & 2 & 10 & 10 \\ 2 & \infty & \infty & \infty & \infty & 5 \\ 3 & \infty & 3 & \infty & 8 & 8 \\ 4 & 6 & 11 & 8 & 16 & 16 \\ 5 & \infty & 7 & 4 & 12 & 12 \end{array}$$

$$D_{(4)} =$$

	1	2	3	4	5
1	16	5	2	10	10
2	∞	∞	∞	∞	5
3	14	3	16	8	8
4	6	11	8	16	16
5	18	7	4	12	12

$$D_{(5)} =$$

	1	2	3	4	5
1	16	5	2	10	10
2	23	12	9	17	5
3	14	3	12	8	8
4	6	11	8	16	16
5	18	7	4	12	12

In this example, the diagonal entries $D[j,j]$ ended up being the length of a shortest cycle through vertex j, a shortest **nontrivial** dipath from vertex j to itself. If they had been initialized as zero, they would stay zero, reflecting the idea that the trivial dipath is a shortest dipath from any vertex to itself.

We will prove the Floyd-Warshall Algorithm is correct; that is, after it is run, the following **post-conditions** are satisfied:

1. For all vertices p and q, if $p \neq q$, then $D[p, q] = \delta[p, q]$.
2. If there is a cycle through vertex p, then $D[p, p]$ is the length of a shortest such cycle, else $D[p, p] = \infty$.

To facilitate the proof, we define an $n \times n$ matrix $\delta 1$ where

$$\delta 1[i,j] = \begin{cases} \text{the length of a } \textbf{\textit{shortest nontrivial}} \text{ dipath from } i \text{ to } j; \\ \infty \quad \text{if there is no such dipath.} \end{cases}$$

Then if $i \neq j$, then $\delta 1[i, j] = \delta[i, j]$. But for each vertex j, $\delta 1[j, j]$ is the length of a shortest cycle through j (if there is a cycle through j); otherwise, $\delta 1[j, j] = \infty$.

// Recall $\delta[j, j] = 0$.

Theorem 6.4.3: The Floyd-Warshall Algorithm is correct; the final matrix D is equal $\delta 1$.

Proof. // in several parts (such a simple algorithm seems to need a complex
         // proof of correctness)

Part 1. At every stage, for all pairs of vertices p and q

if $D[p, q] < \infty$, then $\exists$ a nontrivial dipath from p to q of length $D[p,q]$.   (*)

// We prove this using a variation of Mathematical Induction.

// We could index the iterations of the body if the inner for-loop is from 1 to n^3

// and then use our usual form for MI.

Let x and y be fixed (but arbitrary) vertices. After initializing D to be W, if $D[x, y]$ $< \infty$, then $\exists$ an arc α from x to y with weight equal to $D[x, y]$, so $\pi = (x,\alpha,y)$ is a dipath from x to y of length $D[x, y]$.

Assume that (*) is true up to some point when $D[x,y]$ is revised (downward) when $A = x$, $C = y$ and for some value b of B

$$D[x, y] > D[x, b] + D[b, y]. \qquad\qquad // \ x \neq b \neq y$$

Then $D[x, b]$ must be $< \infty$ and also $D[b, y]$ must be $< \infty$. Hence, from (*)

$$\exists \text{ a nontrivial dipath from } x \text{ to } b \text{ of length } D[x,b],$$

and $\exists$ a nontrivial dipath from b to y of length $D[b,y]$.

These two dipaths can be "concatenated." // traverse the first and then the second Hence, when $D[x, y]$ is revised down to equal $D[x, b] + D[b, y]$,   // which is $< \infty$

$$\exists \text{ a nontrivial dipath from } x \text{ to } y \text{ of length } D[x,y].$$

Part 2. At every stage, for all pairs of vertices p and q,

$$D[p, q] >= \delta 1[p, q]. \qquad // \text{ even when } p = q \text{ or when } D[p,q] = \infty$$

Part 3. Definition of the "**_height_**" of a dipath. // for Mathematical Induction later
    If $\pi = (x_0, \alpha_1, x_1, \alpha_2, x_2, \ldots, \alpha_k, x_k)$ is a nontrivial dipath, we will call $x_1, x_2, \ldots$ x_{k-1} the "intermediate" vertices in π.   // vs. "end" vertices
If π has no intermediate vertices, we say
        $h(\pi) = 0;$
otherwise, $h(\pi) = \max\{x_1, x_2, x_3, \ldots, x_{k-1}\}.$ // So $0 <= h(\pi) <= n$.

Part 4. Let $D_{(k)}$ denote the matrix D after the iteration of the outer for-loop with $B = k$ and let $D_{(0)}$ denote W.

Lemma. If there is a shortest nontrivial path π from a to b with $h(\pi) = k$, then $D_{(k)}[a, b] <= \delta 1[a, b]$.

// Actually, $D_{(k)}[a, b] = \delta 1[a, b]$, but D-values **_may decrease_** so $<=$ will be easier
// to prove.

Proof. // by Mathematical Induction on k

Step 1.   // when $k = 0$

   If there is a shortest nontrivial path π from a to b with $h(\pi) = 0$, then $\pi = (a, \alpha, b)$ where α is an arc from a to b with weight $= W[a, b] = D_{(0)}[a, b]$. So $D_{(0)}[a, b] <= \delta 1[a, b]$.   // Actually, $D_{(0)}[a, b] = \delta 1[a, b]$.

Step 2.   Assume that $\exists$ q where $0 < q <= n$, such that if $0 <= H < q$ then, if there is a shortest path π from a to b with $h(\pi) = H$, then

$$D_{(H)}[a, b] <= \delta 1[a, b].$$

Step 3.   // when $k = q$

   Suppose there is a shortest nontrivial path π from a to b with $h(\pi) = q$, say

$$\pi = \left(a = x_0, \alpha_1, x_1, \ldots, \alpha_j, x_j = q, \alpha_{j+1}, x_{j+1}, \ldots, \alpha_m, x_m = b\right),$$

where the largest intermediate vertex, $q = x_j$.

// Since $q > 0$, there must be some intermediate vertex in π, also $m >= 2$.

Let $\pi 1 = \left(a = x_0, \alpha_1, x_1, \alpha_2, x_2, \ldots, \alpha_j, x_j = q\right)$ and let
         $\pi 2 = \left(q = x_j, \alpha_{j+1}, x_{j+1}, \alpha_{j+2}, x_{j+2}, \ldots, \alpha_m, x_m = b\right)$.

Then $\pi 1$ is a shortest nontrivial path from a to q with $h(\pi 1) = h1 < q$,
and $\pi 2$ is a shortest nontrivial path from q to b with $h(\pi 2) = h2 < q$.
   By the induction hypothesis,   // in Step 2.

$$D_{(h1)}[a, q] <= \delta 1[a, q] \text{ and } D_{(h2)}[q, b] <= \delta 1[q, b].$$

Since the D-values decrease or remain the same, the current D-values when $A = a$ and $C = b$ and $D[a, b]$ are considered for revision, then

$$D[a, q] <= D_{(q-1)}[a, q] <= D_{(h1)}[a, q],$$ // $h1 <= q - 1$.
and $D[q, b] <= D_{(q-1)}[q, b] <= D_{(h2)}[q, b].$ // $h2 <= q - 1$.

After the iteration of the outer for-loop with $B = q$,

$$D_{(q)}[a, b] <= D_{(h1)}[a, q] + D_{(h2)}[q, b] <= \delta 1[a, q] + \delta 1[q, b] = \delta 1[a, b].   \quad \square$$

Part 5. // Putting these parts together.

   If ever $D[a, b]$ becomes $< \infty$, $\exists$ a nontrivial dipath from a to b of length $D[a, b]$, and through all subsequent revisions $D[a, b]$ remains $>= \delta 1[a, b]$.   // by #1 and #2

If there is a nontrivial dipath from a to b, there must be a shortest nontrivial dipath from a to b, and that dipath must have height H where $0 <= H <= n$, so by the Lemma, $D_{(H)}[a, b] <= \delta 1[a, b]$.   // by #4

Then $D_{(n)}[a, q] <= \delta 1[a, b]$ which is $< \infty$ and, when $D_{(n)}[a, q] < \infty$,

$$D_{(n)}[a, q] >= \delta 1[a, b].$$ // by #1

If there is a nontrivial dipath from a to b, then $D_{(n)}[a, q] = \delta 1[a, b]$. If there is no nontrivial dipath from a to b, $D[a, b]$ must remain equal ∞ which equals $\delta 1[a, b]$. Therefore, $\forall$ pairs of vertices, $D_{(n)}[a, b] = \delta 1[a, b]$.   ▯

Recall that a digraph is **strongly connected** means that for any two vertices x and y, there is a dipath from x to y. How can the output from the Floyd-Warshall Algorithm be used to test a digraph for strong connectivity? How can the (input matrix and the) output from the Floyd-Warshall Algorithm be used to determine a shortest dipath from a given vertex y to some other given vertex z? Equations 6.4.1 and 6.4.2 can be used as a basis for an algorithm to recover the shortest path from any particular vertex to any other.

The Most Important Ideas in this Section.
In an arc-weighted digraph, the distance function δ may be defined so that $\delta(x, y) =$ the length of a *shortest* directed path *from x to y* (if there is one).

   Dijkstra's Algorithm finds a shortest dipath from a given vertex to another given vertex, in a simple arc-weighted digraph, using vertex labels to grow a tree of shortest dipaths from a single source.

   The *Floyd-Warshall Algorithm* evaluates the entries in the distance matrix δ and is said to solve the "all-pairs" shortest dipath problem.

6.5   The Maximum Flow Problem

Imagine a group of travelers arrive at the Los Angeles airport wishing to travel to New York City that same day, and that the customer service agent for a particular airline tells them that that no seats are available on any direct flights that day, but she can book them to New York City using seats that are available on connecting flights through several intermediate cities: Denver, Houston, Chicago, and Atlanta. Below is a diagram of these connecting flights with the arcs labelled with the number of available seats on each. *How many travelers can be booked to go from LA to NY that day?*

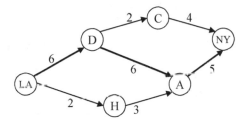

Five people can be sent from LA to D, D to A, and A to NY. In the next diagram, each arc α represents a connecting flight labelled by an ordered pair, "*f/c*" where $f(\alpha)$ is the number of travelers assigned to that flight, and $c(\alpha)$ is the "capacity" of that flight.

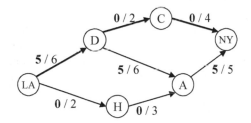

With the graph labelled in this manner, it becomes apparent that it is possible to send one more traveler from LA to D, D to C, and C to NY, as shown below:

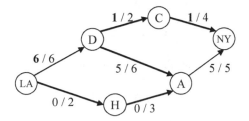

This diagram shows that <u>one</u> more traveler could get from D to NY via C, and <u>two</u> more could get from LA to A via H. If one traveler, *x* say, on the flight from D to A were diverted to go instead to C, and on to NY, then an additional traveler *y* could go from LA to A via H, and then (in what was previously *x*'s seat) from A to NY. This would give the trips displayed in the diagram below, with a total of 7 travelers able to make the trip from LA to NY. *// Is 7 the maximum?*

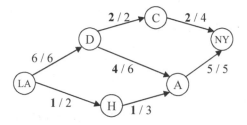

This is indeed the maximum, because any traveler from LA to NY must (at some point) leave the southwest group of cities {LA, D, H, A} on the flight from D to C or the flight from A to NY, and those two flights have a total of only 7 seats available.

The remainder of this section formalizes the ideas in this example and proves the Max-flow/Min-cut Theorem of Ford and Fulkerson from 1962. The proof is based on an algorithm that simultaneously constructs a maximum value flow and a minimum capacity "cut".

A *Flow Network* is a simple digraph $D(V, A)$ with one source s, one sink t, and a *capacity* on each arc given by a function $C: A \to \mathbf{N}$. They model networks that carry some commodity from place to place: travelers on airline flights, oil through pipelines where the capacity might be the number of barrels per day, water in pipes where the capacity might be the number of gallons per hour, manufactured items over a road system where the capacity might be the number of truckloads per week, etc.

In many applications the network is acyclic, but our treatment here allows for the possibility that there may be many cycles present. We assume the vertices in V are given in some order (though not necessarily the order of a topological sort because the flow network is not necessarily acyclic).

A *flow* is a function $F: A \to \mathbf{N}$.   // Flows on arcs are non-negative integers. Because the digraph is simple (having no loops nor strictly-parallel arcs) the arc from a to b can be (uniquely) represented as the ordered pair (a, b). A flow F is said to be *feasible* if it satisfies two sets of constraints:

1. *Capacity constraints*: For all arcs,

$$F(a, b) <= C(a, b).$$

2. *Conservation constraints*: For all intermediate vertices, $i \in V \setminus \{s, t\}$,

$$\sum\{F(i, x): (i, x) \in A\} = \sum\{F(y, i): (y, i) \in A\}.$$

// The total flow out of vertex i equals the total flow into vertex i.
// Flow does not originate at any vertex other than the single source s
// and is not absorbed at any vertex other than the single sink **t**.

The **value** of a *flow* F, written as $||F||$, is the total amount of the commodity that it moves from the source to the sink over the network. The main question is:

"What is the maximum value of a feasible flow over this given network?"

Let G_D denote the underlying, undirected graph obtained from the digraph D by replacing each arc by an undirected edge joining the same two vertices. If

$$\pi = \left(v_0, e_1, v_1, \ldots, v_{j-1}, e_j, v_j, e_{j+1}, v_{j+1}, \ldots, e_k, v_k\right)$$

is a path in G_D and α_j is the arc producing the edge e_j,

α_j is called a "forward arc" if it goes from v_{j-1} to v_j, and
α_j is called a "backward arc" if it goes from v_j back to v_{j-1}.

An **Augmenting Path** for a given flow F is a *simple* path in G_D where

on each forward arc, $F(\alpha) < C(\alpha)$ and
on each backward arc, $0 < F(\alpha)$.

If $\pi = \left(s = v_0, e_1, v_1, \ldots, v_{j-1}, e_j, v_j, \ldots, e_k, v_k = t\right)$

is an *augmenting path* from s to t for a feasible flow F, then a new feasible flow F^+ can be constructed with a **_larger_** value. Let Δ denote

minimum of $\{C(\alpha) - F(\alpha)\colon \alpha \text{ is forward in } \pi\} \cup \{F(\alpha)\colon \alpha \text{ is backward in } \pi\}$.

Then Δ is a positive integer. Now define a new flow on A as follows:

$$F^+(\alpha) = \begin{cases} F(\alpha) + \Delta \text{ if } \alpha \text{ is forward } \text{ in } \pi & \text{// } F(\alpha) + \Delta <= F(\alpha) + C(\alpha) - F(\alpha) \\ F(\alpha) - \Delta \text{ if } \alpha \text{ is backward in } \pi & \text{// } 0 <= F(\alpha) - \Delta <= C(\alpha) \\ F(\alpha) \qquad\qquad \text{ if } \alpha \text{ is not in } \pi \end{cases}$$

Thus F^+ takes non-negative integer values and satisfies all capacity constraints.

Now we must show that F^+ satisfies all the conservation constraints. Let i be any intermediate vertex. We know that $i \in V\backslash\{s, t\}$ and

$$\sum\{F(i, x) : (i, x) \in A\} = \sum\{F(y, i) : (y, i) \in A\}.$$

If i does not appear in π, flow is still conserved at i.   // at i, F-values remain the same
If i does appear in π, it occurs only <u>once</u> as v_j say, where $0 < j < k$. But there are now 4 cases to consider:

1. α_j and α_{j+1} are forward arcs,

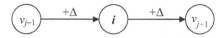

the flow into i and the flow out of i are both increased by Δ so flow is still conserved at i.

2. α_j is a forward arc but α_{j+1} is a backward arc,

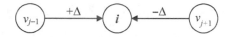

the flow into i from v_{j-1} is increased by Δ and the flow into i from v_{j+1} is decreased by Δ so the total flow into i is unchanged, the total flow out of i is unchanged, and flow is still conserved at i.

3. α_j is a backward arc but α_{j+1} is a forward arc,

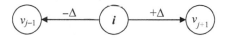

the flow out of i to v_{j-1} is decreased by Δ and the flow out of i to v_{j+1} is increased by Δ so the total flow out of i is unchanged, the total flow into i is unchanged, and flow is still conserved at i.

4. α_j and α_{j+1} are both backward arcs,

the flow out of i and the flow into i are both decreased by Δ so flow is still conserved at i. Thus F^+ is feasible, and the value of the flow has been increased (augmented) by the positive value of Δ.   // $||F^+|| = ||F|| + \Delta$.

A **cut** is a subset K of vertices that contains s but does not contain t. The "capacity" of cut K, written as $C(K)$, is $\sum\{C(a,x) : (a,x) \in A, a \in K$ but $x \in V\backslash K\}$. The set of arcs, $\{(a,x) \in A, a \in K$ but $x \in V\backslash K\}$ is sometimes called the **cut-set** determined by K.

// Now after all these definitions, some theorems:

Theorem 6.5.1: The Flow/Cut Equality
If F is any feasible flow and K is any cut, then
$$||F|| = \text{the value of } F = \text{the } \underline{\textbf{net}} \text{ flow out of } K.$$

Proof. We will use (just) the conservation constraints to prove that

$$||F|| = \sum\{F(s,x): (s,x) \in A\}$$
$$= \sum\{F(a,x): (a,x) \in A, a \in K \text{ but } x \in V\backslash K\} \qquad \text{// the total flow out of } K$$
$$- \sum\{F(y,b): (y,b) \in A, y \in V\backslash K \text{ but } b \in K\}. \qquad \text{// } - \text{the total flow into } K$$

Recall that for all intermediate vertices, $i \in V\backslash\{s, t\}$

$$\sum\{F(i,x): (i,x) \in A\} = \sum\{F(y,i): (y,i) \in A\}.$$

// Any flow introduced at s must be passed on through *all* the intermediate vertices.

For all vertices, $i \in K\backslash\{s\}$

$$\sum\{F(i,x): (i,x) \in A\} - \sum\{F(y,i): (y,i) \in A\} = 0.$$

Furthermore, summing all these differences, we get

$$\sum\{[\sum\{F(i,x): (i,x) \in A\} - \sum\{F(y,i): (y,i) \in A\}] : i \in K\backslash\{s\}\} = 0.$$

That is

$$0 = \sum\{[\sum\{F(i,x): (i,x) \in A\}]: i \in K\backslash\{s\}\}$$
$$- \sum\{[\sum\{F(y,i): (y,i) \in A\}]: i \in K\backslash\{s\}\}.$$

The value of F,

$$||F|| = \sum\{F(s,x): (s,x) \in A\} + 0$$
$$= \sum\{F(s,x): (s,x) \in A\} + \sum\{\sum\{F(i,x):(i,x) \in A\}: i \in K\backslash\{s\}\}$$
$$- \sum\{\sum\{F(y,i): (y,i) \in A\}: i \in K\backslash\{s\}\}$$
$$= \sum\{\sum\{F(a,x): (a,x) \in A\}: a \in K\}$$
$$- \sum\{\sum\{F(y,i): (y,i) \in A\}: i \in K\} \qquad \text{// there is no arc from } y \text{ into } s$$
$$\text{// then changing the variable } i \text{ to } b$$

$$= \sum\{F(a,x): (a,x) \in A, a \in K, x \in K\}$$
$$+ \sum\{F(a,x): (a,x) \in A, a \in K, x \in V\backslash K\}$$
$$- \sum\{F(y,b): (y,b) \in A, b \in K, y \in K\}$$
$$- \sum\{F(y,b): (y,b) \in A, b \in K, y \in V\backslash K\}$$

// The first and third sums both equal the total of all flows on arcs with
// both ends in K, so they "cancel" each other and we get

$$= \sum\{F(a,x): (a,x) \in A, a \in K, x \in V\backslash K\}$$
$$- \sum\{F(y,b): (y,b) \in A, b \in K, y \in V\backslash K\}.$$ ☐

Corollary 6.5.2: The Flow/Cut Inequality

If F is <u>any</u> feasible flow and K is <u>any</u> cut, then

$$\|F\| = \text{the value of } F \leq \text{the capacity of } K = C(K).$$

Proof.

$$\|F\| = \sum\{F(s,x): (s,x) \in A\}$$
$$= \sum\{F(a,x): (a,x) \in A, a \in K \text{ but } x \in V\backslash K\}$$
$$- \sum\{F(y,b): (y,b) \in A, b \in K \text{ but } y \in V\backslash K\}$$
$$\leq \sum\{F(a,x): (a,x) \in A, a \in K, \text{ but } x \in V\backslash K\} \qquad \text{// each } F(y,b) \geq 0$$
$$\leq \sum\{C(a,x): (a,x) \in A, a \in K, \text{ but } x \in V\backslash K\} \qquad \text{// each } F(a,x) \leq C(a,x)$$
$$= C(K) \hspace{7cm} ☐$$

Corollary 6.5.3: The Max-Flow/Min-Cut Inequality

If F is a feasible flow of *maximum* value and K is a cut of *minimum* capacity, then

$$\|F\| = \text{the value of } F \leq \text{the capacity of } K = C(K).$$

Proof. Corollary 6.5.2 applies for this particular flow and this particular cut.   ☐

Theorem 6.5.4: The Max-Flow/Min-Cut Theorem

If F is <u>any</u> feasible flow of maximum value and K is <u>any</u> cut of minimum capacity, then

$$\|F\| = \text{the value of } F = \text{the capacity of } K = C(K).$$

Proof. We will give an algorithm that constructs a feasible flow F^* and a cut K^* such that

$$\|F^*\| = \sum\{F^*(s,x): (s,x) \in A\}$$
$$= \sum\{C(a,x): (a,x) \in A, a \in K^* \text{ but } x \in V\backslash K^*\} = C(K^*).$$

Then,   for *any* flow F, $||F|| <=$ capacity of $K^* =$ the value of $F^* = ||F^*||$

so F^* is a flow of *maximum* value.

Furthermore,   for *any* cut K, $C(K) >=$ the value of $F^* =$ capacity of $K^* = C(K^*)$

so K^* is a cut of *minimum* capacity.   ⬚

Algorithm 6.5.1: The Ford-Fulkerson Algorithm

1. Grow a tree T of <u>simple augmenting paths</u> from the source s, adding new
   vertices one at a time, until

   either (a) the sink, t, enters the tree

   or   (b) the sink is not in the tree, but <u>no</u> new vertex can bc added T.

2. If the sink enters the tree, augment the current flow, and go back and repeat step 1.
When (b) occurs, the vertex set of the final tree is a *cut* that verifies that the current
flow has maximum possible value.

The value of the current flow is augmented by at least one unit in step 2, and is
bounded above by $C(\{s\})$ say, so (b) must eventually occur, and the algorithm must
terminate.

We will describe the algorithm for constructing augmenting paths for <u>any</u> feasible
flow in words, avoiding a description of a particular implementation for the graph
(and, as before, we will follow the algorithm with a walkthrough using diagrams).

But first we make one more (non-standard) definition. We will say an arc α is
"free at v" (free to be added to the tree T of augmenting paths from vertex v)

if   (a) the other end w of α is <u>not</u> already in the tree;

and   (b) if α is a forward arc, then $F(\alpha) < C(\alpha)$;

but   (c) if α is a backward arc, then $0 < F(\alpha)$.

Algorithm 6.5.2: Growing a Tree of Augmenting Paths

```
Begin
    Mark all vertices "unscanned";
    Put s into the tree, T, as the root;
    While (T has an unscanned vertex v) and (the sink, t, is not in T) Do
        If (there is a "free" arc α ending at v) Then
            Let w denote the other end of α;
            Add w and α to T;      // T stays connected and no polygon is created
            P(w) ← v;              // call v the "Parent" of w
        Else                       // now all the free arcs at v are in the tree T
            Mark vertex v as "scanned";
        End;
    End;                           // either t has just been put into T,
End.                               // or t not in T, but all vertices in T have been "scanned"
```

This algorithm provides a great deal of freedom when selecting vertex v and arc α.
We could imitate a ***Breadth-First Traversal*** by using a queue to store the unscanned

Tree
#6

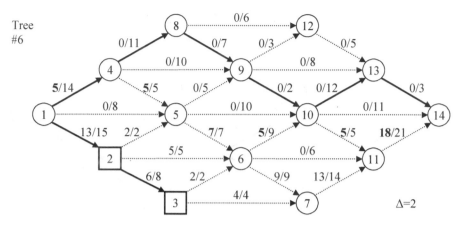

Tree
#7

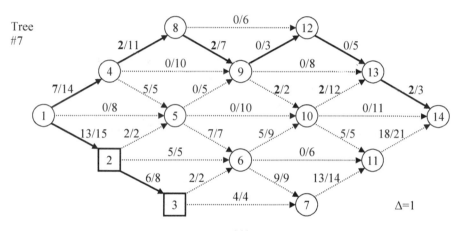

Tree
#8

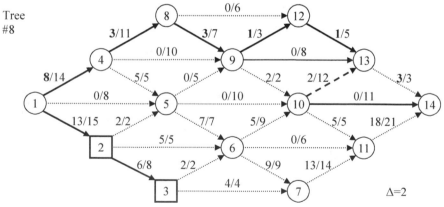

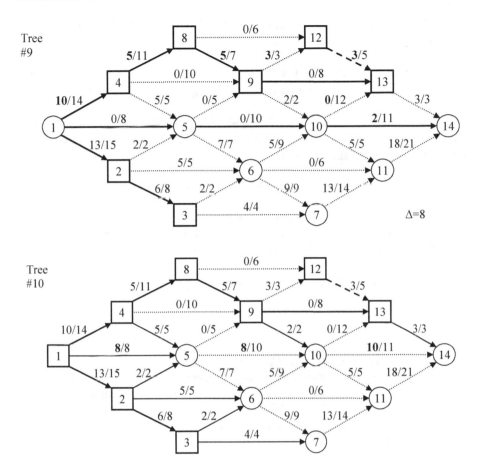

The final tree (#10) shows a feasible flow with value 31 units. The cut that is the vertex set of this final tree has a capacity of 31 units. Therefore this flow is of maximum value and this cut is of minimum capacity.

Theorem 6.5.5: Algorithm 6.5.1, the ***Ford-Fulkerson Algorithm***, constructs a feasible flow F of maximum possible value. It also produces a "certificate of optimality", a cut K where $||F|| = C(K)$.

Proof. Each time the algorithm constructs an augmenting path for the current feasible flow, that flow can be altered to give a new feasible flow with a strictly larger value. We will prove that

if   the algorithm terminates with a tree T of augmenting paths that does not contain the sink t and where no new arc can be added,

then the current flow F has the maximum possible value of a feasible flow and the vertex set of the final tree is a cut K where $||F|| = C(K)$.

Let K denote the vertex set of the tree T. Then K contains s but does not contain t, so K is indeed a "cut". By Theorem 6.5.1

$$||F|| = \sum\{F(a,x)\colon (a,x) \in A, a \in K \text{ but } x \in V \backslash K\} \qquad \text{// total flow out of } K$$
$$-\sum\{F(y,b)\colon (y,b) \in A, y \in V \backslash K \text{ but } b \in K\}. \qquad \text{// } -\text{total flow into } K$$

If there were an arc, (y, b), where vertex y is not in the tree, vertex b is in the tree, *and* $F(y, b) > 0$, then this arc could be added to the tree as a *backward* arc. Since no such arc can be added, $F(y, b)$ must equal 0. Thus,

$$||F|| = \sum\{F(a,x)\colon (a,x) \in A, a \in K \text{ but } x \in V \backslash K\} \qquad \text{// total flow out of } K$$

If there were an arc, (a, x), where vertex x is not in the tree, vertex a is in the tree, *and* $F(a, x) < C(a, x)$, then this arc could be added to the tree as a *forward* arc. Since no such arc can be added, $F(a, x)$ must equal $C(a, x)$. Thus,

$$||F|| = \sum\{C(a,x)\colon (a,x) \in A, a \in K \text{ but } x \in V \backslash K\} \qquad \text{// the capacity of } K$$
$$= C(K) \qquad\qquad\qquad\qquad\qquad\qquad\qquad\qquad\qquad\qquad □$$

The Most Important Ideas in this Section.
Given any commodity Flow Network we can find a feasible flow of maximum possible value, and we can certify its optimality by exhibiting a cut whose capacity equals the value of the flow. The **Max-Flow/Min-Cut Theorem** is proved by the action of the *Ford-Fulkerson Algorithm*.

6.6   Matchings in Bipartite Graphs

In a loopless, undirected graph $G(V, E)$, a *matching* is a subset M of edges where no two have a common end point. Matchings "pair up" vertices like roommates in a dorm, or tasks and workers, or jobs and machines. A matching is said to be *perfect* if every vertex in G is matched with some other vertex. The main problem discussed here is: *How can we find a matching that is as large as possible*? This section will show how matchings in bipartite graphs can be related to flow problems.

A *vertex-cover* of G is a subset W of its vertices such that every edge in G has at least one end in W. V is a vertex-cover, but we would like to find a "small" vertex-cover; as small as possible if we can.

//X Show that W is a **vertex-cover** of G if and only if $V \setminus W$ is an *independent* set of
//X vertices in G.

Theorem 6.6.1: The Matching/Vertex-Cover Inequality

If M is $\underline{any}$ matching and W is $\underline{any}$ vertex-cover, then $|M| <= |W|$.

Proof. Suppose that M is a matching and W is a vertex-cover. Since each edge e in
M has an end-point $f(e)$ in W, and since no two edges in M have an end-point in
common, the set of vertices, $\{f(e): e \in M\}$, is a subset of W that is the same size as
M. Thus $|M| <= |W|$. $\square$

In particular, we have

Corollary 6.6.2: The Max-Matching/Min-Cover Inequality
If M is a matching of *maximum* cardinality, and W is a vertex-cover of *minimum*
cardinality, then $|M| <= |W|$.

If G is a polygon with n vertices (and n edges) then any matching has at most
$\lfloor n/2 \rfloor$ edges.   // all the ends of the matching edges must be different
Any vertex covers exactly 2 edges, so at least $\lceil n/2 \rceil$ vertices are needed to cover all
n edges. If $n = 7$, any matching has at most 3 edges, and any vertex-cover has at
least 4 vertices.

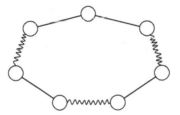

In this example, the inequality of Corollary 6.6.2 is strict.   // it's not an equality
But we will prove that for *bipartite* graphs the inequality of Corollary 6.6.2 is never
strict; it's always an equation.

We will say that an edge "covers" its end points, and a matching M "covers" all
the end points of all the edges in M. For any set of vertices X in any graph, let
$N(X)$ denote all vertices joined by an edge to some vertex in X.
                                        // all neighbours of all vertices in X

Theorem 6.6.3: Suppose that X is an independent set of vertices in a graph G.
If there is a matching M that covers X then $|X| <= |N(X)|$.

Proof. Each vertex v in X is matched by M to a unique vertex $m(v)$. Because X is
independent, each $m(v) \in [N(X) \setminus X]$. Each $m(v)$ is different, so $\{m(v): v \in X\}$ is a

subset of $N(X)$ of the same size as X. Therefore, $|X| <= |N(X)|$. Furthermore, if $|X| > |N(X)|$ then there is no matching M that covers X.   // the contra-positive

□

In particular, if $G(V, E)$ is bipartite with V partitioned into independent sets L and R, then any subset of L (or R) is independent. We will show (near the end of this section) that either   (1) there is a matching in G that covers L;
or   (2) there is a subset X of L where $|X| > |N(X)|$.   // but not both

Example 6.6.1: Consider the matching in the graph below where the 4 matching edges are the ones drawn with "heavy" lines.

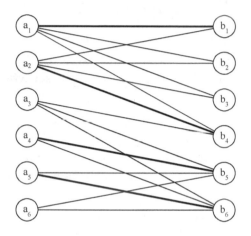

Is this a **maximal** matching? Here we use "maximal" to mean "is not a proper subset of a larger matching". **Yes**, it is maximal. No new edge can be added to it to make a larger matching.   // The 8 ends of these 4 matching edges are a vertex-cover.

Is this a **maximum cardinality** matching? Is it a matching of largest possible **size**? **No**, it is not. We will convert this to a flow problem where we know how to find a flow of maximum possible value in a network.

Suppose that $G(V, E)$ is a bipartite graph where V is partitioned into subsets L and R and every edge joins a vertex in L to one in R. Create a flow network $N(G)$ as follows:

1. Begin with V as the vertex set of $N(G)$.
2. Orient all the edges as arcs from set L on the left to set R on the right.
                // Then all L-vertices are sources, and all R-vertices are sinks.
3. Add a "super-source" s, and an arc *from* it *to* each "left-source".
        // This can be done for any commodity flow network with several sources.
4. Add a "super-sink" t, and an arc *to* it *from* each "right-sink".
        // This can be done for any commodity flow network with several sinks.
5. Let V^* denote $V \cup \{s, t\}$, the vertex set of $N(G)$.

6. Give every arc a capacity of 1 unit.   // But don't put capacities in the diagrams.

Suppose also that **M** is a matching in **G**.   // perhaps the empty set of edges
Create a <u>feasible flow</u> **F**_M in **N(G)** as follows:

> If (l_i, r_j) is an arc that is the orientation of a matching edge, then
> assign a flow of 1 unit to the 3 arcs (s, l_i), (l_i, r_j) and (r_j, t).
> But assign all other arcs a flow of zero.   // The value of this flow equals $|M|$.

Applying this construction to Example 6.6.1, where the arcs are labelled with their
flows, gives:

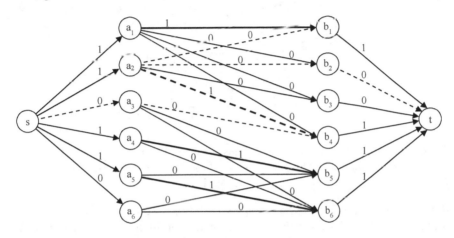

// Is this initial flow of maximum value?
By applying the Ford-Fulkerson Algorithm, exactly as we did in the walkthrough in
the previous section, we obtain a flow with value 5.

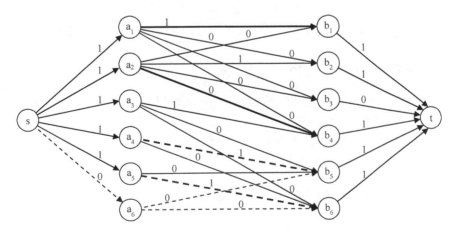

// Is this flow of maximum value?

Applying the Ford-Fulkerson algorithm again, exactly as we did in the walkthrough in the previous section, we obtain a cut with capacity 5. And there is a vertex-cover of G of size 5.

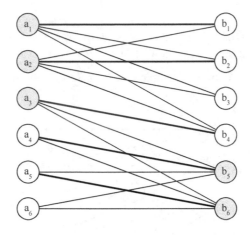

Theorem 6.6.4: If bipartite graph G has a matching M of size p then the flow network $N(G)$ has a feasible flow F_M of value p, and if $N(G)$ has a feasible flow F of value q then G has a matching M_F of size q.

Proof. Let $G(V, E)$ be a bipartite graph where V is partitioned into subsets L and R and every edge joins a vertex in L to one in R. Construct the flow network $N(G)$.

Suppose M is a matching of size p. Construct the feasible flow F_M corresponding to M as we did in the example above. Let K be the cut $\{s\} \cup L$. Then by Theorem 6.5.1 (the Flow/Cut Equality)

$$\|F_M\| = \sum\{F_M(a,x): (a,x) \in A, a \in K \text{ but } x \in V^*\backslash K\} \qquad // \text{ total flow out of } K$$
$$- \sum\{F_M(y,b): (y,b) \in A, y \in V^*\backslash K \text{ but } b \in K\}. \qquad // - \text{ total flow into } K$$

Since there are no arcs into K from $V^*\backslash K$,

$$\|F_M\| = \sum\{F_M(a,x): (a,x) \in A, a \in K \text{ but } x \in V^*\backslash K\} \qquad // \text{ total flow out of } K$$
$$= \sum\{F_M(a,x): (a,x) \in A, a \in L \text{ but } x \in R\}$$
$$= \# \text{ arcs from } L \text{ to } R \text{ given a flow of 1 unit}$$
$$= |M| = p.$$

Thus, if G has a matching of size p then the flow network $N(G)$ has a feasible flow of value p.

Suppose $N(G)$ has a feasible flow F of value q. As above, if K is the cut $\{s\} \cup L$,

$$\|F\| = \sum\{F(a,x): (a,x) \in A, a \in K \text{ but } x \in V^*\backslash K\} \qquad \text{// total flow out of } K$$
$$= \sum\{F(a,x): (a,x) \in A, a \in L \text{ but } x \in R\}$$
$$= \# \text{ arcs from } L \text{ to } R \text{ given a flow of 1 unit}$$
$$= q.$$

Each arc in $N(G)$ from L to R corresponds to an edge in G. Let M_F denote the set of edges in G that correspond to arcs from L to R given a flow of 1 unit. Since no vertex a in L can have 2 arcs out of it with a flow of 1, no 2 edges in M_F meet at vertex a. Since no vertex b in R can have 2 arcs into it with a flow of 1, no 2 edges in M_F meet at vertex b. Thus, M_F is a <u>matching</u> in G of size q.   □

This theorem implies that we can find a maximum cardinality matching M^* in a bipartite graph G by finding a maximum valued flow F^* in the flow network $N(G)$.
// M^* will be M_{F^*}.
// But what about a minimum cut K^* in $N(G)$?
// And what about a vertex-cover W in G?

Let T be the <u>final</u> tree of augmenting paths produced by Algorithm 6.5.1, the Ford-Fulkerson Algorithm, acting on $N(G)$, the flow network associated with bipartite graph G. Also let n denote $|L|$. Either all arcs from s to L have a flow equal 1 unit or there is at least one vertex a^* in L with $F(s,a^*) = 0$.

Case 1. Suppose no arc from s has a flow of zero.
Then there is a flow of 1 unit to every vertex a in L and from every vertex a in L to a unique vertex $m(a)$ in R. The value of the flow will be $|L| = n$ and $\{s\}$ will be a minimum capacity cut. In this case, the edges in G corresponding to the arcs $(a, m(a))$ will be a matching M^* of size n, and L will be a vertex-cover in G of size n. M^* covers L.   // What about the other possibility?
Case 2. Suppose there is at least one vertex a^* in L with $F(s,a^*) = 0$.

Theorem 6.6.5: If T is the final tree produced by the **Ford-Fulkerson Algorithm**, and a is a vertex in $L \cap V(T)$, then all out-neighbours of a are also in T. // in $R \cap V(T)$

Proof.   // Considering $F(s, a) = 0$ and $F(s, a) = 1$ separately.

Suppose $F(s, a) = 0$.   // Whether a is a^* or not.
Because the total flow into a is 0, there can be no arc from a to $b \in R$ with $F(a,b) = 1$. Vertex a would have been put into T when s was being scanned using (s, a) as a forward arc. Then any arc from a to $b \in R$ has $F(a, b) = 0 < C(a, b)$, so this arc would be added to T unless b is already in T. Thus, all out-neighbours b of a are also in T. Also, $a^* \in L \cap V(T)$ but there is no arc from a^* to $b \in R$ with $F(a^*, b) = 1$.

Suppose $F(s, a) = 1$. Then the total flow out of a is 1, and there is a unique vertex $m(a) \in R$ where $F(a, m(a)) = 1 = C(a, m(a))$. Because (s, a) is the only arc into a and $F(s, a) = C(s, a)$, vertex a must have been put into T as the start-end of a <u>backward</u> arc with positive flow. That arc must have been $(a, m(a))$ and $m(a)$ must have been in T. Then any arc from a to $b \in R\backslash\{m(a)\}$ has $F(a,b) = 0 < C(a,b)$, so this arc would be added to T unless b is already in T. Thus, all out-neighbours of a are also in T, including $m(a)$.   □

Therefore, no arcs in $N(G)$ go from a vertex in $L \cap V(T)$ to $R \backslash V(T)$. In fact, $\{L \cap V(T)\} \cup \{R \backslash V(T)\}$ is an <u>independent</u> set of vertices in G. There are several consequences of this theorem.

1. All arcs from the minimum cut, $V(T)$ to its complement in $N(G)$ must go from the source s to a vertex in $L \backslash V(T)$ or from $R \cap V(T)$ to the sink t. All these arcs must have a flow of 1 unit equaling the capacity of the arc. The capacity of the minimum cut must be

$$Q = |L \backslash V(T)| + |R \cap V(T)| \quad = \text{the value of the maximum flow in } N(G)$$
$$= \text{the size of the maximum matching in } G.$$

2. Since no edge in G goes from a vertex in $L \cap V(T)$ to $R \backslash V(T)$,

$$K = \{L \backslash V(T)\} \cup \{R \cap V(T)\} \text{ is a vertex-cover in } G \text{ of size}$$
$$Q = |L \backslash V(T)| + |R \cap V(T)| = \text{the size of the maximum matching in } G.$$

Theorem 6.6.6: If T is the final tree produced by the ***Ford-Fulkerson Algorithm***, and b is a vertex in $R \cap V(T)$, then there is a unique vertex c in $L \cap V(T)$ where $F(c, b) = 1$.

Proof. For all vertices b in R and T, $F(b, t)$ must be 1 since otherwise, this arc could be added to T and the sink t would enter the tree. Therefore the total flow into b must be 1 and so there must be a unique vertex c in L where $F(c, b) = 1$. Then the arc (c, b) could be added to T as a backward arc into b, unless c is already in the tree. Hence, c must be in L and the tree T. However, c is not equal to a^*.   □

Thus every vertex in $R \cap V(T)$ is matched with a vertex in $L \cap V(T)$ by the matching M_F in G determined by the flow F, so $|R \cap V(T)| <= |L \cap V(T)|$. However, a^* is in $|L \cap V(T)|$ but is not matched to any vertex, so $|R \cap V(T)| < |L \cap V(T)|$. Therefore, if $X = L \cap V(T)$, then $N(X) \subseteq R \cap V(T)$ so $|X| > |N(X)|$.

Summarizing these two cases, we have proved what's known as

(Philip) Hall's Marriage Theorem (1935): If $G(V, E)$ is bipartite with V partitioned into independent sets L and R, then
either (1) there is a matching in G that covers L;
or   (2) there is a subset X of L where $|X| > |N(X)|$.   // but not both

// This was called the "Marriage Theorem" because of the setting where L is the set of n
// girls in a certain town, and R the set of n boys, and each girl would only accept a
// proposal from a few boys in some small subset of R. Then ***either*** there is a perfect

// matching (a set of *n* perhaps proposed marriages involving all the boys and all the
// girls) *or* there is a set of picky girls, *X*, all of whom will only accept proposals
// from the boys in a strictly smaller set, *Y*.

> **The Most Important Ideas in this Section.**
> A *matching* is a subset *M* of edges where no two have a common end point.
> A *vertex-cover* of *G* is a subset *W* of vertices such that every edge in *G* has at
> least one end in *W*. If *M* is <u>any</u> matching and *W* is <u>any</u> vertex-cover, then
> $|M| <= |W|$.
>
> If $G(V, E)$ is a *bipartite* graph where *V* is partitioned into subsets *L* and
> *R* and every edge joins a vertex in *L* to one in *R*, we can create a flow network
> $N(G)$ where *G* has a matching of size *q* if and only if $N(G)$ has a feasible flow
> of value *q*. A maximum size matching M^* in *G* is produced by finding a flow
> of maximum value in $N(G)$, using the Ford-Fulkerson algorithm say. That
> algorithm also produces a vertex-cover *W* of *G* where $|W| = |M^*|$, so *W* is as
> small a cover as possible. The section ends with a proof of **Hall's Marriage
> Theorem** that either (1) there is a matching in *G* that covers *L* or (2) there is a
> subset *X* of *L* where $|X| > |N(X)|$.

Exercises

1. Answer the following questions about the graph shown below:
    (a) What are the out-neighbours of v_1?
    (b) What are the out-neighbours of v_3?
    (c) What are the in-neighbours of v_3?
    (d) What are the in-neighbours of v_5?
    (e) What is the set of source vertices?
    (f) What is the set of sink vertices?
    (g) Is the graph simple? Justify your answer.

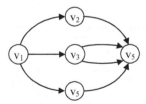

2. A *tournament* with *n* players/teams is an orientation of the complete graph
    K_n. The graph models wins and losses in a round robin competition. Every
    player/team plays one game against each of the others, and the edge between
    vertex *a* and vertex *b* is directed from *a* to *b* if *b* defeats *a*. Assume no ties
    are allowed.
    (a) Does every tournament have a sink? (A champion who defeats all the
        others?)
    (b) Orient K_5 so every player/team wins 2 games and loses 2 games.

(c) Can K_6 be oriented so every player/team wins the same number of games?

(d) Prove that if a has more wins than losses, then there must be a b who has more losses than wins.

(e) Prove that <u>every</u> tournament contains a directed Hamilton Path.

3. Find an orientation that produces a strongly connected digraph for each of the graphs below.

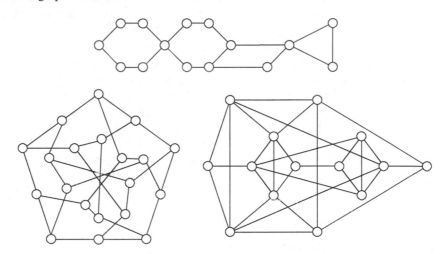

4. Use Algorithm 6.3.1 to topologically sort the vertices in the following graph.

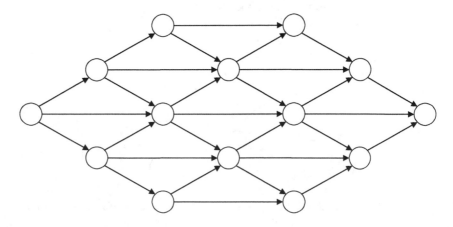

5. Find a topological sorting or find a directed cycle in each of the digraphs below.

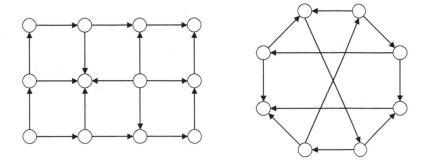

6. When is there only one topological sorting of an acyclic digraph?
7. Orient the graph below to produce an acyclic digraph.

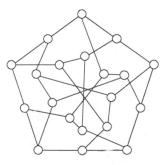

8. Use Algorithm 6.3.2 to count the number of dipaths from v_1 to v_2 in each of the following graphs.

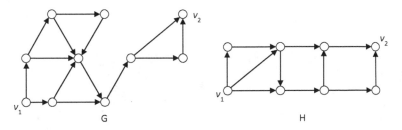

9. Using Algorithm 6.3.3, find the shortest dipath from v_1 to v_2 in each of the following weighted digraphs:

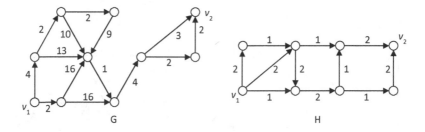

G   H

10. Using Algorithm 6.3.4, find the longest dipath from v_1 to v_2 in each of the weighted digraphs from the previous question.

11. In the acyclic arc-weighted digraph below, find the number of dipaths from a to b, the length of a shortest dipath from a to b, and the length of a longest dipath from a to b.

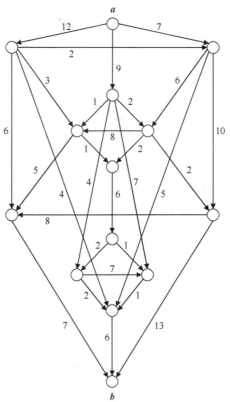

12. Prove that when the jth vertex v_j is added to T in Dijkstra'a Algorithm, $L(v_j) >= L(x)$ for all vertices x in T, and $L(v_j) <= L(w)$ for all vertices w not in T.

13. The *diameter* of a connected graph is the length of a longest shortest path, the *radius* from a given vertex v is the length of a longest shortest path from v, and the *center* of a graph is a vertex with the smallest radius.

                                                    // the shortest longest shortest path?

By referring to <u>directed</u> paths, the notions of diameter, radius and center also apply to strongly connected digraphs.

Find the diameter and center of the directed graph below.

Hint. You could use the Floyd-Warshall Algorithm to get the distance matrix.

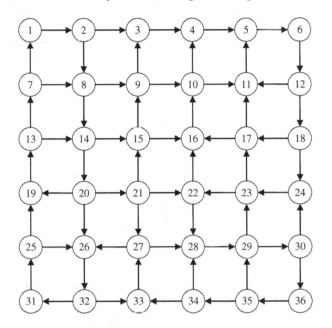

14. Write an algorithm to construct a shortest nontrivial dipath from a given vertex y to another given vertex z using the input and output from the Floyd-Warshall Algorithm.

15. Use the Ford-Fulkerson algorithm to compute the maximum flow from s to t in the following graphs where each arc is labelled with its capacity.

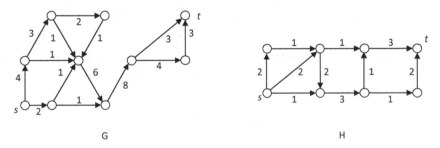

G   H

16. Find a maximum flow in the "standard example" of Sect. 6.5 by applying the Ford-Fulkerson Algorithm but starting with the initial flow given below.

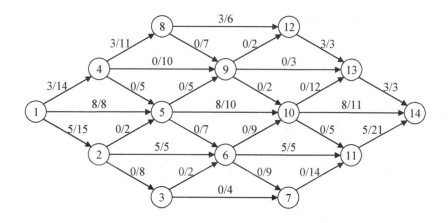

Is the output flow the same as that produced in Sect. 6.5?
Is the output cut the same as that produced in Sect. 6.5?

17. Consider the digraph shown in question 11. Interpret the arc weights as capacities. Find a maximum flow from the single source (vertex **a**) to the single sink (vertex **b**) and a minimum cut.

18. Show that **W** is a *vertex-cover* of **G** if and only if $V \setminus W$ is an *independent* set of vertices in **G**.

19. Find a maximum cardinality matching in the bipartite graph given below.

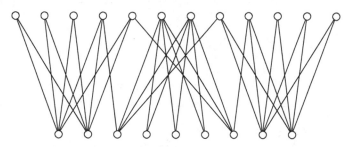

20. If $b(v)$ is a non-negative integer assigned to each vertex v in an undirected graph $G(V,E)$, then a *b-matching* is a subset of edges **M** such that for each vertex v the number of edges in **M** that have v as an end-point is $<= b(v)$. So ordinary matchings are *b*-matchings where all *b*-values are 1.

   (a) If **G** is bipartite, how can a flow network be created so the max-flow algorithm finds a maximum cardinality *b*-matching?

   (b) In the graph given in Exercise 19, find a maximum cardinality *b*-matching if all *b*-values are 2.

   (c) In the graph given in Exercise 19, find a maximum cardinality *b*-matching if all *b*-values are 3.

21. Given a bipartite graph $G(V, E)$ and a positive integer k, use Hall's Marriage Theorem to prove that:
    (a) If all vertices in G have degree k, then G has a perfect matching.
    (b) If all vertices in G have degree k, then E can be partitioned into k perfect matchings.

22. In our reduction of bipartite matching problems to flow problems, a flow can have its value increased if there is an "augmenting path" for the current flow. That path in $N(G)$ corresponds to a path in G itself as illustrated below.

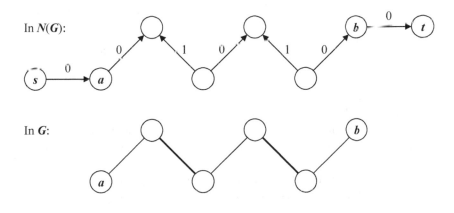

Define an ***augmenting path*** for a matching M to be a path in $G(V, E)$, which may be bipartite or not, as a path of odd length from an "uncovered" vertex a to a different "uncovered" vertex b, and where the edges are alternately not in M and then in M.

(a) Prove that if matching M has such an augmenting path then M is not a maximum cardinality matching.

(b) Prove that if matching M is not a maximum cardinality matching then M has such an augmenting path. Hint: Given two matchings M and M^* where $|M| < |M^*|$, then $|M \backslash M^*| < |M^* \backslash M|$, and so some component of the subgraph whose edge set is $(M \backslash M^*) \cup (M^* \backslash M)$ has more edges from M^* than M.

(c) Find a perfect matching in the graph below.

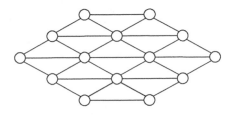

23. Consider the graph **G** below.

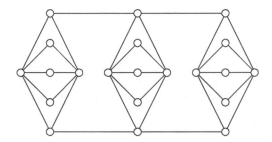

(a) Show that **G** is bipartite by coloring the vertices Red and Blue so every edge joins a Red vertex and a Blue one.

(b) Find a vertex-cover of size 8.

(c) Find a matching of maximum size.

(d) Find an independent set of size 13.

(e) Explain why your independent set has maximum size.

24. (a) Prove that if $G(V, E)$ has a Hamilton Path and $|V|$ is even, then **G** has a perfect matching.

(b) Construct an example of a connected graph $G(V, E)$ where $|V|$ is even and **G** has a perfect matching but where **G** does not have a Hamilton Path.

                                                              // A tree might work.

(c) Prove that if $G(V, E)$ has a Hamilton Circuit and $|V|$ is even, then **G** has two disjoint perfect matchings.

(d) Construct an example of a connected graph $G(V, E)$ where $|V|$ is even and **G** has two disjoint perfect matchings but where **G** does not have a Hamilton Circuit.

                                                              // Try 2 squares joined by an edge.

Relations: Especially on (Integer) Sequences

7

There are two main objectives of this chapter: (1) to provide an "ordering" of sequences of objects so they can be sorted, like individual numbers, and so any finite set of them can be generated (see Chap. 9) in a natural order from the first to the last, and (2) to provide a mechanism for classifying complexity functions by their rates of growth. Relations will be a means to realize both these goals.

7.1 Relations and Representations

The relations in the chapter title don't refer to aunts, uncles, and cousins but more to special or intimate personal relations. A *relation*, R, on some underlying set, S, is some characteristic of certain ordered pairs of elements of S, namely,

on numbers: $a = b$
                  $a < b$
                  $a >= b$

on integers: $a \mid b$

on subsets: $A \subseteq B$
                $|A| = |B|$

on people: *a is married to b*
               a has a crush on b
               a is younger than b
               a is a descendant of b.

© Springer International Publishing AG, part of Springer Nature 2018   299
T. Jenkyns and B. Stephenson, *Fundamentals of Discrete Math for Computer Science: A Problem-Solving Primer*, Undergraduate Topics in Computer Science,
https://doi.org/10.1007/978-3-319-70151-6_7

In every case, R may be identified with the set of ordered pairs connected by the relation. We <u>define</u> a ***relation*** on set S to be <u>any</u> set of ordered pairs of elements of S; that is, any set

$$R \subseteq S \times S.$$

Generalizing the notation of the examples, let's write

$$a \ R \ b \quad \text{to denote} \quad (a, b) \in R.$$

Then "$a \ R \ b$" is a statement and therefore takes a Boolean value. The negation of this is also a Boolean expression, and (as in Java) let's write

$$a \ !R \ b \quad \text{to denote} \quad (a, b) \notin R.$$

Example 7.1.1: Let $S = \{1,2,3,4\}$ and let

$$R = \{(1, 1), (1, 3), (1, 4), (2, 1), (2, 2), (2, 3), (2, 4), (3, 2), (4, 2)\}.$$

Here, $1 \ R \ 1$ but $4 \ !R \ 4$ and $1 \ R \ 3$ but $3 \ !R \ 1$.

Besides listing the pairs in R, there are other ways of representing a relation. We'll look at two of them.

7.1.1   Matrix Representation

Choose <u>any</u> ordering of the set S and use this ordering for the row and column indices of a square matrix M. Then define the entries in M by

$$M[a,b] = \begin{cases} 1 & \text{if } a \ R \ b \\ 0 & \text{if } a \ !R \ b \end{cases}.$$

We could give M Boolean values but it is more conventional to use 0's and 1's in such "characteristic" matrices.

The matrix representation of Example 7.1.1 is then

	1	2	3	4
1	1	0	1	1
2	1	1	1	1
3	0	1	0	0
4	0	1	0	0

7.1.2   Directed Graph Representation

We can construct a directed graph, D, to display the relation R on set S by letting S be the set of vertices of D and then drawing an *arc* (an arrow or a directed edge) from a to b whenever $a\ R\ b$. The directed graph representation of Example 7.1.1 is then

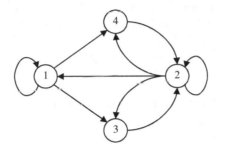

7.1.3   Properties of Relations

Next, we define five properties a relation might have: reflexivity, symmetry, anti-symmetry, transitivity, and comparability. We will use these generic attributes to distinguish certain kinds of relations, especially those indicating similarity and those indicating a ranking (one thing is bigger, better, or sexier than another).

A relation R on S is ***reflexive*** <u>means</u>
  $a\ R\ a\quad \forall\ a \in S.$
                                              // "equals" is reflexive; "<" is not.
A relation R on S is ***transitive*** <u>means</u>
  if $a\ R\ b$ and $b\ R\ c$ then $a\ R\ c\quad \forall a,b,c \in S.$
                               // "<" is transitive; "has a crush on" is not.
A relation R on S is ***symmetric*** <u>means</u>
  if $a\ R\ b$ then $b\ R\ a\quad \forall a,b \in S.$
                               // "is married to" is symmetric; ">" is not.
A relation R on S is ***antisymmetric*** <u>means</u>
  if $a\ R\ b$ and $b\ R\ a$ then $a = b\quad \forall a,b \in S.$
                        // "$\subseteq$" is anti-symmetric; "is married to" is not.
                        // is "<" anti-symmetric?
A relation R on S has the ***comparability property*** <u>means</u>
  either $a\ R\ b$ or $b\ R\ a$ or $a = b\quad \forall a,b \in S.$
  // Both "<" and "<=" have the comparability property; "$\subseteq$" does not.

Symmetry and antisymmetry are not opposites because "equals" has both properties, and some relations, like R in Example 7.1.1, have neither.

// Which properties does R in Example 7.1.1 have?
//   4 !R 4 $\Rightarrow$ R is not reflexive.
//   1 R 3 and 3 R 2 but 1 !R 2 $\Rightarrow$ R is not transitive.
//   1 R 3 but 3 !R 1 $\Rightarrow$ R is not symmetric.

// 2 R 3 and 3 R 2 but 2 $\neq$ 3 $\Rightarrow$ R is not antisymmetric.
// 3 !R 4 and 4 !R 3 and 3 $\neq$ 4 $\Rightarrow$ R does not have the comparability property.

The last three properties are really only of interest when $a \neq b$, so we will rephrase them in that case.

A relation R on S is *symmetric* <u>means</u>
      whenever a and b are distinct elements of S
        if a R b, then b R a.
A relation R on S is *antisymmetric* <u>means</u>
      whenever a and b are distinct elements of S
        if a R b, then b !R a.   // "$<$" <u>is</u> antisymmetric.
A relation R on S has the *comparability property* <u>means</u>
      whenever a and b are distinct elements of S
        if a !R b, then b R a.

Distinct elements x and y are said to be *comparable* (under R) if either x R y or y R x; otherwise, x and y are *incomparable*.

// Which properties do the examples at the beginning of the chapter have?
// How are the properties reflected by the matrix?
// How are the properties reflected by the digraph?

The Most Important Ideas in This Section.
A *relation* R on set S is <u>any</u> set of ordered pairs of elements of S and may be represented as a matrix or as a digraph. But this definition is too general to be of any use. Relations are classified by what properties they possess, particularly the five properties: *reflexivity*, *symmetry*, *antisymmetry*, *transitivity*, and *comparability*.

7.2   Equivalence Relations

In this section, we formalize the idea of "similarity". Any object is similar to itself; if a is similar to b, then b is similar to a, and if a is similar to b and b is similar to c, then a is similar to c.

A relation R is an *equivalence relation* whenever R is *reflexive*, *symmetric*, and *transitive*. Equality (an extreme case of similarity) is an equivalence relation, but there are many others. A fairly general but simple construction is the following.

Example 7.2.1: Let f be any function with domain S. Define a relation R on S by

$$a \ R \ b \ \text{if and only if} \ f(a) = f(b).$$

Because equality is an equivalence relation, no matter how the function f is defined, this R will also be an equivalence relation.

// $f(a) = f(a) \Rightarrow a\,R\,a$ // R is reflexive

// $a\,R\,b \Rightarrow f(a) = f(b)$ $\Rightarrow f(b) = f(a) \Rightarrow b\,R\,a$ // symmetric

// $a\,R\,b$ and $b\,R\,c \Rightarrow f(a) = f(b)$ and $f(b) = f(c) \Rightarrow f(a) = f(c) \Rightarrow a\,R\,c$ // transitive

For instance, suppose S is the set of people in this classroom today and for each person, x, we let $f(x)$ be that person's age (in years, rounded down). Then the relation R partitions the set S into subsets of people of the same age. All 18-year-olds are related to each other but no one else, and the same with all 19-year-olds, 20-year-olds, and so on. This partitioning occurs with every equivalence relation.

If R is an equivalence relation on S, an ***equivalence class*** is a *nonempty* subset E of S such that   (i) every pair of elements of E are related
and   (ii) no element of E is related to an element not in E.

// Some authors define an equivalence class to be a maximal subset with property
// (i), but that's "equivalent" to this definition.

Theorem 7.2.1: If R is an equivalence relation on S, then R partitions S into equivalence classes

Proof. // We must prove that every element of S belongs to one and only one
        // equivalence class.

Let a be any element of S and let $[a]$ be the subset of S defined by

$$[a] = \{x \in S : a\,R\,x\}.$$

Since $a\,R\,a$, $a \in [a]$ and so $[a]$ is nonempty. // Is this subset an equivalence class?

If $x, y \in [a]$, then $a\,R\,x$ and $a\,R\,y$. By symmetry $x\,R\,a$. Then $x\,R\,a$ and $a\,R\,y$, so by transitivity, $x\,R\,y$. Thus, every pair of elements of $[a]$ are related. If $x \in [a]$ and $x\,R\,z$, then $a\,R\,x$ and $x\,R\,z$, so, by transitivity, $a\,R\,z$, and therefore, $z \in [a]$. Thus, no element of $[a]$ is related to any element not in $[a]$, and hence, $[a]$ <u>is</u> an equivalence class.

We now know that every element a of S is in at least one equivalence class, namely, $[a]$.   // Is this the only equivalence class containing element a? Suppose C is some equivalence class <u>containing</u> a. If $z \in C$, then $a\,R\,z$ and so $z \in [a]$. Thus, $C \subseteq [a]$. If $a\,R\,x$, then by (ii), x must be an element of C; thus, $[a] \subseteq C$. The subset C must equal the subset $[a]$, and we have proved the theorem that every element of S is in exactly one equivalence class.   □

7.2.1   Matrix and Digraph of an Equivalence Relation

Example 7.2.2: Let $S = \{0, 1, \ldots, 9\}$ and define a relation R on S by

$$a\,R\,b \text{ if and only if } 3 \mid (a - b).$$

Then R is an equivalence relation.

// Recall $x|y$ means that integer x "divides evenly into" y.
// $3|0$ so $a\ R\ a$ for all a in S.   Thus, R is reflexive.
// If $a\ R\ b$, then $3|(a - b)$ and since $(b - a) = -(a - b)$, $3|(b - a)$ and so $b\ R\ a$.
//   Thus, R is symmetric.
// If $a\ R\ b$ and $b\ R\ c$, then $3|(a - b)$ and $3|(b - c)$,
// so because $(a - c) = (a - b) + (b - c)$, $3|(a - c)$ and $a\ R\ c$.
//   Thus, R is transitive.

With the natural ordering of the elements of S, the matrix representation of this relation is

	0	1	2	3	4	5	6	7	8	9
0	1	0	0	1	0	0	1	0	0	1
1	0	1	0	0	1	0	0	1	0	0
2	0	0	1	0	0	1	0	0	1	0
3	1	0	0	1	0	0	1	0	0	1
4	0	1	0	0	1	0	0	1	0	0
5	0	0	1	0	0	1	0	0	1	0
6	1	0	0	1	0	0	1	0	0	1
7	0	1	0	0	1	0	0	1	0	0
8	0	0	1	0	0	1	0	0	1	0
9	1	0	0	1	0	0	1	0	0	1

However, if the elements of S are ordered somewhat differently, the matrix representation of this relation becomes

	0	3	6	9	1	4	7	2	5	8
0	1	1	1	1	0	0	0	0	0	0
3	1	1	1	1	0	0	0	0	0	0
6	1	1	1	1	0	0	0	0	0	0
9	1	1	1	1	0	0	0	0	0	0
1	0	0	0	0	1	1	1	0	0	0
4	0	0	0	0	1	1	1	0	0	0
7	0	0	0	0	1	1	1	0	0	0
2	0	0	0	0	0	0	0	1	1	1
5	0	0	0	0	0	0	0	1	1	1
8	0	0	0	0	0	0	0	1	1	1

The equivalence classes occur in blocks in this ordering of S, and the ones in the matrix occur in corresponding squares along the main diagonal.

// What will the digraph of this relation look like?

Example 7.2.3: Z_5
Define an equivalence relation R on $\mathbf{Z}$ by

$$a\ R\ b \text{ if and only if } f(a) = f(b),$$

where $f(x) = x$ MOD 5, the nonnegative remainder obtained when x is divided by 5.

Then $f(x)$ takes only five values: 0,1,2,3, or 4. The equivalence classes induced are

```
[0] = {...,-15,-10,-5,0,5,10,15, ...}
[1] = {...,-14, -9,-4,1,6,11,16, ...}          // -9 = 5(-2) + 1
[2] = {...,-13, -8,-3,2,7,12,17, ...}          // -3 = 5(-1) + 2
[3] = {...,-12, -7,-2,3,8,13,18, ...}
[4] = {...,-11, -6,-1,4,9,14,19, ...}.
```

We can define arithmetic on these classes from the ordinary operations, $+$ and $\times$, on Z as follows:

$$[a] \oplus [b] = [a + b] \quad \text{and} \quad [a] \otimes [b] = [a \times b].$$

// The class containing a *plus* the class containing b is the class containing $(a + b)$.
// The class containing a *times* the class containing b is the class containing $(a \times b)$.
//X Is this definition really independent of the representatives of the classes?

The tables for these operations are the following; but inside the table, $[x]$ is written as x.

$\oplus$	[0]	[1]	[2]	[3]	[4]
[0]	0	1	2	3	4
[1]	1	2	3	4	0
[2]	2	3	4	0	1
[3]	3	4	0	1	2
[4]	4	0	1	2	3

$\otimes$	[0]	[1]	[2]	[3]	[4]
[0]	0	0	0	0	0
[1]	0	1	2	3	4
[2]	0	2	4	1	3
[3]	0	3	1	4	2
[4]	0	4	3	2	1

// What will the tables for these operations on Z_6 look like?

Finite algebraic systems like Z_5 turn out to have many applications in computer science, especially in modern cryptography.

The Most Important Ideas in This Section.
A relation R on a set S is an *equivalence relation* whenever R is reflexive, symmetric, and transitive. This formalizes the idea of "similarity". An *equivalence class* is a *nonempty* subset E of S such that:
(1) Every pair of elements in E are related and
(2) No element in E is related to an element not in E.

(continued)

(continued)

> Furthermore, *R partitions* S into **equivalence classes**.
> In Sect. 7.5, we will define an equivalence relation on complexity functions
> of algorithms in a way that will make all linear functions equivalent to **n** and
> put all quadratic functions in the class [**n^2**].

7.3 Order Relations

In this section, we formalize the idea of objects occurring in a certain "preference
ranking". If a is bigger or better than b, then b is <u>not</u> bigger or better than a.
Antisymmetry is a fundamental property of such rankings. But so is transitivity: if a
is bigger or better than b and b is bigger or better than c, then a is bigger or better
than c.

A relation R is an **order relation** whenever R is **antisymmetric** and **transitive**.
               // Some books use the term "quasi-order" for such relations.
An order relation R is a **partial order** whenever R is (also) **reflexive**. A partial order
R is a **total order** whenever R (also) has the **comparability property**. The
distinctions will (perhaps) become clear as we look at some examples.

Let's examine the ten relations on various sets at the start of this chapter and use
fairly obvious abbreviations for the properties:

1. $a = b$:   AS, T and R   so it's a PO.
2. $a < b$:   AS and T   so it's an OR.
3. $a >= b$:   AS, T, R and CP   so it's a TO.
4. $a|b$ on **Z**:   T but not AS   since $(-5)|(+5)$ and $(+5)|(-5)$
                                                   but $(-5) \neq (+5)$.

    $a|b$ on **P**:   AS, T and R   so it's a PO.
5. $A \subseteq B$:   AS, T and R   so it's a PO.
6. $|A| = |B|$:   T but not AS.
7. a is married to b: is not AS.
8. a has a crush on b: is not AS   // Since sometimes, though all too rarely,
                                   // when a has a crush on b, b also has a crush on a.
                     is not T   // Since if a has a crush on b and b has a
                                     // crush on c, then (almost) *never*
                                          // does a have a crush on c.
                     is not S   // Everyone is painfully aware that this
                                          // relation is not symmetric.
9. a is younger than b: AS and T   so it's an OR.
10. a is a descendant of b: AS and T   so it's an OR.

// If a $R= b$ denotes (a R b or $a = b$), then
// whenever R is an order relation, $R=$ is a partial order.

7.3.1 Matrix and Digraph of a Partial Order

Example 7.3.1: Let $S = \{1, 2, \ldots, 12\}$ and consider the partial order, $|$, on S.
// recall that $a|b$ means "a divides evenly into b".

With this natural ordering of the elements of S, the matrix representation of this relation is

	1	2	3	4	5	6	7	8	9	10	11	12
1	1	1	1	1	1	1	1	1	1	1	1	1
2	0	1	0	1	0	1	0	1	0	1	0	1
3	0	0	1	0	0	1	0	0	1	0	0	1
4	0	0	0	1	0	0	0	1	0	0	0	1
5	0	0	0	0	1	0	0	0	0	1	0	0
6	0	0	0	0	0	1	0	0	0	0	0	1
7	0	0	0	0	0	0	1	0	0	0	0	0
8	0	0	0	0	0	0	0	1	0	0	0	0
9	0	0	0	0	0	0	0	0	1	0	0	0
10	0	0	0	0	0	0	0	0	0	1	0	0
11	0	0	0	0	0	0	0	0	0	0	1	0
12	0	0	0	0	0	0	0	0	0	0	0	1

This matrix is said to be "upper triangular" because all the entries below the main diagonal are zeros (and therefore, all the "information" is contained in the upper triangle).

When R is an order relation on S and a and b are distinct elements of S, b is said to **cover** a if $a\,R\,b$, <u>and</u> there is no element $x \in S\backslash\{a,b\}$ where $a\,R\,x$ and $x\,R\,b$.

The graphical representation of a partial order on a finite set S can be simplified to what's known as a **Hasse diagram**, named after Helmut Hasse (1898–1979) where:
1. No loops are drawn – there would be a loop at every vertex so these loops carry no information and just clutter the picture.
2. No arcs are drawn that are implied by transitivity from the ones that are present – draw an arc from a to b if and only if b covers a.
3. No arrowheads are drawn – the vertices are placed on the page so that all arrows would go upward.

// Is this always possible?

The Hasse diagram for Example 7.3.1 is

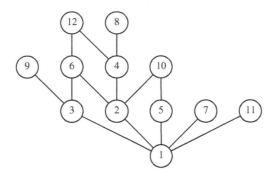

// Here, b covers $a \Leftrightarrow b = a \times p$ where p is a prime.
// $a\,R\,b \Leftrightarrow$ there is a "consistently upward" path in the Hasse diagram from a to b.
// What would the Hasse diagram of a total order look like?

7.3.2 Minimal and Maximal Elements

Suppose that R is an order relation on a set S and that T is a subset of S.

 $q \in T$ is a **minimum** element of T <u>means</u> $\forall x \in T \setminus \{q\}, q\,R\,x$.
    // q is related to every other element of T, like 1 in Example 7.3.1.
      //X T might not have a minimum element,
     // but if T has a minimum element, it is unique.

 $q \in T$ is a **minimal** element of T <u>means</u> $\forall y \in T \setminus \{q\}, y\,!R\,q$.
     // No other element of T is related q, like 1 in Example 7.3.1.
      //X If q is a minimum element, then q is minimal.
        // We'll see that if T is finite,
      // T must have a (at least one) minimal element.
   //X b covers $a \Leftrightarrow b$ is a minimal element of $\{x \in S \setminus \{a\}: a\,R\,x\}$.

 $q \in T$ is a **maximum** element of T <u>means</u> $\forall x \in T \setminus \{q\}, x\,R\,q$.
       // Every other element of T is related to q.
   //X T might not have a maximum element, like Example 7.3.1,
     // but if T has a maximum element, it is unique.

 $q \in T$ is a **maximal** element of T <u>means</u> $\forall y \in T \setminus \{q\}, q\,!R\,y$.
      // q is not related to any other element of T,
     // like 7, 8, 9, 10, 11, and 12 in Example 7.3.1.
    //X If q is a maximum element, then q is maximal.
   //X If T is finite, T must have a (at least one) maximal element.

$z \in S$ is an **upper bound** for T <u>means</u> $\forall x \in T, x\,R\,z$.
  // Every element of T is related to z, like $z = 12!$ in Example 7.3.1
      // or $z = (12)(11)(10)(9)(8)(7)$.
   // The upper bounds for T form another subset U of S
      // and one might ask questions like:
  //X What is the minimum upper bound for the subset in Example 7.3.1?
      // Is it $z = (2^3)(3^2)(5)(7)(11)$?

$z \in S$ is a **lower bound** for T <u>means</u> $\forall x \in T, z\,R\,x$.
     // z is related to every element of T like 1 in Example 7.3.1.
     // The lower bounds for T also form a subset of S.

Suppose R is an order relation on a finite set $S = \{x[1], x[2], \ldots , x[n]\}$. The following algorithm will find a minimal element of S.

Algorithm 7.3.1: Minimal in S

```
Begin
   M ← x[1];
   For j ← 2 To n Do
     If (x[j] R M) Then
        M ← x[j];
     End;
   End; // the for-loop
   Return(M);
End.                              // Is this Algorithm 4.3.1 when R is <?
```

Theorem 7.3.1: This algorithm correctly finds and returns a minimal element of *S* because

$$\text{“M is a minimal element of } S[j] = \{x[1], x[2], \ldots, x[j]\}\text{”}$$

is a loop invariant.   // for the for-loop

Proof. // by Mathematical Induction on $j \in P$

   Before the loop is done the first time, $M = x[1]$, which is a minimal element of $S[1] = \{x[1]\}$.   // No other element of $S[1]$ is related to $x[1]$.

   Assume $\exists q$ where $1 < q <= n$ and that, <u>before</u> the iteration where $j = q$, M is a minimal element of $S[q-1]$. That is, if $y \in S[q-1]\backslash\{M\}$, then y !R M.

   The next iteration of the loop may change the value of M; let M^* denote the value of M after this iteration.   // Is M^* a minimal element of $S[q]$?

   Suppose that $y \subset S[q]\backslash\{M^*\}$.   // We need to show that y !R M^*.

Case 1.   If $x[q]$!R M, then $M^* = M$ and $S[q]\backslash\{M^*\} = \{x[q]\} \cup (S[q-1]\backslash\{M\})$.
         If $y = x[q]$ or $y \in S[q-1]\backslash\{M\}$, then y !R M so y !R M^*.
Case 2.   If $x[q]$ R M then $M^* = x[q]$ and $M^* \neq M$.   // because $M \in S[q-1]$
         The set $S[q]\backslash\{M^*\} = S[q-1] = \{M\} \cup (S[q-1]\backslash\{M\})$ so either $y = M$
         or $y \in S[q-1]\backslash\{M\}$.

   Suppose y R M^*.   // We want to find a contradiction.

   If $y = M$, then, since M^* R M by antisymmetry, we'd have $M = M^*$. But since $M^* \neq M$, we know that $y \neq M$. If $y \in S[q-1]\backslash\{M\}$, then, since M^* R M, by transitivity, we would have y R M, but this contradicts the assumption that M is a minimal element of $S[q-1]$.

   Therefore, y !R M^*, and after the loop is done when $j = q$, the current value of M is a minimal element of $S[q]$.   □

   The elements of S can be indexed as $S = \{y[1], y[2], \ldots, y[n]\}$ so that

$$\text{if } y[i] \ R \ y[j], \text{ then } i <= j$$

by means of the following algorithm.

Algorithm 7.3.2: *R*-indexing of the n-set S

```
Begin
   T ← S;
   For j ← 1 To n Do
   Find a minimal element M of T;
   y[j] ← M;
   T ← T\{M};
   End; // the for-loop
End.
```

Theorem 7.3.2: After this algorithm indexes the elements of set *S*,

$$\text{if } y[i] \ R \ y[j], \text{ then } i <= j.$$

Proof. $y[1]$ is a minimal element of S. Thus, $\forall x \in S \setminus \{y[1]\}$, $x \ !R \ y[1]$.

$$S \setminus \{y[1]\} = \{y[2], y[3], \ldots, y[n]\} \text{ so if } i > 1, \text{ then } y[i] \ !R \ y[1].$$

For each value of j from 2 to $n - 1$, the current set $T = S \setminus \{y[1], \ldots, y[j-1]\}$ and $y[j]$ is a minimal element of T. Thus, $\forall x \in T \setminus \{y[j]\}$, $x \ !R \ y[j]$. But

$$T \setminus \{y[j]\} = (S \setminus \{y[1], \ldots, y[j-1]\}) \setminus \{y[j]\} = \{y[j+1], y[j+2], \ldots, y[n]\}$$

Therefore, if $i > j$, then $y[i] \ !R \ y[j]$. The contrapositive of this statement is

$$\text{if } y[i] \ R \ y[j], \text{ then } i <= j. \qquad \square$$

// Using this ordering of S, the matrix representation of **R** will be upper triangular.
//X Must $y[n]$ be a maximal element of S?

The Most Important Ideas in This Section.
A relation **R** on a set S is an *order relation* whenever it is antisymmetric and transitive. An order relation **R** is a *partial order* whenever **R** is (also) reflexive. A partial order **R** is a *total order* whenever **R** (also) has the comparability property. This section formalized the idea of objects occurring in a preference ranking.

The graphical representation of a Partial Order on a finite set S can be simplified to what's known as a *Hasse Diagram*.

7.4 Relations on Finite Sequences

If $X = (x_1, x_2, \ldots, x_m)$ and $Y = (y_1, y_2, \ldots, y_n)$ are sequences of any objects at all

$$X \text{ \textit{equals} } Y \underline{\text{ means}}$$
$$m = n \text{ and } x_i = y_i \text{ for } i = 1, 2, \ldots, m.$$

While this may be the most fundamental relation between sequences, the main focus of this section is the study of two (other) order relations on sequences of numbers.

7.4.1 Domination

If $X = (x_1, x_2, \ldots, x_m)$ and $Y = (y_1, y_2, \ldots, y_n)$ are sequences of numbers,

$$X \text{ is \textit{dominated by} } Y \text{ [written } X \mathcal{D} Y] \underline{\text{ means}}$$

(1) $m <= n$ // Y is at least as long as X

and   (2) $x_i <= y_i$ for $i = 1, 2, \ldots, m$.   // Each y_i is at least as big as x_i

For instance, $(1,2,3) \mathcal{D} (2,2,4,0)$ and $(2,2,4,0) \mathcal{D} (4,4,4,4)$.

// When $X \mathcal{D} Y$, some books say that X is "majorized" by Y.
// Is this a total order? Which properties does $\mathcal{D}$ have?

Theorem 7.4.1: $\mathcal{D}$ is a partial order but not a total order

Proof. // We must prove that $\mathcal{D}$ is reflexive, transitive, antisymmetric
         // and show it does not have the comparability property.

$\mathcal{D}$ is reflexive because if $X = Y$, then $m = n$, and condition (2) is satisfied so $X \mathcal{D} Y$.

// What about antisymmetry?
   Suppose $X \mathcal{D} Y$ and $Y \mathcal{D} X$.   // Must $X = Y$?
Then

(1) $m <= n$ and $n <= m$ so $m = n$ // $<=$ is anti-symmetric.

and   (2) $x_i <= y_i$ for $i = 1, 2, \ldots, m$ and $y_i <= x_i$ for $i = 1, 2, \ldots, n$ so
         $x_i = y_i$ for $i = 1, 2, \ldots, m = n$.   // $<=$ is anti-symmetric.

Hence, $X = Y$, and therefore, $\mathcal{D}$ is antisymmetric.

// What about transitivity?
   Suppose $X \mathcal{D} Y$ and $Y \mathcal{D} Z$ where $Z = (z_1, z_2, \ldots, z_p)$.   // Must $X \mathcal{D} Z$?
Since $X \mathcal{D} Y$ and $Y \mathcal{D} Z$,

(1) $m <= n$ and $n <= p$ so $m <= p$ // $<=$ is transitive.

and   (2) $x_i <= y_i$ for $i = 1, 2, \ldots, m$ and $y_i <= z_i$ for $i = 1, 2, \ldots, n$ so
         $x_i <= z_i$ for $i = 1, 2, \ldots, m$.   // $<=$ is transitive.

Hence, $X \mathcal{D} Z$, and therefore, $\mathcal{D}$ is transitive.

If $X = (1,2,3)$ and $Y = (3,2,1)$, then $X \, !\mathcal{D} \, Y$ and $Y \, !\mathcal{D} \, X$, and therefore, $\mathcal{D}$ does not have the comparability property.   □

Hasse diagrams of the dominates relation $\mathcal{D}$

In this diagram, $\mathcal{D}$ is restricted to the set of all three sequences on $\{0, 1, 2\}$.

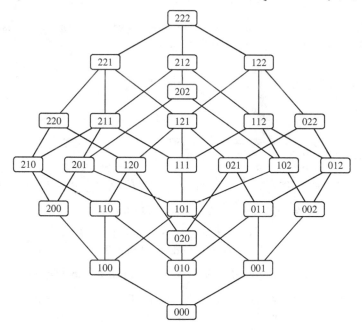

In this diagram, $\mathcal{D}$ is restricted to the set of all increasing three sequences on $\{1..6\}$.

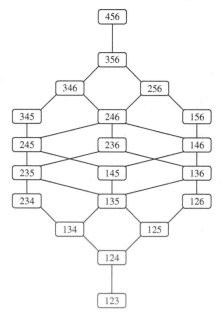

7.4.2 Lexicographic Order

The ordering of words in a dictionary is based on the ordering of the letters in the alphabet. In a dictionary,

KIND   comes before   KINDER   // A prefix comes before an extension.
KINDER   comes before   KINDEST.   // R precedes S in the alphabet.

The basic idea in <u>lexicographic</u> (i.e., dictionary) ordering of words may be used to provide a (total) ordering of finite sequences of numbers as follows:

if $X = (x_1, x_2, \ldots, x_m)$ and $Y = (y_1, y_2, \ldots, y_n)$ are sequences of numbers,

X is *lexicographically less than or equal* Y [written $X \mathcal{L} Y$] <u>means</u>

either   (1) $m <= n$ and $x_i = y_i$ for $i = 1, 2, \ldots, m$;
or   (2) $\exists$ an index j where $1 <= j <= m, n$ such that
            $x_j < y_j$ but if $1 <= i < j$ then $x_i = y_i$.

// Condition (1) asserts that X is a "prefix" of Y.
// Condition (2) asserts that at the first index where $x_i \neq y_i$, we have $x_i < y_i$.
//X If $X \mathcal{D} Y$, then $X \mathcal{L} Y$.   (But the converse is false.)
//X If $X \mathcal{L} Y$, then $x_1 <= y_1$.

// Is $\mathcal{L}$ really a total order? Which properties does $\mathcal{L}$ have?
Since (1,2,3) $\mathcal{L}$ (2,2) but (2,2) $!\mathcal{L}$ (1,2,3), $\mathcal{L}$ is <u>not symmetric</u>.

Theorem 7.4.2: $\mathcal{L}$ is a total order

Proof. // We must prove that $\mathcal{L}$ is reflexive, transitive, antisymmetric, and
         // has the comparability property.

$\mathcal{L}$ is <u>reflexive</u> because if $X = Y$, then $m = n$, and condition (1) is satisfied
so $X \mathcal{L} Y$.

If $X \neq Y$ and $m <= n$, then   // Take X to be the <u>shorter</u> sequence.

either $x_i = y_i$ for $i = 1, 2, \ldots, m$
or $\exists$ an index j where $1 <= j <= m$ such that $x_j \neq y_j$ but if $1 <= i < j$ then
         $x_i = y_i$.

In the first case, m must be $< n$ (if $m = n$, then $X = Y$) so $X \mathcal{L} Y$ and $Y !\mathcal{L} X$.
In the second case, either $x_j < y_j$ or $x_j > y_j$. If $x_j < y_j$, then $X \mathcal{L} Y$ and $Y !\mathcal{L} X$;
if $x_j > y_j$, then $Y \mathcal{L} X$ and $X !\mathcal{L} Y$. Therefore, we have shown that

if $X \neq Y$, then $X \mathcal{L} Y$ or $Y \mathcal{L} X$ but <u>not both</u>,

and this proves both that $\mathcal{L}$ is <u>antisymmetric</u> and has the <u>comparability property</u>.

// But what about transitivity?

Suppose $X \mathcal{L} Y$ and $Y \mathcal{L} Z$ where $Z = (z_1, z_2, \ldots, z_p)$. // Can we show that $X \mathcal{L} Z$?

// It seems that there are several (tedious) cases we *must* consider.

Since $X \mathcal{L} Y$, $x_1 <= y_1$ and since $Y \mathcal{L} Z$, $y_1 <= z_1$. Therefore $x_1 <= z_1$. If $x_1 < z_1$, then $X \mathcal{L} Z$. Otherwise, $x_1 = y_1 = z_1$.

Let j be the length of a longest common prefix of X and Y. Then $1 <= j <= m, n$. If $j = n$, then Y is a prefix of X, so $Y \mathcal{L} X$ by antisymmetry $X = Y$, and therefore, $X \mathcal{L} Z$. Otherwise, $j < n$ and

either $m = j < n$ and X is a prefix of Y

or $j < m$ and $j < n$ and $x_{j+1} < y_{j+1}$.

Let k be the length of a longest common prefix of Y and Z. Then $1 <= k <= n, p$. If $k = p$ then Z is a prefix of Y so $Z \mathcal{L} Y$; by anti-symmetry $Z = Y$, and therefore $X \mathcal{L} Z$. Otherwise, $k < p$ and

either $n = k < p$ and Y is a prefix of Z

or $k < n$ and $k < p$ and $y_{k+1} < z_{k+1}$.

Suppose now that $j < n$ and $k < p$.
If $j > k$, then $x_i = y_i = z_i$ for $i = 1, 2, \ldots, k$ and $x_{k+1} = y_{k+1} < z_{k+1}$, and so $X \mathcal{L} Z$.

Otherwise, $j <= k$ and $x_i = y_i = z_i$ for $i = 1, 2, \ldots, j$.
If $j = m$, then (X is a prefix of Z so) $X \mathcal{L} Z$.
If $j < m$, then $x_{j+1} < y_{j+1} <= z_{j+1}$ and so $X \mathcal{L} Z$.

                                          // If $y_{j+1} \neq z_{j+1}$, then $y_{j+1} < z_{j+1}$.
Thus, $\mathcal{L}$ is transitive.   ▢

The Most Important Ideas in This Section.
Two order relations were defined on finite sequences of numbers:
$$X \text{ is dominated by } Y \; [X \, \mathcal{D} \, Y],$$
    and   *X is lexicographically less than or equal Y* $[X \, \mathcal{L} \, Y]$.
$\mathcal{D}$ is a partial order but not a total order; $\mathcal{L}$ is a total order.

   If $S = \{O[1], O[2], O[3], \ldots, O[n]\}$ is a set of _indexed objects_ of <u>any</u> kind, and T is a sequence of these objects $(O[a], O[b], O[c], O[d])$, T corresponds to the sequence of indices (a, b, c, d). Applying lexicographic ordering to the sequences of indices produces a **total ordering** of the sequences of objects.

   In the next section, we expand the idea of domination to compare complexity functions of algorithms, so we can "rank" their efficiency. In Chap. 9, we develop algorithms for (efficiently) generating sequences of several kinds in "lexicographic" order, from the first to the last.
                                          // the minimum element to the maximum

7.5 Relations on Infinite Sequences

This section concerns relations on complexity functions of algorithms taken as real-valued sequences defined on **P**. The three relations defined in Sect. 7.4 may be extended to infinite sequences. Suppose $f = (x_1, x_2, \ldots)$ and $g = (y_1, y_2, \ldots)$ are sequences of numbers. Then

$$f \; \textbf{\textit{equals}} \; g \quad \underline{\text{means}} \quad x_i = y_i \text{ for } \forall i \in \textbf{P}; \qquad\qquad // f = g$$

$$f \; \textbf{\textit{is lexicographically less than or equal}} \; g \; \underline{\text{means}} \qquad // f \; \pounds \; g$$
$$f = g \text{ or } \exists \text{ an index } j \text{ where } x_j < y_j \text{ but if } 1 <= i < j, \text{ then } x_i - y_i,$$

and f **is dominated by** g <u>means</u> $x_i <= y_i$ for $\forall i \in \textbf{P}.$ $// f \; \mathcal{D} \; g$

As sequences defined on **P**, it's easy to see that

$$n \; \mathcal{D} \; n^2 \quad \text{and} \quad n^2 \; \mathcal{D} \; n^3 \quad \text{and} \quad n^3 \; \mathcal{D} \; n^4 \quad \text{and} \quad n^4 \; \mathcal{D} \; n^5 \ldots$$
$$1^n \; \mathcal{D} \; 2^n \quad \text{and} \quad 2^n \; \mathcal{D} \; 3^n \quad \text{and} \quad 3^n \; \mathcal{D} \; 4^n \quad \text{and} \quad 4^n \; \mathcal{D} \; 5^n \ldots$$

In Chap. 3, we noted that $n \; \mathcal{D} \; 2^n$, and therefore, $lg(n) \; \mathcal{D} \; n$.

Also $\qquad\qquad n \; \mathcal{D} \; [n + lg(n)], [n + lg(n)] \; \mathcal{D} \; 2n, \text{ and } nlg(n) \; \mathcal{D} \; n^2.$

We also saw that $n^2 \; \mathcal{D} \; 3^n$, but $n^2 \; !\mathcal{D} \; 2^n$ because $n^2 <= 2^n$ except when $n = 3$.
$$// \; 3^2 > 2^3$$
In Chap. 4, we proved that $(n!) \; \mathcal{D} \; n^n$ and that $n^n \; \mathcal{D} \; (n!)^2.$

Therefore, $\qquad lg(n!) \; \mathcal{D} \; nlg(n) \text{ and } nlg(n) \; \mathcal{D} \; 2lg(n!).$

However, the main objective of this section is to determine relations between complexity functions of algorithms that will indicate when they are similar and when one is better than another. The relation $\mathcal{D}$ (by itself) does not serve either purpose very well. For instance, n^2 is a much better complexity than 2^n even though $n^2 \; !\mathcal{D} \; 2^n$, and even though $(n - 1) \; \mathcal{D} \; n$, $(n - 1)$ is very similar to n.

We want a fairly rough measure of complexity that will detect <u>large</u> and <u>important</u> changes in complexity. In earlier chapters, we found that (for large values of n)

from prime testing
     $\sqrt{n}$ is much, much better than $n/2$
but $n/2$ is somewhat better than $n - 2$ // but not much, much better

from searching
     $lg(n)$ is much better than $(n + 1)/2;$ // for large n

and from sorting

$n(n-1)/2$ is roughly similar to n^2

but $n \times lg(n)$ is of lower complexity than n^2.

We want a way of comparing and ranking complexity functions that makes these ideas precise.

Complexity functions are nonnegative and usually increasing, since the number of steps it takes to solve a larger problem is (almost always) more than the number of steps it takes to solve a smaller problem. As n gets bigger, $f(n)$ gets bigger.

// These functions are also (almost always) concave up; that is, the extra work in
// doing a problem of size $n + 1$ compared to doing a problem of size n also
// increases (with n).

We will devise a way to *classify* the rates of growth of such functions so that all *linear functions*, $f(n) = An + B$ with $A > 0$, end up in the same class as ***n*** itself, $\Theta(\boldsymbol{n})$, and all *quadratic functions*, $f(n) = An^2 + Bn + C$ with $A > 0$, end up in the same class as $\boldsymbol{n^2}$ itself, $\Theta(\boldsymbol{n^2})$.

We will begin by defining a fairly large set of functions $\mathcal{F}$ that will include the complexity functions of algorithms. Later, we'll define an *equivalence* relation on $\mathcal{F}$ so that all linear functions are equivalent and so that all quadratic functions are equivalent. And finally, we'll define an *order* relation $<<<$ on $\mathcal{F}$ where

$$\sqrt{n} <<< n/2 \text{ and } lg(n) <<< (n+1)/2 \text{ and } n \times lg(n) <<< n^2.$$

Let $\mathcal{F}$ denote the set of all infinite, real-valued sequences (with domain $\mathbf{P}$) that are "eventually positive"; that is, there is an $N \in \mathbf{P}$ such that if $n >= N$, then $f(n) > 0$.

7.5.1   Asymptotic Dominance and Big-Oh Notation

The last section on sorting algorithms listed the time they took to sort a list of length n, $T(n)$. We found that $T(n)$ was roughly proportional to the number of comparisons of array elements, $f(n)$. That is,

$$T(n) \cong K \times f(n),$$

where the constant of proportionality K depended on the machine, operating system, compiler, etc. *The exact value of K was **not** as **important** to us as the fact that it was a constant.* Counting the number of comparisons of array elements gave a rough measure of the "cost" of using each sorting algorithm and also a means of comparing their efficiencies.

Often, we're only interested in the worst case and would be satisfied by an upper bound on the cost which would apply for all large values of n (say $>= M$) like

$$\text{if } n >= M \text{ then } T(n) <= K \times f(n).$$

This relation between $T(n)$ and $f(n)$ is known as *asymptotic dominance*.

If $f(n)$ and $g(n)$ are sequences in $\mathcal{F}$,

f is **asymptotically dominated by** g [written $f \ll g$] <u>means</u>
$\exists\ K \in \mathbf{R}^+$ and $\exists\ M \in \mathbf{P}$ such that if $n >= M$, then $f(n) <= K \times g(n)$.
// $\mathbf{R}^+$ denotes the set of <u>positive</u> real numbers.
Asymptotic domination is a <u>weaker</u> condition than ordinary domination because we can multiply $g(n)$ by an arbitrary positive constant and, even then, the domination inequality does not need to apply until n is sufficiently large. Furthermore, if $f\ \mathcal{D}\ g$, then (taking $K = 1$ and $M = 1$) $f \ll g$. Because $\mathcal{D}$ is reflexive, $\ll$ is also <u>reflexive</u>.

Example 7.5.1: $900n \ll n^2$.

// We need a value for K, a value for M, and an algebraic "argument" that shows
// if $n >=$ our value of M, then $900n <=$ our value of $K \times n^2$.

Let $K = 1$ and let $M = 900$. If $n >= 900$, then $900 \times n <= n \times n = 1 \times n^2$.

// In fact, many different values of K and M will "work", and
// any such pair of values proves asymptotic domination.

Let $K = 900$ and let $M = 1$. If $n >= 1$ then $n <= n \times n$ so $900 \times n < = 900 \times n^2$.
Let $K = 30$ and let $M = 30$. If $n >= 30$ then $30 \times n <= n \times n$ so
$$30 \times n = 30 \times 30n <= 30 \times n^2.$$

// Also, if any particular value of K works, then any larger value of K will work.
// And if any particular value of M works, then any larger value of M will work.

Example 7.5.2: $n^2\ !\ll 900n$.

// Here we need to show that no pair of values "works" to prove asymptotic
// domination. That is, for any value of K (no matter how large)
// and any value of M (no matter how large), the conditional statement
//   "if $n >=$ the value of M then $n^2 <=$ the value of $K \times (900n)$"
// is False.
//
// It's False if we can find a value of n, say n^*, where
// $n^* >=$ the value of M
// but $(n^*)^2 >$ the value of $K \times (900n^*)$.

Let K be any given positive real number and let M be any given positive integer. Now let $n^* = M + \lceil K \rceil \times 900$. Then $n^* >= M$ and, since $n^* > K \times 900$,

$$(n^*)^2 > (K \times 900)(n^*) = K \times (900n^*).$$

In general, if $f(n)$ and $g(n)$ are sequences in $\mathcal{F}$,

f is **not asymptotically dominated by** g [written $f\ !\ll g$] <u>means</u>
$\forall\ K \in \mathbf{R}^+$ and $\forall\ M \in \mathbf{P}$, $\exists\ n^* >= M$ where $f(n^*) > K \times g(n^*)$.

Examples 7.5.1 and 7.5.2 show that $\ll$ is <u>not symmetric</u>. The next example shows it's <u>not antisymmetric</u>.   // What properties does $\ll$ have?

Example 7.5.3: $n(n - 1)/2 \ll n^2$ and $n^2 \ll n(n - 1)/2$

When $n \in \mathbf{P}$, $n(n - 1)/2 <= n(n - 1) < n^2$,   // $n(n - 1)/2 \; \mathcal{D} \; n^2$
Taking $K = 1$ and $M = 1$, we have

if $n >= M$, then $n(n - 1)/2 <= K \times n^2$, and therefore, $n(n - 1)/2 \ll n^2$.

Even though $n(n - 1)/2$ is dominated by n^2, if we multiply it by a sufficiently large constant K, it might then dominate n^2 from some point on.

// Let's try K=4. Then, $n^2 <= K[n(n - 1)/2] = 2n(n - 1) = 2n^2 - 2n$
// $\Leftrightarrow 2n <= n^2 \Leftrightarrow 2 <= n.$

Let $K = 4$ and let $M = 2$. If $n >= 2$, then $n^2 >= 2n$ so

$$n^2 <= n^2 + \left(n^2 - 2n\right) = 2n^2 - 2n = 4[n(n - 1)/2] = K \times [n(n - 1)/2],$$

and therefore, $n^2 \ll n(n - 1)/2$.

Theorem 7.5.1: The relation $\ll$ is *transitive.*

Proof. Suppose $f \ll g$ and $g \ll h$ where f, g, and h are sequences in $\mathcal{F}$.
                                                                    // Is $f \ll h$?
Since $f \ll g$ and $g \ll h$,

$\exists K1 \in \mathbf{R}^+$ and $\exists M1 \in \mathbf{P}$ such that if $n >= M1$, then $f(n) <= K1 \times g(n)$
and $\exists K2 \in \mathbf{R}^+$ and $\exists M2 \in \mathbf{P}$ such that if $n >= M2$, then $g(n) <= K2 \times h(n)$.

Let $K = K1 \times K2$ and let $M = M1 + M2$.   // $K \in \mathbf{R}^+$ and $M \in \mathbf{P}$
Now, if $n >= M$, then

$n >= M2$ so $g(n) <= K2 \times h(n)$
$K1 > 0$ so $K1 \times g(n) <= K1 \times \{K2 \times h(n)\} = \{K1 \times K2\} \times h(n)$
and $n >= M1$ so $f(n) <= K1 \times g(n) <= \{K1 \times K2\} \times h(n) = K \times h(n),$

and therefore, $f \ll h$.   ⬜

Theorem 7.5.2: The relation $\ll$ does <u>not</u> have the *comparability* **property**

Proof. // We must find two sequences, f and g, and show that $f \;!\ll\; g$ and $g \;!\ll\; f$.
        // In fact, we'll construct two incomparable, increasing, integer sequences.

Let $f(n) = (n!) \times (n!)$ and

$$\text{let } g(n) = \begin{cases} (2r)!(2r)!(2r) & \text{if } n = 2r \\ (2r)!(2r)!(2r + 1) & \text{if } n = 2r + 1. \end{cases}$$

Tabulating the first few values of these sequences, we get

n	$f(n)$	$g(n)$
1	1	1
2	4	8
3	36	12
4	576	2304
5	14400	2880

$// \, n = 1 = 2(0) + 1 \text{ so } r = 0$

For any $r \in \mathbf{P}$, we have

$$f(2r) = (2r)!(2r)!$$
$$< g(2r) = (2r)!(2r)!(2r)$$
$$< g(2r+1) = (2r)!(2r)!(2r+1)$$
$$< f(2r+1) = (2r)!(2r)!(2r+1)(2r+1)$$
$$< f(2r+2) = (2r)!(2r)!(2r+1)(2r+1)(2r+2)(2r+2)$$
$$< g(2r+2) = (2r)!(2r)!(2r+1)(2r+1)(2r+2)(2r+2)(2r+2).$$

If $n = 2r$, then $g(n) = f(n) \times n$, and if $n = 2r + 1$, then $f(n) = g(n) \times n$. Recall that $f1 \; !<< \; f2$ means

$$\forall \, K \in \mathbf{R}^+ \text{ and } \forall \, M \in \mathbf{P}, \exists \; n* >= M, \text{where } f1(n*) > K \times f2(n*).$$

Let K be any given positive real number and let M be any given positive integer. Let $r* = M + \lceil K \rceil$. $// \, r* \in \mathbf{P} \text{ and } r* > M, K.$
If $n* = 2r*$, then

$$n* >= M \text{ and } g(n*) = f(n*) \times n* > K \times f(n*) \text{ so } g \, !<< f.$$

If $n* = 2r* + 1$, then

$$n* >= M \text{ and } f(n*) = g(n*) \times n* > K \times g(n*) \text{ so } f \, !<< g. \qquad \square$$

Because we're looking at sequences that are eventually positive, we can give an additional attribute when $f \ll g$. We know $\exists K \in \mathbf{R}^+$ and $\exists M \in \mathbf{P}$ such that if $n >= M$, then $f(n) <= K \times g(n)$, and, since $f \in \mathcal{F}$, $\exists N \in \mathbf{P}$ such that if $n >= N$, then $0 < f(n)$. If we let $M^+ = \max\{M, N\}$, we get

if $f \ll g$, then $\exists K \in \mathbf{R}^+$ and $\exists M^+ \in \mathbf{P}$ such that if $n >= M^+$, then $0 < f(n) <= K \times g(n)$.

If f and g are sequences in $\mathcal{F}$ and $A \in \mathbf{R}^+$, we can create three other sequences in $\mathcal{F}$ as follows: For $\forall \, n \in \mathbf{P}$, let

$$(Af)(n) = A \times f(n)$$
$$(f + g)(n) = f(n) + g(n)$$
and $$(f \times g)(n) = f(n) \times g(n).$$

$// \text{ and } (g + f) = (f + g)$
$// \text{ and } (g \times f) = (f \times g)$

Theorem 7.5.3: For sequences in $\mathcal{F}$

1. $g \ll (f + g)$.
2. If $A \in \mathbf{R}^+$, then $Af \ll f$ and $f \ll Af$.
3. If $f \ll g$ and $A \in \mathbf{R}^+$, then $Af \ll g$.
4. If $f \ll g$, then $(f + g) \ll g$.
5. If $f1 \ll g$ and $f2 \ll g$, then $(f1 + f2) \ll g$.
6. If $f1 \ll g1$ and $f2 \ll g2$, then $(f1 + f2) \ll (g1 + g2)$.
7. If $f1 \ll g1$ and $f2 \ll g2$, then $(f1 \times f2) \ll (g1 \times g2)$.

Proof.
// We'll prove each of these seven assertions but in a slightly different order
// because some are special cases of others.

// First we'll prove 1.
   Because f is eventually positive, $\exists N \in \mathbf{P}$ such that if $n >= N$, then $f(n) > 0$.
Let $K = 1$ and $M = N$. If $n >= M$, then $g(n) <= g(n) + f(n) = (f + g)(n)$;
so $g \ll f + g$.

// Next, we'll prove 3.
   If $f \ll g$, then $\exists K \in \mathbf{R}^+$ and $\exists M \in \mathbf{P}$ such that if $n >= M$, then $f(n) <= K \times g(n)$.
If $A \in \mathbf{R}^+$, then $AK \in \mathbf{R}^+$, and now, if $n >= M$, then

$$(Af)(n) = A \times f(n) <= A \times [K \times g(n)] = (AK) \times g(n)$$

so $Af \ll g$.   // What about 2?
Since $f \ll f$, using $g = f$ in 3, we get $\forall A \in \mathbf{R}^+$, $Af \ll f$, and, in particular,
$f + f = 2f \ll f$. If $A \in \mathbf{R}^+$, then $(1/A) \in \mathbf{R}^+$, and we have $(1/A)(Af) \ll Af$; that
is, $f \ll Af$.

// Next, we'll prove 6.
   Suppose that $f1 \ll g1$ and $f2 \ll g2$. Then

$\exists K1 \in \mathbf{R}^+$ and $\exists M1 \in \mathbf{P}$ such that if $n >= M1$ then $f1(n) <= K1 \times g1(n)$,
and $\exists K2 \in \mathbf{R}^+$ and $\exists M2 \in \mathbf{P}$ such that if $n >= M2$ then $f2(n) <= K2 \times g2(n)$.

Let $K = \max\{K1, K2\}$ and $M = \max\{M1, M2\}$. Suppose $n >= M$. Then

$$(f1 + f2)(n) = f1(n) + f2(n) <= K1 \times g1(n) + K2 \times g2(n)$$
$$<= K \times g1(n) + K \times g2(n) = K \times [g1(n) + g2(n)] = K \times [(g1 + g2)(n)].$$

Thus, $(f1 + f2) \ll (g1 + g2)$.
To prove 5, take $g1 = g2 = g$ in 6, then we get $f1 + f2 \ll g + g \ll g$ (from 2).
To prove 4, take $f1 = f$ and $f2 = g$ in 5, then we get $f + g \ll g$.

// Finally, we'll prove 7.

Suppose that $f1 \ll g1$ and $f2 \ll g2$. Then

$\exists K1 \in \mathbf{R}^+$ and $\exists M1^+ \in \mathbf{P}$ such that if $n >= M1^+$ then $0 < f1(n) <= K1 \times g1(n)$, and

$\exists K2 \in \mathbf{R}^+$ and $\exists M2^+ \in \mathbf{P}$ such that if $n >= M2^+$ then $0 < f2(n) <= K2 \times g2(n)$.

Let $K = K1 \times K2$ and $M = \max\{M1^+, M2^+\}$. Suppose $n >= M$. Then

$$(f1 \times f2)(n) = f1(n) \times f2(n) <= [K1 \times g1(n)] \times f2(n) \qquad // \ f2(n) > 0$$
$$<= [K1 \times g1(n)] \times [K2 \times g2(n)] \qquad // \ g1(n) > 0$$
$$= [K1 \times K2] \times [g1(n) \times g2(n)] = K \times [(g1 \times g2)(n)].$$

Thus, $(f1 \times f2) \ll (g1 \times g2)$. $\qquad\qquad\qquad\qquad\qquad\qquad\qquad\qquad\qquad\qquad$ □

We have used the notation $\ll$ to denote asymptotic domination to stress its properties and similarity to ordinary domination, but other books do not use it. They use what's called <u>Big-Oh notation</u>. When $f \ll g$, it is often said that "f is of order g" or "f is $\mathbf{O}(g)$". In some books, $\mathbf{O}(g)$ is called an "order class" and is defined as a set by

$$\mathbf{O}(g) = \{f : f \ll g\}.$$

From this point on, we will assume that the following are all equivalent assertions:
1. f is asymptotically dominated by g.
2. $f \ll g$.
3. f is of order g.
4. f is $\mathbf{O}(g)$.
5. $f \in \mathbf{O}(g)$.
And all of these mean

$$\exists K \in \mathbf{R}^+ \text{ and } \exists M \in \mathbf{P} \text{ such that if } n >= M, \text{ then } f(n) <= K \times g(n).$$

If an algorithm has "complexity $\mathbf{O}(n^2)$", we now understand that to mean that the cost of running that algorithm on input of size n is bounded by some constant times n^2 when n is large.

Some algorithms are said to have "complexity f which is $\mathbf{O}(1)$".

// What can that mean?

We interpret "**1**" as the sequence $(1,1,1,\ldots)$ and "f is $\mathbf{O}(1)$" to mean that $f \ll 1$; that is,

$$\exists K \in \mathbf{R}^+ \text{ and } \exists M \in \mathbf{P} \text{ such that if } n >= M, \text{ then } f(n) <= K \times 1 = K.$$

Setting $B = \max\{K, f(1), f(2), f(3), f(4),\ldots, f(M-1)\}$, we see $f(n) <= B$ for $\forall n \in \mathbf{P}$. Thus, "f is $\mathbf{O}(1)$" means "f **is bounded**".

7.5.2   Asymptotic Equivalence and Big-Theta Notation

If $f(n)$ and $g(n)$ are sequences in $\mathcal{F}$,

> f is **asymptotically equivalent to** g [written $f \sim g$] <u>means</u>
> $$f \ll g \quad \text{and} \quad g \ll f.$$

We've seen $n(n-1)/2 \sim n^2$.   // But is this really an equivalence relation?

// $f \ll f$ (and $f \ll f$) $\Rightarrow f \sim f$ so $\sim$ is reflexive.
// $f \sim g \Rightarrow f \ll g$ and $g \ll f \Rightarrow g \ll f$ and $f \ll g \Rightarrow g \sim f$, so $\sim$ is symmetric.
// $f \sim g$ and $g \sim h$ $\quad\Rightarrow\quad$ $f \ll g$ and $g \ll f$
//   and $\quad g \ll h$ and $h \ll g$
// $\Rightarrow \quad f \ll h$ and $h \ll f$ (because $\ll$ is transitive)
// $\Rightarrow \quad f \sim h$ and hence $\sim$ is transitive.

//X $\log_b(n) \sim \lg(n)$ for $\forall\, b > 1$
//   that is, all logarithm functions are equivalent no matter what the base is.
//X $f \sim g \Leftrightarrow \exists A, B \in \mathbf{R}^+$ and $\exists\, M \in \mathbf{P}$ such that

$$\text{if } n >= M, \text{ then } A \times f(n) <= g(n) <= B \times f(n).$$

The relation $\sim$ partitions $\mathcal{F}$ into equivalence classes. We will use what's called <u>Big-Theta notation</u> to denote these classes:

$$\Theta(g) = \{f : f \sim g\}.$$

From this point on, we will assume that the following are all equivalent assertions:
1. f is asymptotically equivalent to g.
2. $f \sim g$.
3. f is $\Theta(g)$.
4. $f \in \Theta(g)$.
And all of these mean $f \ll g$ <u>and</u> $g \ll f$.

7.5.2.1   Polynomials

Example 7.5.4: $6n^3 - 10n^2 + 3n - 12 \ll n^3$.
If $n >= 1$, then

$$6n^3 - 10n^2 + 3n - 12 <= 6n^3 + 3n <= 6n^3 + 3n^3 = 9n^3.$$

Thus, if $K = 9$ and $M = 1$, the condition for asymptotic dominance is satisfied.

Example 7.5.5: $n^3 \ll 6n^3 - 10n^2 + 3n - 12$.
If $n >= 1$, then

$$
\begin{aligned}
6n^3 - 10n^2 + 3n - 12 &>= 6n^3 - 10n^2 - 12 \\
&>= 6n^3 - 10n^2 - 12n^2 \\
&= 6n^3 - 22n^2.
\end{aligned}
$$

If $n >= 22$, then $n^3 = n \times n^2 >= 22n^2$, and so

$$6n^3 - 10n^2 + 3n - 12 >= 5n^3 + \left(n^3 - 22n^2\right) >= 5n^3 > n^3.$$

Thus if $K = 1$ and $M = 22$ the condition for asymptotic dominance is satisfied.
// We've shown that $(6n^3 - 10n^2 + 3n - 12) \sim n^3$.

Example 7.5.6: $n^3 \ll (0.4)n^3 - 10n^2 + 3n - 12$.
Let $K = 2/(0.4) = 5$. Then, when $n >= 1$,

$$
\begin{aligned}
K \times \left\{(0.4)n^3 - 10n^2 + 3n - 12\right\} &= 2n^3 - 50n^2 + 15n - 60 \\
&> 2n^3 - 50n^2 - 15n - 60 \\
&> 2n^3 - 50n^2 - 15n^2 - 60n^2 \\
&= 2n^3 - 125n^2 \\
&= n^3 + n^3 - 125n^2 \\
&= n^3 + n^2\{n - 125\}.
\end{aligned}
$$

If $n >= 125$, then $n^3 + n^2\{n - 125\} >= n^3$.

Thus, if $K = 5$ and $M = 125$, the condition for asymptotic dominance is satisfied.

These last few examples are instances of a general theorem about polynomials, which we will prove in detail from the definitions.

Theorem 7.5.4: Suppose $f(n)$ is a polynomial of degree d; that is,
$f(n) = a_d \times n^d + a_{d-1} \times n^{d-1} + \ldots + a_2 \times n^2 + a_1 \times n + a_0$ **where each**
$a_j \in \mathbf{R}$ **and** $a_d \neq 0$.

If $a_d > 0$, then $f(n)$ is a polynomial of degree d in $\mathcal{F}$ and $f \in \Theta(n^d)$.
If $a_d < 0$, then $f(n)$ is eventually negative, so is not in the set $\mathcal{F}$.

Proof. // This argument has three parts to it.
Part 1. Let $K = |a_d| + |a_{d-1}| + \ldots + |a_2| + |a_1| + |a_0|$.   // $K \in \mathbf{R}^+$
If $n >= 1$,

$$
\begin{aligned}
f(n) &= a_d \times n^d + a_{d-1} \times n^{d-1} + \ldots + a_2 \times n^2 + a_1 \times n + a_0 \\
&\leq |a_d|n^d + |a_{d-1}|n^{d-1} + \ldots + |a_2|n^2 + |a_1|n + |a_0| \\
&\leq |a_d|n^d + |a_{d-1}|n^d + \ldots + |a_2|n^d + |a_1|n^d + |a_0|n^d \\
&= \{|a_d| + |a_{d-1}| + \ldots + |a_2| + |a_1| + |a_0|\}n^d \\
&= K \times n^d. \qquad\qquad\qquad\qquad\qquad\qquad\qquad\text{// so } f(n) \ \mathcal{D} \ K \times n^d
\end{aligned}
$$

If f is in $\mathcal{F}$, then $f \ll n^d$.   // But when is f is in $\mathcal{F}$? And when is $n^d \ll f$?
     Let $g(n) = a_{d-1} \times n^{d-1} + \ldots + a_2 \times n^2 + a_1 \times n + a_0$
and $B = |a_{d-1}| + \ldots + |a_2| + |a_1| + |a_0|$.   // $B >= 0$.

Using the same inequalities as above, we have $g(n) <= B \times n^{d-1}$. Then

$$f(n) = a_d \times n^d + g(n) <= a_d \times n^d + B \times n^{d-1} = \{a_d \times n + B\}n^{d-1}.$$

Part 2. If a_d is negative, $a_d = -|a_d|$. Then, for any $n > B/|a_d|$, since $a_d < 0$,

$$a_d \times n < a_d \times (B/|a_d|) = -B.$$

Then $a_d \times n + B < -B + B = 0$

and $f(n) <= \{a_d \times n + B\}n^{d-1} < 0.$ // $f(n)$ is eventually negative

That is, $f(n)$ is not eventually positive and cannot be in $\mathcal{F}$.

Part 3. // Next, we show that if a_d is positive, then f is in $\mathcal{F}$ and $n^d << f(n)$.
Since for any number a, $-|a| <= a$, we have for all $n \in \mathbf{P}$

$$
\begin{aligned}
g(n) \;&= a_{d-1} \times n^{d-1} &&+ \ldots + & a_2 \times n^2 &&+ a_1 \times n &&+ a_0 \\
&>= (-|a_{d-1}|)n^{d-1} &&+ \ldots + & (-|a_2|)n^2 &&+ (-|a_1|)n &&+ (-|a_0|) \\
&>= (-|a_{d-1}|)n^{d-1} &&+ \ldots + & (-|a_2|)n^{d-1} &&+ (-|a_1|)n^{d-1} &&+ (-|a_0|)n^{d-1} \\
&= -\{|a_{d-1}| &&+ \ldots + & |a_2| &&+ |a_1| &&+ |a_0|\}n^{d-1} \\
&= -\{B\}n^{d-1}.
\end{aligned}
$$

Assume that $a_d > 0$. Let $K = 2/a_d$ and let $M = 1 + \lceil KB \rceil$.   // Then $K \in \mathbf{R}^+$ and $M \in \mathbf{P}$.
If $n >= M$, then $n > KB$ and

$$
\begin{aligned}
K \times f(n) &= Ka_d \times n^d + Ka_{d-1} \times n^{d-1} + \ldots + Ka_2 \times n^2 + Ka_1 \times n + Ka_0 \\
&= 2 \times n^d \quad + K \times g(n) \\
&>= 2 \times n^d \quad + K \times \left[-\{B\}n^{d-1} \right] \\
&= n^d \quad + \left[n^d - \{KB\}n^{d-1} \right] \\
&= n^d \quad + [n - KB]n^{d-1} \\
&> n^d.
\end{aligned}
$$
                                                              // and so $f(n) > n^d/K > 0$

Thus, f is eventually positive, f is in $\mathcal{F}$, and $n^d << f$.
We have shown that if $f(n)$ is a polynomial of degree d in $\mathcal{F}$, then $f \sim n^d$ and so
$f \in \Theta(n^d)$.   ◻

For sequences in $\mathcal{F}$ and $A \in \mathbf{R}^+$:
1. If $f \sim g$ and $A \in \mathbf{R}^+$, then $Af \sim g$.
2. If $f << g$, then $(f + g) \sim g$.
3. If $f1 << g$ and $f2 \sim g$, then $(f1 + f2) \sim g$.
4. If $f1 \sim g1$ and $f2 \sim g2$, then $(f1 + f2) \sim (g1 + g2)$.
5. If $f1 \sim g1$ and $f2 \sim g2$, then $(f1 \times f2) \sim (g1 \times g2)$.

// The proofs of these assertions are left as exercises.
// All the work has been done earlier on $<<$ in Theorem 7.5.3.

7.5.3   Asymptotic Ranking

If $f(n)$ and $g(n)$ are sequences in $\mathcal{F}$,

$$f \text{ is } \textbf{of lower order than } g \text{ [written } f <<< g] \underline{\text{means}}$$
$$f << g \text{ but } g !<< f.$$

We've seen $900n <<< n^2$.   // in Examples 7.5.1 & 7.5.2
// $f <<< g$ implies $f << g$ but $f !\sim g$.   // here "but" means "and"

// Is this really an order relation on $\mathcal{F}$?   What properties does $<<<$ have?
The relation $<<<$ is clearly <u>not reflexive</u>; if $f << g$ but $g !<< f$, then $f \neq g$.

Theorem 7.5.5: The relation $<<<$ is an order relation on $\mathcal{F}$

Proof. // We must show that $<<<$ is transitive and antisymmetric.
    Suppose $f <<< g$ and $g <<< h$ where f, g, and h are sequences in $\mathcal{F}$.
                                                    // Is $f <<< h$?
We know that $f << g$ but $g !<< f$ and $g << h$ but $h !<< g$.
Since $f << g$ and $g << h$, we have $f << h$.   // because $<<$ is transitive
If it were the case that $h << f$, then, because $f << g$ and $<<$ is transitive, we'd
have $h << g$; since this contradicts $h !<< g$, we know that $h !<< f$. Because $f << h$
and $h !<< f, f <<< h$. Therefore, $<<<$ is <u>transitive</u>.
    If $f <<< g$, then $f << g$ and $g !<< f$, so it cannot be the case that that $g << f$,
and therefore, it cannot be the case that that $g <<< f$. Thus, $<<<$ is <u>anti-</u>
<u>symmetric</u>.   □

    The relation $<<<$ does <u>not</u> have the *comparability* property. The two sequences,
f and g, given in the proof of Theorem 7.5.2 were incomparable under $<<$; that is,
$f !<< g$ and $g !<< f$. Therefore, they are incomparable under $<<<$; that is, $f !<<< g$
and $g !<<< f$.
    The relation $<<<$ acts (consistently) on whole equivalence classes; that is,

Theorem 7.5.6: If $f1 \sim f$ and $g1 \sim g$ and $f1 <<< g1$ then $f <<< g$

Proof. Suppose that $f1 \sim f$, $g1 \sim g$, and $f1 <<< g1$. Then

$$f << f1 \text{ and } g1 << g \text{ and } f1 << g1 \text{ but } g1 !<< f1.$$

// We want to show that $f << g$ but $g !<< f$.

Since $f << f1$ and $f1 << g1$ and $g1 << g$, and because $<<$ is transitive, we get $\underline{f << g}$.
If it were the case that $g << f$, then

$$g1 << g \text{ and } g << f \text{ and } f << f1 \text{ would give } g1 << f1.$$

Because this contradicts $g1 !<< f1$, we must have $\underline{g !<< f}$. Thus, $f <<< g$.   □

// In fact, what we proved was as follows: If $f \ll f1$ and $f1 \lll g1$ and $g1 \ll g$,
// then $f \lll g$.
// If f <u>is of order</u> $f1$ and $f1$ <u>is of lower order than</u> $g1$ and $g1$ <u>is of order</u> g,
// then f <u>is of lower order than</u> g.

7.5.4 Strong Asymptotic Dominance and Little-Oh Notation

There is yet another relation on $\mathcal{F}$ that is used to rank complexity functions.

  f is *strongly asymptotically dominated by* g [written f S$\mathcal{D}$ g] <u>means</u>
   $\forall K \in \mathbf{R}^+, \exists M(K) \in \mathbf{P}$ such that if $n > M(K)$,   then $K \times f(n) < g(n)$.

// For <u>any</u> positive real number K (no matter how *large*),
// there is a starting point (depending on the value of K) such that
// if n is <u>any</u> integer $> M(K)$, then $K \times f(n)$ is strictly smaller than $g(n)$.

For example, n **S$\mathcal{D}$** n^2.
Suppose K is <u>any</u> given positive real number. Let $M(K) = \lceil K \rceil$.   // $M(K) \in \mathbf{P}$.
If $n > M(K)$, then $n > K$ and so $K \times n < n^2$.

// Some books use "Little-Oh notation" to denote this relation, and it is usually
// defined by f is $\mathbf{o}(g) \Leftrightarrow \lim\{f(n)/g(n)\} = 0$.
// (We discussed limits of sequences in Sect. 2.4).

//X What properties does S$\mathcal{D}$ have? Is S$\mathcal{D}$ an order relation?

Theorem 7.5.7: If f S$\mathcal{D}$ g, then f $\lll$ g   //X The converse is false.

Proof. Suppose that f S$\mathcal{D}$ g // We must show that $f \ll g$ and $g \,!\!\ll f$.
   Then

   $\forall\ K_1 \in \mathbf{R}^+,\ \exists M_1(K_1) \in \mathbf{P}$ such that if $n >= M_1(K_1)$, then $K_1 \times f(n) < g(n)$.

// We use K_1 and $M_1(K_1)$ to distinguish the constants in the definition of S$\mathcal{D}$
// from those in the definitions of that $f \ll g$ and $g \,!\!\ll f$.

   To prove that $f \ll g$, we must show that

   $\exists K_2 \in \mathbf{R}^+$ and $\exists M_2 \in \mathbf{P}$ such that if $n >= M_2$, then $f(n) <= K_2 \times g(n)$.

Let $K_2 = 1$ and $M_2 = 1 + M_1(1)$.   // $K_2 \in \mathbf{R}^+$ and $M_2 \in \mathbf{P}$.
Using $K_1 = 1$, we get

   if $n > M_1(K_1)$,   then $K_1 \times f(n) = 1 \times f(n) < g(n)$; that is,
   if $n >= M_2$,   then $f(n) < g(n) = K_2 \times g(n)$.

Therefore, $f \ll g$.

To prove that that $g \; !<< \; f$, we must show that

$$\forall \; K_3 \in \mathbf{R}^+ \text{ and } \forall \; M_3 \in \mathbf{P}, \exists n^* >= M_3 \text{ where } g(n^*) > K_3 \times f(n^*).$$

Let K_3 be any given positive real number and let M_3 be any positive integer. Using $K_1 = K_3$,

$$\exists M_1(K_3) \in \mathbf{P}, \text{ such that if } n > M_1(K_3), \quad \text{then } K_3 \times f(n) < g(n).$$

Now let $n^* = M_3 + M_1(K_3)$ then $n^* >= M_3$ and $g(n^*) > K_3 \times f(n^*)$.

$$// \; n^* > M_1(K_3)$$

Therefore, $g \; !<< \; f$.   □

The relation $S\mathcal{D}$ does <u>not</u> have the *comparability* property. The two sequences, f and g, given in the proof of Theorem 7.5.2 are incomparable under $<<$; that is, $f \; !<< \; g$ and $g \; !<< \; f$. Therefore, $f \; !<<< \; g$ and $g \; !<<< \; f$. The contrapositive of Theorem 7.5.7 gives us $f \; !S\mathcal{D} \; g$ and $g \; !S\mathcal{D} \; f$.

Theorem 7.5.8: $1 \; S\mathcal{D} \; lg(n)$ and $n \; S\mathcal{D} \; n \times lg(n)$

Proof. // We will show that $\forall K \in \mathbf{R}^+, \exists M(K) \in \mathbf{P}$, such that
    // if $n > M(K)$, then $K \times 1 < lg(n)$, and $K \times n < n \times lg(n)$.

Suppose K is any given positive real number. Let $M(K) = \lceil 2^K \rceil$. If $n > M(K)$, then $n > 2^K$, and so

$$K \times 1 = lg(2^K) < lg(n) \text{ and also } K \times n < n \times lg(n).$$ □

The Most Important Ideas in This Section.

We defined $\mathcal{F}$ to be the set of all infinite, real-valued sequences (with domain $\mathbf{P}$) that are "eventually positive"; that is, there is an $N \in \mathbf{P}$ such that if $n >= N$, then $f(n) > 0$. The set $\mathcal{F}$ contains the complexity functions of algorithms, and we examined several relations on $\mathcal{F}$. If $f(n)$ and $g(n)$ are sequences in $\mathcal{F}$:

1. *f is asymptotically dominated by g* $[f << g \text{ or } f \in \mathbf{O}(g)] \Leftrightarrow$
    $\exists K \in \mathbf{R}^+$ and $\exists M \in \mathbf{P}$ such that if $n >= M$, then $f(n) <= K \times g(n)$.
2. *f is **not** asymptotically dominated by g* $[f \; !<< \; g] \Leftrightarrow$
    $\forall K \in \mathbf{R}^+$ and $\forall M \in \mathbf{P}, \exists \; n^* >= M$ where $f(n^*) > K \times g(n^*)$.
3. *f is asymptotically equivalent to g* $[f \sim g \text{ or } f \in \Theta(g)] \Leftrightarrow$
    $f << g$ and $g << f$.
4. *f is of lower order than g* $[f <<< g] \Leftrightarrow$
    $f << g$ but $g \; !<< \; f$.
5. *f is strongly asymptotically dominated by g* $[f \; S\mathcal{D} \; g \text{ or } f \text{ is } \mathbf{o}(g)] \Leftrightarrow$
    $\forall \; K \in \mathbf{R}^+, \exists M(K) \in \mathbf{P}$ such that if $n > M(K)$, then $K \times f(n) < g(n)$.

(continued)

(continued)

> The relation $\sim$ was shown to be an equivalence relation, so it partitions $\mathcal{F}$ into equivalence classes, where the class containing $f(n)$ is denoted $\Theta(f)$. We proved that if $f(n)$ is a polynomial of degree d in $\mathcal{F}$, then $f \in \Theta(n^d)$.
>
> All three of the relations $\ll$, $\lll$, and $S\mathcal{D}$ are order relations, so they allow ranking of complexity functions. The following list shows many of the common complexity classes in increasing "cost". All of these comparisons can be established using $S\mathcal{D}$ (or using Little-Oh).
>
> $$
> \begin{array}{lll}
> 1 & \lll & \lg(n) \\
> \lg(n) & \lll & n^p & \text{whenever } p > 0 \\
> n^p & \lll & n^q & \text{whenever } q > p > 0 \\
> n^q & \lll & n^{\lg(n)} & \text{whenever } q > 0 \\
> n^{\lg(n)} & \lll & n^{\sqrt{n}} \\
> n^{\sqrt{n}} & \lll & b^n & \text{whenever } b > 1 \\
> b^n & \lll & c^n & \text{whenever } c > b > 0 \\
> c^n & \lll & n! & \text{whenever } c > 0 \\
> n! & \lll & n^n
> \end{array}
> $$
>
> All **bounded** functions are of lower order than all **logarithmic** functions.
> All **logarithmic** functions are of lower order than all **power** functions with $p > 0$.
> All **power** functions (polynomials) are of lower order than $n^{\lg(n)}$.
> The function $n^{\lg(n)}$ is of lower order than $n^{\sqrt{n}}$.
> The function $n^{\sqrt{n}}$ is of lower order than all **exponential** functions with $b > 1$.
> So all **power** functions are of lower order than all **exponential** functions with $b > 1$.
> All **exponential** functions with $b > 1$ are of lower order than the **factorial** function, $n!$.

Exercises

1. Consider the set $S = \{0, 1, 2, 3, 4, 5, 6\}$ and the relations

   $R_1 = \{(a,b): a,b \in S, a > b\}$, and $R_2 = \{(a,b): a,b \in S, \max(a + b, ab) = 3 \text{ or } 6\}$:

   (a) Give a matrix representation for each relation.
   (b) Give a digraph representation for each relation.
   (c) Determine the properties of each relation.
   (d) Which relation is a partial order, an equivalence relation or neither?
2. Let $S = \{1, 2, 3\}$. Define a relation R on S that is reflexive.

3. Let S = {1, 2, 3, 4, 5} and let R be a relation on S where R = {(1, 2), (1, 3), (2, 4), (3, 5), (4, 1), (5, 4)}. What additional pairs must be added to R to make it transitive?

4. Let S = {1, 2, 3}. Define a relation R on S where every integer in S appears in at least one ordered pair in R and R is
   (a) Symmetric
   (b) Antisymmetric

5. Let S = {1, 2, 3, 4}. Define a relation R on S that has the comparability property.

6. Let X = {3, 4, 5, 6, 7} and let R be the relation on X defined by $a \; R \; b \Leftrightarrow ab + b < 30$. Find a "counter-example" to prove each of the following:
   (a) R is NOT reflexive.
   (b) R is NOT transitive.
   (c) R is NOT symmetric.
   (d) R is NOT antisymmetric.

7. Let S denote the set of Boolean expressions:
   (a) The "implies" relation on S is denoted using the symbol "$\Rightarrow$". (Recall from Chap. 3 that $P \Rightarrow Q$ means that the conditional expression, $P \rightarrow Q$, is always True.) What properties does the relation $\Rightarrow$ have?
   (b) In Chap. 3 two Boolean expressions P and Q were said to be **equivalent** [written $P \Leftrightarrow Q$] when they had exactly the same truth tables. Therefore $\Leftrightarrow$ denotes a relation on S. Is this really an equivalence relation? What properties does $\Leftrightarrow$ have?

8. Let $G = (V, E)$ be an undirected graph. Define a relation R on V by

$$v \; R \; w \Leftrightarrow \text{there is a path joining vertices } v \text{ and } w.$$

What properties does R have?

9. Let $D = (V, A)$ be a directed graph. Define a relation R on V by

$$v \; R \; w \Leftrightarrow \text{there is a dipath from vertex } v \text{ to vertex } w.$$

What properties does R have?
If D has no cycles, what properties does R have?

10. Let S = {1, 2, 3, 4} and R be an equivalence relation on S where R = {(1, 1), (1, 4), (2, 2), (2, 3), (3, 2), (3, 3), (4, 1), (4, 4)}. What are the equivalence classes of R?

11. Let S = {1, 2, 3, 4, 5} and R be an equivalence relation on S where R = {(1, 1), (2, 2), (2, 4), (2, 5), (3, 3), (4, 2), (4, 4), (4, 5), (5, 2), (5, 4), (5, 5)}. What are the equivalence classes of R?

12. Let $D = (V, A)$ be a directed graph. Define a relation R on V by

$$v \; R \; w \Leftrightarrow \text{there is a dipath from vertex } v \text{ to vertex } w$$
$$\text{and there is a dipath from vertex } w \text{ to } v.$$

(a) Show that R is an equivalence relation.
(b) An equivalence class X together with all arcs joining vertices in X, is called a "strong component" of D. Prove that all vertices occurring in some cycle C, are in the same strong component.
(c) Consider the digraph $D^* = (V^*, A^*)$ where V^* is the set of equivalence classes of D, and where there is an arc from class A to class B in $D^* \Leftrightarrow$ there is an arc in D from a vertex a in A to a vertex b in class B. Prove that D^* is acyclic.

13. Suppose that R is a relation on set X. Define a second relation, R*, on X by

$$x \; R^* \; y \;\; \Leftrightarrow \;\; x \, R \, y \text{ and } y \, R \, x$$

(a) Prove that R* is symmetric.
(b) Prove that if R is transitive, then R* is transitive.
(c) Prove that if R is reflexive and transitive, then R* is an equivalence relation.

14. Construct the operation tables for Z_6 like the ones given in Example 7.2.3. Does your table show that if $[a] \otimes [b] = [0]$ then either $[a] = [0]$ or $[b] = [0]$?

15. Suppose that k is a given positive integer. Z_k denotes the set of equivalence classes determined by the relation R on Z defined by

$$a \, R \, b \text{ if and only if } f(a) = f(b),$$

where $f(x) = x \text{ MOD } k$, the nonnegative remainder obtained when x is divided by k:
(a) Show $a \, R \, b$ if and only if $b = a + kn$ for some integer n.
(b) Show $a \, R \, b$ if and only if $b - a = kn$ for some integer n.
(c) Show $a \, R \, b$ if and only if $k | (b - a)$.
(d) Show that if $a \, R \, b$ and $c \, R \, d$, then $(a + c) \, R \, (b + d)$.

   // This shows that the addition of classes given by $[a] \oplus [b] = [a + b]$ is
   // "independent of the representatives of the classes".

(e) Show that if $a \, R \, b$ and $c \, R \, d$, then $(a \times c) \, R \, (b \times d)$.

   // This shows that the addition of classes given by $[a] \otimes [b] = [a \times b]$ is
   // "independent of the representatives of the classes".

16. Let $S = \{1, 2, 3, 4, 5\}$ and R be a relation on S where $R = \{(1, 1), (1, 3), (1, 5), (2, 2), (2, 4), (3, 3), (3, 5), (4, 4), (5, 5)\}$.
(a) Is R an order relation on S? Why or why not?
(b) Is R a partial order relation on S? Why or why not?
(c) Is R a total order relation on S? Why or why not?

17. Let $S = \{1, 2, 3, 4, 5\}$ and R be a relation on S where $R = \{(1, 1), (1, 3), (1, 5), (2, 1), (2, 2), (2, 3), (2, 4), (2, 5), (3, 3), (3, 5), (4, 1), (4, 2), (4, 3), (4, 4), (4, 5), (5, 5)\}$.
(a) Is R an order relation on S? Why or why not?
(b) Is R a partial order relation on S? Why or why not?
(c) Is R a total order relation on S? Why or why not?

18. Consider a finite sequence Y = (1, 3, 5, 2, 4).
    (a) Give an example of a finite sequence X such that X is dominated by Y.
    (b) Give an example of a finite sequence Z such that Y is dominated by Z.
19. Consider a finite sequence Y = (1, 3, 5, 2, 4).
    (a) Give an example of a finite sequence X such that X is lexicographically less than Y.
    (b) Give an example of a finite sequence Z such that Y is lexicographically less than Z.
20. Let S denote the set of all five sequences on $\mathbf{Z}$. Consider the set T of ten elements of S:

$$s_1 = (4, 2, 9, 5, 5) \qquad s_2 = (4, 2, 3, 1, 2) \qquad s_3 = (-2, 3, 3, -2, 4)$$
$$s_4 = (7, 6, 8, 3, 9) \qquad s_5 = (0, -1, 5, 1, 3) \qquad s_6 = (3, 1, 3, 0, 3)$$
$$s_7 = (4, 2, 4, 5, 5) \qquad s_8 = (4, -2, 9, 1, 6) \qquad s_9 = (4, 2, 9, -1, 6)$$
$$s_{10} = (4, 2, 9, 5, 6)$$

    (a) Sort the ten elements of T into lexicographic order, smallest to largest.
    (b) Using the relation "is dominated by" on S:
        1. Find an element of S which is dominated by all of these ten.
        2. Find two elements in the ten which are **not** comparable under this relation.
        3. Draw the Hasse diagram of the set T.
        4. Find the least upper bound of the set T in S.
21. Suppose that R is an order relation on a set S and that T is a subset of S. Prove the following:
    (a) If T has a minimum element, it is unique.
    (b) If q is a minimum element, then q is minimal.
    (c) Prove that b covers $a \Leftrightarrow b$ is a minimal element of $\{x \in S \setminus \{a\}: a \, R \, x\}$.
22. Suppose that R is an order relation on a set S and that T is a subset of S. Prove the following:
    (a) If T has a maximum element, it is unique.
    (b) If q is a maximum element, then q is maximal.
    (c) If T is finite, T must have at least one maximal element.
    (d) What is the minimum upper bound for the subset in Example 7.3.1?
23. After Algorithm 7.3.2 has R-indexed the n-set S, is $y[n]$ a maximal element of S?
24. Construct an example where $X \, \mathcal{L} \, Y$ but $X \, !\mathcal{D} \, Y$.
25. Prove that if $X \, \mathcal{L} \, Y$, then $x_1 <= y_1$.
26. Prove that $\log_b(n) \sim \lg(n)$ for $\forall \, b > 1$. Is $\log_b(n) = K \times \lg(n)$ for some particular real number K?
27. Prove that $f \sim g \Leftrightarrow$
    $\exists \, A, B \in \mathbf{R}^+$ and $\exists \, M \in \mathbf{P}$ such that if $n >= M$, then
    $A \times f(n) <= g(n) <= B \times f(n)$.

28. List the following ten complexity classes from lowest complexity to highest: $n\lg(n)$, n^2, 2^n, $n!$, n, 1, n^n, $(\sqrt{5})^n$, $\sqrt{n}$, $\lg(n)$.

29. Suppose that f, $f1$, $f2$, g, $g1$, and $g2$ are sequences in $\mathcal{F}$. Prove each of the following:
    (a) If $f \sim g$ and $A \in \mathbf{R}^+$ then $Af \sim g$.
    (b) If $f \ll g$, then $(f + g) \sim g$.
    (c) If $f1 \ll g$ and $f2 \sim g$, then $(f1 + f2) \sim g$.
    (d) If $f1 \sim g1$ and $f2 \sim g2$, then $(f1 + f2) \sim (g1 + g2)$.
    (e) If $f1 \sim g1$ and $f2 \sim g2$, then $(f1 \times f2) \sim (g1 \times g2)$.

30. Suppose that $f1 = n$, $f2 = n^2$, $g1 = n^3$, and $g2 = 2n^2$. Then $f1 \ll g1$ and $f2 \sim g2$. Prove that $(f1 + f2) \lll (g1 + g2)$ and so $(f1 + f2) \,!\!\sim (g1 + g2)$.

31. Prove that $S\mathcal{D}$ is an order relation.

32. Prove that the converse of Theorem 7.5.7 is False.
    Find two sequences, f and g, where $f \lll g$ and $f \,!S\mathcal{D}\, g$.
    Hint: Consider these two increasing integer sequences

$$f(n) = (n!) \quad \text{and}$$

$$g(n) = \begin{cases} (2r)!(2r) & \text{if } n = 2r \\ (2r)!(2r + 1) & \text{if } n = 2r + 1. \end{cases}$$

Sequences and Series

8

When we considered the Towers of Hanoi in Sect. 2.4, we saw that the number of single disc transfers required to move a tower of $n > 1$ discs, T_n, satisfied the equation

$$T_n = T_{n-1} + 1 + T_{n-1},$$

or
$$T_n = 2T_{n-1} + 1. \tag{8.1.1}$$

This is an example of a recurrence equation − where the generic entry in a sequence is expressed in terms of one or more previous entries. Such equations arise frequently when counting the operations done by an algorithm and in other counting problems.

"*Solving a recurrence equation*" means finding a sequence that satisfies the recurrence equation. Finding a "*general solution*" means finding a formula that describes all possible solutions (all possible sequences that satisfy the equation).

The recurrence equation (8.1.1) tells how the sequence *continues* but doesn't tell us how the sequence starts:

If $T_1 = 1$, then $T = (1, 3, 7, 15, 31, \ldots)$. // if T has domain **P**

Let's assume that T is some sequence defined on the set of positive integers, **P**. Using the recurrence equation alone (not some intrinsic meaning for the entries), we can determine that

If $T_1 = 2$, then $T = (2, 5, 11, 23, 47, 95, \ldots)$.
If $T_1 = 4$, then $T = (4, 9, 19, 39, 79, 159, \ldots)$.
If $T_1 = -1$, then $T = (-1, -1, -1, -1, -1, -1, \ldots)$.

© Springer International Publishing AG, part of Springer Nature 2018
T. Jenkyns and B. Stephenson, *Fundamentals of Discrete Math for Computer Science: A Problem-Solving Primer*, Undergraduate Topics in Computer Science,
https://doi.org/10.1007/978-3-319-70151-6_8

Proof. // by strong Mathematical Induction on n

Step 1. If $n = 2$, then $(1/3)n! = 2/3 < 1 = D_n = (1/2)n!$

and   if $n = 3$, then $(1/3)n! = 6/3 = 2 = D_n < 3 = (1/2)n!$.

Step 2. Assume $\exists k >= 3$ such that if $2 <= n <= k$, then $(1/3)n! <= D_n <= (1/2)n!$.

Step 3. If $n = k + 1$, then $n >= 4$ and

$$D_n = (n - 1)\{D_{n-2} + D_{n-1}\} \text{ where } 2 <= n - 2 < n - 1 <= k.$$

Thus, $D_n >= (n - 1)\{(1/3)[n - 2]! + (1/3)[n - 1]!\} = (1/3)\, n!$,

// as we saw before

and $D_n <= (n - 1)\{(1/2)[n - 2]! + (1/2)[n - 1]!\} = (1/2)\, n!$.

// as we saw before

$\square$

The nicest formula for D_n that we know uses the "nearest integer" function. For any real number x, let $\lceil x \rfloor$ denote the **nearest integer to** x defined as follows: when x is written as $n + f$ where n is the integer $\lfloor x \rfloor$, and f is a fraction where $0 <= f < 1$:

if $0 <= f < \frac{1}{2}$ then $\lceil x \rfloor = n$;

if $\frac{1}{2} <= f < 1$ then $\lceil x \rfloor = n + 1$.   // Is $\lceil x \rfloor = \lfloor x + \frac{1}{2} \rfloor$?

So $\lceil 3.29 \rfloor = 3$, $\lceil -3.78 \rfloor = -4$, $\lceil +3.78 \rfloor = 4$, and $\lceil 3.50 \rfloor = 4$.

Then $D_n = \lceil (n!)/e \rfloor$ where $e = 2.718\ 281\ 828\ 44 \ldots$ is the base of the natural logarithms.   // $(n!)/e$ is never equal to $\lfloor (n!)/e \rfloor + \frac{1}{2}$.

n	D_n	$n!/e$
1	0	0.367 879 441
2	1	0.735 758 882
3	2	2.207 276 647
4	9	8.829 106 588
5	44	44.145 532 94
6	265	264.873 197 6
7	1 854	1 854.112 384
8	14 833	14 832.899 07
9	133 496	133 496.091 6
10	1 334 961	1 334 960.916

// There is another (much less compact) formula for D_n given in the exercises,
// along with an outline of the proof that $D_n = \lceil (n!)/e \rfloor$ (for you to complete).

Example 8.1.2: Ackermann Numbers

In the 1920s, a German logician and mathematician, Wilhelm Ackermann (1896–1962), invented a very curious function, $A: \mathbf{P} \times \mathbf{P} \rightarrow \mathbf{P}$, which we will define recursively using three "rules":

Rule 1. $A(1, n) = 2$ for $n = 1, 2, \ldots,$

Rule 2. $A(m, 1) = 2m$ for $m = 2, 3, \ldots,$

Rule 3.   When both m and n are larger than 1,

$$A(m, n) = A(A(m - 1, n), n - 1).$$

Then $A(2, 2) = A(A(2 - 1, 2) \quad , 2 - 1)$ // Rule 3
$\qquad\qquad = A(A(1, 2) \qquad\quad , 1)$
$\qquad\qquad = A(\quad 2 \qquad\qquad , 1)$ // Rule 1
$\qquad\qquad = 2(2)$ // Rule 2
$\qquad\qquad = 4.$

Also, $A(2, 3) = A(A(2 - 1, 3) \quad , 3 - 1)$ // Rule 3
$\qquad\qquad = A(A(1, 3) \qquad\quad , 2)$
$\qquad\qquad = A(\quad 2 \qquad\qquad , 2)$ // Rule 1
$\qquad\qquad = 4$ // above

In fact,   if $A(2, k) \;\; = 4,$ // for some $k >= 2$
then $A(2, k + 1) = A(A(2 - 1, k + 1) \quad , [k + 1] - 1)$ // Rule 3
$\qquad\qquad\quad = A(A(1, k + 1) \qquad\quad , k)$
$\qquad\qquad\quad = A(\qquad 2 \qquad\qquad , k)$ // Rule 1
$\qquad\qquad\quad = 4.$ // our assumption

Thus, $A(2, \; n) = 4$ for all $n >= 1.$ // by MI

So far the table of Ackermann numbers looks like this:

A	$n = 1$	$n = 2$	3	4	5	6	7	8	9...
$m = 1$	2	2	2	2	2	2	2	2	?
$m = 2$	4	4	4	4	4	4	4	4	4...
3	6								
4	8								
5	10								

// The second *row* is all 4's.
// What's the second *column* like?

$\qquad\qquad A(3, 2) = A(A(3 - 1, 2) \quad , 2 - 1)$ // Rule 3
$\qquad\qquad\qquad = A(A(2, 2) \qquad\quad , 1)$
$\qquad\qquad\qquad = A(\quad 4 \qquad\qquad , 1)$ // second row
$\qquad\qquad\qquad = 2(4)$ // Rule 2
$\qquad\qquad\qquad = 8.$

$\qquad\qquad A(4, 2) = A(A(4 - 1, 2) \quad , 2 - 1)$ // Rule 3
$\qquad\qquad\qquad = A(A(3, 2) \qquad\quad , 1)$
$\qquad\qquad\qquad = A(\quad 8 \qquad\qquad , 1)$ // above
$\qquad\qquad\qquad = 2(8)$ // Rule 2
$\qquad\qquad\qquad = 16.$

// Is the second column the powers of 2?

$$
\begin{array}{lll}
\text{If} & A(k,2) \; = 2^k, & \text{// for some } k >= 2 \\
\text{then} & A(k+1,2) = A(A([k+1]-1,2) \quad ,2-1) & \text{// Rule 3} \\
& \quad\quad\;\; = A(A(k,2) \quad\quad\quad\quad ,1) & \\
& \quad\quad\;\; = A(\quad 2^k \quad\quad\quad\quad\quad ,1) & \text{// our assumption} \\
& \quad\quad\;\; = 2(2^k) & \text{// Rule 2} \\
& \quad\quad\;\; = 2^{k+1}.
\end{array}
$$

Thus, $A(m, 2) = 2^m$ for all $m >= 1$.
// What are the other values like?

$$
\begin{array}{lll}
A(3,3) & = A(A(3-1,3) \quad ,3-1) & \text{// Rule 3} \\
& = A(A(2,3) \quad\quad\;\; ,2) & \\
& = A(\quad 4 \quad\quad\quad\quad ,2) & \text{// second row} \\
& = 2^4 & \text{// second column} \\
& = 16.
\end{array}
$$

$$
\begin{array}{lll}
A(4,3) & = A(A(4-1,3) \quad ,3-1) & \text{// Rule 3} \\
& = A(A(3,3) \quad\quad\;\; ,2) & \\
& = A(\quad 16 \quad\quad\quad ,2) & \text{// above} \\
& = 2^{16} & \text{// second column} \\
& = 65\,536.
\end{array}
$$

$$
\begin{array}{lll}
A(3,4) & = A(A(3-1,4) \quad ,4-1) & \text{// Rule 3} \\
& = A(A(2,4) \quad\quad\;\; ,3) & \\
& = A(\quad 4 \quad\quad\quad\quad ,3) & \text{// second row} \\
& = 65\,536. & \text{// above}
\end{array}
$$

// What is the value of $A(4, 4)$?
// Could you run a simple recursive program to evaluate $A(4, 4)$?

$$
\begin{array}{lll}
A(5,3) & = A(A(5-1,3) \quad ,3-1) & \text{// Rule 3} \\
& = A(A(4,3) \quad\quad\;\; ,2) & \\
& = A(65536 \quad\quad\;\; ,2) & \text{// above} \\
& = 2^{65536} & \text{// second column} \\
& = \text{a very large number.} & \\
& \quad\quad\quad \text{// with about 20,000 digits (in base 10)}
\end{array}
$$

So far, we have

A	$n = 1$	2	3	4	5	6
$m = 1$	2	2	2	2	2	2
2	4	4	4	4	4	4
3	6	8	16	65536	?	
4	8	16	65536	?		
5	10	32	2^{65536}			

// How does the third column continue?

Let $2{\uparrow}k$ denote the value of a "tower" of k 2's, defined recursively by

$$2{\uparrow}1 = 2; \quad \text{and for } k >= 1, \; 2 \uparrow [k+1] = 2^{2^{\uparrow}k}.$$

Then
$$2{\uparrow}2 = 2^{2^{\uparrow}1} = 2^2 = 4,$$
$$2{\uparrow}3 = 2^{2^{\uparrow}2} = 2^4 = 16,$$
and
$$2{\uparrow}4 = 2^{2^{\uparrow}3} = 2^{16} = 65\,536.$$

We can prove (by Mathematical Induction) that $A(m, 3) = 2{\uparrow}m$ for $\forall m \in \mathbf{P}$.
Step 1. If $m = 1$, then by Rule 1, $A(1, 3) = 2$ and $2 = 2{\uparrow}1$.
Step 2. Assume $\exists\; k >= 1$ such that $A(k, 3) = 2{\uparrow}k$.
Step 3. If $m = k + 1$, then by Rule 3

$$
\begin{aligned}
A(k+1, 3) &= A(A([k+1] - 1, 3) \quad, 3 - 1) \\
&= A(A(k, 3) \qquad\qquad, 2) \\
&= A(\qquad 2 \uparrow k \qquad, 2) \qquad && \text{// our assumption} \\
&= 2^{2^{\uparrow}k} && \text{// second column} \\
&= 2 \uparrow (k+1). && \text{// definition of } \uparrow
\end{aligned}
$$

Thus, $A(m, 3) = 2{\uparrow}m$ for all $m >= 1$.   // by MI

                                                                           □

Finally,
$$
\begin{aligned}
A(4, 4) &= A(A(4 - 1, 4) \quad, 4 - 1) && \text{// Rule 3} \\
&= A(A(3, 4) \qquad, 3) \\
&= A(65\,536 \qquad, 3) && \text{// above} \\
&= 2 \uparrow (65\,536).
\end{aligned}
$$

But this is a number so big it could never be written out in decimal digits, not even using all the paper in the world. Its value can never be calculated. Are Ackermann numbers "computable"? On the other hand, let's assume that the sequences we encounter, even those defined by recurrence equations, will be easy to understand and deal with.

is given in two parts:

$$\text{if } a = 1, \qquad S_n = I + nc \qquad\qquad\qquad \text{for } \forall\, n \in \mathbf{N};$$
$$\text{if } a \neq 1, \qquad S_n = a^n A + \frac{c}{1 - a} \qquad\quad \text{for } \forall\, n \in \mathbf{N}.$$

When $a = 1$, any ***particular solution*** is obtained by determining a specific, numerical value for I. In fact, a *particular solution* is determined by a specific, numerical value J for any (particular) entry, S_j. Solving the equation

we get
$$J = I + jc \text{ for } I, \qquad\qquad\qquad // \text{ since } S_j = I + jc$$
$$I = J - jc. \qquad\qquad\qquad\qquad // \text{ where } S_0 = I$$

// One particular "particular solution" has $I = 0$.

When $a \neq 1$, any ***particular solution*** is obtained by determining a specific, numerical value for A; if the starting value I is given, then $A = I - \dfrac{c}{1 - a}$. In fact, a *particular solution* is determined by a specific, numerical value J for any (particular) entry, S_j. Solving the equation

$$J = Aa^j + \frac{c}{1 - a} \quad \text{ for } A,$$

we get
$$A = \frac{1}{a^j}\left[J - \frac{c}{1 - a}\right]. \qquad\qquad // \text{ But what if } a = 0?$$

// One particular "particular solution" has $A = 0$.

Example 8.2.1: The Towers of Hanoi

The recurrence equation for the number of moves in the Towers of Hanoi problem is a first-order linear recurrence equation:

$$T_n = 2T_{n-1} + 1.$$

Here $a = 2$ and $c = 1$, so $\dfrac{c}{1 - a} = \dfrac{1}{1 - 2} = -1$, and <u>any</u> sequence T that satisfies this RE is given by the formula
$$T_n = 2^n[I - (-1)] + (-1)$$
$$= 2^n[I + 1] - 1.$$

Assuming T has domain $\mathbf{N}$ and denoting T_0 by I, we saw at the beginning of this chapter several particular solutions:

if $I = 0$, then	$T = (0, 1, 3, 7, 15, 31, \dots)$;	// $T_n = 2^n[0 + 1] - 1 = 2^n \quad - 1$.
if $I = 2$, then	$T = (2, 5, 11, 23, 47, 95, \dots)$;	// $T_n = 2^n[2 + 1] - 1 = 3 \times 2^n - 1$.
if $I = 4$, then	$T = (4, 9, 19, 39, 79, 159, \dots)$;	// $T_n = 2^n[4 + 1] - 1 = 5 \times 2^n - 1$.
if $I = -1$, then	$T = (-1, -1, -1, -1, -1, \dots)$.	// $T_n = 2^n[-1 + 1] - 1 = \quad -1$.

Example 8.2.2: The three Shipwrecked Pirates

A pirate ship is wrecked in a storm at night. Three of the pirates survive and find themselves on a beach the morning after the storm. They agree to cooperate to ensure their continued survival. They spot a monkey in the jungle near the beach and spend all of that first day collecting a large pile of coconuts and then go to sleep exhausted.

But they are pirates.

The first one sleeps fitfully, worried about his share of the coconuts; he wakes, divides the pile into 3 equal piles, but finds one left over which he throws into the bush for the monkey, buries his third in the sand, heaps the two other piles together, and goes to sleep soundly.

The second pirate sleeps fitfully, worried about his share of the coconuts; he wakes, divides the pile into 3 equal piles, but finds one left over which he throws into the bush for the monkey, buries his third in the sand, heaps the two other piles together, and goes to sleep soundly.

The third one too sleeps fitfully, worried about his share of the coconuts; he wakes, divides the pile into 3 equal piles, but finds one left over which he throws into the bush for the monkey, buries his third in the sand, heaps the two other piles together, and goes to sleep soundly.

The next morning, they all awaken and see a somewhat smaller pile of coconuts which they divide into 3 equal piles but find one left over which they throw into the bush for the monkey.

How many coconuts did they collect on the first day?

Let S_j denote the size of the pile after the jth pirate and let S_0 will be the number they collected on the first day. Then

$$S_0 = 3x + 1 \text{ for some integer } x \text{ and } S_1 = 2x,$$
$$S_1 = 3y + 1 \text{ for some integer } y \text{ and } S_2 = 2y,$$
$$S_2 = 3z + 1 \text{ for some integer } z \text{ and } S_3 = 2z,$$

and
$$S_3 = 3w + 1 \text{ for some integer } w.$$

// Is there a recurrence equation here?

$$S_1 = 2x \text{ where } x = (S_0 - 1)/3, \text{ so } S_1 = (2/3)S_0 - (2/3);$$
$$S_2 = 2y \text{ where } y = (S_1 - 1)/3, \text{ so } S_2 = (2/3)S_1 - (2/3);$$
$$S_3 = 2z \text{ where } z = (S_2 - 1)/3, \text{ so } S_3 = (2/3)S_2 - (2/3).$$

The recurrence equation satisfied by the first few S_j's is

$$S_{j+1} = (2/3)S_j - (2/3). \tag{*}$$

If we now let $S_4 = (2/3)S_3 - (2/3)$, then $S_4 = 2[S_3 - 1]/3 = 2w$ for some integer w.

We want to know what value (or values) of S_0 will produce an <u>even integer</u> for S_4 when we apply the RE (*).
In (*), $a = 2/3$ and $c = -2/3$, so $c/(1 - a) = -2$, and so the general solution of (*) is

$$S_n = (2/3)^n[S_0 + 2] - 2.$$

Hence, $S_4 = (2/3)^4[S_0 + 2] - 2 = (16/81)[S_0 + 2] - 2.$

S_4 will be an integer
  $\Leftrightarrow S_4 + 2$ is an (even) integer
  $\Leftrightarrow 81$ divides into $[S_0 + 2]$
  $\Leftrightarrow [S_0 + 2] = 81k$ for some integer k
  $\Leftrightarrow S_0 = 81k - 2$ for some integer k.
S_0 must be a positive integer, but there are an infinite number of possible answers:

$$79 \text{ or } 160 \text{ or } 241 \text{ or } 322 \text{ or } \ldots$$

// We need more information to determine S_0.
// If we had been told that on the first day the pirates collected
// between 200 and 300 coconuts, we could now say
// "the number they collected on the first day was exactly 241."

Example 8.2.3: Compound Interest

Suppose you are offered two retirement savings plans. In Plan A, you start with $1,000, and each year (on the anniversary of the plan), you are paid 11% simple interest, and you add $1,000. In Plan B, you start with $100, and each month, you are paid one-twelfth of 10% simple (annual) interest, and you add $100. Which plan will be larger after 40 years?   // Can we apply a recurrence equation?
Consider Plan A and let S_n denote the number of dollars in the plan after (exactly) n years of operation. Then $S_0 = \$1,000$ and

$$\begin{aligned}
S_{n+1} &= S_n + \text{interest on } S_n + \$1000 \\
&= S_n + 11\% \text{ of } S_n \quad + \$1000 \\
&= S_n(1 + 0.11) \qquad + \$1000.
\end{aligned}$$

In this RE, $a = 1.11$, $c = 1000$, so $\dfrac{c}{1 - a} = \dfrac{1000}{-0.11}$, and

$$\begin{aligned}
S_n &= (1.11)^n\left[1000 - \frac{1000}{-0.11}\right] + \frac{1000}{-0.11} \\
&= (1.11)^n\left[\frac{1110}{+0.11}\right] - \frac{1000}{+0.11}.
\end{aligned}$$

Hence, $S_{40} = (1.11)^{40}(10\,090.090\,909\ldots)$ $- (9\,090.909\,090\ldots)$
$= (65.000\,867\ldots)(10\,090.090\,909\ldots)$ $- (9\,090.909\,090\ldots)$
$= 655\,917.842\ldots$ $- (9\,090.909\,090\ldots)$
$\cong \$646\,826.$

// Can that be right? You put in \$40,000 and take out $> \$600,000$ in interest.

Now consider Plan B and let T_n denote the number of dollars in the plan after (exactly) n months of operation. Then $T_0 = \$100$ and

$$\begin{aligned} T_{n+1} &= T_n + \text{interest on } T_n &&+ \$100 \\ &= T_n + (1/12) \text{ of } 10\% \text{ of } T_n + \$100 \\ &= T_n[1 + 0.1/12] &&+ \$100. \end{aligned}$$

In this RE, $a = 12.1/12$, $c = 100$, so $\dfrac{c}{1-a} = \dfrac{100}{-0.1/12} = -12000$ and
$$T_n = (12.1/12)^n[100 + 12000] - 12000.$$

Hence, after 40×12 months,

$$\begin{aligned} T_{480} &= (12.1/12)^{480}(12100) &&- (12000) \\ &= (1.008\,333\ldots)^{480}(12100) &&- (12000) \\ &= (53.700\,663\ldots)(12100) &&- (12000) \\ &= 649\,778.023\,4\ldots &&- (12000) \\ &\cong \$637\,778. \end{aligned}$$

Therefore, Plan A has a slightly larger value after 40 years.

The Most Important Ideas in This Section.
A *first-order linear recurrence equation* relates consecutive entries in a sequence by an equation of the form

$$S_{n+1} = aS_n + c \qquad \text{for } \forall\, n \in \mathbf{N}.$$

The *general solution* is given in two parts:

if $a = 1$, $\qquad S_n = A + nc$ $\qquad$ for $\forall\, n \in \mathbf{N}$;
if $a \neq 1$, $\qquad S_n = a^n A + \dfrac{c}{1-a}$ $\qquad$ for $\forall\, n \in \mathbf{N}$.

A *particular solution* is obtained by determining a specific, numerical value for A. In fact, a *particular solution* is determined by a specific, numerical value J for any (particular) entry, S_j.

8.3   The Fibonacci Sequence

Leonardo Fibonacci, an Italian mathematician (c1170–1230), posed the following problem. Imagine a safe enclosure for keeping rabbits. At the beginning of the year, a pair of newborn rabbits is placed in it. Rabbits multiply like rabbits, but (for this example) suppose that a pair will produce a new pair every month as soon as they are old enough, and they are old enough after one month. How many rabbits will be in there after a year?

Let F_n denote the number of pairs of rabbits in the sanctuary after n months. Then

$$
\begin{aligned}
F_0 &= 1, \\
F_1 &= 1, \\
F_2 &= 2, && \text{// an old pair and a new pair} \\
F_3 &= 3, && \text{// 2 old pairs and a new pair} \\
F_4 &= 5, && \text{// 3 old pairs and 2 new pairs} \\
F_5 &= 8. && \text{// 5 old pairs and 3 new pairs}
\end{aligned}
$$

// Is there a recurrence equation at work here?

At the end of any month,
    # of pairs of rabbits = # of pairs of rabbits at the beginning of the month
                    +   # of pairs of rabbits born in that month
    $F_n = F_{n-1}$
            + # of pairs of rabbits old enough to reproduce in that month
        $= F_{n-1}$
            + # of pairs of rabbits alive two months ago.
The "Fibonacci" sequence F satisfies the "Fibonacci recurrence equation",

$$
S_n = S_{n-1} + S_{n-2}. \tag{8.3.1}
$$

Continuing by applying this recurrence, we get

$$
\begin{aligned}
F_6 &= 13, \\
F_7 &= 21, \\
F_8 &= 34, \\
F_9 &= 55, \\
F_{10} &= 89, \\
F_{11} &= 144, \\
F_{12} &= 233.
\end{aligned}
$$

After a year, there will be 233 pairs of rabbits.

// How many will there be after 10 years, if none died of old age?
// Is there a formula for the entries in the Fibonacci sequence?

Example 8.3.1: Certain Subsets of $\{1..n\}$

Call a subset of $\{1..n\}$ a "good" subset if it does not contain two consecutive integers, k and $k + 1$, and let G_n denote the number of such subsets. What are the values of G_n?

If $n = 1$, there are 2 subsets of $\{1\}$, $\emptyset$ and $\{1\}$, and both are good, so $G_1 = 2$.

If $n = 2$, there are 2^2 subsets of $\{1,2\}$; $\emptyset$, $\{1\}$, and $\{2\}$ are good, so $G_2 = 3$.
// The other subset is $\{1,2\}$ and it's not good.

If $n = 3$, there are 2^3 subsets of $\{1,2,3\}$; $\emptyset$, $\{1\}$, $\{2\}$, $\{3\}$, and $\{1,3\}$ are good, so $G_3 = 5$. // The other subsets are $\{1,2\}$, $\{2,3\}$, and $\{1,2,3\}$ and each is bad.

If $n = 4$, there are 2^4 subsets of $\{1,2,3,4\}$; $\emptyset$, $\{1\}$, $\{2\}$, $\{3\}$, and $\{1,3\}$ are good, and so are $\{4\}$, $\{1,4\}$, and $\{2,4\}$; hence, $G_4 = 8$.

// $G = (2, 3, 5, 8, \ldots)$ and looks like the Fibonacci sequence after its first two
// entries.

// Do the G-values satisfy the Fibonacci recurrence equation when $n >= 3$?

If $n >= 3$ and X is a good subset of $\{1..n\}$, then either $n \notin X$ or $n \in X$.

If $n \notin X$, then X is a good subset of $\{1..(n - 1)\}$;

if $n \in X$, then $(n - 1) \notin X$, so X is equal $\{n\} \cup Y$ where Y is a good subset of $\{1..(n - 2)\}$.

In fact, the number of good subsets of $\{1..n\}$ equals

the number of good subsets X of $\{1..(n - 1)\}$ plus

the number of good subsets Y of $\{1..(n - 2)\}$.

That is, if $n >= 3$, then $G_n = G_{n-1} + G_{n-2}$.

Therefore, the sequence of G-values continues like the Fibonacci sequence, and

$$\text{for } n >= 1 \quad G_n = F_{n+1}.$$

//X   In how many n-sequences of flips of a coin are there never two heads in a row?
//   Or how many n-sequences of H's and T's don't contain "HH"?

The first seven rows of *Pascal's triangle* T of the binomial coefficients, $\binom{n}{k}$,

are

// from Chap. 2

		k						
		0	**1**	**2**	**3**	**4**	**5**	**6**
	0	1						
	1	1	1					
	2	1	2	1				
n	**3**	1	3	3	1			
	4	1	4	6	4	1		
	5	1	5	10	10	5	1	
	6	1	6	15	20	15	6	1

If we were to add the numbers on the diagonals slanting upward from southwest down to northeast,

the one from $T[0,0]$ is		1,
the one from $T[1,0]$ is		1,
the one from $T[2,0]$ is	$1+1=$	2,
the one from $T[3,0]$ is	$1+2=$	3,
the one from $T[4,0]$ is	$1+3+1=$	5,
the one from $T[5,0]$ is	$1+4+3=$	8,
the one from $T[6,0]$ is	$1+5+6+1=$	13,
the one from $T[7,0]$ will be	$1+6+10+4=$	21.

These totals are the Fibonacci numbers.   //X so far, but will it continue?

//X   Is F_n equal $\displaystyle\sum_{j=0}^{k} \binom{n-j}{j}$ where $k = \lfloor n/2 \rfloor$?

//   (Do MI on odd & even cases of $k + 1$.)

8.3.1   Algorithms for the Fibonacci Sequence

How can we construct an algorithm to determine the value of F_n? That is, given a value of n, how can we find the value of F_n? We could use recursion, as in

Algorithm 8.3.1. Fib(n)

```
Begin
  If (n < 2) Then
    Return(1);
  Else
    Return(Fib(n - 1) + Fib(n - 2));
  End ;
End.
```

This algorithm is sure to be correct. But is it efficient? The *Tree of Recursive Calls* of **Fib** from an external call with $n = 5$ is

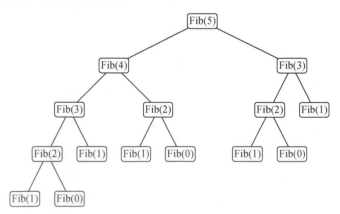

The leaves correspond to calls with no sub-calls, calls where $n = 0$ or 1. Each leaf returns the value 1; each internal vertex just adds the values returned by the two vertices below it. For each input integer n, the *full binary tree* will have F_n leaves and $2F_n - 1$ vertices. If F_n has exponential order, these trees become very large very quickly.

A second shortcoming of the recursive algorithm is that the same value of the parameter n occurs in many sub-calls – when $n = 5$

F_5 is evaluated 1 time,
F_4 is evaluated 1 time,
F_3 is evaluated 2 times,
F_2 is evaluated 3 times,
F_1 is evaluated 5 times,   // are these frequencies always Fibonacci numbers?
and F_0 is evaluated 3 times.

Perhaps we can do better if we remember the F-values as we calculate them – in an array, F say

Algorithm 8.3.2. Fib(n) version #2

```
Begin
  If (n < 2) Then
    Return(1);
  End;
  F[0] ← 1;
  F[1]←1;
  For j ← 2 To n Do
    F[j] ← F[j − 1] + F[j - 2];
  End ;
  Return(F[n]);
End.
```

But we only need the last two entries we evaluated in order to evaluate the next entry; perhaps we can do better by just keeping those two.
                                        // as $A = F[j − 2]$ and $B = F[j − 1]$

Algorithm 8.3.3. Fib(n) version #3

```
Begin
  If (n < 2) Then
    Return(1);
  End;
  A ← 1;
  B ← 1;
  For j ← 2 To n Do
    C ← A + B;
    A ← B;
    B ← C;                          // or B ← A + B; A ← B - A;
  End;
  Return(B);
End.
```

// Might there be another (even better) way, based on a formula for F_n?
// But is there a nice formula?

8.3.2 The Golden Ratio

Let us digress for a moment to introduce you to the "golden ratio". The ancient Greeks used it in their theory of aesthetics (beauty). A rectangle with length L and width W displays the golden ratio when the following occurs:

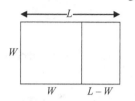

If a square with side length W is drawn inside it, the second smaller rectangle has sides in the same proportions (the same ratio) as the large rectangle. That is,

$$\frac{L}{W} = \frac{W}{L - W} = \gamma, \quad \text{the "golden ratio".}$$

Then

$$\gamma = \frac{W/W}{L/W - W/W} = \frac{1}{\gamma - 1}.$$

So

$$\gamma(\gamma - 1) = 1,$$

or

$$\gamma^2 - \gamma - 1 = 0.$$

Therefore,

$$\gamma = \frac{-(-1) \pm \sqrt{(-1)^2 - 4(1)(-1)}}{2(1)}$$

$$= \frac{1 \pm \sqrt{5}}{2} = \frac{1 \pm 2.236\ 067\ 977...}{2}.$$

That is, $\gamma = +1.618\ 033\ 988...$ or $-0.618\ 033\ 988...$.
But the golden ratio must be positive.   // and >1
So, the golden ratio,

$$\gamma = \frac{1 + \sqrt{5}}{2} = +1.618\ 033\ 988....$$

Let's denote the other root by β; so

$$\beta = \frac{1 - \sqrt{5}}{2} = -0.618\ 033\ 988....$$

Then $\gamma + \beta = 1$, $\gamma \times \beta = -1$, and $\gamma - \beta = \sqrt{5}$.   // are these right?

// Both γ and β satisfy the equation $x^2 = x + 1$, and they are the <u>only</u> solutions.
// But what about the Fibonacci sequence formula?

8.3.3   The Fibonacci Sequence and the Golden Ratio

The Fibonacci numbers grow fairly quickly. Maybe there is a (simple) geometric sequence, $S_n = r^n$, that satisfies the Fibonacci RE,

$$S_{n+2} = S_{n+1} + S_n \quad \text{for } \forall n \in \mathbf{N}. \qquad \text{// Eq. (8.3.1) again}$$

If there were, then $r^{n+2} = r^{n+1} + r^n$ for $\forall n \in \mathbf{N}$.

When $n = 0$, $r^2 = r \quad + 1.$

Hence, r must be either γ or β.
   In fact, we have

Lemma 8.3.1: Any sequence given by $S_n = A\gamma^n + B\beta^n$ for $\forall n \in \mathbf{N}$ satisfies the Fibonacci RE.   // where A and B may be any numbers at all

Proof.

$$\begin{aligned}
S_{n+1} + S_n &= \left[A\gamma^{n+1} + B\beta^{n+1}\right] + \left[A\gamma^n + B\beta^n\right] \\
&= \left[A\gamma^{n+1} + A\gamma^n\right] \quad + \left[B\beta^{n+1} + B\beta^n\right] \\
&= A\gamma^n[\gamma+1] \qquad\quad + B\beta^n[\beta+1] \\
&= A\gamma^n\left[\gamma^2\right] \qquad\qquad + B\beta^n\left[\beta^2\right] \\
&= A\gamma^{n+2} + B\beta^{n+2} \quad = S_{n+2}.
\end{aligned}$$

Furthermore,

Theorem 8.3.2: $S_n = A\gamma^n + B\beta^n$ is the *general solution* of the Fibonacci RE.

Proof. Suppose that T is any *particular* solution of the Fibonacci RE defined on $\mathbf{N}$.

// not necessarily the sequence for counting rabbit pairs
// We will find values for A and B and then prove that $T_n = A\gamma^n + B\beta^n$ for $\forall n \in \mathbf{N}$
// (by MI).

   Let's solve the equations (for A and B) that would guarantee $T_n = A\gamma^n + B\beta^n$ when $n = 0$ and $n = 1$:   // So our sequence starts correctly.

If

$$T_0 = A\gamma^0 + B\beta^0 = A + B \tag{1}$$

and
$$T_1 = A\gamma^1 + B\beta^1 = A\gamma + B\beta \tag{2}$$

then $\gamma T_0 = A\gamma + B\gamma$ // multiplying (1) by γ

and $T_1 = A\gamma + B\beta.$ // (2) again

Subtracting, we obtain

$$\gamma T_0 - T_1 = B\gamma - B\beta \quad = B(\gamma - \beta) \quad = B\sqrt{5}. \qquad // \gamma - \beta = \sqrt{5}$$

So $B = \dfrac{\gamma T_0 - T_1}{\sqrt{5}}.$

Hence, $A = T_0 - B = \dfrac{\sqrt{5}T_0}{\sqrt{5}} - \dfrac{\gamma T_0 - T_1}{\sqrt{5}} = \dfrac{(\sqrt{5} - \gamma)T_0 + T_1}{\sqrt{5}}$

$$= \dfrac{-\beta T_0 + T_1}{\sqrt{5}}.$$

// No matter how the sequence T starts (no matter what the values for T_0 and T_1
// are), there are unique numbers A and B such that $T_n = A\gamma^n + B\beta^n$ for $n = 0$
// and 1.

// Continuing the proof by Mathematical Induction that $T_n = A\gamma^n + B\beta^n$ for $\forall n \in \mathbf{N}$.

Step 1.   If $n = 0$ or 1, then $T_n = A\gamma^n + B\beta^n$, by our "choice" of A and B.
Step 2.   Assume that $\exists \; k \geq= 1$ such that if $0 <= n <= k$, then $T_n = A\gamma^n + B\beta^n$.
Step 3.   If $n = k + 1$, then $n \geq= 2$ so, because T satisfies the Fibonacci RE

$$T_{k+1} = T_k + T_{k-1}$$
$$= \left[A\gamma^k + B\beta^k\right] + \left[A\gamma^{k-1} + B\beta^{k-1}\right] \qquad // \text{ by step 2}$$
$$= A\gamma^{k+1} + B\beta^{k+1}. \qquad // \text{ by Lemma 8.3.1}$$
$$\square$$

This means we can find a formula for the entries in "the Fibonacci sequence"
(for counting rabbits), $F = (1,1,2,3,5,8,\dots)$. This is a *particular solution* to the
Fibonacci RE which must be given by the formula $F_n = A\gamma^n + B\beta^n$ where

$$A = \dfrac{-\beta F_0 + F_1}{\sqrt{5}} \quad = \dfrac{-\beta + 1}{\sqrt{5}} = \dfrac{\gamma}{\sqrt{5}} \qquad // \gamma + \beta = 1$$

and
$$B = \dfrac{\gamma F_0 - F_1}{\sqrt{5}} \quad = \dfrac{\gamma - 1}{\sqrt{5}} = \dfrac{-\beta}{\sqrt{5}}. \qquad // \gamma + \beta = 1$$

Thus, $F_n = \dfrac{\gamma}{\sqrt{5}}\gamma^n + \dfrac{-\beta}{\sqrt{5}}\beta^n = \dfrac{1}{\sqrt{5}}\left[\gamma^{n+1} - \beta^{n+1}\right]. \tag{8.3.2}$

// Can this be so? $\sqrt{5}$, γ, and β are all **irrational** numbers but each F_n is an
// integer!

// Is $F_{12} = 233$ really equal to $\dfrac{1}{\sqrt{5}} \left[\left(\dfrac{1+\sqrt{5}}{2} \right)^{13} - \left(\dfrac{1-\sqrt{5}}{2} \right)^{13} \right]$?

Let's tabulate a few values: // and see?

n	F_n	$\dfrac{1}{\sqrt{5}}\left(\dfrac{1+\sqrt{5}}{2}\right)^{n+1}$	$-\dfrac{1}{\sqrt{5}}\left(\dfrac{1-\sqrt{5}}{2}\right)^{n+1}$
0	1	0.723 606 797 …	+0.276 393 202 …
1	1	1.170 820 39 …	−0.170 820 393 …
2	2	1.894 427 19 …	+0.105 572 808 …
3	3	3.065 247 58 …	−0.065 247 584 0 …
4	5	4.959 674 77 …	+0.040 325 224 5 …
5	8	8.024 922 36 …	−0.024 922 359 3 …
6	13	12.984 597 1 …	+0.015 402 865 1 …
7	21	21.009 519 5 …	−0.009 519 494 16 …
8	34	33.994 116 6 …	+0.005 883 370 94 …
9	55	55.003 636 1 …	−0.003 636 123 20 …
10	89	88.997 752 8 …	+0.002 247 247 72 …
11	144	144.001 389 …	−0.001 388 875 47 …
12	233	232.999 142 …	+0.000 858 372 248 …
13	377	377.000 531 …	−0.000 530 503 223 …

The table is a bit corrupted by round-off error, but it shows that we have

Theorem 8.3.3: $\forall n \in \mathbf{N}, \qquad F_n = \lceil \gamma^{n+1}/\sqrt{5} \rfloor.$ // the nearest integer

Proof. // We'll prove that we always have $F_n - \frac{1}{2} < \gamma^{n+1}/\sqrt{5} < F_n + \frac{1}{2}$.
// Then the nearest integer to $\gamma^{n+1}/\sqrt{5}$ must be F_n.

We've seen that $\forall n \in \mathbf{N}$,

$$F_n = \gamma^{n+1}/\sqrt{5} - \beta^{n+1}/\sqrt{5} \text{ or } \gamma^{n+1}/\sqrt{5} - F_n = \beta^{n+1}/\sqrt{5}.$$

So $\left| \gamma^{n+1}/\sqrt{5} - F_n \right| = \left| \beta^{n+1}/\sqrt{5} \right| = |\beta|^{n+1}/\sqrt{5}.$ // $|xy| = |x| \times |y|$

Since $0 < |\beta| < 1$, for every positive integer k, $0 < |\beta|^k < 1$,   // use MI
and so

$$0 < |\gamma^{n+1}/\sqrt{5} - F_n| = |\beta^{n+1}/\sqrt{5}| = |\beta|^{n+1}/\sqrt{5} < 1/\sqrt{5} < 1/2.$$

If $\gamma^{n+1}/\sqrt{5} > F_n$, then $\gamma^{n+1}/\sqrt{5} - F_n < 1/2$, so $\gamma^{n+1}/\sqrt{5} < F_n + 1/2$.
If $\gamma^{n+1}/\sqrt{5} < F_n$, then $F_n - \gamma^{n+1}/\sqrt{5} < 1/2$, so $F_n - 1/2 < \gamma^{n+1}/\sqrt{5}$.

Thus,

$$F_n - 1/2 < \gamma^{n+1}/\sqrt{5} < F_n + 1/2, \text{ and hence } F_n = \lceil \gamma^{n+1}/\sqrt{5} \rfloor. \qquad \Box$$

// Does this provide a better way to evaluate Fibonacci numbers?

8.3.4   The Order of the Fibonacci Sequence

We have just seen that

$$\begin{aligned} F_n &\cong \left(1/\sqrt{5}\right)\gamma^{n+1} &&= (0.447\ 213\ 595\dots)\gamma^{n+1}, \\ \text{or} \qquad F_n &\cong \left(\gamma/\sqrt{5}\right)\gamma^{n} &&= (0.723\ 606\ 797\dots)\gamma^{n}. \qquad \text{// surely } F_n \text{ is } \Theta(\gamma^n) \end{aligned}$$

Theorem 8.3.4: $\forall n \in \mathbf{N}, \gamma^{n-1} <= F_n <= \gamma^n.$ // when is it $=$?
          // when is it $<$?

Proof. // by mathematical induction
Step 1. If $n = 0$, then $\gamma^{n-1} = \gamma^{-1} = 0.618\ 033\ 988\dots,$ // $1/\gamma = \gamma - 1$

$$F_n = 1, \text{ and } \gamma^n = \gamma^0 = 1;$$

If $n = 1$,   then $\gamma^{n-1} = 1$, $F_n = 1$,   and $\gamma^n = \gamma^1 = 1.618\ 033\ 988\dots.$
If $n = 2$, $\gamma^{n-1} = 1.618\dots$, $F_n = 2$,   and $\gamma^n = \gamma^2 = 2.618\ 033\ 988\dots.$ // $\gamma^2 = \gamma + 1$

Step 2.   Assume that $\exists\ k >= 2$ such that if $0 <= n <= k$, then $\gamma^{n-1} <= F_n <= \gamma^n.$
Step 3.   If $n = k + 1$, then $n >= 3$ so, because T satisfies the Fibonacci RE

$$F_{k+1} = F_k + F_{k-1}.$$

By Step 2 $F_k + F_{k-1} >= \gamma^{k-1} + \gamma^{k-2} = \gamma^k.$ // $=$ by Lemma 8.3.1

Hence, $\gamma^{[k+1]-1} <= F_{k+1}.$

Also by Step 2, $F_k + F_{k-1} <= \gamma^k + \gamma^{k-1} = \gamma^{k+1}.$ // $=$ by Lemma 8.3.1

Hence, $F_{k+1} <= \gamma^{k+1}.$ $\Box$

Since $\gamma/2 = 0.809\ 016\ 994\dots < 1$, we have $\forall n \in \mathbf{N}$

$$(1/2)\gamma^n = (\gamma/2)\gamma^{n-1} < \gamma^{n-1} <= F_n <= \gamma^n$$

and therefore $\qquad\qquad\qquad F_n$ is $\Theta(\gamma^n)$. $\qquad\qquad\qquad$ // see Sect. 7.5

//X Prove that $\forall n \in \mathbf{P}\ (1/3)\gamma^{n+1} < F_n < (1/2)\gamma^{n+1}$ and $(0.6)\gamma^n < F_n < (0.8)\gamma^n$.

8.3.5 The Complexity of Euclid's Algorithm for GCD

$\qquad\qquad\qquad\qquad\qquad\qquad$ // Euclid's Algorithm and the Fibonacci sequence?
Recall from Chap. 1

Algorithm 1.2.5 Euclid's Algorithm for GCD(x,y) $\qquad\qquad$ // for $x, y \in \mathbf{P}$

```
Begin
  A ← x;
  B ← y;
  R ← A MOD B;
  While (R > 0) Do
    A ← B;
    B ← R;
    R ← A MOD B;
  End;
  Output("GCD(", x, ",", y, ")=", B);          // or Return(B)
End.
```

We promised that we would prove: If Euclid's Algorithm requires k iterations of the while-loop, then

$$y \geq \left(\frac{1+\sqrt{5}}{2}\right)^k \text{ and also } k <= \lfloor(3/2) \times \lg(y)\rfloor. \qquad \text{// Usually } k \text{ is } \underline{\text{much}} \text{ smaller}$$

Consecutive Fibonacci numbers never have a common prime factor and provide a worst case for Euclid's Algorithm, which calculates a sequence of remainders, $R[j]$. Consider a

Walkthrough with $x = F_{12} = 233$ and $y = F_{11} = 144$.

#	A	B	R_j	R_j > 0	(3/2)×lg(B)
0	$233 = x$	$144 = y$	$89 = R_0$	T	$10.75\ldots$
1	$144 = y$	$89 = R_0$	$55 = R_1$	T	$9.71\ldots$
2	$89 = R_0$	$55 = R_1$	$34 = R_2$	T	$8.67\ldots$
3	$55 = R_1$	$34 = R_2$	$21 = R_3$	T	$7.63\ldots$
4	$34 = R_2$	$21 = R_3$	$13 = R_4$	T	$6.58\ldots$
5	$21 = R_3$	$13 = R_4$	$8 = R_5$	T	$5.55\ldots$
6	$13 = R_4$	$8 = R_5$	$5 = R_6$	T	4.5
7	$8 = R_5$	$5 = R_6$	$3 = R_7$	T	$3.48\ldots$
8	$5 = R_6$	$3 = R_7$	$2 = R_8$	T	$2.37\ldots$
9	$3 = R_7$	$2 = R_8$	$1 = R_9$	T	1.5
$10 = k$	$2 = R_8$	$1 = R_9$	$0 = R_{10}$	F	0

```
output: GCD(233,144) = 1.
```
$\qquad\qquad\qquad\qquad\qquad\qquad\qquad$ // all positive remainders are Fibonacci #'s

Euclid's Algorithm calculates a *strictly decreasing* sequence of *nonnegative integer* remainders:

$$x = y \times Q[0] + R[0] \qquad \text{where } 0 <= R[0] < y; \text{ // outside the loop}$$
$$\text{// and inside the loop}$$

if $R[0] > 0$, $\quad y = R[0] \times Q[1] + R[1] \quad$ where $0 <= R[1] < R[0]$;

if $R[1] > 0$, $\quad R[0] = R[1] \times Q[2] + R[2] \quad$ where $0 <= R[2] < R[1]$;

if $R[2] > 0$, $\quad R[1] = R[2] \times Q[3] + R[3] \quad$ where $0 <= R[3] < R[2]$;

. . .

if $R[j] > 0$, $\quad R[j-1] = R[j] \times Q[j+1] + R[j+1] \quad$ where $0 <= R[j+1] < R[j]$;

. . . .

// Whenever a remainder is > 0, a next, smaller, nonnegative remainder is
// calculated.

If $R[0]$, $R[1]$, $\ldots$, $R[y]$ were all positive, they would be $y + 1$ distinct numbers in the interval $\{1..(y - 1)\}$. But that (consequent) is impossible. Therefore, for some integer $k < y$, $R[k] = 0$ and the while-loop terminates.

The connection with the Fibonacci sequence is given by the following:

Theorem 8.3.5: If Euclid's Algorithm requires k iterations of the while-loop, then $y >= F_{k+1}$ and k $< (1.5) \times$ lg(y).

Proof. Since the remainders are decreasing,

$$0 = R[k] < R[k-1] < \ldots < R[2] < R[1] < R[0] < y.$$

Here, $R[k - 1]$ to $R[0]$ are k (distinct) positive integers less than y, and so $y >= k + 1$.

// We continue with the cases: $k <= 2$ and $k > 2$ separately.

If $k <= 2$, $\quad$ then $F_{k+1} = k+1$, $\quad$ so $y >= F_{k+1}$. $\quad$ // $F_1 = 1$, $F_2 = 2$ and $F_3 = 3$

Suppose now that $k > 2$. We use Mathematical Induction to prove

$$\text{for } j = 1, 2, .., k, \text{ we have } F_j <= R[k - j].$$

// The remainders, in reverse order, *dominate* the Fibonacci numbers.

Step 1. Since $0 = R[k] < R[k - 1] < R[k - 2]$, $\qquad\qquad$ // $k - 2 > 0$

$$R[k - 1] >= 1 = F_1 \text{ and } R[k - 2] >= 1 + R[k - 1] >= 2 = F_2.$$

Therefore, for $j = 1$ and $j = 2$, we have $F_j <= R[k - j]$.

Step 2.   Assume that $\exists$ q in $\{2..(k-1)\}$ such that if $1 <= j <= q$,
           then $F_j <= R[k-j]$.

Step 3.   If $j = q+1$, then $2 <= q < q+1 <= k$.

Since $(k-q)$ is between 1 and $(k-2)$, $R[(k-q)+1]$ is calculated as the remainder when $R[(k-q)-1]$ is divided by $R[k-q]$.

                                        // and where $R[k-q] < R[(k-q)-1]$

Thus,

$$R[k-(q+1)] = R[(k-q)-1]$$
$$= R[k-q] \times Q[(k-q)+1] + R[(k-q)+1]$$
$$>= R[k-q] \qquad\qquad + R[(k-q)+1]$$
$$\qquad\qquad\qquad\qquad // \ Q[(k-q)+1] >= 1$$
$$>= R[k-q] \qquad\qquad + R[k-(q-1)]$$
$$>= F_q \qquad\qquad +F_{q-1} \qquad\qquad // \text{ by step 2}$$
$$= F_{q+1}.$$

Therefore, for all values of j from 1 to k, we have $F_j <= R[k-j]$.   // by MI
In particular,

$$F_{k-1} <= R[k-(k-1)] = R[1] \quad \text{and} \quad F_k <= R[k-k] = R[0];$$

and $$y = R[0] \times Q[1] + R[1] \qquad\qquad // \text{ where } Q[1] >= 1$$
$$>= R[0] \qquad + R[1]$$
$$>= F_k \qquad + F_{k-1} = F_{k+1}.$$

We saw previously that for $\forall n \in \mathbf{N}$, $\gamma^{n-1} <= F_n <= \gamma^n$.   // Theorem. 8.3.4
Thus, if Euclid's Algorithm requires k iterations of the while-loop, then

$$y >= F_{k+1} >= \gamma^k = \left(\frac{1+\sqrt{5}}{2}\right)^k, \text{ and so} \qquad\qquad // \text{ taking logarithms}$$

$\lg(y) >= k \times \lg(\gamma) = k \times \lg(1.618\ 033\ 988\ 74\ldots) = k \times (0.694\ 241\ 913\ 848\ldots)$.

Hence, $k <= \lg(y)/(0.694\ 241\ 913\ 848\ldots) = \lg(y) \times (1.440\ 420\ 089\ 93\ldots)$
$$< (1.5) \times \lg(y). \qquad\qquad\qquad\qquad\qquad\qquad\qquad \square$$

Because k is an integer, $k <= \lfloor (3/2) \times \lg(y) \rfloor$.

The Most Important Ideas in this Section.

The Fibonacci sequence F satisfies the "Fibonacci recurrence equation,"

$$S_{n+2} = S_{n+1} + S_n \qquad \qquad \text{for } \forall\, n \in \mathbf{N}$$

and the boundary values $F_0 = 1$ and $F_1 = 1$. Three algorithms are considered for evaluating entries in the sequence. But there is a formula for the entries

$$F_n = \lceil \gamma^{n+1}/\sqrt{5} \rfloor \qquad \qquad \text{for } \forall\, n \in \mathbf{N},$$

where $\lceil x \rfloor$ is the nearest integer to the real number x and

$$\gamma = \frac{1 + \sqrt{5}}{2} = +1.618\,033\,988\ldots$$

is the "golden ratio." We proved that F_n is $\Theta(\gamma^n)$.

The Fibonacci sequence occurs in many (surprising) contexts. In particular, we proved that if Euclid's Algorithm for $\mathrm{GCD}(x,y)$ requires k iterations of the while-loop, then $y \ge F_{k+1}$, and from this, $k \le \lfloor (3/2) \times \lg(y) \rfloor$, so Euclid's Algorithm is $\mathbf{O}(\lg(y))$.

The Fibonacci recurrence equation is a "second-order linear recurrence equation", and we'll consider them and their solutions in the next section.

8.4 Solving Second-Order Linear Recurrence Equations

A *second-order linear recurrence equation* relates consecutive entries in a sequence by an equation of the form

$$S_{n+2} = aS_{n+1} + bS_n + c \qquad \text{for } \forall\, n \text{ in the domain of } S. \qquad (8.4.1)$$

But let's assume that the domain of S is $\mathbf{N}$. Let's also assume that ***not both a and b are 0***; otherwise, $S_n = c$ for $\forall n \in \{2..\}$, and the solutions to (8.4.1) are not very interesting.   // What are they?

// First-order RE's are just a special case of second-order RE's where $b = 0$.

When $c = 0$, the RE is said to be ***homogeneous*** (all the terms look the same – a constant times a sequence entry).   // The Fibonacci RE is homogeneous. Let's also restrict our attention (for the moment) to a *second-order linear, homogeneous recurrence equation*

$$S_{n+2} = aS_{n+1} + bS_n \qquad \qquad \text{for } \forall\, n \in \mathbf{N}. \qquad (8.4.2)$$

Just as we did for the Fibonacci recurrence equation, let's suppose there is a geometric sequence, $S_n = r^n$, that satisfies (8.4.2).

If there were,   then $r^{n+2} = ar^{n+1} + br^n$ for $\forall\, n \in \mathbf{N}$.

When $n = 0$, $r^2 = ar + b$.

The "characteristic equation" of (8.4.2) is $x^2 - ax - b = 0$,

$$\text{which has "roots"} \quad r = \frac{-(-a) \pm \sqrt{(-a)^2 - 4(1)(-b)}}{2(1)} = \frac{a \pm \sqrt{a^2 + 4b}}{2}.$$

Let $\Delta = \sqrt{a^2 + 4b}$, $r_1 = \dfrac{a + \Delta}{2}$, and $r_2 = \dfrac{a - \Delta}{2}$.

Then $r_1 + r_2 = a$, $r_1 \times r_2 = -b$, and $r_1 - r_2 = \Delta$.   // are these right?

// The Greek capital letter delta denotes the "difference" in the roots.

// Both r_1 and r_2 satisfy the equation $x^2 = ax + b$, and they are the only solutions.

Example 8.4.1a: If $S_{n+2} = 10S_{n+1} - 21S_n$ for $\forall n \in \mathbf{N}$, the characteristic equation is $x^2 - 10x + 21 = 0$.   // Or $(x - 7)(x - 3) = 0$

Here, $a = 10, b = -21, a^2 + 4b = 100 - 84 = 16, \Delta = 4$, so $r_1 = 7$ and $r_2 = 3$.

Example 8.4.1b: If $S_{n+2} = 3S_{n+1} - 2S_n$ for $\forall n \in \mathbf{N}$, the characteristic equation is $x^2 - 3x + 2 = 0$.   // Or $(x - 2)(x - 1) = 0$

Here, $a = 3, b = -2$, $a^2 + 4b = 9 - 8 = 1$, $\Delta = 1$, so $r_1 = 2$ and $r_2 = 1$.

Example 8.4.1c: If $S_{n+2} = 2S_{n+1} - S_n$ for $\forall n \in \mathbf{N}$, the characteristic equation is $x^2 - 2x + 1 = 0$.   // Or $(x-1)\,(x - 1) = 0$

Here, $a = 2$, $b = -1, a^2 + 4b = 4 - 4 = 0$, $\Delta = 0$, so $r_1 = 1$ and $r_2 = 1$.

// But what about a formula giving the general solution?

Theorem 8.4.2: The *general solution* of the homogeneous RE (8.4.2) is

$$S_n = A(r_1)^n + B(r_2)^n \quad \text{if } r_1 \neq r_2, \qquad\qquad \text{// if } \Delta \neq 0$$
$$\text{and} \quad S_n = A(r)^n + Bn(r)^n \quad \text{if } r_1 = r_2 = r. \qquad\qquad \text{// if } \Delta = 0$$

Proof. Suppose that T is any *particular* solution of the homogeneous RE (8.4.2).

// We deal with the two cases separately.

Case 1.   If $\Delta \neq 0$, then the two roots are distinct (but may be "complex" numbers).

// We'll find values for A and B, then prove that $T_n = A(r_1)^n + B(r_2)^n$ for $\forall n \in \mathbf{N}$
// (by MI). We'll show $A(r_1)^n + B(r_2)^n$ starts correctly for specially chosen values
// of A and B, and then show $A(r_1)^n + B(r_2)^n$ continues correctly.

Let's solve the equations (for A and B) that would guarantee $T_n = A(r_1)^n + B(r_2)^n$ when $n = 0$ and $n = 1$. If

$$T_0 = A(r_1)^0 + B(r_2)^0 = A \quad + B \tag{1}$$
and $\quad T_1 = A(r_1)^1 + B(r_2)^1 = A(r_1) + B(r_2),$ $\qquad\qquad$ (2)

then $\quad (r_1)T_0 = A(r_1) + B(r_1)$ $\qquad\qquad\qquad$ // multiplying (1) by r_1
and $\qquad\quad T_1 = A(r_1) + B(r_2).$ $\qquad\qquad\qquad\qquad$ // (2) again

Subtracting, we obtain

$$(r_1)T_0 - T_1 = B(r_1 - r_2) = B\Delta. \qquad\qquad // \; r_1 - r_2 = \Delta \neq 0$$

So $\quad B = \dfrac{(r_1)T_0 - T_1}{\Delta}.$

Hence, $\;\; A = T_0 - B = \dfrac{\Delta T_0}{\Delta} - \dfrac{(r_1)T_0 - T_1}{\Delta} = \dfrac{(\Delta - r_1)T_0 + T_1}{\Delta} = \dfrac{-(r_2)T_0 + T_1}{\Delta}.$

// No matter how the sequence T starts (no matter what the values for T_0 and T_1
// are), there are unique numbers A and B such that $\mathbf{T}_n = A(r_1)^n + B(r_2)^n$ for $n = 0$
// and 1.

// Continuing the proof by Mathematical Induction that $T_n = A(r_1)^n + B(r_2)^n$ for
// $\forall n \in \mathbf{N}$.

Step 1. If $n = 0$ or 1, then $\mathbf{T}_n = A(r_1)^n + B(r_2)^n$, by our "choice" of A and B.
Step 2. Assume that $\exists \; k >= 1$ such that if $0 <= n <= k$, then $T_n = A(r_1)^n + B(r_2)^n$.
Step 3. If $n = k + 1$, then $n >= 2$ so, because T satisfies the homogeneous RE (8.4.2)

$$\begin{aligned}
T_{k+1} &= aT_k + bT_{k-1} \\
&= a\Big[A(r_1)^k + B(r_2)^k\Big] \quad + b\Big[A(r_1)^{k-1} + B(r_2)^{k-1}\Big] \qquad // \text{ by step 2} \\
&= \Big[aA(r_1)^k + bA(r_1)^{k-1}\Big] + \Big[aB(r_2)^k + bB(r_2)^{k-1}\Big] \\
&= A(r_1)^{k-1}[a(r_1) + b] \quad\;\; + B(r_2)^{k-1}[a(r_2) + b] \\
&= A(r_1)^{k-1}\Big[(r_1)^2\Big] \quad\quad\; + B(r_2)^{k-1}\Big[(r_2)^2\Big] \qquad // \text{ "choice" of } r_1 \text{ and } r_2 \\
&= A(r_1)^{k+1} \quad\qquad\qquad\; + B(r_2)^{k+1}.
\end{aligned}$$

Thus, if $r_1 \neq r_2$, $\;T_n = A(r_1)^n + B(r_2)^n$ for $\forall \, n \in \mathbf{N}$.

Example 8.4.1a: If $S_{n+2} = 10S_{n+1} - 21S_n$ for $\forall n \in \mathbb{N}$,
then $r_1 = 7$ and $r_2 = 3$. Hence, the *general solution* of the RE is $S_n = A7^n + B3^n$.

Example 8.4.1b: If $S_{n+2} = 3S_{n+1} - 2S_n$ for $\forall n \in \mathbb{N}$,
then $r_1 = 2$ and $r_2 = 1$. Hence, the *general solution* of the RE is

$$S_n = A2^n + B1^n = A2^n + B.$$

Case 2. If $\Delta = 0$, then the roots are (both) equal to r where $r = a/2$. Also,
$b = -a^2/4 = -r^2$. If a were 0, then $b = 0$; but we assumed that not both
a and b are 0. Hence, $r \neq 0$.

Let's solve the equations (for A and B) that would guarantee $T_n = A(r)^n + nB(r)^n$ when $n = 0$ and $n = 1$. If

$$T_0 = A(r)^0 + 0B(r)^0 = A \tag{1}$$

and
$$T_1 = A(r)^1 + 1B(r)^1 = Ar + Br, \tag{2}$$

then
$$A = T_0 \text{ and } B = (T_1 - Ar)/r.$$

// No matter how the sequence T starts (no matter what the values for T_0 and T_1
// are), there are unique numbers A and B such that $T_n = A(r)^n + nB(r)^n$ for $n = 0$
// and 1.

// Continuing the proof by Mathematical Induction that $T_n = A(r)^n + nB(r)^n$ for
// $\forall n \in \mathbb{N}$.

Step 1.   If $n = 0$ or 1, then $T_n = A(r)^n + nB(r)^n$, by our "choice" of A and B.
Step 2.   Assume that $\exists\, k \geq 1$ such that if $0 \leq n \leq k$, then $T_n = A(r)^n + nB(r)^n$.
Step 3.   If $n = k + 1$, then $n \geq 2$ so, because T satisfies the RE (8.4.2)

$$
\begin{aligned}
T_{k+1} &= aT_k + bT_{k-1} \\
&= a\Big[A(r)^k + kB(r)^k\Big] + b\Big[A(r)^{k-1} + (k-1)B(r)^{k-1}\Big] && \text{// by step 2} \\
&= \big[aAr^k + bAr^{k-1}\big] \ + \ \big[akBr^k + b(k-1)Br^{k-1}\big] \\
&= Ar^{k-1}[ar + b] \qquad\quad + Br^{k-1}[akr + b(k-1)] \\
&= Ar^{k-1}[ar + b] \qquad\quad + Br^{k-1}[k(ar + b) - b] \\
&= Ar^{k-1}\big[r^2\big] \qquad\qquad\ + Br^{k-1}\big[k(r^2) - b\big] && \text{// } r^2 = ar + b \\
&= Ar^{k+1} \qquad\qquad\qquad + Br^{k-1}\big[k(r^2) + r^2\big] && \text{// } -b = r^2 \\
&= Ar^{k+1} \qquad\qquad\qquad + Br^{k-1}\big[(k+1)r^2\big] \\
&= Ar^{k+1} \qquad\qquad\qquad + (k+1)Br^{k+1}.
\end{aligned}
$$

Thus, if $r_1 = r_2 = r$, $\quad T_n = A(r)^n + nB(r)^n$ for $\forall n \in \mathbb{N}$.   □

Example 8.4.1c: If $S_{n+2} = 2S_{n+1} - S_n$ for $\forall n \in \mathrm{N}$,
then $r_1 = 1$ and $r_2 = 1$. Hence, the *general solution* of the RE is

$$S_n = A1^n + nB1^n = A + nB. \qquad\qquad \text{// an arithmetic sequence?}$$

// But what about a formula giving the general solution to a nonhomogeneous
// case?
   Suppose

$$S_{n+2} = aS_{n+1} + bS_n + c \quad \text{for } \forall n \in \mathrm{N} \text{ where } c \neq 0. \qquad \text{// (8.4.1) again}$$

// Can we somehow use the results for the homogeneous case? Yes.

Suppose that T_n is some particular solution of (8.4.1). If S_n is any (other, particular)
solution of (8.4.1), then
   $W_n = T_n - S_n$ is a solution of the homogenous RE.   // (8.4.2) again

That is, $W_{n+2} = T_{n+2} - S_{n+2}$
$$\begin{aligned}
&= [aT_{n+1} + bT_n + c] - [aS_{n+1} + bS_n + c] &&\text{// } T \ \& \ S \text{ satisfy 8.4.1}\\
&= a[T_{n+1} - S_{n+1}] \quad + b[T_n - S_n] + [c - c] &&\\
&= aW_{n+1} \qquad\qquad\quad + bW_n. &&\text{// } W \text{ satisfies 8.4.2}
\end{aligned}$$

   Therefore, <u>any</u> solution S_n of (8.4.1) may be written as $W_n + T_n$, where T_n is
some particular solution of the nonhomogeneous RE and W_n is a solution of the
corresponding homogeneous RE. Furthermore,

   the **general solution** of (8.4.1), G_n, is $W_n + T_n$,
      where T_n is <u>any</u> **particular solution** of the nonhomogeneous RE and
      W_n is the general solution of the corresponding homogeneous RE.

// But how can we find a *particular solution* of the nonhomogeneous RE?

Particular Solutions

   Suppose $S_{n+2} = aS_{n+1} + bS_n + c \quad \text{for } \forall \, n \in \mathrm{N} \text{ where } c \neq 0.$

// We want to find *simple solutions* like $\qquad\qquad S_n = \mathbf{K},$
// or (when no constant sequence satisfies the RE) $\quad S_n = \mathbf{K}n,$ (a linear function)
// or (when no linear function satisfies the RE) $\qquad S_n = \mathbf{K}n^2.$ (a quadratic)

   If $T_j = K$ for $\forall j \in \mathrm{N}$, then

$$\begin{aligned}
& T_{n+2} = aT_{n+1} &&+\quad bT_n &&+\quad c\\
\Leftrightarrow\ & K \ \ = aK &&+\quad bK &&+\quad c\\
\Leftrightarrow\ & K[1 - (a+b)] &&= \quad c. &&
\end{aligned}$$

Thus, when $a + b \neq 1$, there is a unique constant solution, $T_k = \dfrac{c}{1 - (a+b)}$
$\forall k \in \mathrm{N}$.

// And there is no constant solution when $a + b = 1$.

Example 8.4.2a: The constant sequence $T_k = 2$ for $\forall k \in \mathbf{N}$ is a particular solution of $S_{n+2} = 10S_{n+1} - 21S_n + 24$ for $\forall n \in \mathbf{N}$.

$$// \ 10[2] - 21[2] + 24 = 20 - 42 + 24 = 2$$

The *general solution* of the RE is $S_n = A7^n + B3^n + 2$.

Suppose that $a + b = 1$. If $T_j = Kj$ for $\forall j \in \mathbf{N}$, then

$$
\begin{aligned}
& T_{n+2} && = aT_{n+1} + bT_n && + c \\
\Leftrightarrow\ & K(n+2) && = aK(n+1) + bKn && + c \\
\Leftrightarrow\ & K[(n+2) - \{a(n+1) + bn\}] && = c \\
\Leftrightarrow\ & K[n+2 - an - a - bn] && = c \\
\Leftrightarrow\ & K[2 - a] && = c. && // \ a+b = 1
\end{aligned}
$$

Thus, when $a + b = 1$ but $a \neq 2$, there is a linear solution, $T_k = \dfrac{kc}{2 - a}$ for $\forall k \in \mathbf{N}$.

// But there is no linear solution when $a + b = 1$ and $a = 2$.

Example 8.4.2b: Linear sequence $T_k = 5k$ for $\forall k \in \mathbf{N}$ is a particular solution of $S_{n+2} = 3S_{n+1} - 2S_n - 5$ for $\forall n \in \mathbf{N}$.

$$// \ 3[5(n+1)] - 2[5n] - 5 = 15n + 15 - 10n - 5 = 5[n+2]$$

The *general solution* of the RE is $S_n = A2^n + B + 5n$.

Finally, suppose that $a + b = 1$ and $a = 2$. Then $b = -1$. If $T_j = Kj^2$ for $\forall j \in \mathbf{N}$, then

$$
\begin{aligned}
& T_{n+2} && = aT_{n+1} && + bT_n && + c \\
& && = 2T_{n+1} && + (-1)T_n && + c \\
\Leftrightarrow\ & K(n+2)^2 && = 2K(n+1)^2 && - Kn^2 && + c \\
\Leftrightarrow\ & K[(n+2)^2 - 2(n+1)^2 + n^2] && && = c \\
\Leftrightarrow\ & K[n^2 + 4n + 4 - (2n^2 + 4n + 2) + n^2] && && = c \\
\Leftrightarrow\ & K[4 - 2] && && = c.
\end{aligned}
$$

Thus, when $a + b = 1$ and $a = 2$, there is a quadratic solution, $T_k = \dfrac{k^2 c}{2}$ for $\forall k \in \mathbf{N}$.

Example 8.4.2c: The quadratic sequence $T_k = 3k^2$ for $\forall k \in \mathbf{N}$ is a particular solution of $S_{n+2} = 2S_{n+1} - S_n + 6$ for $\forall n \in \mathbf{N}$.

$$// \ 2\left[3(n+1)^2\right] - [3n^2] + 6 = 6[n^2 + 2n + 1] - 3n^2 + 6$$

$$// = 3n^2 + 12n + 12 = 3[n^2 + 4n + 4] = 3(n+2)^2$$

The *general solution* of the RE is

$$S_n = A + nB + 3n^2.$$

We end this section with a practical reason for knowing the algebraic solution to a recurrence equation.

Example 8.4.3: Round-off errors and recurrence equations

Find S_{10}, given that $S_0 = 1$, $S_1 = 1/9$, and $S_{n+2} = (82/9)S_{n+1} - S_n$.

$$// \text{ for } \forall n \in \mathbf{N}$$

But suppose we attempt this using a machine that rounds numbers to three significant figures.

// This imprecision would be extreme but emphasizes (and exaggerates) our point
// that knowing the algebraic solution to a RE might be useful.

Let the symbol "$\cong$" to denote "is stored as" in the machine. Then

$$S_0 \cong 1.00,$$
$$S_1 \cong 0.111, \quad \text{and } 82/9 \cong 9.11.$$

Using the given RE for calculating later S-values gives

$S_2 = 9.11\times$	0.111	$-$	(1.00)	$= 1.01121$	-1.00	$= 0.01121$	so $S_2 \cong 0.0112$
$S_3 = 9.11\times$	0.0112	$-$	(0.111)	$= 0.102032$	-0.111	$= -0.008968$	so $S_3 \cong -0.00897$
$S_4 = 9.11\times$	-0.00897	$-$	(0.0112)	$= -0.0817167$	-0.0112	$= -0.0929167$	so $S_4 \cong -0.0929$
$S_5 = 9.11\times$	-0.0929	$-$	(-0.00897)	$= -0.846319$	$+0.00897$	$= -0.837349$	so $S_5 \cong -0.837$
$S_6 = 9.11\times$	-0.837	$-$	(-0.0929)	$= -7.62507$	$+0.0929$	$= -7.53217$	so $S_6 \cong -7.53$
$S_7 = 9.11\times$	-7.53	$-$	(-0.837)	$= -68.5983$	$+0.837$	$= -67.7613$	so $S_7 \cong -67.8$
$S_8 = 9.11\times$	-67.8	$-$	(-7.53)	$= -617.658$	$+7.53$	$= -610.128$	so $S_8 \cong -610$
$S_9 = 9.11\times$	-610	$-$	(-67.8)	$= -5557.1$	$+67.8$	$= -5489.3$	so $S_9 \cong -5490$
$S_{10} = 9.11\times$	-5490	$-$	(-610)	$= -50013.9$	$+610$	$= -49403.9$	so $S_{10} \cong -49400$

// How much damage did round-off error do?
// Is the correct value of S_{10} close to $-49,400$?

The characteristic equation of the RE

$$S_{n+2} = (82/9)S_{n+1} - S_n \quad \text{for } \forall n \in \mathbf{N}$$

is

$$x^2 - (82/9)x + 1 = (x - 1/9)(x - 9) = 0.$$

Therefore, the *general solution* of the RE is $G_n = A(1/9^n) + B9^n$.

The calculated sequence appears to be approximating a *particular solution* of the form $T_n = B9^n$ where B is negative. In the last few rows of the table, the entries <u>are</u> negative, each entry is about nine times the previous entry.

$$// \text{ in fact, } 5490/610 = 9$$

But the *particular solution* that begins $S_0 = 1$, $S_1 = 1/9$ is $S_n = 1/9^n$. Therefore, the correct value of S_{10} is

$$1/9^{10} = 1/(3\,486\,784\,401) = 2.867\,971\,991\ldots \times 10^{-10}$$
$$= 0.000\,000\,000\,286\,797\,199\,1\ldots.$$

// $-49,400$ is just a truly awful approximation of this. What went wrong?

The Most Important Ideas in This Section.

A *second-order linear recurrence equation* relates consecutive entries in a sequence by an equation of the form

$$S_{n+2} = aS_{n+1} + bS_n + c \quad \text{for } \forall\, n \in \mathbf{N}, \tag{$*$}$$

where not both a and b are 0. When $c = 0$, the RE $(*)$ is said to be *homogeneous*.

We first dealt with the homogeneous form   // with $c = 0$

$$S_{n+2} = aS_{n+1} + bS_n \quad\quad \text{for } \forall\, n \in \mathbf{N}. \tag{$**$}$$

The *characteristic equation* of $(**)$ is $x^2 - ax - b = 0$ which has roots

$$r_1 = \frac{a + \Delta}{2} \quad \text{and} \quad r_2 = \frac{a - \Delta}{2} \quad \text{where } \Delta = \sqrt{a^2 + 4b}.$$

// these roots may be complex numbers

The *general solution* of $(**)$ is

$$S_n = A(r_1)^n + B(r_2)^n \quad \text{if } r_1 \neq r_2, \quad\quad \text{// if } \Delta \neq 0$$
$$\text{and} \quad\quad S_n = A(r)^n + Bn(r)^n \quad \text{if } r_1 = r_2 = r. \quad\quad \text{// if } \Delta = 0$$

The *general solution* of $(*)$, the nonhomogeneous equation, is $G_n = W_n + T_n$, where W_n is the *general solution* of the corresponding homogeneous RE $(**)$ and where T_n is <u>any</u> *particular solution* of the nonhomogeneous RE $(*)$. When $a + b \neq 1$, there is a unique constant solution to $(*)$,

$$T_k = \frac{c}{1 - (a + b)} \quad\quad \text{for } \forall k \in \mathbf{N}.$$

When $a + b = 1$ but $a \neq 2$, there is a linear solution to $(*)$,

$$T_k = \frac{kc}{2 - a} \quad\quad \text{for } \forall k \in \mathbf{N}.$$

When $a + b = 1$ and $a = 2$, there is a quadratic solution $(*)$,

$$T_k = \frac{k^2 c}{2} \quad\quad \text{for } \forall k \in \mathbf{N}.$$

8.5   Infinite Series

Can you add up an infinite number of numbers? How? Why would anyone want to?
And if you could somehow add up an infinite number of positive numbers, could
the total ever be finite? Zeno of Elea (c490–430 BC) who founded the Stoic school
of philosophy in Athens didn't think so.

8.5.1   Zeno's Paradoxes

Achilles and the Tortoise. Achilles was the champion sprinter of his time, and the
Tortoise was notoriously slow. Zeno said that if they raced each other, Achilles
(being a fair-minded athlete) would agree to give the Tortoise a head start. But then
Zeno pointed out that when Achilles reached the point where the Tortoise started,
the Tortoise would have moved on; and when Achilles reached that point (where
the Tortoise was), the Tortoise would have moved on; and when Achilles reached
that new point where the Tortoise was, the Tortoise would have moved on:

>   *At any time that the Tortoise is ahead of Achilles,*
>   *when Achilles reaches the point where the Tortoise was,*
>   *the Tortoise would have moved on, and would still be ahead of Achilles.*

From this, Zeno concluded that **Achilles can never overtake the Tortoise**.

// So even if Achilles can cover 10 meters per second, and the Tortoise (at top
// speed) covers 1 meter in 10 seconds, when the Tortoise starts ahead of Achilles,
// *will Achilles never overtake the Tortoise?*

Zeno and the Archer. Suppose an Archer is about to launch an arrow at Zeno.
Zeno says that before the arrow can strike me, it must reach the point halfway to me.
And if it reaches that point, it must still reach the point halfway to me. And when it
reaches that point, it must still reach the point halfway to me:

>   *At any point in its flight, the arrow must reach the point halfway to me*
>   *before it reaches me. It must reach an infinite number of halfway points,*
>   *before it reaches me. Therefore, it can never reach me.*

// Is that conclusion correct? Will Zeno's "argument" protect him from arrows?

Suppose the arrow flies at about 30 meters per second and that Zeno is standing
60 meters from the Archer. The arrow will fly for 2 seconds and then strike Zeno.
                                                    // But what about all those midpoints?
It will take 1 second to reach the first midpoint.
From there, it will take half a second to reach the next midpoint.
From there, it will take a quarter of a second to reach the next midpoint.
From there, it will take an eighth of a second to reach the next midpoint.

$$1 + \tfrac{1}{2} \qquad\qquad\qquad = 2 - \tfrac{1}{2}$$
$$1 + \tfrac{1}{2} + \tfrac{1}{4} \qquad\qquad = 2 - \tfrac{1}{4}$$
$$1 + \tfrac{1}{2} + \tfrac{1}{4} + \tfrac{1}{8} \qquad = 2 - \tfrac{1}{8}$$
$$1 + \tfrac{1}{2} + \tfrac{1}{4} + \ldots + 1/2^n \;= 2 - 1/2^n \qquad \text{// using MI or Theorem 3.6.8}$$

The arrow does reach all these midpoints; it reaches the $n + 1^{st}$ after $(2 - 1/2^n)$ seconds. The flight of the arrow may be partitioned into an infinite number of (nonoverlapping) sub-flights, but the total time the arrow is in the air remains 2 seconds.

   Maybe we can resolve these paradoxes of Zeno by devising a theoretical method for adding an infinite number of numbers. It cannot be done by repeated (ordinary) addition. Maybe we can give a useful idea to describe the sum of entries in an infinite sequence – at least sometimes.

8.5.2 Formal Definitions of Convergence of Sequences and Series

In Chap. 2, when we defined infinite sequences, we also wrote that
   a sequence S **converges** to $L \in \mathbf{R}$ [written $S_n \to L$] *means*
      for any number $\delta > 0$, there is an integer M such that
         if $(n > M)$, then $|S_n - L| < \delta$.
And we explained this definition by referring to the sequences generated by the Bisection Algorithm of Chap. 1.

   We all know that when a fraction is raised to higher and higher powers, the values tend to zero. Let us prove that in a formal sense.

Theorem 8.5.1: If $-1 < r < +1$ and $S_n = Ar^n + C$ where A and C are any real numbers, then $S_n \to C$.

Proof. // This may seem obvious, but we want to apply the formal definition. Suppose that δ is some given positive number.

// We must determine a positive integer M (which may depend on the value of δ)
// and then demonstrate that
//   if $(n > M)$, then $|S_n - C| < \delta$.

We use the fact that $\qquad |S_n - C| = |Ar^n| = |A| \times |r|^n$.

Case 1.   If $A = 0$ or $r = 0$, then taking $M = 1$, if $(n > M)$, then $|S_n - C| = 0 < \delta$.

// Any constant sequence converges in the formal sense.

From now on, suppose that $A \neq 0$ and also that $r \neq 0$:

$$|S_n - C| = |A| \times |r|^n \; < \; \delta \Leftrightarrow |r|^n \; < \; \delta/|A|.$$

Case 2.   If $1 <= \delta/|A|$, then since $0 < |r| < 1$, we have $|r|^n < 1 <= \delta/|A|$ $\forall n \in \mathbf{P}$. Then, taking $M = 1$, if $(n > M)$, then $|S_n - C| < \delta$.

// But what if δ is very small?

Case 3.   If $\delta/|A| < 1$, then since $0 < \delta/|A| < 1$, $\lg(\delta/|A|)$ is defined but is $< \lg(1) = 0$.

Since $0 < |r| < 1$, $\lg(|r|)$ is also negative.

Let $B = \dfrac{\lg(\delta/|A|)}{\lg(|r|)}$ and let $M = \lceil B \rceil$.   // Then $M \in \mathbf{P}$.

If $(n > M)$, then

$n > B$ so	$n \times \lg(|r|) < B \times \lg(|r|)$	// $\lg(|r|)$ is negative
and	$n \times \lg(|r|) < \lg(\delta/|A|)$.	
Then	$\lg(|r|^n) < \lg(\delta/|A|)$.	// $\lg(a^b) = b \times \lg(a)$.

Raising 2 to these powers gives

$$|r|^n < \delta/|A|.$$

Therefore, if $(n > M)$, then $|S_n - C| < \delta$.   □

The formal definition of an infinite sum is given as follows:

An infinite **series** S **converges** to $L \in \mathbf{R}$ [written $\displaystyle\sum_{j=0}^{\infty} S_j = L$] *means*

the sequence of partial sums $T_n = \displaystyle\sum_{j=0}^{n} S_j$ converges to L.

A Formula for (Convergent) Infinite Geometric Series

Theorem 8.5.2: If $-1 < r < +1$ and $S_n = Ar^n$ where A is any real number, then $\displaystyle\sum_{j=0}^{\infty} S_j = \dfrac{A}{1 - r}$.

Proof.   Let $T_n = \displaystyle\sum_{j=0}^{n} S_j = Ar^0 + Ar^1 + Ar^2 + \ldots Ar^n.$

Then $T_n = A \times \dfrac{r^{n+1} - 1}{r - 1}$ // Theorem 3.6.8

$$= \left(\dfrac{A}{r - 1}\right)r^{n+1} - \left(\dfrac{A}{r - 1}\right) = \left(\dfrac{Ar}{r - 1}\right)r^n + \dfrac{A}{1 - r}.$$

Then by Theorem 8.5.1,

$$T_n \to \dfrac{A}{1 - r}, \text{ so by our definition } \sum_{j=0}^{\infty} S_j = \dfrac{A}{1 - r}.$$ □

// Now what? We want to show that this is consistent with our past experience
// and resolves Zeno's paradoxes.

Recall   // from grade school

$$0.\overline{3} = 0.333\,333\,333\ldots$$ // which means

$$= 0.3 + 0.03 + 0.003 + 0.0003 + 0.00003 + \ldots$$

$$= \text{the sum of an infinite geometric sequence with } A = 0.3 \text{ and } r = 0.1$$

$$= \frac{A}{1 - r} = \frac{0.3}{1 - (0.1)} = \frac{0.3}{0.9} = \frac{1}{3}.$$

Resolving Zeno's Paradoxes

Zeno and the Archer. Letting T_n denote the time the arrow takes to reach the $n + 1^{\text{st}}$ midpoint,

$$T_n = 1 + \tfrac{1}{2} + \tfrac{1}{4} + \ldots + 1/2^n = 2 - 1/2^n,$$ // Theorem 3.6.8

and T_n converges to 2.   // Theorem 8.5.1

Therefore, the total time the arrow takes to reach **all** these midpoints is

$$1 + \tfrac{1}{2} + \tfrac{1}{4} + \ldots = \sum_{j=0}^{\infty} \left(\frac{1}{2}\right)^j = 2 \text{ seconds.}$$ // which we knew

Achilles and the Tortoise. Suppose that A gives T a head start of 30m. Let $D_0 = 30m$:

A covers D_0 in $t_0 = \dfrac{30m}{10m/s} = 3$ seconds. But in that time,

T has covered $D_1 = t_0 \times (1/10)m/s = 3s/(10m/s) = 0.3m.$

A covers D_1 in $t_1 = \dfrac{(0.3)m}{10m/s} = \dfrac{t_0 \times (1/10)m/s}{10m/s} = \dfrac{t_0}{100} = 0.03$ seconds.

If at some moment, T is D_n meters ahead of A,

A covers D_n in $t_n = D_n/10$ seconds. But in that time,

T has covered $D_{n+1} = t_n/10m.$

A covers D_{n+1} in $t_{n+1} = \dfrac{D_{n+1}\, m}{10\, m/s} = \dfrac{t_n}{100} s.$

The time intervals form a geometric sequence with common ratio $r = 1/100 = 0.01$. Adding up **all** the time intervals when T is ahead of A,

$$t_0 + t_1 + t_2 + t_3 + \ldots = 3 + 0.03 + 0.0003 + \ldots$$

$$= \frac{A}{1 - r} = \frac{3}{1 - 0.01} = \frac{3}{0.99} = \frac{300}{99} = \frac{100}{33} = 3 + (1/33) \text{ seconds.}$$

// Now, T is **not** ahead of A.

Therefore, A overtakes T in exactly $3 + (1/33)$ seconds.

If $T[2^k] > 1 + k/2$, then $// \ k >= 2$

$$T[2^{k+1}] = \{1 + 1/2 + 1/3 + \ldots + 1/2^k\}$$
$$+ \{1/(2^k+1) + 1/(2^k+2) + 1/(2^k+3) + \ldots + 1/(2^k+2^k)\}$$

$$= T[2^k] \quad + \{1/(2^k+1) + 1/(2^k+2) + 1/(2^k+3) + \ldots + 1/(2^k+2^k)\}$$
$$> T[2^k] \quad + \{1/(2^k+2^k) + 1/(2^k+2^k) + 1/(2^k+2^k) + \ldots + 1/(2^k+2^k)\}$$
$$= T[2^k] \quad + 2^k/(2^k+2^k)$$
$$= T[2^k] \quad + 1/2$$
$$> 1 + k/2 \quad + 1/2 \qquad\qquad = 1 + (k+1)/2.$$

Therefore (by MI) $T[2^n] \quad > 1 + n/2$ for $\forall n \in \{2..\}$,

and $T[2^{2M}] > 1 + (2M)/2 > M$ for $\forall M \in \mathbf{P}$.

In particular, $T[2^6] \quad > 1 + 6/2 \quad > 3,$

                               $T[2^{20}] > 1 + 20/2 \ > 10,$

and $T[2^{200}] > 1 + 200/2 > 100.$

   Thus, the harmonic <u>sequence</u> converges to 0, but the harmonic <u>series</u> does not converge. The converse of Theorem 8.5.4 is not true. The condition that $S_n \to 0$ is necessary but not sufficient for the convergence of the series $S_1 + S_2 + S_3 + \ldots$.

$// $ Using calculus and the fact that $\displaystyle\int_1^n \frac{1}{x}\,dx = \ln(n)$, you can show that

$// \ 1/2 + 1/3 + \ldots + 1/n < \ln(n) < 1 + 1/2 + 1/3 + \ldots + 1/(n-1)$, and then

$// \qquad\qquad 1/n + \ln(n) \ < T[n] < 1 + \ln(n).$

$// $ So $\qquad\qquad \ln(n) \ \ < T[n] < 2 \times \ln(n)$ (when $n > e = 2.718\ldots$).

$// $ Thus, $T[n]$ is $\Theta(\ln(n))$, and $T[n]$ is $\Theta(\lg(n))$. $// $ see Sect. 7.5.2

The Most Important Ideas in This Section.
We gave a formal definition of the sum of an infinite series to resolve Zeno's paradoxes: *Achilles and the Tortoise* and *Zeno and the Archer*. Recall from Chap. 2 that

a sequence S *converges* to $L \in \mathbf{R}$ [written $S_n \to L$] <u>means</u>
for any number $\delta > 0$, there is an integer M such that
if $(n > M)$, then $|S_n - L| < \delta$.

Then a formal definition of an infinite sum was given.

An infinite *series*, S *converges* to $L \in \mathbf{R}$ [written $\displaystyle\sum_{j=0}^{\infty} S_j = L$] <u>means</u>

the sequence of partial sums $\displaystyle T_n = \sum_{j=0}^{n} S_j$ converges to L.

We proved a formula for (convergent) infinite geometric series:
If $-1 < r < +1$ and $S_n = Ar^n$ where A is any real number, then

(continued)

(continued)

$$\sum_{j=0}^{\infty} S_j = \frac{A}{1-r}.$$

If the terms in a sequence S_n can be added up to give a finite total, then the entries in S must get smaller and smaller. In fact,

$$\text{If } \sum_{j=0}^{\infty} S_j = L \in \mathbf{R}, \text{ then } S_n \to 0.$$

Finally, we proved that the converse of this is not true, by showing that the harmonic <u>sequence</u> converges to 0, but the harmonic <u>series</u> does not converge; it eventually grows larger than <u>any</u> given number.

Exercises

1. Suppose E_n is defined recursively on $\mathbf{P}$ by

$$E_0 = 0, \ E_1 = 2, \text{and } E_{n+1} = 2n\{E_n + E_{n-1}\} \text{ for } n >= 1.$$

Determine the value of E_{10}.

2. Suppose that the function f is defined recursively on $\mathbf{P}$ by

$$f(n) = \begin{cases} 1 & \text{if } n = 2^k \text{ for some } k \in \mathbf{N} \\ f(n/2) & \text{if } n \text{ is even but not a power of 2} \\ f(3n+1) & \text{if } n \text{ is odd.} \end{cases}$$

Then $f(3) = f(10)$ because 3 is odd

$ = f(5)$ because $10 = 2 \times 5$

$ = f(16)$ because 5 is odd

$ = 1$ because $16 = 2^4$.

    (a) Show that $f(11)$ also equals 1.
    (b) Show that $f(9)$, $f(14)$, and $f(25)$ all equal $f(11)$ and so all equal 1.
    (c) Write a program to find $f(27)$.

    // Do you think this function will always evaluate to 1? No matter what n you
    // start with? Look up "Collatz's conjecture" or the "Hailstone problem" on
    // the web.

3. We could define a derangement as an n-permutation S of $\{1..n\}$ where each $S_j \neq j$ and then define $\mathbf{D}_n$ to be the number of derangements of $\{1..n\}$. Then $\mathbf{D}_n$ is the unique sequence satisfying the recurrence equation

$$D_n = (n-1)\{D_{n-1} + D_{n-2}\} \quad \text{for } n = 3,4,5,\ldots \qquad // \; 8.1.3$$

and initial conditions $\qquad D_1 = 0$ and $D_2 = 1$.

(a) Show that $D_2 = (2)(D_1) + (-1)^2$.

(b) Use Mathematical Induction to prove that for all integers $n >= 2$,

$$D_n = (n)(D_{n-1}) + (-1)^n.$$

4. Use Mathematical Induction and (8.1.3) to prove that

$$\text{for all positive integers } n, \; D_n = n! \sum_{j=0}^{n} \frac{(-1)^j}{j!}.$$

5. Assume that (or look up these two calculus results)

   A. for all real numbers x, $e^x = \sum_{j=0}^{\infty} \frac{x^j}{j!}$, and so $e^{-1} = \sum_{j=0}^{\infty} \frac{(-1)^j}{j!}$,

   and B. for any positive integer n,

$$e^{-1} = \sum_{j=0}^{n} \frac{(-1)^j}{j!} + E_n \quad \text{where } |E_n| < \left| \frac{(-1)^{n+1}}{(n+1)!} \right| = \frac{1}{(n+1)!}.$$

(a) Use the result of the previous question to show

$$\frac{n!}{e} = D_n + n! E_n \quad \text{where } |n! E_n| < \frac{n!}{(n+1)!} = \frac{1}{n+1} <= 1/2.$$

(b) Explain why $D_n - \frac{1}{2} <= n!/e <= D_n + \frac{1}{2}$.

(c) Is $\lceil n!/e \rfloor = D_n$?

6. *Ackermann's function* is sometimes defined recursively in a slightly different form.

   Rule 1. $B(0,n) = n+1$ $\qquad\qquad\qquad$ for $n = 0,1,2,\ldots$,

   Rule 2. $B(m,0) = B(m-1,1)$ $\qquad\qquad$ for $m = 1,2,3,\ldots$,

   and   Rule 3. $B(m,n) = B(m-1, B(m,n-1))$ when both m and n are positive.

(a) Use MI to prove $B(1,n) = n+2$ $\qquad\qquad$ for all $n \in \mathbf{N}$.

(b) Use MI to prove $B(2,n) = 3+2n$ $\qquad\qquad$ for all $n \in \mathbf{N}$.

(c) Use MI to prove $B(3,n) = 2^{3+n} - 3$ $\qquad\quad$ for all $n \in \mathbf{N}$.

(d) Use MI to prove $B(4,n) = (2 \uparrow [3+n]) - 3$ for all $n \in \mathbf{N}$.

(e) Give an expression using the symbol "$\uparrow$" for the values of $B(5,1)$ and $B(5,2)$.

7. Suppose that A is a set of $2n$ objects. Let P_n denote the number of different ways that the objects in A may be "paired up" (the number of different partitions of A into 2-subsets).   // Assume n is a positive integer.

If $n = 2$, then A has four elements, $A = \{x_1, x_2, x_3, x_4\}$.

The three possible pairings are   1. x_1 with x_2 and x_3 with x_4,

2. x_1 with x_3 and x_2 with x_4,

3. x_1 with x_4 and x_2 with x_3.   // So $P_2 = 3$

(a)   Show that if $n = 3$ and $A = \{x_1, x_2, x_3, x_4, x_5, x_6\}$, there are 15 possible pairings by listing them all:

1. x_1 with x_2 and x_3 with x_4 and x_5 with x_6

2. ...   // So $P_3 = 15$.

(h)   Show that P_n must satisfy the RE $P_n = (2n - 1)P_{n-1}$ for $\forall\, n \geq 2$.

(c)   Use this recurrence equation and Mathematical Induction to prove

$$P_n = (2n)!/[2^n \times n!] \text{ for } \forall\, n \geq 1.$$

8. Show that $y_n = \dfrac{n(n-1)}{2} + c$ for $n > 0$ is a solution of the recurrence relation

$$y_{n+1} = y_n + n.$$

9. Suppose that a sequence is defined by:

$$f(0) = 5 \text{ and}$$
$$f(n+1) = 2 \times f(n) + 1 \quad \text{for } n = 0, 1, 2, \ldots$$

(a)   Find the value of $f(10)$

(b)   Prove that the sequence is neither an arithmetic sequence nor a geometric sequence.

10. (a) Find the General Solution of the recurrence equation

$$S_n = 3S_{n-1} - 10 \quad \text{for } n = 1, 2, \ldots.$$

(b)   Determine the particular solution where $S_0 = 15$.

(c)   Use the formula in (b) to evaluate S_6 and check your answer using the recurrence equation itself.

11. Suppose $s_0 = 60$ and $s_{n+1} = (1/5)s_n - 8$ for $n = 0, 1, \ldots$.

(a)   Find s_1, s_2, and s_3.

(b)   Solve the recurrence relation to give a formula for s_n.

(c)   Does this sequence converge? If so, what is the limit?

(d)   Does the corresponding series converge? If so, what is the limit?

12. Suppose $s_0 = 75$ and $s_{n+1} = (1/3)s_n - 6$ for $n = 0, 1, \ldots$.

(a)   Find s_1, s_2, and s_3.

(b)   Solve the recurrence relation to give a formula for s_n.

(c)   Does this sequence converge? If so, what is the limit?

(d)   Does the corresponding series converge? If so, what is the limit?

13. (a) Show that $f_n = A \times 3^n + B \times 2^n$ satisfies the recurrence equation

$$f_n = 5f_{n-1} - 6f_{n-2} \quad \text{for } n >= 2.$$

(b) Find the particular solution (values for A and B) so that

$$f_0 = 4 \text{ and } f_1 = 17.$$

14. In how many n-sequences of flips of a coin is there never two heads in a row? Or how many n-sequences of H's and T's do not contain "HH"? For each positive integer n, let $f(n)$ denote the number of such sequences. The list of all two-sequences of H's and T's that don't contain "HH" is

$$\text{HT} \qquad \text{TH} \qquad \text{TT,} \qquad\qquad\qquad\qquad \text{so } f(2) = 3.$$

(a) List all three-sequences of H's and T's that don't contain "HH".
(b) List all four-sequences of H's and T's that don't contain "HH".
(c) Find a recurrence equation satisfied by the sequence f.

15. Suppose that $\alpha^2 = \alpha + 1$ and suppose F_n denotes the Fibonacci sequence.
(a) Show that $\alpha^3 = 2\alpha + 1$, $\alpha^4 = 3\alpha + 2$, and $\alpha^5 = 5\alpha + 3$.
(b) Prove that for $n >= 2$, $\alpha^n = (F_{n-1}) \times \alpha + (F_{n-2})$.

16. Suppose F_n denotes the Fibonacci sequence.

Is F_n equal $\sum_{j=0}^{k} \binom{n-j}{j}$ where $k = \lfloor n/2 \rfloor$?

Hint: Use Mathematical Induction but separate the cases of $(k + 1)$ being odd or even.

17. Using your favourite programming language (Python, Java, C++, or something else), write a program that uses Algorithm 8.3.1 to compute and display several Fibonacci numbers of modest size (perhaps Fib(20), Fib(30) and Fib(40)). Write a second program that uses Algorithm 8.3.3 to compute and display the same Fibonacci numbers. What do you notice about the running time for the programs?

18. Prove $|x + y| <= |x| + |y|$ for all real numbers x and y.

19. Prove $|xy| = |x| \times |y|$ for all real numbers x and y.

20. Let γ denote the "golden ratio" and let F_n denote the Fibonacci sequence. Prove that $\forall n \in \mathbf{P}$

$$(1/3)\gamma^{n+1} < F_n < (1/2)\gamma^{n+1},$$
and $\qquad\qquad (0.6)\gamma^n < F_n < (0.8)\gamma^n.$

21. For $\forall n \in \mathbf{N}$, let $T_n = 1 + F_0 + F_1 + \ldots + F_n$ tabulate the first few values of T_n. Formulate a conjecture about a formula for T_n, and prove that your formula is correct (by Mathematical Induction).

22. Suppose that $S_{n+2} = 13S_{n+1} + 48S_n$ for $\forall n \in \mathbf{N}$.
    (a) Find the General Solution of this recurrence equation.
    (b) Find the particular solution where $S_0 = 1$ and $S_1 = 5$.
23. Suppose that $S_{n+2} = 22S_{n+1} - 121S_n$ for $\forall n \in \mathbf{N}$.
    (a) Find the General Solution of this recurrence equation.
    (b) Find the particular solution where $S_0 = 1$ and $S_1 = 5$.
24. Suppose that $S_{n+2} = S_{n+1} + S_n + 2$ for $\forall n \in \mathbf{N}$.
    (a) Find the General Solution of this recurrence equation.
    (b) Find the particular solution where $S_0 = 1$ and $S_1 = 5$.
25. Suppose that $S_{n+2} = 2S_{n+1} - S_n + 3$ for $\forall n \in \mathbf{N}$.
    (a) Find the General Solution of this recurrence equation.
    (b) Find the particular solution where $S_0 = 1$ and $S_1 = 5$.
26. Suppose that $S_{n+2} = -S_{n+1} + 2S_n + 7$ for $\forall n \in \mathbf{N}$.
    (a) Find the General Solution of this recurrence equation.
    (b) Find the particular solution where $S_0 = 1$ and $S_1 = 5$.
27. (a) Is $0.29999999\ldots = 0.300000000\ldots$?
    Hint : $0.29999999\ldots = 0.2 + 0.09 + 0.009 + 0.0009 + 0.00009 + \ldots$
    $$= 0.2 + A \quad + Ar \quad + Ar^2 \quad + Ar^3 \quad + \ldots$$
    $$\text{where } A = 0.09 \text{ and } r = 0.1$$
    $$= 0.2 + A/(1 - r) \qquad \text{by Theorem 8.5.2}$$
    (b) Is $0.9999999\ldots = 1$?
    (c) Is $99.99999\ldots = 100$?
    (d) Does every rational number have two different positional representations in base 10? Or just those with a "terminating" positional representation?
28. In Example 1.3.3 in Chap. 1, when converting fractions from decimal notation to other bases, we asserted that

    $$0.7\{10\} = 0.1\overline{0110}\{2\}.$$

    Then $0.7\{10\} = 0.1\,0110\,0110\,0110\,0110\ldots\{2\}$
    $$= 0.1 + 0.00110 + 0.000000110 + 0.0000000000110 + \ldots\{2\}$$
    $$= 0.1\{2\} + A + Ar + Ar^2 + Ar^3 + \ldots$$
    $$= \qquad\qquad \text{where } A = 0.00110\{2\} \text{ and } r = 0.0001\{2\}$$
    $$= 0.1\{2\} + A/(1 - r). \qquad \text{by Theorem 8.5.2}$$

    (a) Convert 0.1, A, and r to base 10.
    (b) Check that in base 10, $0.1\{2\} + A/(1 - r)$ really is equal 0.7.

29. Assume that for all x, $e^x = \sum_{j=0}^{\infty} \frac{1}{j!} x^j$.   // Look up infinite Series Expansions.

    Consider the case that $x = 1/3$.

    (a) Use a machine to calculate $e^{1/3}$
    (b) Add the terms of the infinite series until the last term added is < 0.0005.
    (c) Does that partial sum give 3 decimal places of accuracy?
    (d) Repeat (a), (b) and (c) for $x = -1/3$.

                              // Look up convergence of Alternating Series.

Generating Sequences and Subsets

9

The Knapsack Problem is a famous problem with many applications. Suppose you have (or a cat-burglar finds) a set of n objects $U = \{O_1, O_2, ..., O_n\}$ where each object O_j has a positive "weight" W_j and a positive "value" V_j. Suppose also that there is a positive value B equal to the total weight you (or the burglar) are willing to carry in your knapsack:

> Find a subset X of U with maximum total value
> subject to the constraint that the total weight of X is $<= B$.

One method for solving this problem is to consider ***all subsets*** of objects, S, in turn, and "Process" the subset S as follows:

> Find the total weight of S.
> If that total weight is $<= B$, then find the total value of S.
> If that total value is larger than the largest so far,
>   then remember S as the solution so far.

We might start with the empty subset; its total weight is $< B$, its total value is zero, but that's the largest so far, so remember $\varnothing$ as the solution so far. After we've looked at all the (nonempty) subsets, our solution so far will be a solution to the Knapsack Problem.

```
// But how do we generate all the nonempty subsets "in turn"?
// As we saw in Chap. 2, subsets have characteristic vectors, n-sequences of 0's
// and 1's where
//        X[j] = 1        if Oⱼ is in the subset
// and    X[j] = 0        if Oⱼ is not in the subset.
//
// Can these n-sequences of bits be generated "in turn"?
```

Actually, the Knapsack Problem is the collection of all "instances" of a Knapsack Problem where an instance is determined by specific values for the parameters n and B and specific values for the entries in the two arrays V and W. A ***solution***

© Springer International Publishing AG, part of Springer Nature 2018
T. Jenkyns and B. Stephenson, *Fundamentals of Discrete Math for Computer Science: A Problem-Solving Primer*, Undergraduate Topics in Computer Science, https://doi.org/10.1007/978-3-319-70151-6_9

to the Knapsack Problem is an algorithm that determines an optimal subset of objects in every instance.

Another famous problem is the **Traveling Salesman's Problem**. A traveling salesman wishes to visit each one of n towns (or prospective customers). However, there is a "cost" (in time or distance) for moving from one town to another town. The salesman wants to minimize the total cost of his "tour" of visits, including the trip from his home base to the first town and from the last town back to his home base.

And there is another "constraint": he sells a product he claims will cure everything – from measles to melancholia and from insomnia to ingrown toenails. But the immediate effect of his "snake oil" is to cause blurred vision, a rash, and diarrhea; so he never wants to return to a town he's already visited, not even to pass through.

If his home base is $t[0]$ and the other towns are $t[1]$, $t[2]$, to $t[n]$ and the (nonnegative) cost of going directly from $t[j]$ to $t[k]$ is $C[j,k]$ for all distinct pairs j and k in $\{0..n\}$, then the **TSP** can be expressed as follows:

Every permutation $S = (s_1, s_2, s_3, \ldots, s_n)$ of $\{1..n\}$ corresponds to a **tour** for the salesman:

$$T = (t[0],\ t[s_1],\ t[s_2],\ \ldots,\ t[s_n],\ t[0]),$$

where the total cost of this tour is

$$C(T) = C[0, s_1] + C[s_1, s_2] + C[s_2, s_3] + \ldots + C[s_{n-1}, s_n] + C[s_n, 0].$$

Find a tour T* whose total cost is as small as possible.
The **TSP** in this form has many important theoretical and practical applications. It would be nice if there were an efficient algorithm to solve it.

// An efficient algorithm to solve it would bring *fame* and quite probably *fortune*.

The "total enumeration" algorithm solves it but has order $n!$:
        Generate all the tours T in some order.
        Find the cost of T.
        Keep track of the best tour so far.
When all the tours have been considered, the best tour so far is the best tour of all, T^*.   // But how do we generate all the permutations and tours "in some order"?

The Traveling Salesman's Problem, like the Knapsack Problem, is really the collection of all "instances" of a Traveling Salesman's Problem where an instance is determined by specific values for the parameters n and transition-cost matrix C. A **solution** to the **TSP** would be an algorithm that determines an optimal salesman's tour in every instance.

In this chapter, we develop a strategy for listing all sequences in some finite set $\mathbb{S}$. We want to do it in a manner so we can be sure that every sequence occurs, and it occurs exactly once. If these are sequences of integers, we can use Lexicographic-Order. In particular, we want simple algorithms to generate the following: **all k-sequences** on $\{1..n\}$; all nonempty, **increasing sequences**; all **increasing k-sequences**; all **permutations**; and all **k-permutations**.

We will use these algorithms to illustrate ***average-case complexity*** and show that the average-case complexity may be "of lower order" than the worst-case complexity.

9.1 Generating Sequences in Lexicographic-Order

We used a systematic method with a tree diagram in Chap. 2 to generate integer sequences in Lexicographic-Order. Here let's try to do it without drawing a tree (which may be huge). And let's begin with an example.

// We want to formulate a general strategy for listing sequences in Lexicographic
// Order.

Example 9.1.1: Some Strange Sequences

Let A denote all 5-sequences $S = (x_1, x_2, x_3, x_4, x_5)$ on $\{1..5\}$ such that
(1) x_1 and x_3 are odd.   // that is, 1 or 3 or 5
(2) x_2 and x_4 are even.   // that is, 2 or 4
(3) $x_1 <= x_4$.   // So $x_1 \neq 5$.
(4) x_5 is not equal x_1 or x_2 or x_3 or x_4.   // (3,2,3,4,1) is in A.
List all the elements of A.   // How many of them are there? (Is there no formula?)

Before creating the list, we want to determine how the list will start, how to continue a partial list and how to know when to stop.

What Would Be the Lexicographic-First Sequence, F?

All the sequences that begin with 1 will come before those that begin with 3, (and those that begin with 3 would come before those that begin with 5). So in F, $x_1 = 1$.
Among all the sequences that begin with $x_1 = 1$, those with $x_2 = 2$ will come before those with $x_2 = 4$. So in F, $x_1 = 1$ and $x_2 = 2$.
Among all the sequences that begin with $x_1 = 1$ and $x_2 = 2$, those with $x_3 = 1$ will come before those with $x_3 = 3$ and those with $x_3 = 5$. So in F, $x_3 = 1$.
Then, the smallest possible value of x_4 (that can follow 1,2,1...) is 2;
                                                             so in F, $x_4 = 2$.
Then, the smallest possible value of x_5 (that can follow 1,2,1,2,...) is 3;
                                                             so in F, $x_5 = 3$.
Thus, the Lexicographic-First sequence is $F = (1, 2, 1, 2, 3)$.

What Would Be the Lexicographic-Next Sequence After F?

// If we were looking at 5-sequences of letters that might be words in a dictionary,
// after ***ababc***, we would look for ***ababd***. We'd make a change at the right-hand
// end and increase the last entry as little as possible.

The Lexicographic-Next sequence after $F = (1, 2, 1, 2, 3)$ is $(1, 2, 1, 2, \mathbf{4})$.
The Lexicographic-Next sequence after $(1, 2, 1, 2, 4)$ is $(1, 2, 1, 2, \mathbf{5})$.

But now we cannot increase the rightmost entry.

// If we were looking at 5-sequences of letters that might be words in a dictionary,
// after **abczz**, we would look for **abdaa**. We'd make a change as far to the right-
// hand end as possible; increase that entry that can be increased by as little as
// possible, and then reset the following entries to the smallest possible values (just
// as we did to get F).

To get the Lexicographic-Next sequence after $(1, 2, 1, 2, 5)$, we cannot increase x_5,
so we take one step to the left. We can increase x_4 from 2 to 4. Then, the smallest
possible value for $x_5 = 3$. Hence, the Lexicographic-Next sequence after
$(1, 2, 1, 2, 5)$ is $(1, 2, 1, \mathbf{4}, \mathbf{3})$.

How Do We Know When to Stop?

// If we were looking at 5-sequences of letters that might be words in a dictionary,
// after *zzzzz*, we would stop because <u>no</u> entry can be increased.

In the Lexicographic-Last sequence, <u>no</u> entry can be increased; all are as large as
possible. We will discover, as we work from right to left, that no entry can be
increased. Our method of finding the Lex-Next sequence itself will tell us that
we've reached the Lex-Last sequence.   // and should stop

// This implies that x_1 now has the largest value that it can take, and
// in all sequences with that value of x_1, the current value of x_2 is now has the
// largest value that it can take. And so on, for all the entries.

Example 9.1.1: The List of the Sequences

	x_1	x_2	x_3	x_4	x_5	
1.	**1**	**2**	**1**	**2**	**3**	// taking the smallest possibility in each place in turn
2.	1	2	1	2	**4**	// increasing the rightmost place as little as possible
3.	1	2	1	2	**5**	// "
4.	1	2	1	**4**	**3**	// finding the rightmost place we can increase …
5.	1	2	1	4	**5**	
6.	1	2	**3**	**2**	**4**	
7.	1	2	3	2	**5**	
8.	1	2	3	**4**	**5**	
9.	1	2	**5**	**2**	**3**	
10.	1	2	5	2	**4**	
11.	1	2	5	**4**	**3**	
12.	1	**4**	**1**	**2**	**3**	
13.	1	4	1	2	**5**	
14.	1	4	1	**4**	**2**	
15.	1	4	1	4	**3**	
16.	1	4	1	4	**5**	

```
17.  1  4  3  2  5
18.  1  4  3  4  2
19.  1  4  3  4  5
20.  1  4  5  2  3
21.  1  4  5  4  2
22.  1  4  5  4  3
23.  3  2  1  4  5      // Remember x₄ must be >= x₁.
24.  3  2  3  4  1
25.  3  2  3  4  5
26.  3  2  5  4  1
27.  3  4  1  4  2
28.  3  4  1  4  5
29.  3  4  3  4  1
30.  3  4  3  4  2
31.  3  4  3  4  5
32.  3  4  5  4  1
33.  3  4  5  4  2
```

Now no entry can be increased, so we're done. We've generated the Lexicographic-Last sequence in A.   // Now, we know A has 33 sequences.

The method that worked in that example was:

Step 1.   Generate the Lex-First sequence in S by taking the smallest possibilities for the entries, working from left to right.

Step 2.   Having just generated sequence $S = (x_1, x_2, x_3, \ldots x_k)$, find the Lex-Next sequence T:

> By finding the largest index j where x_j can be increased, and increasing x_j by the smallest possible amount to $x'_j = x_j + q$, and then resetting the entries following x'_j to their smallest possible values.

But stop when we reach a sequence S where no entry can be increased. When all sequences in S have the same length k, we can prove that this method does indeed generate all sequences in S in Lexicographic-Order.

The Most Important Ideas in This Section.
Lexicographic-Order is a total order, so the elements of a finite set of sequences, S, can be sorted and listed from the first to the last. The general method to do this is:

Step 1.   Generate the Lex-First sequence in S.

Step 2.   Having just generated sequence $S = (x_1, x_2, x_3, \ldots x_k)$, find the Lex-Next sequence T. Stop when sequence S is the Lex-Last in S.

9.2   Generating All *k*-Sequences on {1..*n*}

Let's first look at the example with $k = 4$ and $n = 3$.

Example 9.2.1: All 4-Sequences on {1, 2, 3}

// The list will have $3^4 = 81$ sequences.

x_1	x_2	x_3	x_4	
1	**1**	**1**	**1**	// The Lex-First has the smallest possibility in each place.
1	1	1	**2**	// increasing the rightmost place as little as possible
1	1	1	**3**	// "
1	1	**2**	**1**	// finding the rightmost place we can increase and resetting…
1	1	2	**2**	
1	1	2	**3**	
1	1	**3**	**1**	
1	1	3	**2**	
1	1	3	**3**	// This is the 9th.
1	**2**	**1**	**1**	
1	2	1	**2**	
1	2	1	**3**	
1	2	**2**	**1**	
1	2	2	**2**	
1	2	2	**3**	
1	2	**3**	**1**	
1	2	3	**2**	
1	2	3	**3**	// This is the 18th.
1	**3**	**1**	**1**	
…				
1	3	3	**3**	// This is the 27th.
2	**1**	**1**	**1**	
…				
2	3	3	**3**	// This is the 54th.
3	**1**	**1**	**1**	
…				
3	3	**3**	**1**	
3	3	3	**2**	
3	3	3	**3**	// This is the 81st, the Lex-Last.

9.2.1   Average-Case Complexity

Let $c(T)$ denote the number of entries in T that are assigned a (new) value when T is constructed. Then $c(T)$ ranges from 1 up to k.

// $c(T)$ is (roughly) the "cost" of T.

What Is the Average Cost of Constructing a Sequence?   (in example 9.2.1)

$c(T) = 1$ if in the previous sequence, $x_4 < 3$;
$c(T) = 2$ if in the previous sequence, $x_4 = 3$, but $x_3 < 3$;
$c(T) = 3$ if in the previous sequence, $x_4 = 3$, $x_3 = 3$, but $x_2 < 3$;
$c(T) = 4$ if in the previous sequence, $x_4 = 3$, $x_3 = 3$, $x_2 = 3$, but $x_1 < 3$
          or T is the Lex-First sequence.

Now let $f(a)$ denote the number of sequences T where $c(T) = a$, for $a = 1, 2, 3, 4$.
                                                                          // 1 to k

Then

$f(1) = $ the number of 4-sequences on {1,2,3} where $x_4 <= 2$
      $= (\text{\# choices for } x_1) \times (\text{\# for } x_2) \times (\text{\# for } x_3) \times (\text{\# for } x_4)$
      $= (3) \times (3) \times (3) \times (2) = 54$;

$f(2) = $ the number of 4-sequences on {1,2,3} where $x_3 <= 2$ and $x_4 = 3$
      $= (\text{\# choices for } x_1) \times (\text{\# for } x_2) \times (\text{\# for } x_3) \times (\text{\# for } x_4)$
      $= (3) \times (3) \times (2) \times (1) = 18$;

$f(3) = $ the number of 4-sequences on {1,2,3} where $x_2 <= 2$ and $x_3 = x_4 = 3$
      $= (\text{\# choices for } x_1) \times (\text{\# for } x_2) \times (\text{\# for } x_3) \times (\text{\# for } x_4)$
      $= (3) \times (2) \times (1) \times (1) = 6$;

and

$f(4) = 1(\text{for the Lex-First})$
          $+ $ the number of 4-sequences on {1, 2, 3} where
              $x_1 <= 2$ but $x_2 = x_3 = x_4 = 3$
      $= 1 + (\text{\# choices for } x_1) \times (\text{\# for } x_2) \times (\text{\# for } x_3) \times (\text{\# for } x_4)$
      $= 1 + (2) \times (1) \times (1) \times (1) = 1 + 2 = 3$.

// All the sequences have been counted; $f(1) + f(2) + f(3) + f(4)$
// $= 54 + 18 + 6 + 3 = 81$.

The average cost of constructing a sequence

        $= (\text{the sum of all the costs})/(\text{the number of sequences})$.

The 54 sequences that cost 1 assignment will contribute 54 to the sum of the costs;
the 18 sequences that cost 2 assignments will contribute $18 \times 2 = 36$ to the sum;
the 6 sequences that cost 3 assignments will contribute $6 \times 3 = 18$ to the sum; and
the 3 sequences that cost 4 assignments will contribute $3 \times 4 = 12$ to the sum.

The average cost of constructing a sequence

$$= \text{(the sum of all the costs)}/\text{(the number of sequences)}.$$
$$= (54 + 36 + 18 + 12)/(81) = 120/81 = 1.481481\ldots < 2.$$

// Would you believe that for every k and every $n \neq 1$, the average cost is < 2
// assignments? The exercises will lead you through a proof that

$$// \quad \text{The average cost of a sequence} = \frac{n + n^2 + \ldots + n^k}{n^k} = \left(\frac{n}{n-1}\right)\left(\frac{n^k - 1}{n^k}\right).$$

The following pseudo-code combines the search for the index j (where an increase may be made) and the "resetting" part of our strategy. But in this implementation, the Lex-Last sequence is not detected in the search for index j; instead, the sequences are counted (indexed by the for-loop control variable) as they are constructed.

Algorithm 9.2.1: Generating All k-Sequences on $\{1..n\}$ in Lex-Order

```
Begin
  For j ← 1 To k Do                          // Lex-First = (1, 1, ..., 1).
    S[j] ← 1;
  End;
                                             // Print or process sequence S.
  For index ← 1 To (n^k − 1) Do              // Get the Lex-Next.
    j ← k;
    While (S[j] = n) Do
      S[j] ← 1;
      j ← j − 1;
    End;     // the while-loop
    S[j] ← S[j] + 1;
                                             // Print or process sequence S.

  End;       // the for-loop
End.         // of Algorithm 9.2.1
```

Modifying this algorithm slightly, we can generate all 4-sequences on $\{0,1\}$ in Lexicographic-Order. Interpreting these sequences as characteristic vectors, we get a list of all subsets of $\{1..4\}$ as follows:

x_1	x_2	x_3	x_4	Subset
0	0	0	0	$\varnothing$
0	0	0	1	$\{\qquad 4\}$
0	0	1	0	$\{\quad 3 \quad \}$
0	0	1	1	$\{\quad 3,4\}$
0	1	0	0	$\{\ 2 \qquad \}$
0	1	0	1	$\{\ 2, \ \ 4\}$
0	1	1	0	$\{\ 2,3 \ \}$
0	1	1	1	$\{\ 2,3,4\}$

1	0	0	0	{1 }
1	0	0	1	{1, 4}
1	0	1	0	{1, 3 }
1	0	1	1	{1, 3,4}
1	1	0	0	{1,2 }
1	1	0	1	{1,2, 4}
1	1	1	0	{1,2,3 }
1	1	1	1	{1,2,3,4}

// Aren't these the binary representations of the integers from 0 to 15?

Processing these subsets, perhaps for the Knapsack Problem, could be made easier and faster if, as we go from one subset S in our list to the next subset T, a "minimal change" is made to S. Would it be possible to list these 16 subsets so that in going from S in our list to the next subset T,

> **either**   one new element is added to S
> **or**   one old element is removed from S?

x_1	x_2	x_3	x_4	Subset
0	0	0	0	∅
0	0	0	1	{ 4}
0	0	1	1	{ 3,4}
0	0	1	0	{ 3 }
0	1	1	0	{ 2,3 }
0	1	1	1	{ 2,3,4}
0	1	0	1	{ 2, 4}
0	1	0	0	{ 2 }
1	1	0	0	{1,2 }
1	1	0	1	{1,2, 4}
1	1	1	1	{1,2,3,4}
1	1	1	0	{1,2,3 }
1	0	1	0	{1, 3 }
1	0	1	1	{1, 3,4}
1	0	0	1	{1, 4}
1	0	0	0	{1 }

Any listing of all the n-sequences of bits with minimal changes is known as a "Gray code." They say that these are named after Frank Gray, who, in the 1950s, applied for a patent on a device that produced such a list.

//X   Can you develop an algorithm (or computer program) to list all n-sequences
//   of bits as a Gray code for any value of n, that is, with minimal changes?

The Most Important Ideas in This Section.
All k-sequences on $\{1..n\}$ may be generated in Lexicographic-Order fairly easily by Algorithm 9.2.1, with an average cost of < 2 assignments. Modifying this algorithm slightly, we can generate all k-sequences on $\{0,1\}$ in Lexicographic-Order, and interpreting these sequences as characteristic vectors, we get a list of all subsets of $\{1..k\}$.

A "minimal change" to a subset, S, is

> **either** one new element is added to S
> **or** one old element is removed from S.

A **Gray code** is any listing of all the objects in some class $\mathbb{S}$ with minimal changes between consecutive entries in the list. There are a number of algorithms to list all k-sequences on $\{1..n\}$ and all k-subsets on $\{1..n\}$ in Gray codes.

9.3 Generating Subsets of {1..n} as Increasing Sequences

If $n = 7$, the subset $\{5, 2, 3, 6\}$ of $\{1..7\}$ may be rewritten in many ways by changing the order of the elements in the braces:

$$\{5,2,3,6\} = \{5,6,2,3\} = \{2,3,5,6\} = \{6,2,3,5\} = \ldots \qquad \text{// 24 ways?}$$

The most natural listing is $\{2, 3, 5, 6\}$ where the elements occur in the list between the braces in **increasing order**. // in their usual order
If we were to always list them that way, then every nonempty subset of integers corresponds to a unique increasing sequence of integers.

We want an algorithm to list all (nonempty) increasing sequences on $\{1..n\}$ in Lexicographic-Order. But these are sequences of varying lengths. Therefore, we'll have to revise our general method to incorporate the fact that prefixes come before extensions:

Step 1. Generate the Lex-First sequence in $\mathbb{S}$ by taking smallest possible values for the entries x_1, x_2, x_3, and so on until we have a sequence in $\mathbb{S}$

Step 2. Having generated sequence $S = (x_1, x_2, x_3, \ldots x_k)$, find the Lex-Next sequence T:
(a) By *extending* S by finding smallest possible values
for the entries x_{k+1}, x_{k+2}, and so on
until we have a sequence in $\mathbb{S}$;
or when S cannot be extended,

(b) By finding the largest index j where x_j can be increased, and then
      increasing x_j by the smallest possible amount to $x'_j = x_j + q$, and then
      (if necessary) *extending* $(x_1, x_2, x_3, \ldots x'_j)$ by finding smallest
      possible values for the entries x_{j+1}, x_{j+2}, and so on
      until we have a sequence in $\mathbb{S}$.
   But stop when we reach a sequence S that cannot be extended and
   where no entry can be increased.
This method can be proven to generate all sequences in $\mathbb{S}$ in Lexicographic-Order.
                                                                        // very tediously
   Let's look at the example of $n = 4$. The (nonempty) subsets range in size
from 1 to 4, so the sequences will range in length k from $k = 1$ to $k = 4$. In
Lexicographic-Order, prefixes come before extensions; so the sequence (2) will
come before (2, 3, 4). We need all the sequences

$$S = (x_1, x_2, \ldots, x_k) \text{ on } \{1..4\} \text{ such that } 1 <= x_1 < x_2 < \ldots < x_k <= 4.$$

Example 9.3.1: All Nonempty Increasing Sequences on {1..4}
                                     // The list will have $2^4 - 1 = 15$ sequences.

	x_1	x_2	x_3	x_4	k	
1.	1				1	// the Lex-First
2.	1	2			2	// extending the previous (as little as possible)
3.	1	2	3		3	// "
4.	1	2	3	4	4	// "
5.	1	2	4		3	// finding the rightmost place we can increase
6.	1	3			2	// finding the rightmost place we can increase
7.	1	3	4		3	
8.	1	4			2	
9.	2				1	
10.	2	3			2	
11.	2	3	4		3	
12.	2	4			2	
13.	3				1	
14.	3	4			2	
15.	4				1	// the Lex-Last

   Here, the Lex-Last sequence is easy to detect; it's the only sequence that begins
with n. The following pseudo-code gives the algorithm.

Algorithm 9.3.1: Generating All (Nonempty) Subsets of {1..n}
as Increasing Sequences (S[1]..S[k]) in Lex-Order

```
Begin
  S[1] ← 1;                          // Lex-First = (1)
  k ← 1;

                                     // Print or process sequence (S[1])
  While (S[1]< n) Do                 // Get the Lex-Next
    If (S[k]< n) Then
      S[k + 1] ← S[k] + 1;
      k ← k + 1;                     // The length goes up
    Else // S[k] = n
      S[k − 1] ← S[k−1] + 1;
      k ← k − 1;                     // The length goes down
    End;  // the if-statement

                                     // Print or process sequence (S[1]..S[k])
  End;   // the while-loop
End.     // of Algorithm 9.3.1
```

This algorithm correctly implements the general strategy: it generates the Lex-First increasing sequence, and whenever the current sequence S is not the Lex-Last, it produces the Lex-Next increasing sequence after S.

With this algorithm, each new sequence costs exactly one assignment to one entry in S.   // But is this a "minimum-change" order? Is this a Gray code?

Generating All k-Subsets of {1..n}

We will begin with an example.

Example 9.3.2: If $n = 7$ and $k = 4$, then there are exactly

$$\binom{7}{4} = \frac{7!}{4!3!} = \frac{7 \times 6 \times 5}{3 \times 2 \times 1} = 35 \text{ subsets.}$$ // and increasing sequences

They are listed below as increasing sequences on {1..7} written in Lex-Order along with the j-value for generating the Lex-Next.

// j is the largest index where $S[j]$ may be increased and a is the number
// of entries assigned a new value as the sequence was constructed.

	x_1	x_2	x_3	x_4	j	a	
1.	1	2	3	4	4	4	// the Lex-First
2.	1	2	3	5	4	1	
3.	1	2	3	6	4	1	
4.	1	2	3	7	3	1	
5.	1	2	4	5	4	2	
6.	1	2	4	6	4	1	
7.	1	2	4	7	3	1	

```
 8.  1  2  5  6     4     2
 9.  1  2  5  7     3     1
10.  1  2  6  7     2     1
11.  1  3  4  5     4     3
12.  1  3  4  6     4     1
13.  1  3  4  7     3     1
14.  1  3  5  6     4     2
15.  1  3  5  7     3     1
16.  1  3  6  7     2     1
17.  1  4  5  6     4     3
18.  1  4  5  7     3     1
19.  1  4  6  7     2     1
20.  1  5  6  7     1     1
21.  2  3  4  5     4     4
22.  2  3  4  6     4     1
23.  2  3  4  7     3     1
24.  2  3  5  6     4     2
25.  2  3  5  7     3     1
26.  2  3  6  7     2     1
27.  2  4  5  6     4     3
28.  2  4  5  7     3     1
29.  2  4  6  7     2     1
30.  2  5  6  7     1     1
31.  3  4  5  6     4     4
32.  3  4  5  7     3     1
33.  3  4  6  7     2     1
34.  3  5  6  7     1     1
35.  4  5  6  7     -     1
```

// Now no entry can be increased, so we've generated the Lexicographic-Last.

We want an algorithm which generates all sequences $S = (x_1, x_2, \ldots, x_k)$ on $\{1..n\}$ such that $\qquad 1 <= x_1 < x_2 < \ldots < x_k <= n.$

// How can we design it? What properties of these sequences can we exploit?
// In particular, what are the upper and lower limits on each entry, x_i?

Lemma 9.3.1: If $S = (x_1, x_2, \ldots, x_k)$ is an increasing sequence on $\{1..n\}$, then
$$i <= x_i <= n - k + i \quad \text{for } i = 1, 2, \ldots, k$$

Proof. When p and q are integers and $p <= q$, the interval

$$\{p..q\} = \{p+0, p+1, p+2, \ldots, p+(q-p)\}, \text{ so } |\{p..q\}| = (q-p)+1.$$

When $1 <= p <= q <= k$, $\{x_p, x_{p+1}, \ldots, x_q\}$ is a set of $q - p + 1$ distinct integers in the interval $\{x_p .. x_q\}$, so

$$q - p + 1 <= x_q - x_p + 1$$

and $$q - p <= x_q - x_p.$$

If $p = 1$ and $q = i$, then $$i - 1 <= x_i - x_1 <= x_i - 1 \quad \text{// since } 1 <= x_1$$
so $$i <= x_i.$$

If $p = i$ and $q = k$, then $$k - i <= x_k - x_i,$$
so $$k - i + x_i <= x_k <= n.$$
Hence, $$x_i <= n - k + i.$$ □

We could summarize this as follows:

$(x_1, x_2, \ldots, x_{k-1}, x_k)$ is dominated by $(n - k + 1, n - k + 2, \ldots, n - 1, n)$,
and $(1, 2, \ldots, k - 1, k)$ is dominated by $(x_1, x_2, \ldots, x_{k-1}, x_k)$.

// When $n = 7$ and $k = 4$, $(x_1, x_2, x_3, x_4) \, \mathcal{D} \, (4, \; 5, \; 6, \; 7)$,   // See Chap. 7.
// and $(1, \; 2, \; 3, \; 4) \, \mathcal{D} \, (x_1, x_2, x_3, x_4)$.

Note also that if $p = i$ and $q = i + t$ then from Lemma 9.3.1,   // taking $t >= 0$

$$t = q - p <= x_{i+t} - x_i,$$
or $$x_i <= x_{i+t} - t.$$

Now if $$x_i = n - k + i,$$ // $=$ the upper limit for x_i
then $$n - k + i <= x_{i+t} - t$$ // for $t = 0, 1, 2, \ldots (k - i)$
                                       $$n - k + i + t <= x_{i+t},$$
so $$x_{i+t} = n - k + (i + t). \; \text{// } = \text{the upper limit for } x_{i+t}$$

This implies that   if $x_i < n - k + i,$ // $<$ the upper limit for x_i
then for $p = 1, 2, \ldots, (i - 1)$ $x_p < n - k + p.$ // $<$ the upper limit for x_p

The facts which follow from these results that we may then use in the construction of our algorithm are:
1. There is only one sequence with $x_1 = n - k + 1$, and it's the Lex-Last.

// We can easily detect the Lex-Last sequence by looking at $S[1]$ alone.
// (Also, the Lex-First equals the Lex-Last if and only if $n = k$).

2. If x_j can be increased, then so can x_{j-1}.

// If x_j is increased to its upper limit $n - k + j$, then x_{j-1} can be increased,
// but no x_q with $j < q$ can be increased.
// The next j-value needed by the algorithm is the current j-value minus 1.

The following pseudo-code generates the sequences in Lex-Order, <u>and</u> with each sequence S (except the Lex-Last), it gives the largest index j where $S[j]$ may be increased.

Algorithm 9.3.2: Generating All k-Subsets of {1..n}
as Increasing Sequences (S[1]..S[k]) in Lex-Order

```
Begin
  For j ← 1 To k Do                    // Lex-First = (1, 2, ..., k)
    S[j] ← j;
  End;

                                       // Print or process sequence S
  j ← k;
  While (S[1] < n − k + 1) Do          // Get the Lex-Next, T
    S[j] ← S[j] + 1;
    If (S[j] = n − k + j) Then
      j ← j − 1;                       // One change to S gives T
    Else                               // Reset all entries after S[j]
      While (j < k) Do                 // to the smallest possibility
        j ← j + 1;
        S[j] ← S[j − 1] + 1;
      End;     // the inner while-loop  // and now, j = k.
    End;       // the if-statement
                                       // Print or process sequence S
  End;         // the outer while-loop
End.           // of Algorithm 9.3.2
```

This algorithm correctly implements the general strategy: it generates the Lex-First increasing sequence, and whenever the current sequence S is not the Lex-Last, it produces the Lex-Next increasing sequence after S.

// Is Lexicographic-Order a minimal-change list? A Gray code? Is there such a
// list? What would be a "minimum change"?

Average-Case Complexity of Algorithm 9.3.2

Let $c(T)$ denote the number of entries in T that are assigned a (new) value when T is constructed. Then $c(T)$ ranges from 1 up to k.

// $c(T)$ is (roughly) the "cost" of T.

What is the *average cost of constructing a sequence*?

If $k = n$, then there is only 1 sequence which costs k assignments, so the average cost is k assignments. From this point on, assume that $1 <= k < n$.

// Then $n − k − 1 >= 0$.

But first, let's look at the j-values. For each,

$$S[j] < n − k + j.$$ // because $S[j]$ may be increased

So $S[j] <= (n − k + j) − 1,$ and

if $j < q <= k,$ $S[q] = n − k + q.$ // the upper limit for $S[q]$

The number of such sequences   // with a given value of j

$$= \text{the number of increasing } j\text{-sequences on } \{1..(n-k+j-1)\}$$

$$= \binom{n-k+j-1}{j}.$$

// In Example 9.3.2 where $n = 7$, $k = 4$, and $n - k - 1 = 2$.

// The number of times $j = 4$ equals $\binom{2+4}{4} = \binom{6}{4} = 15$.

// The number of times $j = 3$ equals $\binom{2+3}{3} = \binom{5}{3} = 10$.

// The number of times $j = 2$ equals $\binom{2+2}{2} = \binom{4}{2} = 6$.

// The number of times $j = 1$ equals $\binom{2+1}{1} = \binom{3}{1} = 3$.

These numbers occur in Pascal's triangle.

```
                        1
                      1   1
                    1   2   1
                  1   3   3   1
                1   4   6   4   1
              1   5  10  10   5   1
            1   6  15  20  15   6   1
          1   7  21  35  35  21   7   1
```

Every sequence, *except the last one*, has a j-value associated with it, and

$$1 + 3 + 6 + 10 + 15 = 35.$$ // and 35 is also in the triangle

We can generalize this equation.

Lemma 9.3.2: For all nonnegative integers q and t,

$$\sum_{j=0}^{t} \binom{q+j}{j} = \binom{q}{0} + \binom{q+1}{1} + \binom{q+2}{2} + \ldots + \binom{q+t}{t} = \binom{q+t+1}{t}.$$

Proof. // by Mathematical Induction on t

Step 1.   If $t = 0$, then LHS $= \binom{q}{0} = 1$, and RHS $= \binom{q+1}{0} = 1$.

                                                    // even if $q = 0$

Step 2.   Assume $\exists$ a nonnegative integer r such that

$$\sum_{j=0}^{r} \binom{q+j}{j} = \binom{q+r+1}{r}.$$

Step 3.   If $t = r + 1$, then

$$\text{LHS} = \sum_{j=0}^{r+1}\binom{q+j}{j} = \sum_{j=0}^{r}\binom{q+j}{j} + \binom{q+r+1}{r+1}$$

$$= \binom{q+r+1}{r} + \binom{q+r+1}{r+1} \qquad \text{// by the inductive hypothesis}$$

$$= \binom{(q+r+1)+1}{r+1} \qquad \text{// by the Bad Banana Theorem (Example 2.3.3)}$$

$$= \binom{q+(r+1)+1}{r+1} = \text{RHS.} \qquad\qquad □$$

// What about the average cost of constructing a sequence?
// The average cost of constructing a sequence
// = (the sum of all the costs)/(the number of sequences).

Let's look again at the sequences with a certain (fixed) j-value.

$$S[j] \ < \ n - k + j, \qquad\qquad \text{// because } S[j] \text{ may be increased}$$
so
$$S[j] <= (n - k + j) - 1, \quad \text{and}$$
if $j < q <= k$, $\quad S[q] \ = \ n - k + q.$ \qquad\qquad // the upper limit for $S[q]$

Case 1

If $S[j] = (n - k + j) - 1,$ then only 1 assignment would be done
to construct the Lex-Next sequence, T.

Case 2

If $S[j] <= (n - k + j) - 2,$ then T is constructed by replacing
$S[j], S[j+1]\ldots S[k]$.

In the first case $c(T) = 1$, and in the second case, $c(T) = k - (j - 1) = (k + 1) - j$.
The number of sequences in the first case,   // when $S[j] = (n - k + j) - 1$

$$= \text{the number of increasing } (j - 1)\text{-sequences on } \{1..(n - k + j - 2)\}$$

$$= \binom{n - k + j - 2}{j - 1}.$$

The number of sequences in the second case,   // when $S[j] <= (n - k + j) - 2$

$\qquad$ = the number of increasing j-sequences on $\{1..(n - k + j - 2)\}$

$$= \binom{n - k + j - 2}{j}.$$

The total number of sequences S falling into Case 1, setting $Q = n - k - 1$, is

$$\sum_{j=1}^{k} \binom{Q + (j - 1)}{j - 1} = \binom{Q + 0}{0} + \binom{Q + 1}{1} + \ldots + \binom{Q + (k - 1)}{k - 1}$$

$$= \binom{Q + (k - 1) + 1}{k - 1} \qquad\qquad \text{// by Lemma 9.3.2}$$

$$= \binom{n - 1}{k - 1}. \qquad\qquad \text{// as } Q = n - k - 1$$

For $a = 1, 2, \ldots, k$, let $f(a)$ denote the number of sequences T such that $c(T) = a$. Then

$f(1) =$ the number of sequences in Case 1

$\qquad$ + the number of sequences in Case 2 where $a = 1 = (k + 1) - j$

$$= \binom{n - 1}{k - 1} + \binom{R + j}{j} \quad \text{where } R = n - k - 2 \text{ and } j = k$$

$$= \binom{n - 1}{k - 1} + \binom{R + k}{k}. \quad \text{// if } n = 7 \text{ and } k = 4, \ R = 1 \text{ and } f(1) = 20 + 5$$

For $1 < a < k$, $f(a) =$ the number of sequences in Case 2 where $j = (k + 1) - a$

$$= \binom{R + j}{j}. \qquad\qquad \text{// } R = n - k - 2.$$

Also, $f(k) = 1$ (for the Lex-First sequence)

$\qquad$ + the number of sequences in Case 2 where $j = (k + 1) - a = 1$

$$= 1 + \binom{R + j}{j} \qquad\qquad \text{// } R = n - k - 2.$$

$$= \binom{R + 0}{0} + \binom{R + 1}{1}.$$

Finally, the sum of all the costs

$$= \sum_{a=1}^{k} a \times f(a)$$

$$= 1 \times f(1) + 2 \times \binom{R+k-1}{k-1} + 3 \times \binom{R+k-2}{k-2} + \ldots + (k-1)$$

$$\times \binom{R+2}{2} + k \times f(k).$$

// This looks quite ugly but ... using Lemma 9.3.2 and the hints given in the
// exercises, you (too) will be able to prove the remarkable fact that

$$(k+1) \times \binom{R+0}{0} + k \times \binom{R+1}{1} + \ldots + 3 \times \binom{R+k-2}{k-2} + 2$$

$$\times \binom{R+k-1}{k-1} + 1 \times \binom{R+k}{k} = \binom{n}{k}.$$

It then follows that the sum of all the costs $= \binom{n}{k} + \binom{n-1}{k-1} - 1.$

Since $\binom{n-1}{k-1} = \dfrac{(n-1)!}{(k-1)!([n-1]-[k-1])!} = \dfrac{(n-1)!}{(k-1)!(n-k)!} \times \dfrac{n}{n} \times \dfrac{k}{k}$

$$= \dfrac{n!}{k!(n-k)!} \times \dfrac{k}{n} \qquad\qquad = \dfrac{k}{n} \times \binom{n}{k},$$

the sum of all the costs $= \binom{n}{k} + \dfrac{k}{n} \times \binom{n}{k} - 1 < \left\{1 + \dfrac{k}{n}\right\}\binom{n}{k}.$

Therefore, the average cost is $< 1 + \dfrac{k}{n}$ which is $<= 2.$

// no matter how large n and k are
// so long as $k < n$

// In Example 9.3.2 where $k = 4$ and $n = 7$,
// the average was 54/35 and $1 + k/n = 11/7 = 55/35.$

The Most Important Ideas in This Section.
All nonempty subsets of {1..n} may be generated as increasing sequences in Lexicographic-Order fairly easily by Algorithm 9.3.1, where exactly one entry in the sequence is changed to get the next subset, but the subset sizes go up or down by 1.

All k-subsets of {1..n} may be generated as increasing sequences in Lexicographic-Order fairly easily by Algorithm 9.3.2, with an average cost of < 2 assignments.

(Continued)

(Continued)

> A **Gray code** is a listing of all the subsets with minimal changes between consecutive entries in the list. There are a number of algorithms to list all subsets of $\{1..n\}$, and all k-subsets of $\{1..n\}$ in Gray codes. Algorithms 9.3.1 and 9.3.2 do <u>not</u> produce Gray codes.
>
> For all nonnegative integers q and t,
>
> $$\sum_{j=0}^{t}\binom{q+j}{j} = \binom{q}{0} + \binom{q+1}{1} + \binom{q+2}{2} + \ldots + \binom{q+t}{t}$$
>
> $$= \binom{q+t+1}{t}.$$

9.4 Generating Permutations in Lexicographic-Order

We would like a simple algorithm to generate all $n!$ permutations of $\{1..n\}$ in Lexicographic-Order. The Lex-First will be $(1, 2, 3, \ldots, n-1, n)$, and the Lex-Last will be $(n, n-1, n-2, \ldots, 2, 1)$.

If $n = 4$, then there are $4! = 24$ (full) permutations of $\{1..n\}$. They are listed below in Lex-Order along with the j-value for generating the Lex-Next and with $a =$ the number of entries assigned a new value as the sequence was constructed.

Example 9.4.1: All Permutations of $\{1..4\}$

	x_1	x_2	x_3	x_4	j	a	
1.	1	2	3	4	3	4	// the Lex-First
2.	1	2	4	3	2	2	
3.	1	3	2	4	3	3	
4.	1	3	4	2	2	2	
5.	1	4	2	3	3	3	
6.	1	4	3	2	1	2	
7.	2	1	3	4	3	4	
8.	2	1	4	3	2	2	
9.	2	3	1	4	3	3	
10.	2	3	4	1	2	2	
11.	2	4	1	3	3	3	
12.	2	4	3	1	1	2	
13.	3	1	2	4	3	4	
14.	3	1	4	2	2	2	
15.	3	2	1	4	3	3	
16.	3	2	4	1	2	2	

17.	3	**4**	1	2	3	3
18.	3	4	**2**	1	1	2
19.	**4**	1	2	3	3	4
20.	4	1	**3**	2	2	2
21.	4	**2**	1	3	3	3
22.	4	2	**3**	1	2	2
23.	4	**3**	1	2	3	3
24.	4	3	**2**	1	–	2

// Now no entry can be increased, so we've generated the Lexicographic-Last.

// Is Lexicographic-Order a minimal-change list?
// What would be a minimal change between two permutations?
// Is there a minimal-change list?

// How do we determine the Lex-Next permutation, *T*?

What will be the Lex-Next after $S = (2, 6, 9, 1, 5, 8, 7, 4, 3)$?   // Here $n = 9$.
We must find the largest index j where $S[j]$ may be increased.

$S[n] = 3$ cannot be increased because all integers larger than 3 in $\{1..9\}$
            occur to the left of $S[n]$.
            // This happens in every permutation so j is always $< n$.
$S[n-1] = 4$ cannot be increased because all larger values are left of $S[n-1]$.
$S[n-2] = 7$ cannot be increased because all larger values are left of $S[n-2]$.
$S[n-3] = 8$ cannot be increased because all larger values are left of $S[n-3]$.

But
$S[n-4] = 5$ ***can be increased*** because some integer larger than 5 in $\{1..9\}$
            does not occur to the left of $S[n-3]$.
                        // 7 and 8 occur to the **right**
Thus, $j = n - 4$.
// If we were to continue following our general strategy to find the Lex-Next
// sequence, *T*
//
// $S = (2, 6, 9, 1, \mathbf{5}, 8, 7, 4, 3)$,   so $T = (2, 6, 9, 1, ?, , , ,)$.
//
// $S[j] = 5$ cannot be increased to 6
//   because 6 occurs to the left, but
// $S[j]$ can be increased to 7,   so $T = (2, 6, 9, 1, 7, ?, , ,)$.
//
//   The smallest value that can follow (2, 6, 9, 1, **7**,
//   is not 1 nor 2, but is 3,   so $T = (2, 6, 9, 1, \mathbf{7}, \mathbf{3}, ?, ,)$.
//
//   The smallest value that can follow (2, 6, 9, 1, 7, **3**,
//   is not 1, 2, or 3, but is 4,   so $T = (2, 6, 9, 1, \mathbf{7}, \mathbf{3}, \mathbf{4}, ?,)$.
//
//   The smallest value that can follow (2, 6, 9, 1, 7, 3, **4**,
//   is not 1, 2, 3, or 4, but is 5,   so $T = (2, 6, 9, 1, \mathbf{7}, \mathbf{3}, \mathbf{4}, \mathbf{5}, ?)$.
//
//   The smallest (and only) value that can follow (2, 6, 9, 1, 7, 3, 4, **5** is 8,

//   so $T = (2, 6, 9, 1, \mathbf{7, 3, 4, 5, 8})$.

If $S[n-1] < S[n]$, then $j = n - 1$.   // This will happen half the time.

If $S[n-1] > S[n]$ but $S[n-2] < S[n-1]$,   then $j = n - 2$.

                                    // This happens a third of the time.

If $S[n-2] > S[n-1] > S[n]$ but $S[n-3] < S[n-2]$,   then $j = n - 3$.

We need to find the largest index j where

$$S[j+1] > S[j+2] > \ldots > S[n-2] > S[n-1] > S[n] \text{ but } S[j] < S[j+1].$$

And we can do this in a loop where we start with $k = n - 1$:

```
While (S[k] > S[k + 1]) Do
   k ← k - 1;
End.
```

When this loop terminates $S[k] < S[k + 1]$, so

$S[k]$ is less than at least one later entry, namely, $S[k + 1]$, but
there is no larger index j, where $S[j]$ is less than a later entry.

Therefore, we can set j to be the final k-value. In fact, we can even use the variable j instead of k in the loop.

    There is one **snag** though, if S is the Lex-Last permutation, our while-loop will go "off the end" of S when k (or j) becomes zero. One **remedy** for this is to set $S[0]$ to be zero at the beginning of the algorithm and never change it. Then when k (or j) becomes 0, $S[k]$ must be $< S[k + 1]$. Furthermore, the final value of k (or j) will be zero **if and only if** S is the Lex-Last permutation.

    Now we know how to determine j where $S[j]$ may be increased for any S (except the Lex-Last). How do we find the smallest value larger than the current value of $S[j]$ such that $S[j]$ may be increased to it? "It" will be the **smallest** value following $S[j]$ in S that's larger than the current value of $S[j]$ where

$$S[j+1] > S[j+2] > \ldots S[n-2] > S[n-1] > S[n] \text{ but } S[j] < S[j+1].$$

We can start from $S[n]$ and go down the sequence until we find the first entry $> S[j]$. So starting with $m = n$, we can use the loop

```
While (S[j] > S[m]) Do
   m ← m - 1;
End.
```

This loop is sure to terminate because if m reaches $j + 1$, we know that $S[j] < S[m]$. We will change $S[j]$ to $S[m]$.

// How do we reset the entries in $S[j + 1]$, $S[j + 2]$, and so on to the smallest
// possible value?

    The entries in S after position j are decreasing

$$S[j+1] > S[j+2] > \ldots S[m-1] > S[m] > S[m+1] > \ldots > S[n-1] > S[n].$$

The smallest of these is $S[n]$ so it can be put into position $j + 1$.
The next smallest of these is $S[n - 1]$ so it can be put into position $j + 2$.
Maybe we should simply reverse the order of these entries.
But we cannot use the value of $S[m]$ again, and we must use the old value of $S[j]$.

We know that $S[m] > S[j] > S[m + 1]$.
If we *interchange* the values of $S[m]$ and $S[j]$, we then have

$$S[j+1] > S[j+2] > \ldots S[m-1] > S[j] > S[m+1] > \ldots > S[n-1] > S[n].$$

<u>Now</u>, we can reset the entries in $S[j + 1]$, $S[j + 2]$ ... $S[n]$ to their smallest possible values by simply *reversing* the order of these entries.

The algorithm given below in pseudo-code generates each permutation S and then the j-value of that sequence; it terminates after it generates the Lex-Last which is the only one with j-value $= 0$.

Algorithm 9.4.1: Generating All (Full) Permutations of $\{1..n\}$
               as Sequences (S[1]..S[n]) in Lex-Order

```
Begin
    For i ← 0 To n Do                    // Lex-First = (1, 2, ..., n)
        S[i] ← i;
    End;

                                          // Print or process sequence S

    j ← n - 1;
    While (j > 0) Do                      // Get the Lex-Next
                                          // S[j] may be increased

        X ← S[j];
        m ← n;
        While (X > S[m]) Do
            m ← m - 1;
        End;  // the inner while-loop   // and now Sm > Sj > Sm+1
        S[j] ← S[m];                      // Interchange Sm and Sj
        S[m] ← X;

        p ← j + 1;
        q ← n;                            // Reset all entries after S[j]
        While (p < q) Do                  // by reversing them
            X ← S[p];
            S[p] ← S[q];
            S[q] ← X;
            p ← p + 1;
            q ← q - 1;
        End;  // the second inner while-loop

                                          // Print or process sequence S
        j ← n - 1;                        // Find next index j in S
        While (S[j] > S[j + 1]) Do
            j ← j - 1;
        End;
    End;     // the outer while-loop
End.         // of Algorithm 9.4.1
```

Walk through the next four iterations of the body of the loop constructing the Lex-Next permutation assuming that $n = 10$, and at this point,

$$S = (9, 1, 4, 10, 8, 7, 6, 5, 2, 3). \qquad\qquad // \text{ with } j\text{-value} = 9$$

1. On the next iteration, $j = 9$ and so X will be $S[9] = 2$.
   (a) Find m and interchange $S[m]$ and $S[j]$:

m	S[m]	X > S[m]	the permutation S
10	3	F	9,1,4,10,8,7,6,5,**3,2**

   (b) Reverse the order of $S[j + 1]..S[n]$:

p	q	p < q	the permutation S
10	10	F	9,1,4,10,8,7,6,5,**3,2**

// After $S = (9, 1, 4, 10, 8, 7, 6, 5, 2, 3)$, the Lex-Next is $(9, 1, 4, 10, 8, 7, 6, 5, \mathbf{3}, \mathbf{2})$.

   (c) Find the j-value of this new current permutation:

j	S[j]	S[j] > S[j + 1]	the permutation S
9	3	T	9,1,4,10,8,7,6,5,3,2
8	5	T	
7	6	T	
6	7	T	
5	8	T	
4	10	T	
3	4	F	

2. On the next iteration, $j = 3$ and so X will be $S[3] = 4$.
   (a) Find m and interchange $S[m]$ and $S[j]$:

m	S[m]	X > S[m]	the permutation S
10	2	T	9,1,4,10,8,7,6,5,3,2
9	3	T	
8	5	F	9,1,**5**,10,8,7,6,**4**,3,2

   (b) Reverse the order of $S[j + 1]..S[n]$:

p	q	p < q	the permutation S
			9,1,5,10,8,7,6,4,3,2
4	10	T	9,1,5, **2**,8,7,6,4,3,**10**
5	9	T	9,1,5, 2,**3**,7,6,4,**8**,10
6	8	T	9,1,5, 2,3,**4**,6,**7**,8,10
7	7	F	

// After $S = (9, 1, 4, 10, 8, 7, 6, 5, 3, 2)$, the Lex-Next is $(9, 1, \mathbf{5}, \mathbf{2}, \mathbf{3}, \mathbf{4}, \mathbf{6}, \mathbf{7}, \mathbf{8}, \mathbf{10})$.

(c) Find the j-value of this new current permutation:

j	S[j]	S[j] > S[j + 1]	the permutation S
9	8	F	9,1,5,2,3,4,6,7,8,10

3. On the next iteration, $j = 9$ and so X will be $S[9] = 8$.

(a) Find m and interchange $S[m]$ and $S[j]$:

m	S[m]	X > S[m]	the permutation S
10	10	F	9,1,5,2,3,4,6,7,**10**,**8**

(b) Reverse the order of $S[j + 1]..S[n]$:

p	q	p < q	the permutation S
10	10	F	9,1,5,2,3,4,6,7,**10**,**8**

// After $S = (9, 1, 5, 2, 3, 4, 6, 7, 8, 10)$, the Lex-Next is $(9, 1, 5, 2, 3, 4, 6, 7, \mathbf{10}, \mathbf{8})$.

(c) Find the j-value of this new current permutation:

j	S[j]	S[j] > S[j + 1]	the permutation S
9	10	T	9,1,5,2,3,4,6,7,10,8
8	7	F	

4. On the next iteration, $j = 8$ and so X will be $S[8] = 7$.

(a) Find m and interchange $S[m]$ and $S[j]$:

m	S[m]	X > S[m]	the permutation S
10	8	F	9,1,5,2,3,4,6,**8**,10,**7**

(b) Reverse the order of $S[j + 1]..S[n]$:

p	q	p < q	the permutation S
9	10	T	9,1,5,2,3,4,6,8,10,7
10	9	F	

// After $S = (9, 1, 5, 2, 3, 4, 6, 7, 10, 8)$, the Lex-Next is $(9, 1, 5, 2, 3, 4, 6, \mathbf{8}, \mathbf{7}, \mathbf{10})$.

(c) Find the j-value of this new current permutation:

j	S[j]	S[j] > S[j + 1]	the permutation S
9	10	F	9,1,5,2,3,4,6,8,7,10

// On the next iteration, $j = 9$ and so X will be $S[9] = 7$.

This algorithm correctly implements the general strategy: it generates the Lex-First permutation, and whenever the current sequence S is not the Lex-Last, it produces the Lex-Next permutation after S.

Average-Case Complexity of Algorithm 9.4.1

Let $c(T)$ denote the number of entries in T that are assigned a (new) value when T is constructed. Then $c(T)$ ranges from 2 up to n.

// $c(T)$ is (roughly) the "cost" of T.
// We never just change 1 entry (if $n > 1$).

What is the ***average cost of constructing a sequence***?

// The average cost of constructing a sequence
// = (the sum of all the costs)/(the number of sequences).
// We need a formula for the sum of all the costs.

For $A = 1, 2, \ldots, n$, let $f(A)$ denote the number of sequences T such that $c(T) = A$; that is, in the formation of the Lex-Next permutation T, the last A entries are assigned new values. Then $(j - 1) + A = n$.

For each possible (positive) j-value, how many permutations S are there where

$$S[j+1] > S[j+2] > \ldots > S[n-2] > S[n-1] > S[n] \text{ but } S[j] < S[j+1]?$$

We can construct such a permutation S in 5 steps:
1. Choose A values from $\{1..n\}$ to put at the end of the sequence.
2. Put the largest of these in position $j + 1$.
3. Select 1 of the remaining values to put these in position j.
4. Arrange the remaining $A - 2$ values in decreasing order in positions $j + 2$ to n.
5. Arrange the "un-chosen" $n - A$ values in any order in positions 1 to $j - 1$.

Then, the number of sequences constructed in this way

$$= \binom{n}{A} \times (1) \times (A - 1) \times (1) \times (n - A)! \qquad \text{// The "steps" are independent.}$$

$$= \frac{n!}{A!(n - A)!} \times (A - 1) \times (n - A)!$$

$$= n! \times \left(\frac{A - 1}{A!} \right).$$

Every permutation except the Lex-Last (where $j = 0$) can be constructed (using these 5 steps), so adding these counts should total $n! - 1$.

Because $\dfrac{A - 1}{A!} = \dfrac{A}{A!} - \dfrac{1}{A!} = \dfrac{1}{(A - 1)!} - \dfrac{1}{A!},$ \qquad // the series "telescopes"

$$\sum_{A=1}^{n} \left(\frac{1}{(A - 1)!} - \frac{1}{A!} \right)$$

$$= \left(\frac{1}{0!} - \frac{1}{1!} \right) + \left(\frac{1}{1!} - \frac{1}{2!} \right) + \ldots + \left(\frac{1}{(n - 2)!} - \frac{1}{(n - 1)!} \right) + \left(\frac{1}{(n - 1)!} - \frac{1}{n!} \right)$$

$$= \frac{1}{0!} + \left(-\frac{1}{1!} + \frac{1}{1!} \right) + \ldots + \left(-\frac{1}{(n - 1)!} + \frac{1}{(n - 1)!} \right) - \frac{1}{n!}$$

$$= 1 - \frac{1}{n!}.$$

Hence, $\displaystyle\sum_{A=1}^{n} n! \times \left(\frac{A-1}{A!}\right) = n! \times \sum_{A=1}^{n} \left(\frac{A-1}{A!}\right) = n! \times \sum_{A=1}^{n} \left(\frac{1}{(A-1)!} - \frac{1}{A!}\right)$

$$= n! \times \left(1 - \frac{1}{n!}\right) \qquad = n! - 1.$$

//X Prove that if x_0, x_1, x_2, ..., x_n is <u>any</u> sequence, then $\displaystyle\sum_{j=1}^{n}(x_{j-1} - x_j) = x_0 - x_n$.

We can now evaluate the frequencies, $f(A)$.

For $1 <= A < n$, $f(A) = n! \times \left(\dfrac{A-1}{A!}\right)$,

and $\qquad\qquad f(n) = 1$ (for the Lex-First sequence)

$$+ n! \times \left(\frac{n-1}{n!}\right) = 1 + (n-1) = n.$$

Then $\quad f(1) = 0$ $\qquad\qquad\qquad$ // We always have to change at least 2 entries.
$\qquad f(2) = n! \times \frac{1}{2}$ $\qquad\qquad$ // Half the time we interchange the last 2 entries.
$\qquad f(3) = n! \times \frac{1}{3}$ $\qquad\qquad$ // One third of the time we change the last 3 entries.

Finally, the average cost

$$= \frac{1}{n!}\sum_{A=1}^{n} A \times f(A)$$

$$= \frac{1}{n!}\sum_{A=2}^{n} A \times f(A) \qquad\qquad\qquad\qquad\qquad // f(1) = 0.$$

$$= \frac{1}{n!}\left\{\sum_{A=2}^{n-1} A \times n! \times \left(\frac{A-1}{A!}\right) \quad + n \times (n)\right\} \qquad // f(n) = n.$$

$$= \sum_{A=2}^{n-1} A \times \left(\frac{A-1}{A!}\right) \qquad\qquad\qquad + \frac{n^2}{n!}$$

$$= \sum_{A=2}^{n-1} \frac{1}{(A-2)!} \qquad\qquad\qquad\qquad + \frac{n}{(n-1)!}$$

$$= \frac{1}{0!} + \frac{1}{1!} + \frac{1}{2!} + \ldots + \frac{1}{(n-3)!} + \frac{n-1}{(n-1)!} + \frac{1}{(n-1)!}$$

$$= \frac{1}{0!} + \frac{1}{1!} + \frac{1}{2!} + \ldots + \frac{1}{(n-3)!} + \frac{1}{(n-2)!} + \frac{1}{(n-1)!}.$$

// When n was 4, the average cost was $64/24 = 8/3$, and
// $1/0! + 1/1! + 1/2! + 1/3! = 1 + 1 + 1/2 + 1/6$
// $= (6 + 6 + 3 + 1)/6 = 16/6 = 8/3 = 2.\overline{6}$.

The next theorem proves that the average cost is always < 3.

Theorem 9.4.1: For all $n \in \mathbf{P}$, $\dfrac{1}{0!} + \dfrac{1}{1!} + \dfrac{1}{2!} + \ldots + \dfrac{1}{n!} <= 3 - \dfrac{1}{n}$,

and equality holds only for $n <= 3$.

Proof. // by Mathematical Induction

Step 1.

> If $n = 1$, then LHS $= 1 + 1 = 2$ and RHS $= 3 - 1 = 2$.
> If $n = 2$, then LHS $= 2 + 1/2 = 5/2$ and RHS $= 3 - 1/2 = 5/2$.
> If $n = 3$, then LHS $= 5/2 + 1/6 = 16/6$ and RHS $= 3 - 1/3 = 8/3$.
> If $n = 4$, then LHS $= 8/3 + 1/24 = 65/24$ and RHS $= 3 - 1/4 = 66/24$.

Step 2.   Assume $\exists$ an integer $k >= 4$ such that

$$\frac{1}{0!} + \frac{1}{1!} + \frac{1}{2!} + \ldots + \frac{1}{k!} < 3 - \frac{1}{k}.$$

Step 3.   If $n = k + 1$, then

$$(k+1)! = (k+1) \times (k) \times (k-1)! >= (k+1) \times (k) \times (3)! > (k+1) \times (k),$$

// since $k >= 4$

and so $\dfrac{1}{(k+1)!} < \dfrac{1}{k(k+1)}$.

Hence,

$$
\begin{aligned}
\text{LHS} &= \frac{1}{0!} + \frac{1}{1!} + \frac{1}{2!} + \ldots + \frac{1}{k!} \quad + \frac{1}{(k+1)!} \\
&< 3 - \frac{1}{k} \quad\quad\quad\quad\quad\quad\quad + \frac{1}{(k+1)!} && \text{// by Step 2} \\
&< 3 - \frac{1}{k} \quad\quad\quad\quad\quad\quad\quad + \frac{1}{k(k+1)} && \text{// as we just saw} \\
&= 3 - \left[\frac{1}{k} - \frac{1}{k(k+1)}\right] \quad = 3 - \frac{(k+1)-1}{k(k+1)} \\
&= 3 - \frac{1}{k+1}.
\end{aligned}
$$

☐

// We get a better result if we use the series expansion for e^x given in calculus
// books: $e^x = \displaystyle\sum_{j=0}^{\infty} \frac{x^j}{j!}$ for $\forall x \in \mathbf{R}$. The average cost is $< \displaystyle\sum_{j=0}^{\infty} \frac{1}{j!} = e^1 = 2.718281\ldots$.

9.4.1 Generating All *k*-Permutations of {1..*n*} in Lex-Order

Suppose that $k = 5$ and $n = 9$.
What will be the next 5-permutation after $S = (8, 1, 5, 9, 7)$?
We must find the largest index j where $S[j]$ may be increased.

$S[k] = 7$ cannot be increased because all integers larger than 7 in {1..9} occur to the left of $S[k]$.

$S[k - 1] = 9$ cannot be increased.

But $S[k - 2] = 5$ ***can be increased*** because some integer larger than 5 in {1..9} does not occur to the left of $S[k - 2]$.

                           // 6 does not occur at all in S.

Thus, $j = 3 = k - 2$.

// If we continue following our general method to find the Lex-Next sequence, T
//
// $S = (8, 1, \mathbf{5}, 9, 7)$,   so $T = (8, 1, ?, ?, ?)$.
//
// $S[j] = 5$ can be increased to 6
//   because 6 does not occur in S,   so $T = (8, 1, \mathbf{6}, ?, ?)$.
//
//   The smallest value that can follow (8, 1, **6**,
//   is (not 1, but is) 2,   so $T = (8, 1, \mathbf{6}, \mathbf{2}, ?)$.
//
//   The smallest value that can follow (8, 1, 6, **2**,
//   is 3,   so $T = (8, 1, \mathbf{6}, \mathbf{2}, \mathbf{3})$.
//

For each k-permutation S of {1..n}, define S^+ to be the unique full permutation of {1..n} obtained by extending S by adding the integers not already there, in descending order. For example, when $k = 5$ and $n = 9$,

$$\text{if } S = (8, 1, 5, 9, 7), \quad \text{then} \quad S^+ = (8, 1, 5, 9, 7, \mathbf{6}, \mathbf{4}, \mathbf{3}, \mathbf{2}).$$

When we generated all full permutations, we were able to utilize the right-hand end of S to determine j, the largest index where $S[j]$ may be increased, and also the smallest increase that can be made to $S[j]$. We'll do that again now. We'll maintain a full permutation S where $S[k + 1] > S[k + 2] > \ldots > S[n - 1] > S[n]$. When we find the largest j where $S[j] < S[j + 1]$, that j will be $<= k$. And when we find the smallest entry $S[m] > S[j]$ to the right of $S[j]$, we will interchange $S[j]$ and $S[m]$.

Looking again at the example above, how would the construction of the next full permutation of {1..9} after

$$S = (8, 1, 5, 9, 7, \mathbf{6}, \mathbf{4}, \mathbf{3}, \mathbf{2})$$

begin? We know $j = 3$ and $S[j] = 5$. Then we find $m = 6$ and $S[m] = 6$ and interchange $S[j]$ and $S[m]$ and get

$$S1 = (8, 1, 6, \underline{9, 7, 5, 4,} \ 3, 2).$$

The smallest possible values to reset $S[4]$ and $S[5]$ to are 3 and 2 at the right-hand end of $S1$. Let's copy them into an auxiliary array R in reverse order, so $R = (2, 3, ..)$ – later we'll place them in $S[4]$ and $S[5]$. Now let's shift "9, 7, 5, 4" two places right in S. Finally, place the two values from R into $S1$ to produce

$$T = (8, 1, 6, 2, 3, \underline{9, 7, 5, 4}).$$

Then the first 5 entries in T constitute the Lex-Next 5-permutation of $\{1..9\}$ after $S = (8, 1, 5, 9, 7)$, and T is a full permutation of $\{1..9\}$ with the final $n - k$ entries in decreasing order.

We can modify Algorithm 9.4.1 to generate all k-permutations of $\{1..n\}$ in Lexicographic-Order, using the n-sequences just described and assuming the precondition that $0 < k < n$.

Algorithm 9.4.2: Generating All k-Permutations of {1..n} in Lex-Order as (S[1]..S[k]), the Beginning of an n-Permutation S

```
Begin
    For i ← 0 To k Do                        // Lex-First = (1, 2, ..., k)
        S[i] ← i;
    End;
    For j ← 1 To (n − k) Do
        S[k + j] ← (n + 1) − j;
    End;
                                             // First S⁺ = (1, 2,..., k, n, n − 1, ..., k + 1)
                                             // Print or process the k-permutation (S[1]..S[k])
    j ← k;
    While (j > 0) Do                         // Get the Lex-Next
                                             // S[j] may be increased

        X ← S[j];
        m ← n;
        While (X > S[m]) Do
            m ← m − 1;
        End;        // the inner while-loop       // And now Sₘ > Sⱼ > Sₘ₊₁
        S[j] ← S[m];                              // Interchange Sₘ and Sⱼ
        S[m] ← X;

        If (j < k) Then                       // Reset all entries after S[j]
            t ← k − j;
            For i ← 1 To t Do
                R[i] ← S[n + 1 − i];
            End;
```

```
      q ← n;
      While (q - t > j) Do          // shift n - k entries to the end of S
        S[q] ← S[q - t];
        q ← q - 1;
      End;   // the inner while-loop

      For i ← 1 To t Do
        S[j + i] ← R[i];
      End;
    End;     // the if statement
                              // Print or process the k-permutation (S[1]..S[k])
    j ← k;                    // Find largest index j where S[j] may be increased
    While (S[j] > S[j + 1]) Do
      j ← j - 1;
    End;
  End;       // the outer while-loop
End.         // of Algorithm 9.4.2
```

Walkthrough the next four iterations of the body of the main while-loop in Algorithm 9.4.2 constructing the Lex-Next k-permutation on $\{1..n\}$ where $k = 5$ and $n = 9$, and at this point,

$$S = (8, 1, 5, 9, 7, 6, 4, 3, 2). \qquad\qquad // \text{ with } j\text{-value} = 3$$

1. On the next iteration, $j = 3$ and so X will be $S[3] = 5$.
   (a) Find m and interchange $S[m]$ and $S[j]$:

m	S[m]	X > S[m]	the permutation S
9	2	T	8,1,5,9,7,6,4,3,2
8	3	T	
7	4	T	
6	6	F	8,1,**6**,9,7,**5**,4,3,2

   Since $j < k$, more revisions of S are done.
   (b1) Calculate $t = k - j = 5 - 3 = 2$.
   Reverse the order of the last t entries of S and copy them into **R**:

i	n + 1 - i	S[n + 1 - i]	array R
1	9	2	(**2**, ...)
2	8	3	(**2,3**, ...)

   (b2) Shift the block of $n - k$ entries in S t-steps to the right:

q	q - t	q - t > j	the permutation S
9	7	T	8,1,**6**,9,7,**5**,4,3,**4**
8	6	T	8,1,**6**,9,7,**5**,4,**5**,4
7	5	T	8,1,**6**,9,7,**5**,**7**,**5**,4
6	4	T	8,1,**6**,9,7,**9**,**7**,**5**,4
5	3	F	

(b3) Copy the t entries from R into $S[j + 1]..S[k]$:

i	j + i	R[i]	the permutation S
1	4	2	8,1,**6**,**2**,7,**9**,**7**,**5**,**4**
2	5	3	8,1,**6**,**2**,**3**,**9**,**7**,**5**,**4**

(c) Find the j-value of this new current k-permutation:

j	S[j]	S[j] > S[j + 1]	the permutation S
5	3	F	8,1,**6**,**2**,**3**,**9**,**7**,**5**,**4**

// After $S = $ (8, 1, 5, 9, 7, 6, 4, 3, 2), the next S is (8, 1, **6**, **2**, **3**, **9**, **7**, **5**, **4**).
// After   (8, 1, 5, 9, 7), the Lex-Next 5-permutation is (8, 1, **6**, **2**, **3**).

2. On the next iteration, $j = 5$ and so X will be $S[5] = 3$.
    (a) Find m and interchange $S[m]$ and $S[j]$:

m	S[m]	X > S[m]	the permutation S
9	4	F	8,1,6,2,**4**,9,7,5,**3**

(b) Since "$j < k$" is false, no further revisions of S are made.
(c) Find the j-value of this new current permutation:

j	S[j]	S[j] > S[j + 1]	the permutation S
5	5	F	8,1,6,2,**4**,9,7,5,**3**

// After $S = $ (8, 1, 6, 2, 3, 9, 7, 5, 4), the next S is (8, 1, 6, 2, **4**, 9, 7, 5, **3**).
// After   (8, 1, 6, 2, 3), the Lex-Next 5-permutation is (8, 1, 6, 2, **4**).

3. On the next iteration, $j = 5$ and so X will be $S[5] = 4$.
    (a) Find m and interchange $S[m]$ and $S[j]$:

m	S[m]	X > S[m]	the permutation S
9	3	T	8,1,6,2,4,**9**,**7**,**5**,**3**
8	5	F	8,1,6,2,**5**,9,7,**4**,3

(b) Since "$j < k$" is false, no further revisions of S are made.
(c) Find the j-value of this new current permutation:

j	S[j]	S[j] > S[j + 1]	the permutation S
5	5	F	8,1,6,2,**5**,9,7,**4**,3

// After $S = $ (8, 1, 6, 2, 4, 9, 7, 5, 3), the next S is (8, 1, 6, 2, **5**, 9, 7, **4**, 3).
// After   (8, 1, 6, 2, 4), the Lex-Next 5-permutation is (8, 1, 6, 2, **5**).

4. On the next iteration, $j = 5$ and so X will be $S[5] = 5$.
    (a) Find m and interchange $S[m]$ and $S[j]$:

m	S[m]	X > S[m]	the permutation S
9	3	T	8,1,6,2,5,9,7,4,3
8	4	T	8,1,6,2,5,9,7,4,3
7	7	F	8,1,6,2,**7**,9,**5**,4,3

(b) Since "$j < k$" is false, no further revisions of S are made.

(c) Find the j-value of this new current permutation:

j	S[j]	S[j] > S[j + 1]	the permutation S
5	7	F	8,1,6,2,**7**,9,**5**,4,3

// After S = (8, 1, 6, 2, 5, 9, 7, 4, 3), the next S is (8, 1, 6, 2, **7**, 9, **5**, 4, 3).
// After   (8, 1, 6, 2, 5), the Lex-Next 5-permutation is (8, 1, 6, 2, **7**).

After this walkthrough, it might appear that $j = k$ fairly often.   // How often?
We know that the number of k-permutations on $\{1..n\}$ is

$$(\# \text{ possible values of } x_1)(\# \text{ for } x_2)\ldots(\# \text{ for } x_{k-1})(\# \text{ for } x_k)$$
$$= (n)(n-1)\ldots(n-[k-2])(n-[k-1])$$
$$= (n)(n-1)\ldots(n-k+2) \times (n-k+1).$$

The number of k-permutations on $\{1..n\}$ where the last entry is not the largest possible value is

$$(\# \text{ possible values of } x_1)(\# \text{ for } x_2)\ldots(\# \text{ for } x_{k-1})(\# \text{ for } x_k)$$
$$= (n)(n-1)\ldots(n-[k-2])(\{n-[k-1]\}-1)$$
$$= (n)(n-1)\ldots(n-k+2) \times (n-k).$$

Thus, the fraction of k-permutations on $\{1..n\}$ where x_k may be increased is

$$\frac{n-k}{n-k+1}.$$

When $k = 5$ and $n = 9$, this fraction is 4/5; so 80% of the iterations have that $j = k$.
// and only two entries in the sequence are changed.
// Recall that when $k = n$, it never happens that $j = k$.

The Most Important Ideas in This Section.
All (full) permutations on $\{1..n\}$ may be generated in Lexicographic-Order by Algorithm 9.4.1. All k-permutations on $\{1..n\}$ may be generated in Lexicographic-Order by Algorithm 9.4.2, when $0 < k < n$.

We also looked at the "average cost" of generating the full permutations and found that on average, < 3 entries are changed. A worst case requires changing all n entries. The next chapter looks at average-case complexity in some detail.

Exercises

1. You are given the 5-set of letters {a,b,c,d,e} and the requirement that any ordering in (i) and (ii) which follow must have the form (consonant, vowel, consonant, vowel, consonant).
    (i) Find the next three 5-sequences in Lexicographic-Order following (b,a,d,e,c).
    (ii) Find the next three permutations in Lexicographic-Order following (b,a,c,e,d).
2. Consider the 5-sequences on {1..7}.
    (a) How many are there?
    (b) Which is the Lex-First?
    (c) Which is the Lex-Last?
    (d) Which is the Lex-Next after (2, 3, 4, 6, 7)?
3. Consider the increasing sequences on {1..7} of length $>= 1$.
    (a) How many are there?
    (b) Which is the Lex-First?
    (c) Which is the Lex-Last?
    (d) Which is the Lex-Next after (2, 3, 4, 6, 7)?
4. Consider the increasing 5-sequences on {1..7}.
    (a) How many are there?
    (b) Which is the Lex-First?
    (c) Which is the Lex-Last?
    (d) Which is the Lex-Next after (2, 3, 4, 6, 7)?
5. Consider the 5-permutations on {1..7}.
    (a) How many are there?
    (b) Which is the Lex-First?
    (c) Which is the Lex-Last?
    (d) Which is the Lex-Next after (2, 3, 4, 6, 7)?
    (e) Which is the Lex-Next after (2, 3, 4, 7, 6)?
6. Using Lex-Order,
    (a) Find the next twelve 7-permutations on {1..9} after (8, 1, 3, 9, 7, 6, 4).
    (b) Find the next 7 full permutations on {1..12} after
        (8, 11, 10, 1, 12, 5, 3, 9, 7, 6, 4, 2).
7. List all <u>decreasing</u> 3-sequences on {1..6} in Lexicographic Order.
8. **The Knapsack Problem** and the **Greedy** Algorithm.
    An instance of **the Knapsack Problem** is a set of n objects $U = \{O_1, O_2, O_3, ..., O_n\}$ where each object O_j has a positive weight W_j and a positive value V_j; together with a positive value B, a bound on the total weight you are willing to carry in your knapsack.
    (a) The usual Greedy Algorithm takes the objects in order of their values until no more can be taken. (This would solve the problem if all weights were equal.) Construct an instance where it fails to find an optimum solution.

(b) A second Greedy Algorithm takes the objects in order of their "value-density," the ratio V_j/W_j, until no more can be taken.

// It's better to take diamonds than TVs.

(This would solve the problem if all values were equal).

Construct an instance where this version fails to find an optimum solution.

9. Telescoping Series

(a) Prove that if $x_0, x_1, x_2, \ldots, x_n$ is any sequence and $0 < a <= b < n$, then

$$\sum_{j=a}^{b} (x_{j-1} - x_j) = x_{a-1} - x_b.$$

(b) Express $\sum_{j=a}^{b} (x_j - x_{j-1})$ as a difference of two x's.

(c) Express $\sum_{j=a}^{b} (x_{j+1} - x_j)$ as a difference of two x's.

(d) Express $\sum_{j=a}^{b} n^j(n - 1)$ as a difference of two terms.

(e) Express $\sum_{j=a}^{b} n^{k-j}(n - 1)$ as a difference of two terms.

10. The average cost for **Algorithm** 9.2.1 generating all k-sequences on $\{1..n\}$ in Lexicographic-Order.

For each k-sequence T, let $c(T)$ be the number of assignments done to produce T. Then $1 <= c(T) <= k$, and for $a = 1,2,\ldots k$, let $f(a) =$ the number of sequences T with $c(T) = a$.

Then $f(1) = n^{k-1}(n - 1)$,

$\quad f(2) = n^{k-2}(n - 1)$,

$\quad \ldots,$

$\quad f(k - 1) = n^{k-(k-1)}(n - 1) = n(n - 1)$,

and $\quad f(k) = 1(\text{for the Lex-First}) + (n - 1) = n$.

Prove that $f(1) + f(2) + \ldots + f(k) = n^k$ // the # of k-sequences on $\{1..n\}$

$$// \sum_{a=1}^{k} f(a) = \sum_{a=1}^{k-1} n^{k-a}(n - 1) + n = \sum_{a=1}^{k-1} (n^{k-a+1} - n^{k-a}) + n.$$

// And the series "telescopes".

The total cost for all the sequences,

$$\text{TC} = \sum_{a=1}^{k} a \times f(a) = \sum_{a=1}^{k-1} a \times n^{k-a}(n - 1) \qquad\qquad\qquad + k \times n$$

$$= (n - 1) \times 1 \times n^{k-1} + (n - 1) \times 2 \times n^{k-2} + (n - 1) \times 3 \times n^{k-3} +$$

$$\ldots + (n - 1) \times (k - 2) \times n^2 + (n - 1) \times (k - 1) \times n^1 \qquad\qquad + kn$$

$$= (n - 1) \times (k - 1) \times n^1 + (n - 1) \times (k - 2) \times n^2 + (n - 1) \times (k - 3) \times n^3 +$$

$$\ldots + (n - 1) \times (2) \times n^{k-2} + (n - 1) \times (1) \times n^{k-1} \qquad\qquad + kn.$$

// This series can be evaluated by "Triangulation":

//   The term "$(n-1) \times n^1$" could be added $(k-1)$ times to give $(n-1) \times (k-1) \times n^1$;

//   write them in the first column.

//   The term "$(n-1) \times n^2$" could be added $(k-2)$ times to give $(n-1) \times (k-2) \times n^2$;

//   write them in the second column. And so on.

//

// TC $= (n-1) \times n^1 + (n-1) \times n^2 + (n-1) \times n^3 + \ldots + (n-1) \times n^{k-2} + (n-1) \times n^{k-1}$

// $+ (n-1) \times n^1 + (n-1) \times n^2 + (n-1) \times n^3 + \ldots + (n-1) \times n^{k-2}$

// $+ \ldots$

// $+ (n-1) \times n^1 + (n-1) \times n^2 + (n-1) \times n^3$

// $+ (n-1) \times n^1 + (n-1) \times n^2$

// $+ (n-1) \times n^1$ $+ kn.$

//

// now adding the rows, which can be written as telescoping series

//

// TC $= n^k - n$

// $+ n^{k-1} - n$

// $+ \ldots$

// $+ n^4 - n$

// $+ n^3 - n$

// $+ n^2 - n$ $+ kn$

//

// $= n^k + n^{k-1} + n^{k-2} + \ldots + n^4 + n^3 + n^2 - (k-1)n$ $+ kn$

// $= n + n^2 + n^3 + \ldots + n^{k-2} + n^{k-1} + n^k.$

(a)   Prove that when $n > 1$, the average cost of a sequence is

$$\frac{n + n^2 + \ldots + n^k}{n^k} = \left(\frac{n}{n-1}\right)\left(\frac{n^k - 1}{n^k}\right).$$

(b)   When is this average < 2?

11. Explain why the following method generates all sequences in $\mathbb{S}$ in Lexicographic-Order.

   Stage 1.   Generate the Lex-First sequence in $\mathbb{S}$
                   by taking smallest possible values for the entries x_1, x_2, x_3, and so on
                   until we have a sequence in $\mathbb{S}$.

   Stage 2.   Having generated sequence $S = (x_1, x_2, x_3, \ldots x_k)$, find the Lex-Next
                   sequence T:

   (a)   By *extending* S by finding smallest possible values
         for the entries x_{k+1}, x_{k+2}, and so on
         until we have a sequence in $\mathbb{S}$;

   or when S cannot be extended,

   (b)   By finding the largest index j where x_j can be increased, and then
         increasing x_j by the smallest possible amount to $x'_j = x_j + q$,
         and then (if necessary) *extending* $(x_1, x_2, x_3, \ldots x'_j)$ by finding
         smallest possible values for the entries x_{j+1}, x_{j+2}, and so on until
         we have a sequence in $\mathbb{S}$.

   But stop when we reach a sequence S that cannot be extended and
   where no entry can be increased.

12. **Hamiltonian Graphs** and the **TSP**

Suppose **G** is an undirected graph with vertex set $V = \{x_1, x_2, \ldots, x_n\}$.
Define a transition-cost matrix for a Traveling Salesman's Problem by

$$C[i,j] = \begin{cases} 1 & \text{If } x_i \text{ is adjacent to } x_j \\ 2 & \text{Otherwise} \end{cases}.$$

(a) Prove that **G** has a Hamilton Circuit if there is a traveling salesman's tour with total cost $\le n$.

(b) Add a new vertex x_0 to **G** and join x_0 to all the other vertices in **G** and define a new transition-cost matrix **C*** so that **G** has a Hamilton Path if there is a traveling salesman's tour with total cost $\le n + 1$.

(c) Are there corresponding theorems for directed graphs?

13. Algorithm 9.3.2 generates all k-subsets of $\{1..n\}$.

(a) What would be a "minimum change" between two k-subsets?

   // At least 1 new element must be added and the same number removed.

(b) Is Lexicographic-Order a minimum-change list? (A Gray code?)

(c) Is there such a list when $k = 2$ and $n = 4$?

14. Suppose that R and k are certain non-negative integers. Prove that

$$(k+1) \times \binom{R+0}{0} + k \times \binom{R+1}{1} + (k-1) \times \binom{R+2}{2} + \ldots + 2 \times \binom{R+k-1}{k-1} + 1 \times \binom{R+k}{k}$$
$$= \binom{R+k+2}{k}.$$

Hint. Consider using the following triangular pattern and show that

$$\text{LHS} = \binom{R+0}{0} + \binom{R+0}{0} + \binom{R+0}{0} + \ldots + \binom{R+0}{0} + \binom{R+0}{0} + \binom{R+0}{0}$$

// $(k+1)$ times

$$+ \binom{R+1}{1} + \binom{R+1}{1} + \binom{R+1}{1} + \ldots + \binom{R+1}{1} + \binom{R+1}{1}$$

// k times

$$+ \binom{R+2}{2} + \binom{R+2}{2} + \binom{R+2}{2} + \ldots + \binom{R+2}{2}$$

// $(k-1)$ times

$$+ \ldots$$
$$+ \ldots$$
$$+ \ldots$$

$$+ \binom{R+k-1}{k-1} + \binom{R+k-1}{k-1}$$

// 2 times

$$+ \binom{R+k}{k}.$$

// 1 time

Then use Lemma 9.3.2 to add each of the $(k + 1)$ columns in the pattern. Then use Lemma 9.3.2 again to add the $(k + 1)$ column-sums.

15. Algorithm 9.4.1 generates all full permutations of $\{1..n\}$.
    (a) What would be a "minimum change" between two full permutations?
                // At least 2 entries must change, two values could be interchanged.
                        // Perhaps two consecutive values could be interchanged.
    (b) Is Lexicographic-Order a minimum-change list? (A Gray code?)
    (c) Is there such a list when $n = 3$?

16. **Ringing the Changes** is a type of bell ringing that is peculiarly English, producing a music all its own, developed in the seventeenth century. It consists of producing all the note sequences of a set of n bells tuned to the notes of the major scale. The "peal" corresponds to a minimum-change list of all permutations of $\{1..n\}$ where moving from one permutation to the next interchanges two adjacent values.

// For 4 bells, this takes about a minute; for 6 bells, this takes ~ 30 minutes.
// For 7 bells, this takes > 3 hours; for 8 bells, this takes a day plus ~ 4 hours.
// For 12 bells, it is estimated this would take ~ 38 years.

A minimum-change list of all permutations of $\{1..2\}$ is

12   $\rightarrow$ 21
             $\downarrow$ // Now if 2 moves up we return to 12

A minimum-change list of all permutations of $\{1..3\}$ is

123   $\rightarrow$ 132 $\rightarrow$ 312   // **3** moves "down" the sequence.
                           $\downarrow$ // now make **2** move down its sequence.
213   $\leftarrow$ 231 $\leftarrow$ 321   // then make **3** move "up" the sequence.
$\downarrow$ // Now if 2 moves up, we return to 123.

A minimum-change list of all permutations of $\{1..4\}$ is

1234 $\rightarrow$ 1243 $\rightarrow$ 1423 $\rightarrow$ 4123 // **4** moves "down" the sequence.
                                      $\downarrow$ // Now make the change in the peal for $n = 3$.
1324 $\leftarrow$ 1342 $\leftarrow$ 1432 $\leftarrow$ 4132 // Then make **4** move "up" its sequence.
$\downarrow$ // Now make the change in the peal for $n = 3$.
3124 $\rightarrow$ 3142 $\rightarrow$ 3412 $\rightarrow$ 4312 // Then make **4** move "down" its sequence.
                                      $\downarrow$ // Now make the change in the peal for $n = 3$.
3214 $\leftarrow$ 3241 $\leftarrow$ 3421 $\leftarrow$ 4321 // Then make **4** move "up" its sequence.
$\downarrow$ // Now make the change in the peal for $n = 3$.
2314 $\rightarrow$ 2341 $\rightarrow$ 2431 $\rightarrow$ 4231 // Then make **4** move "down" its sequence.
                                      $\downarrow$ // Now make the change in the peal for $n = 3$.
2134 $\leftarrow$ 2143 $\leftarrow$ 2413 $\leftarrow$ 4213 // Then make **4** move "up" its sequence.
$\downarrow$ // Now make the change in the peal for $n = 3$.
                                                   // And if 2 moves up, we return to 1234.

Show that this list and its reverse are not the only minimum-change lists for
4 bells.

Is there a minimum-change list of all permutations of $\{1..n\}$ for any n?

Is there an inductive proof of this?

Could you write a program to give a minimum-change list of all permutations
of $\{1..n\}$?

Would you use recursion?

17. The ***Traveling Salesman's Problem.***

Suppose that transition-cost matrix C is given by the following array:

	0	1	2	3	4	5
0	*	12	13	9	8	7
1	12	*	5	16	31	8
2	13	30	*	7	32	33
3	9	20	24	*	21	18
4	8	6	20	22	*	15
5	7	18	9	17	19	*

The tour given by the Lex-First permutation of $\{1..5\}$, $S = (1, 2, 3, 4, 5)$ has a
total cost of 67 units. It's very unlikely that this is the best tour.

(a)   The "nearest-neighbor" tour is constructed by going to the nearest neighbor
from $t[0]$, then the nearest-neighbor to that town (not already visited), and
so on to the nearest neighbor (not already visited) until a complete tour has
been constructed. Find the nearest-neighbor tour and show its total cost is
82 units.

// This tour starts well but ends up taking some very costly transitions.

(b)   The "Greedy" Tour is constructed by taking the cheapest transition we can
take until a complete tour has been constructed.

// The best transition is from $t[1]$ to $t[2]$ with a cost of 5 units, (the
// smallest entry in C). Construct a tour that utilizes this smallest possible
// cost; that is,
//   after $t[1]$ go to $t[2]$ (and before $t[2]$ go to $t[1]$).
// The next smallest transition we can use is from $t[4]$ to $t[1]$ with a cost
// of 6 units. Construct a tour that <u>also</u> utilizes this second smallest
// possible cost; that is,
//   after $t[4]$ go to $t[1]$ (and before $t[1]$ go to $t[4]$).

(c)   Find the Greedy Tour and show its total cost is 53 units.   // Is this T^*?

(d)   Determine the best possible tour T^*.

18. Run an implementation or walk through a small case of the execution of

Discrete Probability and Average-Case Complexity

10

Suppose you flip a coin until you get two heads in a row. That might happen on the first two flips, but it might take 17 flips. In fact, there's no limit to the number of flips it might take. If everyone in the class each flipped a coin until they got two heads in a row and counted how many flips were done, the best case might be 2 and the worst case might be a fairly large number.

// Could there be someone who takes forever?
// Could a "reasonable" prediction be made of how many flips are needed *on*
// *average*? Would you believe 6 flips?

Probability is an *idea* (an intellectual construct) used to do just that – make reasonable assertions about how a process behaves *on average*. In many algorithms, like searching and sorting, the number of steps depends on the individual input instance itself (and not just on the size of the input). For such algorithms, we would like a way to determine the *average-case complexity*. In particular, we want to show that on average, QuickSort makes $O(n\lg(n))$ key comparisons when sorting a list of length n. In a worst case, QuickSort makes $n(n - 1)/2$ key comparisons, which is $\Theta(n^2)$.

10.1 Probabilistic Models

Consider flipping a coin into the air and letting it fall and then observing the top face — is it "heads" or "tails"? In engineering or classical physics, you imagine that if you had enough information about exactly how the coin was sitting on your finger, how much force was applied by your thumb, and how and where on the coin the force was applied, you could predict precisely the path of the coin, its spin as it flew, its landing point, and exactly how it would come to rest. Everyone imagines that if the coin were flipped again, in exactly the same way it would travel along the same path as before and land with the same side up as before. If you built a machine

© Springer International Publishing AG, part of Springer Nature 2018   421
T. Jenkyns and B. Stephenson, *Fundamentals of Discrete Math for Computer Science: A Problem-Solving Primer*, Undergraduate Topics in Computer Science, https://doi.org/10.1007/978-3-319-70151-6_10

to flip a coin, exactly the same way every time, the outcome would be exactly the same every time. This describes a "deterministic" model where the past (application of a certain force to a certain point on the coin) *determines* the future, "heads" or "tails". With enough data, you can predict the outcome.

On the other hand, when I flip a coin, I have no idea what the outcome will be. Except that it will be heads or tails – never both and never neither. Sometimes heads, sometimes tails, but no obvious pattern at all. So on any particular flip, it's just as likely to be heads as tails, and after a long sequence of flips, I would expect about the same number of heads as tails. A ***probabilistic model*** is a description of a part of reality that expresses this sort of uncertainty.

// and is almost useless for predicting individual outcomes
The rest of this section introduces the terms we use for probabilistic models and gives a few examples.

10.1.1   Sample Spaces

An ***experiment*** is a process that produces an "outcome"; a ***sample space*** for an experiment is a set of outcomes such that every time the experiment is repeated, exactly one outcome in the sample space is produced.

Example 10.1.1: Experiments and Sample Spaces
1. A sample space for the experiment: "Pick a number between 1 and 10" might be the set $\{1, 2, 3, \ldots, 10\}$.   // It could also be all of **P**.
2. A sample space for the experiment: "Flip a coin" might be the set $\{H, T\}$, where H denotes the outcome "the coin came to rest with head-side up" and T denotes the outcome "the coin came to rest with tail-side up."
3. A sample space for the experiment: "Register in MATH 140" might be the set $\{A, B, C, D, F\}$ where these letters are the final grade you might obtain as an outcome.
4. A sample space for the experiment: "Flip a coin until you get 2 heads in a row" might be the set of all sequences on $\{H, T\}$ that end with HH where there is no other occurrence of two consecutive Hs.

10.1.2   Probability Functions

The *main* element of a probabilistic model is a function P defined on the sample space, S. If O_j is some outcome, we would like $P(O_j)$ to be

> The proportion (or fraction) of outcomes equal to O_j that
> a "reasonable person" would expect when
> the experiment is repeated a large number of times.

Then, we would have $\qquad$ $0 <= P(O_j) <= 1.$

Let's interpret $\qquad$ "$0 = P(O_j)$"   to mean O_j ***never*** occurs

and $\qquad$ "$P(O_j) = 1$"   to mean O_j ***always*** occurs.

We would also have
$$\sum_{O_j \in S} P(O_j) = 1.$$

The sum of all those proportions accounts for all the outcomes obtained, so it equals 1.

However, in math courses (this one included), we can study the models themselves without worrying much about their connection to the real world (unlike engineering or physics). We can discuss probability in an abstract form (without any reference to a "reasonable person").

A ***probability function*** is a real-valued function P defined on some nonempty, discrete set S satisfying the two probability axioms:

I. $0 <= P(O_j)$ $\qquad\qquad$ for all outcomes O_j in the sample space S

II. $\displaystyle\sum_{O_j \in S} P(O_j) = 1.$

Sets of possible outcomes are called ***events***, and we calculate the "probability of an event" as follows: For $A \subseteq S$,

$$prob(A) = \sum_{O_j \in A} P(O_j). \qquad\qquad\qquad \text{// So } prob(S) = 1.$$

Example 10.1.2: Sample Spaces and Probability Functions

1. For the experiment: "Pick a number between 1 and 10" with $S = \{1, 2, \ldots, 10\}$.
   We could let $P(j) = 1/10$ for each $j \in S$. $\qquad$ // Axioms I and II then hold.

   // But are people less likely to pick 1 or 10 than 3 or 4 or 5 or 7?
   // Are all ten possibilities equally likely to be chosen?

2. For the experiment: "Flip a coin" with $S = \{H, T\}$.
   We could let $P(H) = ½$ and $P(T) = ½$. $\qquad$ // Axioms I and II then hold.

   // I think this is the model that best describes what happens when I flip a coin.

3. For the experiment: "Register in MATH 140" with $S = \{A, B, C, D, F\}$.
   Do you think that if you took this course, <u>you</u> have an equal chance of getting an A or a B or an F? Could we construct a "reasonable" model if we assume that the *proportions* of outcomes for this year will be similar to last year when

   $P(A) = 40\%$ $P(B) = 20\%$ $P(C) = 15\%$ $P(D) = 10\%$ $P(F) = 15\%$?

   // Do axioms I and II then hold?

4. For the experiment: "Flip a coin until you get 2 heads in a row" with S equal to the (*infinite*) set of all sequences on {H, T} that end with HH but have no other occurrence of two consecutive Hs. Here, it is more difficult to find any function that satisfies both axioms, let alone a "reasonable" or "realistic" one.

                                      // But we will near the end of this chapter.

// Do you think it's "easier" (and "more likely") to obtain the outcome-
// sequence TTHH than to get
// THTHTHTHTHTHTHTHTHTHTHTHTHTHTHTHTHTHTHTHTHHH
// (which alternates T and H twenty times and then there's another H)?

10.1.3   The Special Case of Equally Likely Outcomes

By far the simplest case is when S consists of a *finite number of equally likely* outcomes; that is,

$$S = \{O_1, O_2, \ldots, O_n\} \text{ and}$$

$$P(O_j) = 1/n \qquad \text{for all } O_j \in S. \qquad \text{// Axioms I and II hold.}$$

When all outcomes in S are *equally likely*, for any event $A \subseteq S$,

$$\boldsymbol{prob}(A) = \sum_{O_j \in A} P(O_j) = |A| \times (1/n) \qquad \text{where } n = |S|.$$

Here, the probability of event A is

$$\boldsymbol{prob}(A) = |A|/|S|,$$

the number of outcomes in set A divided by the total number of outcomes.

    To indicate that this probability function (and the assumption of *equally likely* outcomes) is being used, we often use the word "***random.***"

    This basic description of probability has been used since the beginning of probability theory (for the study of gambling games in the seventeenth century by Pascal and Fermat, and later by Gauss and Laplace). But this is the crudest model; it uses no information about the different outcomes and treats all outcomes the same, and it provides no information about predicting the next outcome. Surprisingly, it is very useful, and it is a very suitable model for many, many (real world) processes.

Example 10.1.3: The 2 DBs

    Suppose that beside your bed you keep 12 books on a shelf of which 2 are what your mother would call "dirty books" – DBs. Suppose also that she pays a surprise

visit to your room and you worry that if the 2 DBs are side by side, she will notice
them and be embarrassed. **What is the probability that they are side by side?**

// To answer that question, we need a probabilistic model:
// an *experiment* and a *sample space* S together with
// a *probability function* that reflects the implicit idea in the question that
// the 2 DBs may be anywhere among the other ten books.
//
// The case we're interested in is "the 2 DBs are side by side on the shelf"; this
// will be *event A*. But A must be expressed as a subset of outcomes of the
// experiment.
//
// We'll give you three solutions.

Solution #1

    Experiment: The 12 books were placed in ***random*** order on your shelf.
    Sample space: All possible orderings of the 12 books
    Event A: The orderings with the 2 DBs side by side

$|S| = 12!$, the number of (full) permutations of 12 objects.   // $= 479{,}001{,}600$
To count the number of orderings where the 2 DBs are side by side, suppose we
place them side by side, tape them together, and then arrange the 11 objects on
the shelf. So   // by the product rule

$|A| =$ (# ways the 2 DBs may be placed side by side and taped together)
        $\times$ (# ways 11 objects can be ordered) $= 2! \times 11!$. // $= 2 \times 39{,}916{,}800$

Thus, $prob(A) = \dfrac{|A|}{|S|} = \dfrac{2 \times 11!}{12 \times 11!} = \dfrac{1}{6}.$

In Solution #1, the arrangement of the 10 "other" books was irrelevant. Perhaps
we can describe a new experiment which disregards the order of the books.

Solution #2

    Experiment: 2 places on the shelf are chosen ***at random*** (for the DBs).
    Sample space: All possible selections of 2 places out of 12
    Event A: The 2 places are side by side

$|S| = \dbinom{12}{2} = 66$, the number of 2-subsets in a 12-set.

If the two selected places are side by side, we know they are places 1 and 2,
or 2 and 3, or 3 and 4, or … 11 and 12. Therefore
$|A| = 11.$ // We counted the possibilities without a formula.

Thus, $prob(A) = \dfrac{11}{66} = \dfrac{1}{6}.$ // again!

Finally, let's construct a solution which reflects what would happen if the last
book you put on the shelf happened to be one of the DBs.

Solution #3

Experiment: The last book was a DB and was placed *at random* on the shelf already containing the other 11 books.

Sample space: All possible positions for the 12th book

Event A: The 2 positions beside the DB already on the shelf

$|S| = 12$,   because the last book may be put to the left of the first book already on the shelf, to the left of the second, ..., to the left of the 11th, or to the right of the 11th book already on the shelf.

$|A| = 2$.   // There are 2 positions beside the other DB.

Thus, $prob(A) = \dfrac{2}{12} = \dfrac{1}{6}$.   // again!

When the outcomes are equally likely, probability questions become counting questions. Sometimes we count permutations, sometimes we count combinations, and sometimes we simply count possibilities. We count whatever the experiment produces and use whichever formula or method counts what we want to count.

The Most Important Ideas in This Section.

A *probabilistic model* consists of three things: an *experiment* which is a process that produces an "outcome"; a *sample space* which is a set S of outcomes such that every time the experiment is repeated, exactly one outcome in S is produced; and a *probability function* which is a real-valued function P defined on S satisfying the two probability axioms:

I. $0 <= P(O_j)$ for all outcomes O_j in the sample space S.

II. $\displaystyle\sum_{O_j \in S} P(O_j) = 1$.

Sets of possible outcomes are called *events*, and the probability of an event A is

$$prob(A) = \sum_{O_j \in A} P(O_j).$$

When S consists of a *finite number of equally likely* outcomes, for any event $A \subseteq S$,

$$prob(A) = |A|/|S|,$$

the number of outcomes in set A divided by the total number of outcomes. When this probability function (and the assumption of *equally likely* outcomes) is being used, we will use the word *random*.

(continued)

(continued)

> Probability values can <u>only</u> be calculated in the context of some model, that is, some assumed experiment, some assumed sample space S, and especially, some assumed probability function on S.

10.2 Conditional Probability

Suppose that we have a friend, Mr. X, who's not politically correct and is in fact a male chauvinist pig. He classifies the women he meets according to two main criteria: Are they coeds or townies, and are they innocent or otherwise? (By "innocent," he might mean "politically naïve.") Below is a "contingency table" showing the number of women he knows in the four possible classifications.

	Coeds	Townies
Innocent	10	30
Otherwise	140	20

Suppose also that you meet one of Mr. X's friends at a party. **What is the probability that she is a coed? She is innocent? She is an innocent coed?**

To answer those questions, you must construct a *probabilistic model* that suits this situation:

Experiment: You meet one of the women Mr. X has met and classified.

Sample space: All the women Mr. X has met and classified

Probability function: We will assume that the woman you meet has been selected at *random* from the sample space, that is, each possibility is equally likely.

Event A: She is a coed.

Event B: She is innocent.

Event C: She is an innocent coed.

For each of these events, we need to know the size of the corresponding subset of the sample space. Let's add the "marginal totals" to the contingency table — the totals of the rows, totals of the columns, and the "grand" total.

	Coeds	Townies	
Innocent	10	30	40
Otherwise	140	20	160
	150	50	200

Now we can see that

$$prob(\text{she's a coed}) \quad\quad = \frac{150}{200} = \frac{3}{4} = 0.75 = 75\%,$$

$$prob(\text{she's innocent}) \quad\quad = \frac{40}{200} = \frac{1}{5} = 0.20 = 20\%,$$

$$prob(\text{she's an innocent coed}) = \frac{10}{200} = \frac{1}{20} = 0.05 = 5\%.$$

10.2.1   Combinations of Events

Often, events are taken to be the sentences that describe subsets of the sample space. Here, event C could be given as "A *and* B" − she is a coed *and* she is innocent. In general, for any pair of events A and B,

$$prob(A \text{ and } B) = \sum_{x \in A \cap B} P(x)$$

and
$$prob(A \text{ or } B) = \sum_{x \in A \cup B} P(x).$$

When $x \in A \cap B$, the value of $P(x)$ is added into $prob(A)$ and into $prob(B)$, so

$$prob(A \text{ or } B) = prob(A) + prob(B) - prob(A \text{ and } B). \quad\quad (10.2.1)$$

Returning to the example of Mr. X's friends,

$$prob(\text{she is a coed } or \text{ she is innocent})$$
$$= prob(A \textbf{ or } B)$$
$$= prob(A) + prob(B) - prob(A \textbf{ and } B).$$
$$= \frac{150}{200} + \frac{40}{200} - \frac{10}{200} \quad\quad = \frac{180}{200}$$
$$= 0.90 = 90\%.$$

// All the women except the 20 townies who are otherwise are counted.

In general, for any pair of events,

$$prob(A \text{ and } B) <= prob(A) <= prob(A \text{ or } B) <= prob(A) + prob(B).$$

10.2.2   Conditional Probability

If in talking to the friend of Mr. X that you met, you find out that she's a coed, does that provide some information as to whether or not she is innocent? Can we calculate the probability that she is innocent **given** the extra information that she is a coed? We want what's called the **conditional probability of B given A**, which is written

$$prob(B \mid A).$$

// The vertical stroke in this context is read "given" – **not** "divides evenly into".
// We want the probability that event B occurs given that event A has occurred.
// So $prob(A)$ must be > 0.

Looking back at the contingency table, if we know she's a coed, we know she's one of the 150 coeds Mr. X knows, so our probabilistic model can be modified:

Experiment: You meet one of the coeds Mr. X has met and classified, selected at *random*.

Sample space*: The 150 coeds Mr. X has met and classified

Event B*:   She is innocent: // She is an innocent coed?

$$prob(B^*) = \frac{|B^*|}{|S^*|} = \frac{10}{150} = \frac{1}{15} = 0.0\bar{6} = 6^{2/3}\%.$$

But B^* is the event A *and* B in the old sample space and S^* is the event A in the old sample space, so in terms of the original model

$$prob(B^*) = \frac{10/200}{150/200} = \frac{prob(A \; and \; B)}{prob(A)} = 0.0\bar{6} = 6^{2/3}\%.$$

// We asked earlier "if you find out that she's a coed, does that provide some
// information as to whether or not she is innocent?"
// We now know she's *less likely* to be innocent:
// $prob$(she's innocent) $= 20\%$,
// $prob$(she's innocent | she's a coed) $= 6^{2/3}\%$.

   In general, the **conditional probability of B given A** is only defined when $prob(A) > 0$ and is given by

$$prob(B \mid A) = \frac{prob(A \; and \; B)}{prob(A)}. \qquad (10.2.2)$$

There are many practical and theoretical applications of conditional probabilities; we will use conditional probabilities to determine average-case complexity of algorithms.

   For all $x \in A$, we define $P(x \mid A)$ to be $prob(\{x\} \mid A)$. Then

$$P(x \mid A) = \frac{prob(A \; and \; \{x\})}{prob(A)} = \frac{P(x)}{prob(A)},$$

and $P(x \mid A)$ is a probability function on A.

$$//X \; P(x \mid A) >= 0 \text{ and } \sum_{x \in A} P(x \mid A) = 1.$$

The **specificity** of the test is the probability that it produces a negative result when the disease is not present. This should be close to 1 showing that positive results are "specific" to (and characterized by) the presence of disease D. This test's *specificity* is

$$prob(- \mid \sim D) = 95\%.$$

// The test sometimes produces a positive result in people that don't have D; such
// results are called **false positives**. For this test, the probability of a *false positive*
// result
// $prob(+ \mid \sim D) = 1 - prob(- \mid \sim D) = 5\%.$

Now suppose **you** take the diagnostic test for D and get a positive result – what is the probability that (it is a true positive and) you do have the disease? **What is the value of *prob(D $\mid$ +)*?**

Solution #1
Let's imagine the experiment to be as follows: A person is selected at *random* from some population S, and the diagnostic test is performed on that person, producing a positive result or a negative result (i.e., this test is never "inconclusive").

"*D*" represents the subset of the population that has the disease;
"$\sim D$" represents the subset of the population that does not have the disease;
"$+$" represents the subset of the population that tests positive; and
"$-$" represents the subset of the population that tests negative.

// By definition, $prob(D \mid +) = prob(+ \ and \ D)/prob(+).$ // See 10.2.2.
// So we need to evaluate both $prob(+ \ and \ D)$ and $prob(+)$:

$$prob(+ \ and \ D) = prob(+ \mid D) \times prob(D) \qquad \text{// by 10.2.3}$$
$$= 0.98 \times 0.005 = 0.0049,$$

and $prob(+ \ and \sim D) = prob(+ \mid \sim D) \times prob(\sim D)$ // by 10.2.3
$$= [1 - prob(- \mid \sim D)] \times [1 - prob(D)] \qquad \text{// by 10.2.5}$$
$$= [1 - 0.95] \times [1 - 0.005]$$
$$= 0.05 \times 0.995 = 0.04975.$$

Since events "$+$ and D" and "$+$ and $\sim D$" partition the event "$+$", by Theorem 10.2.1,

$$prob(+) = prob(+ \ and \ D) + prob(+ \ and \sim D)$$
$$= 0.0049 + 0.04975 = 0.05465.$$

Thus, $prob(D|+) = prob(+ \ and \ D)/prob(+)$
$$= 0.0049/0.05465 = 0.089661482 \ldots < 9\%.$$

// Does that make sense? That when you get a positive result, the chances of
// actually having the disease is less than 10%.

Solution #2

Let's look at this again by constructing a contingency table. And to make things simpler still, let's assume the population S has 100,000 people.

	D	~D	
+			
−			
			100000

Since D is a disease that occurs in ½ of 1% of the population, $prob(D)=0.005$ and in the population of 100,000 people, we would "expect" 500 have the disease.

// and 99,500 do not

	D	~D	
+			
−			
	500	99500	100000

Of the 500 who have the disease, 98% will test positive, and 2% will test negative.
// 98% of 500 $= 0.98 \times 500 = 490$, and 2% of 500 $= 10$.

	D	~D	
+	490		
−	10		
	500	99500	100000

Of the 99,500 who do not have the disease, 95% will test negative, and 5% positive.
// 95% of 99,500 $= 0.95 \times 99,500 = 94,525$, and 5% of 99,500 $= 4,975$.

	D	~D	
+	490	4975	
−	10	94525	
	500	99500	100000

Adding the rows gives

	D	$\sim D$	
+	490	4975	5465
−	10	94525	94535
	500	99500	100000

Of the 5,465 people who got a positive result, most of them (4,975) got a *false positive*, and only 490 (got a *true positive* because they really) had the disease. Therefore,

$$prob(D \mid +) = \frac{490}{5465}$$
$$= 0.0049/0.05465 = 0.089\ 661\ 482.\ldots \qquad \text{// again!}$$

// Because the incidence rate of this disease (and most diseases) is so small,
// a large percentage of those with the disease is much smaller
// than a small percentage of those without the disease, so in this case most of the
// positive test results were false positives.

The Most Important Ideas in This Section.
Events may be taken to be the sentences that describe subsets of the sample space. In general, for any pair of events A and B,

$$prob(A \text{ and } B) \;\; = \sum_{x \in A \cap B} P(x)$$
$$\text{and} \quad prob(A \text{ or } B) \;\; = \sum_{x \in A \cup B} P(x).$$

For any pair of events, A and B,

$$prob(A \text{ or } B) = prob(A) + prob(B) - prob(A \text{ and } B)$$
$$\text{and} \quad prob(A \text{ and } B) <= prob(A) <= prob(A \text{ or } B) <= prob(A) + prob(B).$$

The *conditional probability of B given A* is only defined when $prob(A) > 0$ and is given by

$$prob(B \mid A) = \frac{prob(A \text{ and } B)}{prob(A)}.$$

Events A and B are **independent** <u>means</u> $prob(A \text{ and } B) = prob(A) \times prob(B)$.
Events A and B are **mutually exclusive** <u>means</u> $prob(A \text{ and } B) = 0$.

(continued)

(continued)

> So A and B are mutually exclusive if and only if
>
> $$prob(A \text{ or } B) = prob(A) + prob(B).$$
>
> A *set* of events $\{A_1, A_2, A_3, \ldots, A_k\}$ is said to be *mutually exclusive* if no two events in the set can occur together. If $\{A_1, A_2, A_3, \ldots, A_k\}$ is a partition of some subset B of the sample space, then $prob(B) = \sum_{j=1}^{k} prob(A_j)$.

10.3 Random Variables and Expected Values

We want to calculate average outcomes when we count the number of operations algorithms perform. To do this, we need outcomes that are numbers. As a first step in that direction, we define a **random variable** to be a function from the sample space of some experiment into the real numbers.   // Some $X: S \to \mathbf{R}$.

Example 10.3.1: Rolling 2 Dice

Consider the *experiment* of rolling two ordinary dice, a *red* die and a *green* die.
                                                                        // A die is one dice.
There are 36 *outcomes* from this experiment.
Let X denote the sum of the top two faces. Then, X takes values in the set $\{2, 3, \ldots, 12\}$.

// Are those <u>values</u> equally likely? Do they each occur 1/11th of the time?

We can represent the sample space as a 6×6 table and the X-values as entries:

		result on the green die				
	1	**2**	**3**	**4**	**5**	**6**
1	2	3	4	5	6	7
2	3	4	5	6	7	8
3	4	5	6	7	8	9
4	5	6	7	8	9	10
5	6	7	8	9	10	11
6	7	8	9	10	11	12

result on the red die

The sample space for this experiment is all ordered pairs, (r, g), where both r and g are in $\{1..6\}$. With ordinary, unbiased dice all 36 outcomes are equally likely. The event "$X=4$" corresponds to 3 of these outcomes: $r = 1$ and $g = 3$, $r = 2$ and $g = 2$, and $r = 3$ and $g = 1$. Thus,

$$prob(X = 4) = 3/36 = 1/12.$$

We can tabulate the possible values v of X and the probability that $X = v$.

v	$prob(X = v)$
2	1/36
3	2/36
4	3/36
5	4/36
6	5/36
7	6/36
8	5/36
9	4/36
10	3/36
11	2/36
12	1/36
	36/36

// How can we estimate the average X-value when 2 fair dice are rolled?

10.3.1 Expected Frequency

Imagine rolling these dice 3,600 times. Because $prob(X = 4) = 1/12$, we should "expect" the event "$X = 4$" to occur one twelfth of the time. One twelfth of 3,600 times is 300 times.

In general, when an experiment is repeated N times, the **expected frequency** of event A is defined by

$$Ef(A) = N \times prob(A).$$

Continuing with the "imagined experiment" of rolling these dice 3,600 times, the average outcome would be equal to the sum of all the outcomes divided by 3,600. We expect 300 of the outcomes to be 4, and adding up all these 4s would contribute $4 \times 300 = 1,200$ to the "sum of all the outcomes." Let's extend the table with columns for Ef and $v \times Ef$ when $N = 3,600$.

v	$prob(X = v)$	$Ef(X = v)$	$v \times Ef$
2	1/36	100	200
3	2/36	200	600
4	3/36	300	1,200
5	4/36	400	2,000
6	5/36	500	3,000
7	6/36	600	4,200
8	5/36	500	4,000
9	4/36	400	3,600
10	3/36	300	3,000
11	2/36	200	2,200
12	1/36	100	1,200
	36/36	3,600	25,200
	$= 1$	$= N$	

Then, the average value of X in this idealized process of repeating the experiment 3,600 times would be

$$\overline{X} = \frac{25,200}{3,600} = 7.$$

In an "imagined experiment" of rolling these dice N times, the "average outcome" would be given by

$$\frac{\displaystyle\sum_{v=2}^{12} v \times N \times prob(X = v)}{N} = \sum_{v=2}^{12} v \times prob(X = v),$$

which is independent of N. This *theoretical average* value will be called the "expected value of X" and is defined formally next.

10.3.2 Expected Values

Suppose that X is a random variable defined on the sample space S of some probabilistic model with its experiment and probability function. Then for each outcome O_j, the probability $P(O_j)$ may be found. The *expected value* of X is defined by

$$E(X) = \sum_{O_j \in S} X(O_j) \times P(O_j).$$

// $E(X)$ equals the sum over all possible outcomes of the value of X for that
//   outcome times the probability of that outcome.
//X If for each j, $X(O_j) = C$, then $E(X) = C$.
//X If for each j, $X(O_j) <= Y(O_j)$, then $E(X) <= E(Y)$.

//   Where are we going with all this probability theory?

For an algorithm A: O_j will be some possible, individual input instance, $X(O_j)$ will be the number of steps A takes when O_j is the input, and $E(X)$ will be the average-case complexity of algorithm A.

Theorem 10.3.1: If X and Y are <u>any</u> two random variables defined on the same sample space S, then $E(X + Y) = E(X) + E(Y)$.

Proof.
$$E(X+Y) = \sum_{O_j \in S} (X+Y)(O_j) \times P(O_j)$$

$$= \sum_{O_j \in S} [X(O_j) + Y(O_j)] \times P(O_j)$$

$$= \sum_{O_j \in S} X(O_j) \times P(O_j) + \sum_{O_j \in S} Y(O_j) \times P(O_j)$$

$$= E(X) + E(Y).$$ □

When function X maps S into $\mathbf{R}$, we say that X "takes values" in the set

$$V = \{X(O_j): O_j \in S\}. \qquad\qquad // \text{ the "range" of the function}$$

Furthermore, the function $X{:}S \to \mathbf{R}$ partitions the sample space S into sets A_v where for each $v \in V$,

$$A_v = \{O_j \in S : X(O_j) = v\}.$$

The event "$X = v$" corresponds precisely to the set A_v; hence, we may define an associated function also denoted by P but having the set V as its domain:

$$P(X = v) = prob(A_v). \qquad\qquad // = \sum_{O_j \in A_v} P(O_j)$$

These functions can be studied without any reference to an experiment in an abstract form (as in the next subsection), and the expected value of X can be expressed in terms of this new function:

$$E(X) = \sum_{O_j \in S} X(O_j) \times P(O_j)$$

$$= \sum_{v \in V} \sum_{O_j \in A_v} X(O_j) \times P(O_j)$$

$$= \sum_{v \in V} \sum_{O_j \in A_v} v \times P(O_j)$$

$$= \sum_{v \in V} \left\{ v \times \sum_{O_j \in A_v} P(O_j) \right\}.$$

So
$$E(X) = \sum_{v \in V} v \times P(X = v).$$

10.3.3   Probability Distributions

Suppose that X is a random variable taking values in some discrete set $V \subset \mathbf{R}$. A **probability distribution** for X is a function $P{:}V \to \mathbf{R}$ such that

I. $P(X = v) >= 0$ for all $v \in V$;   and

II. $\sum_{v \in V} P(X = v) = 1.$

Furthermore, the **expected value** of X is defined by

$$E(X) = \sum_{v \in V} v \times P(X = v).$$

The Most Important Ideas in This Section.
We want to calculate the "expected" number of operations algorithms perform. To do this, we need outcomes that are numbers. A *random variable* is a function from the sample space of some experiment into the real numbers.

When an experiment is repeated N times, the *expected frequency* of event A is $Ef(A) = N \times prob(A)$. The *expected value* of a random variable X is defined by

$$E(X) = \sum_{O_j \in S} X(O_j) \times P(O_j).$$

For an algorithm A: O_j will be some possible, individual input instance, $X(O_j)$ will be the number of steps A takes when O_j is the input, and $E(X)$ will be the *average-case complexity* of algorithm A.

If X and Y are any two random variables defined on the same sample space S, then $E(X + Y) = E(X) + E(Y)$.

A function $X{:}S \rightarrow \mathbf{R}$ partitions the sample space S into sets A_v, where for each $v \in V, A_v = \{O_j \in S{:} X(O_j) = v\}$. The event "$X = v$" corresponds to the set A_v; hence, $P(X = v) = prob(A_v)$.

Suppose X is a random variable taking values in some discrete set $V \subset \mathbf{R}$. A *probability distribution* for X is a function $P{:}V \rightarrow \mathbf{R}$ such that

I. $P(X = v) >= 0$ for all $v \in V$;   and

II. $\displaystyle\sum_{v \in V} P(X = v) = 1.$

The *expected value* of X is $E(X) = \sum_{v \in V} v \times P(X = v)$.

Expected values can only be calculated in the context of some probability model, that is, some defined random variable X, with some set of possible values V for X, and especially some assumed probability distribution on V.

10.4   Standard Distributions and Their Expected Values

In this section, we present a few standard distributions with an underlying experiment and random variable.

10.4.1   The Uniform Distribution

	// like rolling a "fair" n-sided die
$V = \{v_1, v_2, v_3, \ldots, v_n\},$	// any nonempty, finite set
$P(X = v) = 1/n$ for all $v \in V,$	// all values of X are equally likely

so $E(X) = v_1(1/n) + v_2(1/n) + \ldots + v_n(1/n)$

$$= \frac{v_1 + v_2 + \ldots + v_n}{n} \qquad\qquad \text{// the ordinary "average" value}$$

Example 10.4.1: Roll a Fair Six-Sided Die

If X denotes the number on the top face of the die when it comes to rest, then

$$V = \{1, 2, 3, 4, 5, 6\}, \qquad\qquad\qquad \text{// Here, does } V = S?$$

$$P(X = v) = 1/6 \text{ for all } v \in V, \qquad \text{// All values of } X \text{ are equally likely.}$$

so $E(X) = \dfrac{1 + 2 + 3 + 4 + 5 + 6}{6} = \dfrac{21}{6} = 3\frac{1}{2}.$

// The average roll is not a possible value. This usually happens, and expected
// values are <u>not</u> rounded to a possible value – better information is conveyed by
// writing fractions as rationals or decimal numbers with several significant
// figures.
//
// Note also that if you roll a red die and a green die, by Theorem 10.3.1,
//
// $E\big(X_{red} + X_{green}\big) = E\big(X_{red}\big) + E\big(X_{green}\big) = 3\frac{1}{2} + 3\frac{1}{2} = 7.$ // as we saw before

A **binomial experiment** consists of: // five things?
1. **n** identical "trials" where   // like flipping a coin n times
2. Each trial results in one of two outcomes, a "success" or a "failure";
3. The probability of success on any trial is a fixed value, **p**;
                // So the probability of failure on any trial is a fixed value, $q = 1 - p$.
4. The trials are independent; and
                // trials with the properties 2, 3, and 4 are called "Bernoulli trials"
5. We're interested in the number of successes, X.
                // not the order of the successes <u>or</u> which trials are successful

Example 10.4.2: The Marksman

Suppose that a certain marksman fires 5 times in competition. His score is how
many "bull's eyes" are hit in those five shots. If in practice shooting, he hits the
bull's eye 80% of the time, **what will be his expected score in competition?**

// Is this a binomial experiment?

A *trial* is a single shot fired at a target, the **n** $= 5$ shots are (nearly) identical.
A *success* would be hitting the bull's eye, and missing it would be a *failure*.
If we assume that every shot is like his practice shots, we could say that the
probability of success on *any* trial is a fixed value, **p** $= 80\%$.

(If I were shooting and missed the entire target on my first shot, I'd be even more nervous on the later shots — my shots would not be independent. But let's assume that he is very used to the stress of competition, and each shot (trial) is independent of the others).

We are interested in his score, $X =$ the number of successes in the five trials, and not in which trials are successful.

// A binomial experiment is a fairly realistic model for this process.

What's the Probability that $X = 2$?

Even though we're not interested in which trials are successful, let's decompose the event "$X = 2$" by looking at all the cases where there are 2 hits and 3 misses. Let H_j denote "he hits on shot number j" and M_j denote "he misses on shot number j". Then

$$
\begin{aligned}
P(X = 2) = P[\quad &(H_1 \ \& \ H_2 \ \& \ M_3 \ \& \ M_4 \ \& \ M_5) \\
\text{or } &(H_1 \ \& \ M_2 \ \& \ H_3 \ \& \ M_4 \ \& \ M_5) \\
\text{or } &(H_1 \ \& \ M_2 \ \& \ M_3 \ \& \ H_4 \ \& \ M_5) \\
\text{or } &(H_1 \ \& \ M_2 \ \& \ M_3 \ \& \ M_4 \ \& \ H_5) \\
\text{or } &(M_1 \ \& \ H_2 \ \& \ H_3 \ \& \ M_4 \ \& \ M_5) \\
\text{or } &(M_1 \ \& \ H_2 \ \& \ M_3 \ \& \ H_4 \ \& \ M_5) \\
\text{or } &(M_1 \ \& \ H_2 \ \& \ M_3 \ \& \ M_4 \ \& \ H_5) \\
\text{or } &(M_1 \ \& \ M_2 \ \& \ H_3 \ \& \ H_4 \ \& \ M_5) \\
\text{or } &(M_1 \ \& \ M_2 \ \& \ H_3 \ \& \ M_4 \ \& \ H_5) \\
\text{or } &(M_1 \ \& \ M_2 \ \& \ M_3 \ \& \ H_4 \ \& \ H_5)].
\end{aligned}
$$

Because this is a list of mutually exclusive alternatives,   // See (10.2.4).

$$
\begin{aligned}
P(X = 2) = \ &P(H_1 \ \& \ H_2 \ \& \ M_3 \ \& \ M_4 \ \& \ M_5) \\
+ &P(H_1 \ \& \ M_2 \ \& \ H_3 \ \& \ M_4 \ \& \ M_5) \\
+ &P(H_1 \ \& \ M_2 \ \& \ M_3 \ \& \ H_4 \ \& \ M_5) \\
+ &P(H_1 \ \& \ M_2 \ \& \ M_3 \ \& \ M_4 \ \& \ H_5) \\
+ &P(M_1 \ \& \ H_2 \ \& \ H_3 \ \& \ M_4 \ \& \ M_5) \\
+ &P(M_1 \ \& \ H_2 \ \& \ M_3 \ \& \ H_4 \ \& \ M_5) \\
+ &P(M_1 \ \& \ H_2 \ \& \ M_3 \ \& \ M_4 \ \& \ H_5) \\
+ &P(M_1 \ \& \ M_2 \ \& \ H_3 \ \& \ H_4 \ \& \ M_5) \\
+ &P(M_1 \ \& \ M_2 \ \& \ H_3 \ \& \ M_4 \ \& \ H_5) \\
+ &P(M_1 \ \& \ M_2 \ \& \ M_3 \ \& \ H_4 \ \& \ H_5).
\end{aligned}
$$

Because the trials are independent,

$$
\begin{aligned}
P(X = 2) = \;& P(H_1)P(H_2)P(M_3)P(M_4)P(M_5) \\
& + P(H_1)P(M_2)P(H_3)P(M_4)P(M_5) \\
& + P(H_1)P(M_2)P(M_3)P(H_4)P(M_5) \\
& + P(H_1)P(M_2)P(M_3)P(M_4)P(H_5) \\
& + P(M_1)P(H_2)P(H_3)P(M_4)P(M_5) \\
& + P(M_1)P(H_2)P(M_3)P(H_4)P(M_5) \\
& + P(M_1)P(H_2)P(M_3)P(M_4)P(H_5) \\
& + P(M_1)P(M_2)P(H_3)P(H_4)P(M_5) \\
& + P(M_1)P(M_2)P(H_3)P(M_4)P(H_5) \\
& + P(M_1)P(M_2)P(M_3)P(H_4)P(H_5).
\end{aligned}
$$

Because $P(H_j) = 0.8$ and $P(M_j) = 0.2$ for any j,

$$
\begin{aligned}
P(X = 2) = \;& (0.8)(0.8)(0.2)(0.2)(0.2) \\
& + (0.8)(0.2)(0.8)(0.2)(0.2) \\
& + (0.8)(0.2)(0.2)(0.8)(0.2) \\
& + (0.8)(0.2)(0.2)(0.2)(0.8) \\
& + (0.2)(0.8)(0.8)(0.2)(0.2) \\
& + (0.2)(0.8)(0.2)(0.8)(0.2) \\
& + (0.2)(0.8)(0.2)(0.2)(0.8) \\
& + (0.2)(0.2)(0.8)(0.8)(0.2) \\
& + (0.2)(0.2)(0.8)(0.2)(0.8) \\
& + (0.2)(0.2)(0.2)(0.8)(0.8).
\end{aligned}
$$

Because each term has exactly 2 hits and 3 misses, each term is $(0.8)^2(0.2)^3$. The number of terms equals the number of ways to choose 2 out of 5 trials to be successes. Thus

$$
prob(X = 2) = \binom{5}{2}(0.8)^2(0.2)^3 = 10(0.64)(0.008) = 0.0512.
$$

The number of bull's eyes he hits ranges from 0 to 5, and for each value v in that range,

$$
\begin{aligned}
prob(X = v) = \;& \text{(the number of ways to choose } v \text{ out of 5 trials to be successes)} \\
& \times \text{(the success-probability raised to the number of successes)} \\
& \times \text{(the failure-probability raised to the number of failures)} \\
= \;& \binom{5}{v} p^v q^{n-v}.
\end{aligned}
$$

So for the marksman, where $p = 0.8$ and $q = 0.2$, we have

v	$prob(X = v)$		$\binom{5}{v}p^v q^{n-v}$	$v \times prob(X = v)$
0	0.00032	=	$1(1)(0.00032)$	0
1	0.00640	=	$5(0.8)(0.0016)$	0.0064
2	0.05120	=	$10(0.64)(0.008)$	0.1024
3	0.20480	=	$10(0.512)(0.04)$	0.6144
4	0.40960	=	$5(0.4096)(0.2)$	1.6384
5	0.32768	=	$1(0.32768)(1)$	1.6384
	1.00000	=		$4.0000 = E(X)$

// That expected value for X is right. If we think of one trial as the experiment,
// "success" is one possible outcome and has probability $p = 80\%$.
// If this (one trial) experiment is repeated $N = 5$ times,
// the expected frequency of success is $Np = 80\%$ of $5 = 4$.

Let's return to a general binomial experiment with **n** trials and success-probability **p**.   // Binomial experiments have two "parameters", n and p.
An outcome of this experiment is a sequence of trial results:

$$\sigma = (R_1, R_2, R_3, \ldots R_n).$$

Because the trial results are independent, we must have

$$P(\sigma) = (P(R_1))(P(R_2))(P(R_3)) \ldots (P(R_n)).$$

If k of these results are successes, each of those results has probability p; the other $n - k$ results are failures, each of which has probability q. Thus,

$$P(\sigma) = p^k q^{n-k}. \tag{10.4.1}$$

// Outcomes like σ from a binomial experiment would be *equally likely* if and
// only if $p = q = \frac{1}{2}$ and then $P(\sigma) = (\frac{1}{2})^n$.

For each k in the range 0 to n, the number of output sequences with exactly k successes is equal to the number of subsets of size k in the set $\{1..n\}$.

10.4.2   The Binomial Distribution

For a binomial experiment with $n \in \mathbf{P}$ trials and success-probability p, where X is the number of successes,

$$V = \{0, 1, \ldots, n\},$$

$$P(X = v) = \binom{n}{v}p^v q^{n-v} \text{ for all } v \in V, \qquad\qquad // q = 1 - p.$$

and $E(X) = np.$

// Is there a proof that this is indeed a probability distribution and $E(X) = np$?

Since p is a probability, $0 <= p <= 1$. Then $0 <= q <= 1$, and for all $v \in V$, $p^v q^{n-v} >= 0$ and so $P(X = v) >= 0$. Furthermore,

$$\sum_{v=0}^{n} P(X = v) = \sum_{v=0}^{n} \binom{n}{v} p^v q^{n-v}$$

$$= (p+q)^n \qquad \text{// by the Binomial Theorem (3.8.1)}$$

$$= (1)^n$$

$$= 1.$$

And $\quad E(X) = \sum_{v \in V} v \times P(X = v) = \sum_{v=0}^{n} v \times \binom{n}{v} p^v q^{n-v} = \sum_{v=1}^{n} v \times \binom{n}{v} p^v q^{n-v}.$

But $\quad v \times \binom{n}{v} = v \times \dfrac{n!}{v!(n-v)!} = \dfrac{v \times n(n-1)!}{v(v-1)!([n-1]-[v-1])!} = n \times \binom{n-1}{v-1}.$

$$\text{// } v > 0.$$

Thus, $\quad E(X) = \sum_{v=1}^{n} n \times \binom{n-1}{v-1} p \times p^{v-1} q^{[n-1]-[v-1]}$

$$= np \sum_{w=0}^{n-1} \binom{n-1}{w} p^w q^{[n-1]-w} \qquad \text{// setting } w = v - 1$$

$$= np(p+q)^{n-1} \qquad \text{// Binomial Theorem (again)}$$

$$= np.$$

10.4.3   The Geometric Distribution

$\qquad\qquad\qquad\qquad\qquad\qquad$ // like flipping a coin until the first H

A geometric experiment is repeating a Bernoulli trial with success-probability p until the first success; X is the number of trials required. We assume that $0 < p < 1$: $\qquad\qquad\qquad$ // What happens when $p = 0$? When $p = 1$?

$$V = \{1, 2, \dots\}, \qquad\qquad\qquad\qquad\qquad\qquad \text{// } V = \boldsymbol{P}$$

$$P(X = v) = q^{v-1} p \qquad \text{for all } v \in V, \qquad\qquad \text{// } q = 1 - p$$

and $\qquad E(X) = 1/p$

The outcome sequences from a geometric experiment have exactly one success in them and it's at the end. Each value of X corresponds to exactly one such sequence. Therefore, by formula (10.4.1),

$$P(X = v) = q^{v-1} p \qquad \text{for all } v \in V.$$

Since $0 < p < 1$; $0 < q < 1$, and for all $v \in V$, $P(X = v) = q^{v-1}p >= 0$. Furthermore, when $0 < q < 1$, we may apply Theorem 8.5.2 and obtain

$$\sum_{v=1}^{\infty} P(X = v) = q^0 p + q^1 p + q^2 p + q^3 p + q^4 p + \ldots$$

$$= \left\{ \frac{p}{1-q} \right\} = 1.$$

This is known as "the geometric distribution" because the probabilities form an infinite geometric sequence.   // Geometry is not involved.

// Is there a proof that $E(X) = 1/p$?

// If you roll a die until you get a 5, <u>on average</u> it should take $\dfrac{1}{1/6} = 6$ rolls.

// If you flip a coin until you get an H, <u>on average</u> it should take $\dfrac{1}{1/2} = 2$ flips.

$$E(X) = \sum_{v \in V} v \times P(X = v) = \sum_{v=1}^{\infty} v \times q^{v-1}p$$

$$= 1q^0 p + 2q^1 p + 3q^2 p + 4q^3 p + 5q^4 p + \ldots$$

// But does the infinite series really have a finite value?
// Do the terms $nq^{n-1} \times p$ converge to zero? (q^n tends to zero, but n gets larger and
// larger). Does this product tend to zero?

Consider the partial sums T_n where $n \in \mathbf{P}$ and

$$T_n = 1q^0 p + 2q^1 p + 3q^2 p + \ldots + nq^{n-1}p. \qquad \text{// Does } T_n \to 1/p?$$

Then $qT_n = \qquad 1q^1 p + 2q^2 p + \ldots + (n-1)q^{n-1}p + nq^n p.$

So $T_n - qT_n = 1q^0 p + 1q^1 p + 1q^2 p + \ldots + \quad 1q^{n-1}p \quad - nq^n p$

$(1 - q)T_n = q^0 p \ + q^1 p \ + q^2 p \ + \ldots + \quad q^{n-1}p \quad - nq^n p$

$$= \qquad\qquad p\frac{(1-q^n)}{(1-q)} \qquad\qquad - nq^n p.$$

// by Theorem 3.6.8

Since $p = 1 - q$, we have

$$pT_n = (1 - q^n) - nq^n p,$$

$$T_n = 1/p - q^n/p - nq^n = 1/p - q^n\{1/p - n\} = 1/p + q^n\{n - 1/p\},$$

and $\qquad\qquad T_n - 1/p = q^n\{n - 1/p\} = \{n - 1/p\}q^n.$

Thus, $\qquad\qquad |T_n - 1/p| = |\{n - 1/p\}q^n| = \{n - 1/p\}q^n < nq^n.$

The fact that T_n converges to $1/p$ follows from

Theorem 10.4.1: If k is *any* positive integer and q is any real number where $0 < q < 1$, then $n^k q^n \to 0$.

Proof. // Warning: This proof appears to rely on algebraic trickery.
$\qquad\qquad\qquad\qquad$ // (One might apply l'Hopital's Rule from calculus).

// We must show that for every $\varepsilon > 0$, $\exists$ M $\in$ **P** such that
// if $n >= $ M, then $|n^k q^n - 0| = n^k q^n < \varepsilon.$

$\qquad$ Suppose some $\varepsilon > 0$ is given. Let $b = 1/q;$ $\qquad$ // Then $b > 1$ and $\lg(b) > 0.$
let $B = 1 + max\{4, \lceil (k+1)/\lg(b) \rceil\};$ and
$\qquad\qquad\qquad\qquad\qquad$ // Then $B >= 5$ and $B - 1 >= (k+1)/\lg(b).$
let $M = max\{\lceil 1/\varepsilon \rceil, 2^B\}.$ $\qquad\qquad$ // Then $M >= 1/\varepsilon$ and $M >= 2^B.$
If $n >= $ M, then, setting $t = \lfloor \lg(n) \rfloor,$ $\qquad$ // So $t <= \lg(n) < t + 1.$
we have:

1. $2^B <= n = 2^{\lg(n)}$ so $B <= \lg(n).$
2. Since B is an integer, $B <= t.$
3. $t >= 5$ and $t - 1 >= (k + 1)/\lg(b).$
4. $\lg(n) \times (k + 1)/\lg(b) < (t + 1)(t - 1) = t^2 - 1 < t^2 < 2^t.$ $\quad$ // by Theorem 3.6.1
5. $\lg(n) \times (k+1)/\lg(b) \ < 2^{\lg(n)} = n.$ $\qquad\qquad\qquad$ // $t <= \lg(n)$
6. $\lg(n) \times (k+1) \qquad\qquad < n \times \lg(b).$ $\qquad\qquad\qquad$ // $\lg(b) > 0$
7. $\lg(n^{k+1}) \qquad\qquad\qquad < \lg(b^n).$ $\qquad\qquad$ // $\lg(x^y) = y \times \lg(x)$
8. $n^{k+1} \qquad\qquad\qquad\quad < b^n.$ $\qquad\qquad\qquad$ // $\lg$ is increasing
9. $n \times n^k \qquad\qquad\qquad\ < (1/q)^n.$
10. $(1/\varepsilon) \times n^k \qquad\qquad\ < (1/q)^n.$ $\qquad\qquad$ // $n >= M >= 1/\varepsilon$
11. $n^k \qquad\qquad\qquad\qquad < (1/q)^n \times \varepsilon.$
Hence, $\quad n^k q^n \qquad\qquad\ < \varepsilon.$ $\qquad\qquad\qquad\qquad\qquad\qquad$ $\Box$

The Most Important Ideas in This Section.
Three standard distributions were discussed:
1. The Uniform Distribution $\qquad\qquad\qquad\qquad$ // like rolling a "fair" n-sided die

$V = \{v_1, v_2, v_3, \ldots, v_n\}$ and $P(X = v) = 1/n$ for all $v \in V.$

$E(X) = \dfrac{v_1 + v_2 + \ldots + v_n}{n}$ $\qquad\qquad\qquad$ // the ordinary "average" value

2. The Binomial Distribution $\qquad\qquad\qquad\qquad$ // like flipping a coin n times
$\quad$ A *binomial experiment* consists of:
$\quad$ 1. n identical "Bernoulli trials" where
$\quad$ 2. Each trial results in a "success" or a "failure";
$\quad$ 3. The probability of success on any trial is a fixed value, p;
$\qquad\qquad\qquad\qquad\qquad\qquad\qquad\qquad$ // $q = 1 - p$

(continued)

(continued)

    4. The trials are independent; and
    5. We're interested in the number of successes, X.

$$V = \{0, 1, \ldots, n\} \text{ and } P(X = v) = \binom{n}{v} p^v q^{n-v} \text{ for all } v \in V.$$

$E(X) = np.$

3. The Geometric Distribution   // like flipping a coin until the first H
    A geometric experiment is repeating a Bernoulli trial with success-probability p until the first success, X is the number of trials required. We assume that $0 < p < 1$.

$$V = \{1, 2, \ldots\} \text{ and } P(X = v) = q^{v-1} p \text{ for all } v \in V.$$
$$E(X) = 1/p.$$

If k is *any* positive integer and $0 < q < 1$, then $n^k q^n \to 0$.

10.5   Conditional Expected Values

Let's return to the example that began this chapter.

Example 10.5.1: Flipping Until HH
    Suppose you flip a coin until you get two heads in a row, and suppose that it takes exactly X flips. At the beginning of this chapter, we asked "Could a 'reasonable' prediction be made of how many flips are needed on average?" Now we can answer that question with "yes" because we can calculate the expected number of flips, E(X).

// But how do we find $prob(X = n)$?

    Let's say that a sequence on {H, T} is "*good*" if it ends HH and that is the first and only occurrence of HH. Each repetition of the *experiment*, "flip a coin until you get two heads in a row," produces a *good* sequence as its outcome.

// Every good sequence is finite, but there are an infinite number of good
// sequences. The sample space for this experiment is the infinite set of all *good*
// sequences.

For all integers $n >= 2$, define

    $G(n)$ to be the set of all *good* **n**-sequences, and $g(n)$ to be $|G(n)|$.

A *good* sequence σ must end in HH; if its length is > 2, it must end in THH and cannot begin with HH. Then

$$G(2) = \{\mathbf{HH}\}$$ so $g(2) = 1,$
$$G(3) = \{\mathbf{THH}\ \}$$ so $g(3) = 1,$
$$G(4) = \{\mathbf{HTHH}, \mathbf{TTHH}\ \}$$ so $g(4) = 2,$
$$G(5) = \{\mathbf{THTHH}, \mathbf{TTTHH}, \mathbf{HTTHH}\ \}$$ so $g(5) = 3,$
$$G(6) = \{\mathbf{TTHTHH}, \mathbf{TTTTHH}, \mathbf{THTTHH},$$
$$\qquad\quad \mathbf{HTHTHH}, \mathbf{HTTTHH}\ \}$$ so $g(6) = 5.$

// The sequence $g = (1, 1, 2, 3, 5, \ldots)$ begins just like the Fibonacci sequence.
// Is this just coincidence or does it continue like the Fibonacci sequence?

A *good* sequence either begins with a T or begins with an H. But when $n > 2$, if it begins with an H it must begin with HT.   // It can't begin HH.
Therefore, a *good* sequence of length $n + 1$ is either:

    "T" followed by a *good* sequence of length $n,$

or   "HT" followed by a *good* sequence of length $n - 1.$

In fact, for $n > 2$, $g(n + 1) = g(n) + g(n - 1).$ // Can you prove this?
Because $g(n)$ begins and continues like the Fibonacci sequence, for $n >= 2$, we know it is the Fibonacci sequence   // "shifted" by 2 units

$$g(n) = F_{n-2}.$$ // see Sect. 8.3.

// But how do we find $prob(X = n)$?

When we assume that the coin is "fair" (or that each flip is equally likely to be H or T), and that coin flips are independent, the probability that n flips produces any particular sequence on $\{T, H\}$ equals $(\frac{1}{2})^n.$ // from (10.4.1)

$prob(X = n) =$ the probability that the experiment produces a *good*
                 sequence of length n
                      // $X = n$ corresponds to a subset of outcomes.
          $=$ the probability that the experiment produces a
                 sequence in $G(n)$

$$= \sum_{\sigma \in G(n)} prob(n \text{ flips produce sequence } \sigma)$$

$$= \sum_{\sigma \in G(n)} \left(\tfrac{1}{2}\right)^n = |G(n)| \times \left(\tfrac{1}{2}\right)^n \qquad\qquad // \text{ this is} >= 0.$$

$$= \frac{g(n)}{2^n} = \frac{F_{n-2}}{2^n} \qquad\qquad\qquad\qquad (10.5.1)$$

$$= \left[\frac{1}{\sqrt{5}} \left(\frac{1+\sqrt{5}}{2} \right)^{n-1} - \frac{1}{\sqrt{5}} \left(\frac{1-\sqrt{5}}{2} \right)^{n-1} \right] \left(\frac{1}{2^n} \right) \quad \text{// Eq. 8.3.2}$$

$$= \frac{1}{\sqrt{5}} \left[\left(\frac{1+\sqrt{5}}{2} \right)^{n-1} \left(\frac{1}{2} \right)^{n-1} - \left(\frac{1-\sqrt{5}}{2} \right)^{n-1} \left(\frac{1}{2} \right)^{n-1} \right] \left(\frac{1}{2} \right)$$

$$= \frac{1}{2\sqrt{5}} \left[\left(\frac{1+\sqrt{5}}{4} \right)^{n-1} - \left(\frac{1-\sqrt{5}}{4} \right)^{n-1} \right].$$

$$(10.5.2)$$

//X Is $prob(X = 1) = 0$? $Prob(X = 2) = 1/4$? $Prob(X = 3) = 1/8$?
//X Is $prob(X = 4) = 2/16$? $Prob(X = 5) = 3/32$?
//   Is this really a probability function? defined on all of **P**?

For each n, $prob(X = n) = g(n)/2^n >= 0$ and   // Probability Axiom I holds.
                                                         // Does Axiom II also hold?

$$\sum_{n=1}^{\infty} prob(X = n) = \sum_{n=1}^{\infty} \frac{1}{2\sqrt{5}} \left[\left(\frac{1+\sqrt{5}}{4} \right)^{n-1} - \left(\frac{1-\sqrt{5}}{4} \right)^{n-1} \right]$$

$$= \frac{1}{2\sqrt{5}} \left[\sum_{n=1}^{\infty} \left(\frac{1+\sqrt{5}}{4} \right)^{n-1} - \sum_{n=1}^{\infty} \left(\frac{1-\sqrt{5}}{4} \right)^{n-1} \right].$$

// Recall that when $|r| < 1, 1 + r + r^2 + r^3 + r^4 + \ldots = \dfrac{1}{1-r}$.   (Theorem 8.5.2)
We can also prove

Theorem 10.5.1: If $|r| < 1$, then $1r^0 + 2r^1 + 3r^2 + 4r^3 + 5r^4 + \ldots = \left(\dfrac{1}{1-r} \right)^2$.

Proof. Consider the partial sums T_n where $n \in \mathbf{P}$ and

$$T_n = 1r^0 + 2r^1 + 3r^2 + \ldots + nr^{n-1}.$$

Then $rT_n = \qquad 1r^1 + 2r^2 + \ldots + (n-1)r^{n-1} + nr^n,$

so $T_n - rT_n = 1r^0 + 1r^1 + 1r^2 + \ldots + 1r^{n-1} \qquad - nr^n$

$(1-r)T_n = r^0 + r^1 + r^2 + \ldots + r^{n-1} \qquad - nr^n$

$$= \frac{(1-r^n)}{(1-r)} \qquad - nr^n. \quad \text{// by Theorem 3.6.8}$$

Because r^n converges to 0   // by Theorem 8.5.1
and nr^n converges to 0,   // by Theorem 10.4.1

$(1-r) T_n$ converges to $\dfrac{1}{(1-r)}$.   Therefore, T_n converges to $\left(\dfrac{1}{1-r} \right)^2$.   □

Since $2 < \sqrt{5} < 3$,

$$-1 < \frac{1-3}{4} < \frac{1-\sqrt{5}}{4} < 0 < \frac{1+\sqrt{5}}{4} < \frac{1+3}{4} = +1.$$

If $r = \dfrac{1 \pm \sqrt{5}}{4}$, then $\dfrac{1}{1-r} = \dfrac{4}{4-4r} = \dfrac{4}{4-(1\pm\sqrt{5})} = \dfrac{4}{3\mp\sqrt{5}}$

and $\left(\dfrac{1}{1-r}\right)^2 = \left(\dfrac{4}{3\mp\sqrt{5}}\right)^2 = \dfrac{16}{9\mp6\sqrt{5}+5} = \dfrac{16}{14\mp6\sqrt{5}} = \dfrac{8}{7\mp3\sqrt{5}}.$

Thus,

$$\sum_{n=1}^{\infty} prob(X = n) = \frac{1}{2\sqrt{5}}\left[\sum_{n=1}^{\infty}\left(\frac{1+\sqrt{5}}{4}\right)^{n-1} - \sum_{n=1}^{\infty}\left(\frac{1-\sqrt{5}}{4}\right)^{n-1}\right]$$

$$= \frac{1}{2\sqrt{5}}\left[\frac{4}{3-\sqrt{5}} - \frac{4}{3+\sqrt{5}}\right] = \frac{4}{2\sqrt{5}}\left[\frac{(3+\sqrt{5})-(3-\sqrt{5})}{(3-\sqrt{5})(3+\sqrt{5})}\right]$$

$$= \frac{4}{2\sqrt{5}}\left[\frac{2\sqrt{5}}{(9-5)}\right] = 1.$$

Also,

$$E(X) = \sum_{n=1}^{\infty} n \times prob(X = n)$$

$$= \sum_{n=1}^{\infty}\frac{n}{2\sqrt{5}}\left[\left(\frac{1+\sqrt{5}}{4}\right)^{n-1} - \left(\frac{1-\sqrt{5}}{4}\right)^{n-1}\right]$$

$$= \frac{1}{2\sqrt{5}}\left[\sum_{n=1}^{\infty} n\left(\frac{1+\sqrt{5}}{4}\right)^{n-1} - \sum_{n=1}^{\infty} n\left(\frac{1-\sqrt{5}}{4}\right)^{n-1}\right]$$

$$= \frac{1}{2\sqrt{5}}\left[\left(\frac{4}{3-\sqrt{5}}\right)^2 - \left(\frac{4}{3+\sqrt{5}}\right)^2\right]$$

$$= \frac{8}{2\sqrt{5}}\left[\left(\frac{1}{7-3\sqrt{5}}\right) - \left(\frac{1}{7+3\sqrt{5}}\right)\right]$$

$$= \frac{4}{\sqrt{5}}\left[\frac{(7+3\sqrt{5})-(7-3\sqrt{5})}{7^2-(3\sqrt{5})^2}\right]$$

$$= \frac{4}{\sqrt{5}}\left[\frac{6\sqrt{5}}{49-9\times5}\right] = \frac{4}{\sqrt{5}}\left[\frac{6\sqrt{5}}{4}\right] = 6.$$

// We will derive this expected value again using a recurrence equation.

Recall that if $n > 2$, then

$$g(n+1) = g(n) + g(n-1),$$

so

$$\frac{g(n+1)}{2^{n+1}} = \frac{g(n)}{2 \times 2^n} + \frac{g(n-1)}{4 \times 2^{n-1}}.$$

Therefore, $n \geq 2$ and   // since $prob(X = 1) = 0$

$$prob(X = n+1) = \frac{1}{2}prob(X = n) + \frac{1}{4}prob(X = n-1), \qquad (10.5.3)$$

and so $E(X) = \sum_{n=1}^{\infty} n \times prob(X = n)$

$$= 1 \times prob(X = 1) + 2 \times prob(X = 2) + \sum_{n=3}^{\infty} n \times prob(X = n)$$

$$= 1 \times 0 + 2 \times \frac{1}{4} + \sum_{n=3}^{\infty} n \left\{ \frac{1}{2}prob(X = n-1) + \frac{1}{4}prob(X = n-2) \right\}$$

$$= \frac{1}{2} + \sum_{n=3}^{\infty} n \frac{1}{2}prob(X = n-1) + \sum_{n=3}^{\infty} n \frac{1}{4}prob(X = n-2)$$

$$= \frac{1}{2} + \frac{1}{2} \sum_{j=2}^{\infty} (j+1)prob(X = j) + \frac{1}{4} \sum_{k=1}^{\infty} (k+2)prob(X = k)$$

$$= \frac{1}{2} + \frac{1}{2} \left\{ \sum_{j=2}^{\infty} j \times prob(X = j) + \sum_{j=2}^{\infty} prob(X = j) \right\}$$

$$\quad + \frac{1}{4} \left\{ \sum_{k=1}^{\infty} k \times prob(X = k) + 2 \sum_{k=1}^{\infty} prob(X = k) \right\}$$

$$= \frac{1}{2} + \frac{1}{2} \{ E(X) + 1 \} + \frac{1}{4} \{ E(X) + 2 \}.$$

Multiplying both sides of the equation by 4 gives

$$4E(X) = 2 + 2\{ E(X) + 1 \} + E(X) + 2 = 3E(X) + 6.$$

Subtracting $3E(X)$ from both sides of the equation gives

$$E(X) = 6. \qquad \qquad \square$$

10.5.1 Conditional Expectation

We just saw that

$$E(X) = \frac{1}{2} + \frac{1}{2}\{E(X)+1\} + \frac{1}{4}\{E(X)+2\}. \qquad (10.5.4)$$

This equation may be interpreted in the context of "conditional expected values".

// and derived much more easily using conditional expectation.

If A is an event with $prob(A) > 0$ and X is a random variable taking values in set V, then $prob(X = v \mid A)$ is a probability function, and we can define **the expected value of X conditional on A** as follows:

$$E(X \mid A) = \sum_{v \in V} v \times prob(X = v \mid A).$$

This formula calculates the expected value of X (the theoretical average value of X) when condition A holds.

Let us try to explain how in Example 10.5.1 we eventually obtained

$$\begin{aligned}
E(\# \text{ flips}) &= E(\# \text{ flips} \mid \sigma \text{ starts HH}) \times prob(\sigma \text{ starts HH}) \\
&\quad + E(\# \text{ flips} \mid \sigma \text{ starts HT}) \times prob(\sigma \text{ starts HT}) \\
&\quad + E(\# \text{ flips} \mid \sigma \text{ starts T}) \times prob(\sigma \text{ starts T})
\end{aligned}$$

$$\begin{aligned}
&= (2) \times {}^{1}\!/_{4} \\
&\quad + (2 + E(\# \text{ flips})) \times {}^{1}\!/_{4} \\
&\quad + (1 + E(\# \text{ flips})) \times {}^{1}\!/_{2}.
\end{aligned}$$

Each outcome of the experiment is a *good* sequence σ on {H, T}. So the sample space S for the experiment is all *good* sequences on {H, T}. The random variable, $X(\sigma)$, is the length of σ (the number of flips done). The sample space S may be partitioned into three parts:

1. *Good* sequences that begin with HH.
2. *Good* sequences that begin with HT.
3. *Good* sequences that begin with T.

If σ is a *good* sequence that begins with HH, its length *must be* 2 so the "expected" length of such sequences equals 2.   *// $E(\# \text{ flips} \mid \sigma \text{ starts HH}) = 2$*

If σ is a *good* sequence that begins with HT, σ must be HT followed by a *good* sequence τ, and the "expected" length of σ must be 2 plus the "expected" length of τ.   *// $E(\# \text{ flips} \mid \sigma \text{ starts HT}) = 2 + E(\# \text{ flips} \mid \tau \in S) = 2 + E(\# \text{ flips})$*

If σ is a *good* sequence that begins with T, σ must be T followed by a *good* sequence τ, and its "expected" length must be 1 plus the "expected" length of τ.

// $E(\#$ flips $\mid \sigma$ starts with T$) = 1 + E(\#$ flips $\mid \tau \in S) = 1 + E(\#$ flips$)$
The equation we obtained as (10.5.4) is an instance of

Theorem 10.5.2: If $\{A_1, A_2, A_3, \ldots, A_k\}$ is a partition of the sample space for a random variable X where each $prob(A_j) > 0$, then

$$E(X) = \sum_{j=1}^{k} E(X \mid A_j) \times prob(A_j).$$

Proof. The proof is easier to follow if we imitate the last example and emphasize the role of the underlying experiment, which produces an outcome that we'll denote by σ. We'll also denote the *event* "A_j" by "$\sigma \in A_j$" and the value of X on σ by $X(\sigma)$. And we'll assume that X takes values in set V. Then,

$$\sum_{j=1}^{k} E(X \mid A_j) \times prob(A_j) \qquad\qquad\qquad \text{// may be rewritten as}$$

$$= \sum_{j=1}^{k} E(X(\sigma) \mid \sigma \in A_j) \times prob(\sigma \in A_j)$$

$$= \sum_{j=1}^{k} \left\{ \sum_{v \in V} v \times prob(X(\sigma) = v \mid \sigma \in A_j) \right\} \times prob(\sigma \in A_j)$$

$$= \sum_{j=1}^{k} \left\{ \sum_{v \in V} v \times \frac{prob(X(\sigma) = v \, and \, \sigma \in A_j)}{prob(\sigma \in A_j)} \right\} \times prob(\sigma \in A_j) \qquad \text{(by 10.2.2)}$$

$$= \sum_{j=1}^{k} \left\{ \sum_{v \in V} v \times prob(X(\sigma) = v \, and \, \sigma \in A_j) \right\}$$

$$= \sum_{v \in V} v \times prob(X(\sigma) = v \, and \, \sigma \in A_1) + \sum_{v \in V} v \times prob(X(\sigma) = v \, and \, \sigma \in A_2)$$

$$+ \ldots + \sum_{v \in V} v \times prob(X(\sigma) = v \, and \, \sigma \in A_k)$$

$$= \sum_{v \in V} v \times prob[(X(\sigma) = v \, and \, \sigma \in A_1) \, or \, (X(\sigma) = v \, and \, \sigma \in A_2) \, or \, \ldots$$

$$or \, (X(\sigma) = v \, and \, \sigma \in A_k)]$$

$$= \sum_{v \in V} v \times prob[X(\sigma) = v \, and \, (\sigma \in A_1 \, or \, \sigma \in A_2 \, or \, \ldots \, or \, \sigma \in A_k)]$$

$$= \sum_{v \in V} v \times prob[X(\sigma) = v] = E(X). \qquad\qquad\qquad\qquad □$$

> **The Most Important Ideas in This Section.**
> We showed that if a coin is flipped until the first occurrence of two heads in a row, then the expected number of flips is 6. Our demonstration used conditional expectation, and we proved that if $\{A_1, A_2, A_3, \ldots, A_k\}$ is a partition of the sample space for a random variable X where each $prob(A_j) > 0$, then $E(X) = \sum_{j=1}^{k} E(X \mid A_j) \times prob(A_j)$. In the next section, we use this equation to examine algorithm performance on average.

10.6   Average-Case Complexity

Sometimes, the number of steps an algorithm requires depends on the input instance, as we saw in prime testing using the method of trial divisors, in searching, and in sorting with BetterBubbleSort or QuickSort. For such cases, how can we determine the average-case complexity?

There seem to be at least four methods:

1. Run the algorithm on all possible inputs and average.

   // But there may be too many cases.

2. Run the algorithm on one *random* input instance.

   // But how do you determine a *random* input instance?

3. Run the algorithm on a *random* sample of input instances and average.

   // But how do you determine a *random* sample of input instances?

4. Use the theory of expected values.

In this section, we apply strategy #4.

10.6.1   Applying Expectation to Linear Search

Suppose we are searching array $A[1]..A[n]$ for a target value T. The simplest probabilistic model for searches is: when T is in A, all positions are equally likely to contain T, but T is not always in A. That is,

$$prob(T \text{ is in the list}) = p \qquad \text{where } 0 <= p <= 1$$
$$\text{and} \quad prob(T = A[j] \mid T \text{ is in the list}) = 1/n \quad \text{for } j = 1, 2, \ldots, n.$$

The cost of searching is usually taken to be the number of probes, so we want

$$E(\# \text{ of probes}) = E(\# \text{ of probes} \mid T \text{ is in the list}) \times p +$$
$$E(\# \text{ of probes} \mid T \text{ is not in the list}) \times q \quad \text{where } q = (1 - p).$$

For *Linear Search*, determining that $T = A[j]$ when T is in the list, costs exactly j probes, and determining that T is not in the list costs exactly n probes, so

$$E(\text{\# of probes} \mid T \text{ is in the list}) = \sum_{j=1}^{n} j \times \frac{1}{n}$$

$$= \frac{1}{n}\left(\sum_{j=1}^{n} j\right) = \frac{1}{n}\left(\frac{n(n+1)}{2}\right) = \frac{n+1}{2},$$

and hence, $E(\text{\# of probes}) = \left(\dfrac{n+1}{2}\right)p + nq.$

10.6.2 Applying Expectation to QuickSort

We assume that the input is a random ordering of the entries in $A[1]..A[n]$ and that all the entries are distinct.   // A is a random permutation of its entries.

In Chap. 4, QuickSort was given as a recursive algorithm.

Algorithm 4.5.1: QuickSort(p,q)

```
Begin
  If (p < q) Then
    M ← A[q];                      // "partition" A using pivot-value M
    j ← p;
    For k ← p to (q - 1) Do
      If (A[k] < M) Then
        x ← A[j];
        A[j] ← A[k];
        A[k] ← x;
        j ← j + 1;
      End;    // the inner if-statement
    End;      // the for-k loop
    A[q] ← A[j];                    // This is the end of the "partitioning".
    A[j] ← M;
    QuickSort(p, j - 1);           // the first "recursive" sub-call
    QuickSort(j + 1, q);           // the second "internal" sub-call
  End;        // the outer if-statement
End.          // the recursive algorithm
```

Let $X(A)$ denote the number of comparisons of key values done by QuickSort applied to input A, a list of length n. That is, $X(A)$ is the number of times the Boolean Expression "$A[k] < M$" is evaluated inside the for-loop in the main invocation of QuickSort (where parameter p equals 1 and parameter q equals n) and in all the recursive sub-calls.

QuickSort works on sublists of the form $A[p] .. A[q]$ by:

1. Setting M to be $A[q]$;   // our "estimate" of the Median
2. "Partitioning" the entries in $A[p] .. A[q]$

   so that after the first partition, we have an index j where

$$A[p], A[p+1], \ldots, A[j-1] < M = A[j]$$

and $M <= A[j+1], A[j+2] \ldots, A[q];$ and then,

3. Sorting $A[p] .. A[q]$ is completed by
   sorting $A[p] .. A[j - 1]$ and sorting $A[j + 1] .. A[q]$.   // by QuickSort too
   The number of comparisons done in partitioning $A[p] .. A[q]$ is equal to the number
of k-values in the for-loop, namely, $q - p$. The length of sublist $A[p] .. A[j - 1]$ is
$j - p$, and the length of the sublist $A[j + 1] .. A[q]$ is $q - j$. But **no** comparisons
of key values will be done unless p is strictly less than q.   // And $q - p >= 1$.
   Thus,

$$X(A) = \text{the cost of the partition}$$
$$+ \text{ cost of sorting the first sublist, } A[1] .. A[j - 1] \text{ of length } j - 1$$
$$+ \text{ cost of sorting the second sublist, } A[j + 1] .. A[n] \text{ of length } n - j$$
$$= (n - 1) + X(A[1] .. A[j - 1]) + X(A[j + 1] .. A[n]).$$

The cost of sorting depends on the outcome of the partition(s). The most costly
cases are

$$j = 1 \text{ and } X(A) = (n - 1)   + X(A[j + 1] .. A[n])$$
$$\text{and } j = n \text{ and } X(A) = (n - 1) + X(A[1] .. A[j - 1]).$$

// Recall a worst case for QuickSort occurs when A is already sorted.
// Then $X(A) = (n - 1) + (n - 2) + \ldots + 2 + 1 = n(n - 1)/2$ like MinSort
// and BubbleSort.

   However, we want to calculate the expected value of X where the sample space S
we're interested in is all possible input sequences, A, of length n. It appears that
we'll need to divide S into cases, one for each j-value produced by the partition
portion of QuickSort. We can set

$$S_j = \{A \in S: \text{the partition part of QuickSort generates index } j\}$$

for $j = 1$ to n. Then

$$prob(A \in S_j) = prob(M \text{ is the } j^{\text{th}} \text{ smallest entry in } A)$$
$$= prob(M \text{ is } j^{\text{th}} \text{ entry when } A \text{ is sorted})$$
$$= 1/n$$

since M is equally likely to end up in any position $A[j]$ after A is sorted. Each
$prob(S_j) > 0$, so applying Theorem 10.5.2, we obtain

$$E(X) = \sum_{j=1}^{k} E(X \mid S_j) \times prob(S_j)$$

$$= \sum_{j=1}^{n} E(X \mid M \text{ ends up in } A[j]) \times prob(M \text{ ends up in } A[j]).$$

$$= \sum_{j=1}^{n} E(X \mid M \text{ ends up in } A[j]) \times \left(\frac{1}{n}\right).$$

Let $E[n]$ denote the expected value of $X(A)$ over all inputs A where

$$X(A) = (n - 1) + X(A[1]..A[j - 1]) + X(A[j + 1]..A[n]).$$

But if $j = 1$, $X(A) = (n - 1) + \qquad\quad 0 \qquad + X(A[j + 1]..A[n]);$
and if $j = n$, $X(A) = (n - 1) + X(A[1]..A[n - 1]) + 0.$

Setting $E[0] = 0$, we get

$$E(X \mid M \text{ ends up in } A[j]) = (n - 1) + E[j - 1] + E[n - j].$$

Hence,

$$E[n] = \sum_{j=1}^{n} \{(n - 1) + E[j - 1] + E[n - j]\}\frac{1}{n}$$

$$= \frac{1}{n}\left\{\sum_{j=1}^{n}(n - 1) + \sum_{j=1}^{n}E[j - 1] + \sum_{j=1}^{n}E[n - j]\right\}$$

$$= (n - 1) + \frac{1}{n}\{E[0] + E[1] + E[2] + \ldots + E[n - 1]\}$$

$$\qquad\quad + \frac{1}{n}\{E[n - 1] + E[n - 2] + \ldots + E[2] + E[1] + E[0]\}$$

$$= (n - 1) + \frac{2}{n}\{E[0] + E[1] + E[2] + \ldots + E[n - 1]\}.$$

Then, $E[0] = 0$ // the default value
$E[1] = 0$
$E[2] = (2 - 1) + (2/2)\,\{E[0] + E[1]\}$
$\quad\; = (1) \qquad + (1) \quad \{0 + 0\}$
$\quad\; = 1$ // $< 2\lg(2) = 2$

$E[3] = (3 - 1) + (2/3)\,\{E[0] + E[1] + E[2]\}$
$\quad\; = (2) \qquad + (2/3)\,\{0 + 1\}$
$\quad\; = 8/3$ // $< 3\lg(3) = 4.754$

$E[4] = (4 - 1) + (2/4)\,\{E[0] + E[1] + E[2] + E[3]\}$
$\quad\; = (3) \qquad + (1/2)\,\{ \qquad\qquad\quad 1\; + 8/3\}$
$\quad\; = 29/6$ // $< 4\lg(4) = 8$

$E[5] = (5 - 1) + (2/5)\{E[0] + E[1] + E[2] + E[3] + E[4]\}$
$\quad\; = (4) \qquad + (2/5)\{ \qquad\qquad\qquad 11/3 + 29/6\}$
$\quad\; = 37/5.$ // $< 5\lg(5) = 11.609$

We have $E[n] = (n-1) + (2/n)\{E[0] + E[1] + E[2] + \ldots + E[n-1]\}$.
Multiplying by n gives

$$nE[n] = n(n-1) + (2)\{E[0] + E[1] + E[2] + \ldots + E[n-2] + E[n-1]\}.$$

That same equation holds when n is replaced by $(n-1)$ // when $n > 1$

$$(n-1)E[(n-1)] = (n-1)(n-2) + (2)\{E[0] + E[1] + E[2] + \ldots + E[n-2]\}.$$

Subtracting produces a first order recurrence equation (but having variable coefficients)

$$
\begin{aligned}
nE[n] - (n-1)E[n-1] &= n(n-1) - (n-1)(n-2) + (2)\{E[n-1]\} \\
nE[n] &= (n-1)[n - (n-2)] \qquad + (2)E[n-1] + (n-1)E[n-1] \\
&= (n-1)[2] \qquad\qquad\quad + [(2+(n-1)])E[n-1] \\
&= 2(n-1) \qquad\qquad\quad + (n+1)E[n-1].
\end{aligned}
$$

When $n > 1$ $\qquad E[n] = \dfrac{2(n-1)}{n} + \dfrac{n+1}{n}E[n-1]$ $\qquad\qquad$ (10.6.1)

$$
\begin{aligned}
E[2] &= 2(2-1)/2 + [(2+1)/2]E[1] \\
&= (1) \qquad + (3/2)\{0\} \\
&= 1 \qquad\qquad\qquad\qquad\qquad // < 2\lg(2) = 2.
\end{aligned}
$$

$$
\begin{aligned}
E[3] &= 2(3-1)/3 + [(3+1)/3]E[2] \\
&= (4/3) \qquad + (4/3)\{1\} \\
&= 8/3 \qquad\qquad\qquad\qquad\quad // < 3\lg(3) = 4.754.
\end{aligned}
$$

$$
\begin{aligned}
E[4] &= 2(4-1)/4 + [(4+1)/4]E[3] \\
&= (3/2) \qquad + (5/4)\{8/3\} \\
&= 29/6 \qquad\qquad\qquad\qquad\quad // < 5\lg(5) = 11.609.
\end{aligned}
$$

$$
\begin{aligned}
E[5] &= 2(5-1)/5 + [(5+1)/5]E[4] \\
&= (8/5) \qquad + (6/5)\{29/6\} \\
&= 37/5. \qquad\qquad\qquad\qquad\quad // < 5\lg(5) = 11.609.
\end{aligned}
$$

// It appears that $E[n] < n\lg(n)$.

We will prove that the average-case complexity of QuickSort is $O(n\lg(n))$ by proving

Theorem 10.6.1: For $\forall\, n \in \{2..\,\},$ $\qquad E[n] < 2 \times n\lg(n).$

Proof. // by Mathematical Induction

Step 1. If $n = 2$, then $E[2] = 1$ but $2 \times n\lg(n) = 2 \times 2 \times \lg(2) = 2 \times 2 \times 1 = 4$.
Step 2. Assume that $\exists\, k \in \{2..\,\}$, where $E[k] < 2 \times k\lg(k)$.

Step 3. If $n = k+1$, then

$$E[k+1] = \frac{2([k+1]-1)}{k+1} + \frac{[k+1]+1}{k+1}E[k] \qquad \text{// by (10.6.1)}$$

$$= \frac{2k}{k+1} + \frac{k+2}{k+1}E[k]$$

$$< \frac{2k}{k+1} + \frac{k+2}{k+1} \times 2 \times k\lg(k) \qquad \text{// by Step 2}$$

$$= \frac{2k}{k+1} + \frac{(k+2)k}{k+1} \times 2 \times \lg(k)$$

$$< \frac{2k}{k+1} + (k+1)2 \times \lg(k) \qquad \text{// } (k+1)^2 = k^2 + 2k + 1.$$

$$< \frac{2(k+1)}{k} + (k+1)2 \times \lg(k) \qquad \text{// } \frac{k}{k+1} < 1 < \frac{k+1}{k}.$$

$$= 2(k+1)\left\{\frac{1}{k} + \lg(k)\right\}.$$

// The theorem will be proven if we show that $\frac{1}{k} + \lg(k)$ is $< \lg(k + 1)$.

If $n \geq 2$ and $x > 0$, applying the Binomial Theorem $\qquad$ (Theorem 3.8.1)

$$(1+x)^n = \sum_{r=0}^{n}\binom{n}{r}x^r(1)^{n-r} = 1 + nx + \sum_{r=2}^{n}\binom{n}{r}x^r > 1 + nx.$$

//X Prove $(1 + x)^n > 1 + nx$ directly using Mathematical Induction on n.

Then,

$$\left(\frac{k+1}{k}\right)^k = \left(1 + \frac{1}{k}\right)^k > \left(1 + k\frac{1}{k}\right) = 2.$$

Therefore, $\qquad\qquad \lg(2) < \lg\left[\left(\frac{k+1}{k}\right)^k\right]; \qquad$ // lg is increasing.

that is, $\qquad 1 < k \times \lg\left(\frac{k+1}{k}\right) = k \times \{\lg(k+1) - \lg(k)\}. \qquad$ // See Chap. 2

So $\qquad \frac{1}{k} < \lg(k+1) - \lg(k)$

and $\qquad \frac{1}{k} + \lg(k) < \lg(k+1).$ $\qquad\qquad\qquad$ $\square$

The Most Important Ideas in This Section.
If we search array $A[1]..A[n]$ for a target value T, a natural probabilistic model is:

$$prob(T \text{ is in the list}) = p \qquad\qquad \text{where } 0 <= p <= 1$$
$$\text{and}\quad prob(T = A[j] \mid T \text{ is in the list}) = 1/n \quad \text{for } j = 1, 2, \ldots, n.$$

Then, for ***Linear Search***, $E(\# \text{ of probes}) = \left(\frac{n+1}{2}\right)p + nq$.

If the input to QuickSort is a random ordering of the entries in $A[1]..A[n]$, all the entries are distinct, and $X(A)$ denotes the number of comparisons of key values done; we proved (with some effort): For $\forall\, n \in \{2.. \}$, $E(\#$ key comparisons$)$ $<$ $2 \times n\lg(n)$. Therefore, the ***average-case complexity*** of QuickSort is $O(n\lg(n))$.

Average-case complexity values can <u>only</u> be calculated in the context of some probability model, that is, some complexity measure given as a random variable X, some set of values for X, and especially, some assumed probability distribution for X.

Exercises

1. Consider two women, Alice and Barbara, who are each about to have a baby.
   (a) What is the probability that Alice has a girl?
   (b) What is the probability that Barbara has a girl?
   (c) What is the probability that both Alice and Barbara have girls?
   (d) If Alice gives birth to a girl before Barbara has her baby, what is the probability that Barbara has a girl?
   (e) Are the outcomes Alice has a girl and Barbara has a girl independent events? Justify your answer.

2. Consider the following pairs of events. For each pair of events, indicate whether the events are independent, mutually exclusive, both independent and mutually exclusive, or neither independent nor mutually exclusive.
   (a) An eight is drawn from a standard deck of cards. A coin toss results in heads.
   (b) A king is the first card received from the dealer when playing a hand of Blackjack. A king is the second card received from the dealer when playing the same hand of Blackjack.
   (c) A 4 is rolled on a 6-sided die. A second 4 is rolled on the same 6-sided die.
   (d) The sum of two random integers, x and y, is greater than x and greater than y. The product of the same two integers, x and y, is zero.

3. Suppose that some "experiment" selects a number Q at random from $\{9, 10, \ldots, 88\}$.
   Let **A** be the event that $3 \mid Q$.   // 3 divides evenly into Q.
   Let **B** be the event that $7 \mid Q$.   // 7 divides evenly into Q.

(a) Find $prob(A)$, $prob(B)$, $prob(A \text{ and } B)$, $prob(A \text{ or } B)$, $prob(A \mid B)$, and $prob(B \mid A)$.

(b) Are **A** and **B** mutually exclusive? Explain why or why not.

(c) Are **A** and **B** independent? Explain why or why not.

4. An "experiment" produces a number n from 1 to 31 inclusive at random. Let **A** be the event that n is prime, and let **B** be the event that n is odd.

(a) Find $prob(A)$, $prob(B)$, $prob(A \text{ and } B)$, $prob(A \text{ or } B)$, $prob(A \mid B)$, and $prob(B \mid A)$.

(b) Are **A** and **B** mutually exclusive? Explain why or why not.

(c) Are **A** and **B** independent? Explain why or why not.

(d) Find $E(n)$.

5. A small delegation of k students is chosen by lot (at random) from a class of size n to complain about a certain aspect of the course. Assume that Flora and Mike are in the class and that $1 < k < n$.

(a) Is the probability that Mike is in the delegation $1/k$? Or $1/n$? Or something else?

(b) What's the probability that Flora is in the delegation?

(c) Are the events "Mike is in" and "Flora is in" independent?

6. Suppose Mr. X flips coins in some peculiar way so that $prob$ (Mr. X gets H) $= p \neq \frac{1}{2}$, but the $prob$(you get H) $= \frac{1}{2}$. If you both flip independently, what is the probability that your coins "match," that is, you both get H or you both get T?

7. Suppose A is an event with $prob(A) > 0$. For all $x \in A$, define $P(x \mid A)$ by

$$P(x \mid A) = \frac{prob(A \text{ and } \{x\})}{prob(A)} = \frac{P(x)}{prob(A)}.$$

Prove $P(x \mid A)$ is a probability function on A; that is, $P(x \mid A) \geq 0$ and $\sum_{x \in A} P(x \mid A) = 1$.

8. Find an expression for the expected number of probes made during a Binary Search of an array of length n, when all the entries are different and all are equally likely to be the target of the search. When is this $< \lg(n)$?

9. If certain entries in an array, X, are much, much more likely than others to be the target of a search, it has been suggested that the entries should be sorted from most likely to least likely into an array Y, and then Y should be searched using Linear Search. Suppose $X_1 < X_2 < \ldots < X_{10}$ and past history indicates the following probabilities that each item will be the target of the search:

X-values	X_{10}	X_7	X_4	X_9	X_6	X_3	X_1	X_8	X_2	X_5
$Prob(T = X_i)$	0.40	0.22	0.10	0.07	0.06	0.05	0.04	0.03	0.02	0.01

(a) Calculate the expected number of probes for a Linear Search of array Y.

(b) Calculate the expected number of probes for a Binary search of array X.

(c) Has this example been constructed so that Linear Search is better on average?

10. Suppose that over the past several years, 4,000 students have taken a certain course, which has a midterm test. Suppose also that

85% passed the test,

95% of those who passed the test also passed the course, and

92% of those who failed the test also failed the course.

(a) Construct a contingency table.

(b) If one of those students were selected at random, calculate to 5 decimal places:

(i) *prob*(he failed the test | he passed the course)

(ii) *prob*(he passed the test | he failed the course)

11. A die is weighted so that

(i) 3 comes up 7 times as often as 4,

(ii) 4 comes up twice as often as 1, 2, 5 and 6, and

(iii) 1, 2, 5, and 6 are equally likely.

Let X denote the outcome on any roll and let q denote $prob(X = 4)$.

(a) Determine the probability distribution for X and the value of q.

(b) Find E(X), the expected value of X.

12. While dice with 6 sides are, perhaps, the most common, dice are also available with 4 and 8 sides (among others).

(a) Create a table, similar to the one in Section 10.3 that shows the possible outcomes of an experiment where a 4-sided die is rolled with an 8-sided die.

(b) The totals that can occur when rolling such a pair of dice range between 2 and 12. Construct a table that shows the probability that each total will occur.

(c) How do the probabilities of the totals for two 6-sided dice compare to the probabilities of the totals for a 4-sided die and an 8-sided die?

(d) Is the expected value for the total of two 6-sided dice the same as the expected value for the total of a 4-sided die and an 8-sided die?

13. A bankcard "password" is a sequence of 4 digits.

(a) How many passwords are there?

(b) How many have no repeated digit?

(c) How many have the digit "5" repeated j times for all possible values of j?

(d) If a password were generated at random, what would be the expected number of 5s?

14. (a) Suppose that X is a "random variable" defined on a sample space S but $X(O_j) = C$ for every outcome. Prove $E(X) = C$.
    (b) Suppose X and Y are two random variables defined on the same sample space S. Prove that if for each outcome in S, $X(O_j) <= Y(O_j)$, then $E(X) <= E(Y)$.

15. It is known that 85% of individuals who purchase a particular make of laptop do not make any claims on their guarantee. Suppose 43 customers buy that make of laptop from a dealer. Compute the probability that at least 3 owners will make a claim on their guarantee. Hint: Will this be 1 – the probability that fewer than 3 owners will make a claim?

16. A certain network component receives "encoded messages" from other components over very "noisy" data channels where 20% of the messages received contain errors. Assuming that the errors are independent, calculate:
    (a) *prob*(exactly 3 out of 13 messages contain errors).
    (b) *prob*(fewer than 3 out of 13 messages contain errors).
    (c) *prob*(more than 3 out of 13 messages contain errors).
    (d) The expected number of error-free messages in 13 transmissions.
    (e) Suppose also that the component can detect errors, and when a message is received that contains an error, the component requests that the message be sent again. Calculate *prob*(a message is sent exactly 6 times before it is correctly received). What is the expected number of times any message must be sent?

17. If 38% of computer science students in North America last year were women,
    (a) What is the probability that the top mark was earned by a woman?
    (b) What is the probability that exactly 6 of the 10 top marks were earned by women?
    (c) How many women would you expect in the top 10?

18. Suppose the probability of passing a driving test is 65%.
    (a) On a certain afternoon 15 people are tested.
        (i) What is the expected number of people who pass?
        (ii) What is the probability that exactly 11 people pass?
    (b) A certain teenager is determined to get a driving license no matter how many times he has to take the test until he finally passes.
        (i) What is the probability that he passes on his third try?
        (ii) What is the expected number of tries it takes to pass?

19. Suppose that in a certain board game player A will win if she rolls a 4 on either or both of two unbiased dice.
    (a) What is the probability she will win on her next roll?
    (b) How many rolls should she expect to use to win?

20. For the experiment of flipping a coin until the first occurrence of HH after X flips in Example 10.5.1
    (a) Show $prob(X = 2) = 1/4$ using formula (10.5.2).
    (b) Show $prob(X = 3) = 1/8$.
    (c) Show $prob(X = 4) = 2/16$.
    (d) Show $prob(X = 5) = 3/32$.

21. If you flip a quarter and a dime together until both are heads, each trial consists of two flips where

$$prob[(\text{quarter is H) and (dime is H)}] = (\tfrac{1}{2})(\tfrac{1}{2})$$

because the two flips are independent. Therefore, on average it should take

$$\frac{1}{1/4} = 4 \text{ trials; that is, you'll flip 8 coins.}$$

Explain why this is **not** the same as flipping **one** coin until you get 2 heads in a row.

22. Suppose some "experiment" produces an integer $n \in \{2, 3, \ldots, 65\}$ at random. Let X be a random variable associated with this experiment where

when n is the integer produced, then
$$X(n) = \lfloor \sqrt{n} \rfloor - 1 \quad \text{if } n \text{ is prime;}$$
otherwise, $X(n) = p - 1,$ where p is the smallest prime factor of n.

   (a) Determine the possible values for X.
   (b) Determine the probability distribution for X.
   (c) Find the expected value of X.
   (d) How is this connected to the Algorithm 1.2.3 for prime testing?

23. Imagine a gambling game played with an ordinary deck of 52 playing cards. You pay $2.50 to play and you play by shuffling the deck and then turning over the top card. You win different amounts depending on which card you turn over.

If the top card is a diamond,   you win $4.50;
If the top card is a picture card, you win $4.50,
                                   but J, Q, or K of diamonds wins $7.00;
If the top card is an ace,   you win $5.00,
                                   but the ace of diamonds wins $10.00; and
If the top card is any other card, you win nothing.

   (a) Let f be your "payoff", that is, your win minus $2.50. Then f has a value for every outcome (card turned up) but only 5 different values. Determine the probability distribution for these values.
   (b) Calculate the expected value of f.
   (c) Why is this average payoff per play negative? Does that make sense?

24. **InsertionSort** appeared in the exercises following Chap. 4. While it can be implemented in a variety of ways, the general step assumes that $A[1] \ldots A[k-1]$ are in nondecreasing order, and now we must "**insert**" $Q = A[k]$ into its correct position among the previous entries.

(i) If $Q < A[1]$, then all the entries from $A[1]$ to $A[k-1]$ must be shifted up one position and then the value of Q may be put into $A[1]$.

(ii) Otherwise, $A[1] <= Q$, and the value of Q may be inserted after the rightmost entry $A[j]$ which is $<= Q$, provided that all the entries from $A[j+1]$ to $A[k-1]$ have been shifted up one position (because they are larger than Q).

(a) Assume the input array A is a random permutation of some n-set $B = \{b_1, b_2, \ldots, b_n\}$. Prove that after the first $k-1$ entries have been sorted so that $A[1] <= A[2] <= \ldots <= A[k-1]$,

$$prob(A[k] \text{ must be inserted at position } j) = 1/k.$$

// The number of different input arrays $= n!$. To construct (and later
// count) the input sequences where $A[k]$ must be inserted at position j:
//   1.   Choose k elements of B to put into $A[1] .. A[k]$.
//   2.   Put the jth smallest of these into $A[k]$.
//   3.   Arrange the other $k-1$ chosen elements in any order in
// $A[1] \ldots A[k-1]$.
//   4.   Arrange the "un-chosen" $n-k$ elements in any order in
// $A[k+1] \ldots A[n]$.

(b) **Algorithm InsertionSort**

// Rearranges the entries in array $A[1] \ldots A[n]$ so that
// $A[1] <= A[2] <= \ldots <= A[n]$.

```
Begin
    For k ← 2 To n Do            // Insert A[k] into the sorted sub-array.
                                 // A[1] ... A[k − 1]

        Q ← A[k];
        j ← k − 1;
        If (Q < A[1]) Then
            While (j > 0) Do
                A[j + 1] ← A[j];
                j ← j − 1;
            End;                 // the while-loop, now j = 0
            A[1] ← Q;
        Else                     // A[1] <= Q.
            While (A[j] > Q) Do
                A[j + 1] ← A[j];
                j ← j − 1;
            End;                 // the while-loop, now A[j] <= Q
            A[j + 1] ← Q;
        End;                     // the if-then-else statement
    End;                         // the for-loop
End.                             // of Algorithm InsertionSort
```

Show that the expected number of key comparisons for InsertionSort is

$$\left(\frac{(n-1)(n+4)}{4} \right) \text{ which is } O(n^2).$$

(c) **Algorithm InsertionSort#2**
// Rearranges the entries in array A[1]...A[n] so that
// A[1] <= A[2] <= ... <= A[n].

Begin
   For k ← 2 **To** n **Do**   // Insert A[k] into the sorted sub-array.
                                            // A[1]...A[k − 1]

    Q ← A[k];
    j ← k − 1;
    A[0] ← Q;
    While (A[j] > Q) **Do**
      A[j + 1] ← A[j];
      j ← j − 1;
    End;   // the while-loop, now A[j] <= Q even if j = 0
    A[j + 1] ← Q
   End;   // the for-loop
End.   // of Algorithm InsertionSort#2

Determine the Expected number of key comparisons for InsertionSort#2. Should this run about twice as fast (half as slowly) as BubbleSort?

Turing Machines

11

At the International Congress of Mathematicians held in Paris in 1900, David Hilbert, a very eminent German professor, proposed a list of 23 problems that he felt should be undertaken in the new century. The one related to this chapter is the tenth problem.

> Is there a mechanical procedure for determining whether or not
> a given polynomial equation with several variables and integer
> coefficients has a solution in integers?

// For instance, are there integer values for x, y and z so that
//
// $6x^3yz^2 + 3xy^2 - 4z^3 - 12 = 0$?
//   (try $x = 1$, $y = 2$, $z = 3$)
// Such equations are named Diophantine equations after
// Diophantus of Alexandria (c 250 AD).

In the 1920s, Hilbert asked a much more general question:

> Is there a mechanical procedure for determining whether
> a given mathematical statement is True or False?

This became known as the "*Entscheidungsproblem*" – German for "decision problem".

By a mechanical procedure, Hilbert meant a **general method** for applying the deductive rules of logic to the axioms, definitions, and known theorems of algebra and arithmetic to answer the question. But more importantly, the general method must be a sequence of simple "operations", unambiguous steps that could be carried out by an unthinking automaton.

This seems to describe the essential character of a programmable algorithm, so let us reconsider the question posed in Sect. 1.1.

© Springer International Publishing AG, part of Springer Nature 2018   467
T. Jenkyns and B. Stephenson, *Fundamentals of Discrete Math for Computer Science: A Problem-Solving Primer*, Undergraduate Topics in Computer Science,
https://doi.org/10.1007/978-3-319-70151-6_11

11.1   What Is an Algorithm?

A number of mathematicians and logicians began to work on Hilbert's problems. Among them were Alan Turing and Alonzo Church. The remainder of this chapter is a description of several of Turing's ideas. These originated in a paper Turing published in 1936 entitled "On Computable Numbers, with an Application to the Entscheidungsproblem".

Before either of Hilbert's questions could be answered, the idea of what constitutes a "mechanical procedure" had to be clarified and formalized in precise (mathematical) language.

// How could one possibly demonstrate that the answer is "No, there is no such
// procedure."? If the answer to Hilbert's second question were "yes", then
// mathematicians could be replaced by unthinking automatons.

Turing first had to "invent" an unthinking automaton that could carry out a sequence of simple, unambiguous steps – his invention is now called a "Turing Machine". (It was invented long before any electronic computer had been invented).

Turing thought about a human "computer" working with paper and pencil when doing arithmetic and tried to abstract the essential elements. First, the sheet of paper could be imagined as a long strip of paper divided into squares that would contain individual digits of the calculation; second, the digit in any square could be erased and replaced by some other digit (or a digit could be written in a blank square), and then the person could move on to an adjacent square and continue the calculation.

The "hardware" of a Turing Machine consists of:
1. A "tape" divided into squares where each square may contain any one of a finite number of symbols from the machine's "alphabet".
   // This tape is a medium for input and output *and* serves as memory.
2. A "read-write head" (RWH) that is positioned on the tape at some square and can read the symbol in that square, overwrite that symbol with a symbol, and then move one square left or one square right.
3. A finite set of "states" (like gears in a car), one of which is designated the starting state, which we denote by s.

The actions of the machine are determined by a finite set of "quintuples" of the following form:

  $(p, x: y, dir, q)$ where

  p and q are states of the machine
  x and y are symbols in the machine's alphabet, and
  dir   is a direction the RWH might move
                      either **L** for left or **R** for right.

However, no two quintuples begin with the same two entries p and x.

The way in which the quintuples control the action of the machine is this: Suppose the machine is in state p and the RWH is positioned at a square containing symbol x.

If the set of quintuples contains one like $(p, x: y, dir, q)$, then:
1. The machine changes x on the tape to y.
2. The RWH moves one square along the tape in the direction dir.
3. The machine changes to state, q.

But if no quintuple begins "$(p, x\ldots$ " then the machine halts.   // or "crashes"

The machine's alphabet is assumed to contain a special symbol, called the blank, which we will indicate by the symbol $\square$.   // so we can see it
At the start of any computation by the machine, the tape must contain only a finite number of nonblank symbols, and the RWH is positioned at the leftmost of these nonblank symbols.   // if there is one
And the machine starts in its special "starting" state, s.

// This will make more sense when we do a walkthrough.

Example 11.1.1: A decimal successor Turing Machine

Input: A nonnegative integer N in decimal notation   // base 10
Output: $N + 1$ in decimal notation

Quintuples:

$(s, \quad 0: 0, R, s)$ // In state s, it moves to RHE, *the right-hand-end*,
$(s, \quad 1: 1, R, s)$ // of the string of nonblank symbols.
$(s, \quad 2: 2, R, s)$
$(s, \quad 3: 3, R, s)$
$(s, \quad 4: 4, R, s)$
$(s, \quad 5: 5, R, s)$
$(s, \quad 6: 6, R, s)$
$(s, \quad 7: 7, R, s)$
$(s, \quad 8: 8, R, s)$
$(s, \quad 9: 9, R, s)$
$(s, \quad \square: \square, L, t)$ // When it goes past the RHE, it steps left and goes to state t

$(t, \quad 0: 1, L, h)$ // In state t, it adds 1 to current digit and halts.
$(t, \quad 1: 2, L, h)$
$(t, \quad 2: 3, L, h)$
$(t, \quad 3: 4, L, h)$
$(t, \quad 4: 5, L, h)$
$(t, \quad 5: 6, L, h)$
$(t, \quad 6: 7, L, h)$
$(t, \quad 7: 8, L, h)$
$(t, \quad 8: 9, L, h)$
$(t, \quad 9: 0, L, t)$ // But, if the current digit is 9, it "carries one".
$(t, \quad \square: 1, L, h)$ // For when the input is all 9's.

// State h is a "halting" state because no quintuple begins with h.
// What happens if the input tape is all blanks?

Walkthrough when $N = 360499$

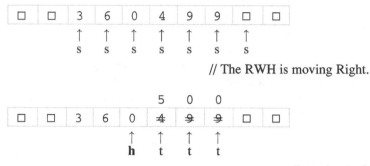

// The RWH is moving Right.

// moving Left

A ***palindrome*** is a word like *racecar* or *level* that is the same when written backwards. Sometimes this is extended to phrases or sentences, but then spaces, capitals, and punctuation are ignored, as in

Was it a car or a cat I saw?   Never odd or even.

Marge lets Norah see Sharon's telegram.   Drab as a fool, aloof as a bard.

A man, a plan, a canal: Panama.   No lemons, no melon.

or perhaps the world's first pick-up line – Madam, I'm Adam.

If a string of letters w is given, how easy is it to decide whether or not w is a palindrome? Is it so easy a Turing Machine could do it? The following example makes it even easier by restricting the alphabet to just two letters.

Example 11.1.2: An {a,b}-palindrome Turing Machine

Input: A string w of a's and b's

Output: *Will be expressed by the halting state*:

            If w is a palindrome, the Turing Machine halts in state "yes", but

            if w is not a palindrome, the Turing Machine halts in state "no".

Strategy: If w has only one letter, then w is a palindrome.

            If $w = aXb$ or bXa, then w is not a palindrome.

            And $w = aXa$ or bXb is a palindrome

            $\Leftrightarrow X$ is a palindrome (or X has no letters at all).

Quintuples:

   $(s, a: \square, \text{R}, 1)$ // In state s, it scans the LHE, ***the left-hand-end***.
                                       // If w starts with an a, does it end in an a?

   $(1,\ a: a, \text{R}, 1)$ // In state 1, it moves to RHE
   $(1,\ b: b, \text{R}, 1)$
   $(1, \square: \square, \text{L}, 2)$ // and goes to state 2.

$(2, \boldsymbol{a}: \square, \text{ L}, 3)$ // Instate 2, it checks that RHE = $\boldsymbol{a}$
$(2, \boldsymbol{b}: \boldsymbol{b}, \text{ R, no})$
$(2, \square: \square, \text{ L, yes})$ // was $\boldsymbol{w} = \boldsymbol{a}$?   [or was $X = \boldsymbol{a}$?]

$(3, \boldsymbol{a}: \boldsymbol{a}, \text{ L}, 3)$ // In state 3, it moves back to LHE
$(3, \boldsymbol{b}: \boldsymbol{b}, \text{ L}, 3)$
$(3, \square: \square, \text{ R, s})$ // and starts again.

$(\text{s}, \boldsymbol{b}: \square, \text{ R}, 4)$ // If w starts with a $\boldsymbol{b}$, does it end in a $\boldsymbol{b}$?

$(4, \boldsymbol{a}: \boldsymbol{a}, \text{ R}, 4)$ // In state 4, it moves to RHE.
$(4, \boldsymbol{b}: \boldsymbol{b}, \text{ R}, 4)$
$(4, \square: \square, \text{ L}, 5)$

$(5, \boldsymbol{a}: \boldsymbol{a}, \text{ R, no})$ // In state 5, it checks that RHE = $\boldsymbol{b}$
$(5, \boldsymbol{b}: \square, \text{ L}, 3)$
$(5, \square: \square, \text{ L, yes})$ // was $w = \boldsymbol{b}$?   [or was $X = \boldsymbol{b}$?]

$(\text{s}, \square: \square, \text{ R, yes})$ // Is the word with no letters a palindrome?

Walkthrough when $w = $ abbba

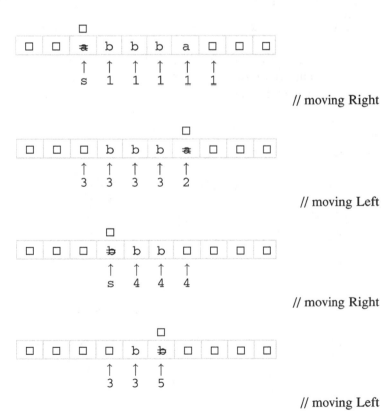

// moving Right

// moving Left

// moving Right

// moving Left

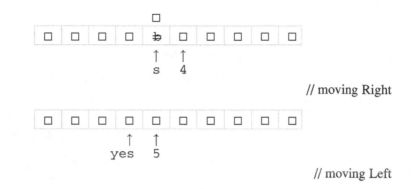

// moving Right

// moving Left

// What about computer operations?

Example 11.1.3: A binary adder

Input: A string of *0*s and *1*s,   // the summand
           then "+",
           then another string of *0*s and *1*s   // the addend
Output: A string of string of *0*s and *1*s that is the sum

Strategy: We must add corresponding bits together (and sometimes "carry" 1):
           the units bits, then the 2s bits, then the 4s bits, ...
           Let's work from the RHE; find the next bit *x* to be added
           but, let's erase bits from the addend as we use them.
           To indicate how far to the left we've done, let's replace *0* by *a* and *1* by *b*.
           Then, we move past the *a*'s and *b*'s in the summand to the corresponding
           bit *y*, where we do the addition.

Quintuples:

        (s, *0*: *0*, R, s)   // In state *s*, it moves to RHE.
        (s, *1*: *1*, R, s)
        (s, *a*: *a*, R, s)   // Later, there will be *a*'s and *b*'s in the summand.
        (s, *b*: *b*, R, s)
        (s, +: +, R, s)
        (s, □: □, L, 1)

        (1, *0*: □, L, 2)   // In state 1, it finds *x* = 0.
                               // We'll do the other cases for *x* later.

        (2, *0*: *0*, L, 2) // In state 2, it moves left across 0 and 1 to +.
        (2, *1*: *1*, L, 2)
        (2, +: +, L, 3)

$(3, \boldsymbol{a}: \boldsymbol{a}, L, 3)$ // In state 3, it moves left across $\boldsymbol{a}$ and $\boldsymbol{b}$ to $\boldsymbol{y}$,
$(3, \boldsymbol{b}: \boldsymbol{b}, L, 3)$
$(3, \boldsymbol{0}: \boldsymbol{a}, R, s)$ // $\boldsymbol{y} = 0$, and $\boldsymbol{x} + \boldsymbol{y} = 0$, but this is recorded as $\boldsymbol{a}$.
                  // Then, we move right to get the next $\boldsymbol{x}$.
$(3, \square: \boldsymbol{a}, R, s)$ // When there is no corresponding bit $\boldsymbol{y}$ because the
                  // summand is shorter than the addend, we take 0 for
                  // $\boldsymbol{y}$ and record an $\boldsymbol{a}$ and move right to get the next $\boldsymbol{x}$.
$(3, \boldsymbol{1}: \boldsymbol{b}, R, s)$ // $\boldsymbol{y} = 1$ and $\boldsymbol{x} + \boldsymbol{y} = 1$, but this is recorded as $\boldsymbol{b}$.
                  // Then, we move right to get the next $\boldsymbol{x}$.

$(1, \boldsymbol{1}: \sqcup, L, 4)$ // In state 1, it finds $\boldsymbol{x} = 1$.

$(4, \boldsymbol{0}: \boldsymbol{0}, L, 4)$ // In state 4, it moves left across 0 and 1 to $+$.
$(4, \boldsymbol{1}: \boldsymbol{1}, L, 4)$
$(4, +: +, L, 5)$

$(5, \boldsymbol{a}: \boldsymbol{a}, L, 5)$ // In state 5, it moves left across $\boldsymbol{a}$ and $\boldsymbol{b}$ to $\boldsymbol{y}$.
$(5, \boldsymbol{b}: \boldsymbol{b}, L, 5)$
$(5, \boldsymbol{0}: \boldsymbol{b}, R, s)$ // $\boldsymbol{y} = 0$ and $\boldsymbol{x} + \boldsymbol{y} = 1$, but this is recorded as $\boldsymbol{b}$.
                  // Then, we move right to get the next $\boldsymbol{x}$.
$(5, \square: \boldsymbol{b}, R, s)$ // When there is no corresponding bit $\boldsymbol{y}$, take 0 for $\boldsymbol{y}$
                  // and record a $\boldsymbol{b}$ and move right to get the next $\boldsymbol{x}$.
$(5, \boldsymbol{1}: \boldsymbol{a}, L, 6)$ // $\boldsymbol{y} = 1$ and $\boldsymbol{x} + \boldsymbol{y} = 10$, but this is recorded as $\boldsymbol{a}$.
                  // Then, we move to state 6 and carry the one.

$(6, \boldsymbol{0}: \boldsymbol{1}, R, s)$
$(6, \square: \boldsymbol{1}, R, s)$
$(6, \boldsymbol{1}: 0, L, 6)$

$(1, +: \square, L, 7)$ // In state 1, it finds no $\boldsymbol{x}$.

$(7, \boldsymbol{a}: \boldsymbol{0}, L, 7)$ // In state 7, it replaces $\boldsymbol{a}$'s by $\boldsymbol{0}$s,
$(7, \boldsymbol{b}: \boldsymbol{1}, L, 7)$ // and it replaces $\boldsymbol{b}$'s by $\boldsymbol{1}$s,
$(7, \boldsymbol{0}: \boldsymbol{0}, L, 7)$ // moves to the LHE,
$(7, \boldsymbol{1}: \boldsymbol{1}, L, 7)$
$(7, \square: \square, R, \boldsymbol{h})$ // and halts.

Walkthrough   // $29 + 10 = ?$

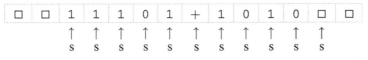

                                              // moving R

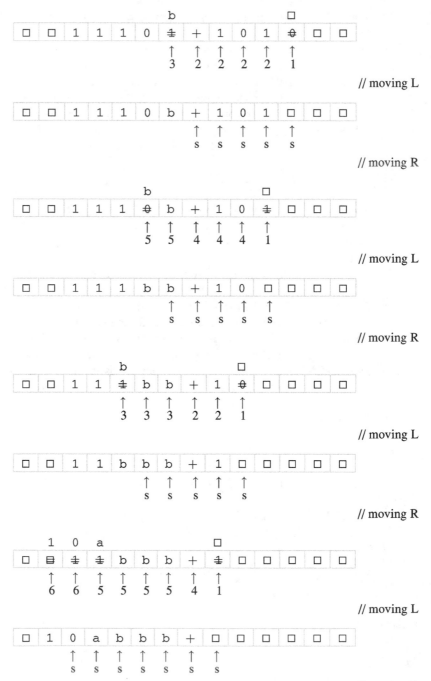

// moving L

// moving R

// moving L

// moving R

// moving L

// moving R

// moving L

// moving R

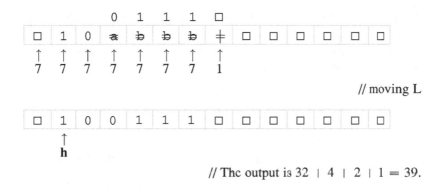

// moving L

// The output is 32 | 4 | 2 | 1 = 39.

// What about other computer operations?
// What machine language instructions can be performed by a Turing Machine?
// Can every machine language instruction be performed by a Turing Machine?

11.1.1 The Church-Turing Thesis

Throughout this text, we have written about algorithms – methods for doing computations or processing data.

// not algorithms to tie your shoes

We used an intuitive and informal "definition" of an algorithm (in Chap. 1) as a "step-by-step process," where a "step" was a relatively easy sub-task, but we never had a formal definition of a "step." This clouded our idea of complexity, the number of steps required for the algorithm to complete its job, as a function of the input size, n. (We never even thought about proving that no algorithm could be devised for some task.)

All that has changed now that we've considered Turing Machines, where a step is the execution of one quintuple, and the input size is the number of nonblank symbols on the tape when the machine begins. Turing Machines provide an objective model of what can be computed and how efficiently the computation may be done. This assertion is known as the Church-Turing Thesis: Any process that is an algorithm (in the informal sense) can be realized as a Turing Machine.

The Church-Turing Thesis: Algorithms EQUAL Turing Machines.

// This is not called a theorem because "algorithm" is not defined precisely in
// formal mathematical terms; it remains an informal idea.

This thesis is universally accepted, in part, because several other interpretations of a "mechanical procedure" were all proved equivalent to Turing Machines.

11.1.2   Universal Turing Machine: As a Computational Model

We've walked through several examples of the operation of a Turing Machine, **M**. It was a fairly simple process:

> Given the list of the quintuples of **M**,
>> the current contents of **M**'s tape,
>> the current state **p** of **M**, and
>> the current symbol **x** being scanned
>> we searched through the quintuples for one that began (**p**, **x** . . . .

> If we found the sought for quintuple (**p**, **x**: **y**, **dir**, **q**), then we
>> changed **x** on the tape to **y**,
>> moved one square along the tape in the direction **dir**,
>> found the new current symbol **z**, and
>> changed to the new current state, **q**.
> Otherwise, no quintuple began with (**p**, **x** . . . so the walkthrough halted.

// Could this process be described as an algorithm?
// Could this process be done by a computer program?
// Could this process be done by a Turing Machine, with enough states and tape-
// symbols?

The answer is "yes" to all three questions. Turing himself described how to construct a **Universal Turing Machine**, **U**. The input to **U** is a description <**M**> of (the quintuples of) a Turing Machine **M** together with a description of the input tape <**w**> for **M**. Then, **U** simulates the action of **M** on its own tape.

> // until **M** halts, or forever if **M** never halts

This Universal Turing Machine is a model of modern, stored program computers; they accept programs as input and simulate the action of each program on its own data.

11.1.3   The Halting Problem

When you write programs in some language like Java, part of debugging is making sure the program never enters a loop from which it never exits, an infinite loop. You carefully analyze the program and remove any such loop, or program carefully enough to never introduce any such loop. The Halting Problem is about using a program to detect infinite loops.

Some programs have a text-file as input, like word-processors. You might write one to count the symbols, words, sentences, commas, or whatever in a certain block of text. Your program could also work on a program, input as a text-file, and count the symbols, words, sentences, commas, or whatever. Compilers and interpreters are programs that have programs as input.

Now imagine a debugging program D that has as its input two text-files: the first is a program P, and the second is a data-file X to be used as input for P. The debugger D analyzes the program P and the data X carefully enough to determine whether or not

> P acting on X gets into an infinite loop
>   or not.   // crashes, or terminates "normally"

Suppose the output from D is a Boolean variable H // for "halts"
so that

> H is *False* means P acting on X gets into an infinite loop and never halts.
> H is *True* means P acting on X eventually does halt.

// Besides evaluating H, D might issue some appropriate diagnostic message.

Determining whether or not P halts on X would be an instance of *the Halting Problem*. The program D would *solve* the Halting Problem (for <u>any</u> program and <u>any</u> data set).

Alan Turing showed that no Turing Machine could solve the Halting Problem for Turing Machines. We will prove that no program in language J can solve the Halting Problem for programs written in language J. The debugging program we imagined simply *cannot exist*; it cannot be constructed no matter how smart the programming team tasked to write it is and no matter how much money and brain power is invested in writing it. The Halting Problem cannot be solved.

Theorem 11.3.1: The debugging program D (described above) cannot exist.

Proof. // We use an indirect argument.

Suppose that there were such a debugging program D that works correctly on *any* pair of text-files: the first containing a program P and the second containing a text-file X used as input for P. That is, for every input-pair P and X:

1. D analyzes the operation of P on input X.
2. D assigns a value to a Boolean variable H, where:
   > H is set to *True* if P acting on X eventually halts.
   > H is set to *False* if P acting on X never halts.
3. D itself halts, immediately after assigning a value to H.

If we have program D, we can make a small modification to it and construct a new program E: immediately after D has given a value to H, insert one statement:

```
While (H) Do
   H ← H;
End;                         // or the equivalent form in language J
```

// Then, E enters an infinite loop if H is True, but halts, just as D did, if H is False.

If program **D** exists, then so does program **E**.

But now, suppose we take *X* to be equal to **P**, and we use **D** to determine what will happen if we ran program **P** with its own description as input: // **P** might crash.

> If program **P** acting on itself halts, **D** sets *H = **True***,
> and then **E** enters an infinite loop.

> If program **P** acting on itself never halts, **D** sets *H = **False***,
> and then **E** halts immediately.

But what happens if as input for **E** we take *X* equal to **E**, and **P** equal to **E** itself?

> If program **E** acting on itself halts, **D** sets *H = **True***,
> and then **E** enters an infinite loop.

> If program **E** acting on itself never halts, **D** sets *H = **False***,
> and then **E** halts immediately.

We have a contradiction:

> If program **E** acting on itself halts, then **E** acting on itself does not halt.
> If program **E** acting on itself does not halt, then **E** acting on itself does halt.

Therefore, program **E** cannot exist.

Therefore, program **D** cannot exist.   ☐

Alan Turing's paper gave a basis for proving that no Turing Machine could solve the Halting Problem for Turing Machines. This solved Hilbert's Entscheidungsproblem.

> Is there (a mechanical procedure) *a Turing Machine*
> for determining whether a given mathematical statement, say
> *"Turing Machine M with input tape w eventually halts"*
> is True or False?

The answer is "no".

The Church-Turing Thesis says, in effect, that any task that can be done by a computer (program) can be done by a Turing Machine. Therefore, any task that cannot be done by a Turing Machine cannot be done by a computer. Turing Machines not only provide a context for ascertaining algorithmic complexity, they determine what is possible to be computed and what numbers are "computable".

One final idea we want to give you is that if the restriction that "no two quintuples of a Turing Machine may begin with the same two entries *p* and *x*" is removed, "nondeterministic" Turing Machines are produced. Problems that can be solved in polynomial time by nondeterministic Turing Machines form the class NP; problems that can be solved in polynomial time by deterministic Turing Machines form the class P. The question "Is P = NP?" is thought by some to be the most important question in theoretical computer science. And it is worth $1,000,000 to the Clay Mathematics Institute.

> **The Most Important Ideas in This Section.**
> We said an algorithm is a "step-by-step process" but we never defined a "step". Our idea of complexity was the number of steps required by the algorithm, as a function of the input size, n.
>
> Turing Machines provide an objective model of what can be computed, and how efficiently the computation may be done. In a Turing Machine, a step is execution of a quintuple, and the input size is the number of nonblank symbols on the tape when the machine begins.
>
> The Church-Turing Thesis asserts any process that is an algorithm can be realized as a Turing Machine. Any process that can be done by a computer can be done by a Turing Machine, and vice versa.
>
> Turing gave a basis for proving that no Turing Machine could solve the Halting Problem for Turing Machines, and thereby proved
> ***mathematicians cannot be replaced by unthinking automatons.***

Exercises

1. Let T be the Turing Machine whose quintuples are:

$(s, \ 0: \ 0, R, \ s)$ $(2, \ 0: 1, L, 1)$ $(4, \ 0: 0, L, 3)$
$(s, \ 1: \ 1, R, \ s)$ $(2, \ 1: 0, L, 4)$ $(4, \ 1: 1, L, 4)$
$(s, \square: \ \square, L, 1)$ $(2, \square: 1, L, h)$ $(4, \square: 0, L, 3)$

$\qquad\qquad (1, \ 0: \ 0, L, 1)$ $(3, \ 0: 1, L, 1)$
$\qquad\qquad (1, \ 1: \ 1, L, 2)$ $(3, \ 1: 0, L, 4)$
$\qquad\qquad (1, \square: \ \square, L, h)$ $(3, \square: 1, L, h)$

   (a) Walk through the operation of machine T applied to the following input tapes. Assume the RWH starts in state **s** reading the leftmost nonblank character.

   (i)   … $\square\square\square$101$\square\square\square$…
   (ii)   … $\square\square\square$110$\square\square\square$…
   (iii) … $\square\square\square$101101$\square\square\square$…

   (b) What function does this machine evaluate? That is, if the input tape contains the binary representation of the integer n, what is left on the tape when T halts? {Hint: look at the input and output strings in decimal.}

2. Let **T** be the Turing Machine whose quintuples are:

$(s, 0: 0, R, s)$ $(1, 0: \square, L, 1)$ $(2, 0: \square, L, 2)$ $(3, 0: \square, L, 3)$
$(s, 1: 1, R, s)$ $(1, 1: \square, L, 2)$ $(2, 1: \square, L, 3)$ $(3, 1: \square, L, 1)$
$(s, 2: 2, R, s)$ $(1, 2: \square, L, 3)$ $(2, 2: \square, L, 1)$ $(3, 2: \square, L, 2)$
$(s, 3: 0, R, s)$ $(1, \square: 0, L, h)$ $(2, \square: 1, L, h)$ $(3, \square: 2, L, h)$
$(s, 4: 1, R, s)$
$(s, 5: 2, R, s)$
$(s, 6: 0, R, s)$
$(s, 7: 1, R, s)$
$(s, 8: 2, R, s)$
$(s, 9: 0, R, s)$
$(s, \square: \square, L, 1)$

(a) Walk through the operation of machine T applied to the following tape. Assume the RWH starts in state s reading the leftmost nonblank character.
$\ldots \square\square\square 4207\square\square\square \ldots$

(b) What does the machine leave on the tape when applied to an input tape positive integer n written in decimal notation?
// Try a few small examples like 42 and 20 and 07.

3. Let **T** be the Turing Machine whose quintuples are:

$(0, 0: 0, R, 0)$ $(2, 0: 0, R, 4)$ $(4, 0: 1, R, 3)$
$(0, 1: 0, R, 1)$ $(2, 1: 1, R, 0)$ $(4, 1: 1, R, 4)$

$(1, 0: 0, R, 2)$ $(3, 0: 1, R, 1)$
$(1, 1: 0, R, 3)$ $(3, 1: 1, R, 2)$

(a) Walk through the operation of machine T applied to the following input tapes. Assume the RWH starts in state $s = 0$ reading the leftmost nonblank character.
   (i) $\ldots \square\square\square 110011\square\square\square \ldots$
   (ii) $\ldots \square\square\square 111110\square\square\square \ldots$
   (iii) $\ldots \square\square\square 1001010\square\square\square \ldots$
   (iv) $\ldots \square\square\square 0011001\square\square\square \ldots$

(b) What function does this machine evaluate? That is, if the input tape contains the binary representation of the integer n, what is left on the tape when T halts?   // Hint: look at the input and output strings in decimal.

(c) The final state is 0, 1 etc. Interpret this value as a function of the input integer n.

4. Construct the quintuples of a Turing Machine M that operates on (an input tape containing) a string of a's and b's, and halts in a "yes" state if and only if there are an integer number of a's followed by the same number of b's. The machine is allowed to crash in any state on other input strings.

5. Construct the quintuples of a Turing Machine M that operates on (an input tape containing) a string of a's, b's, and c's, and halts in a "yes" if and only if there are an integer number of a's followed by the same number of b's followed by the same number of c's. The machine is allowed to crash in any state on other input strings.

Hint: If the input tape has no nonblank symbols, halt in the "yes" state. Otherwise, move from left to right checking that the input string is some a's followed by some b's followed by some c's. Then, remove 1 c from the right-hand end, move left to the first b, replace that b by a c, go to the right-hand end, and remove another c. Then, go back to the left-hand end and remove an a, move right one square, and begin again in state s.

6. Look up Hilbert's tenth problem on the Internet. When was it solved? Who solved it? What is the answer to Hilbert's question?

7. Look up Hilbert's problems on the Internet. Were all 23 solved in the twentieth century? Are any of them on the list of million dollar problems given by the Clay Mathematics Institute for solution in this century?

Appendix A: Solutions to Selected Exercises

A.1 Solutions from Chapter 1

1. Using the conversion technique described in Section 1.3, or RPM described in Section 1.1:

$2015 = 2\ (1007) + 1$	$2015 =$	1024	
$1007 = 2\ (\ 503) + 1$		$+\ 512$	1536
$503 = 2\ (\ 251) + 1$		$+\ 256$	1792
$251 = 2\ (\ 125) + 1$		$+\ 128$	1920
$125 = 2\ (\ \ 62) + 1$		$+\ \ 64$	1984
$62 = 2\ (\ \ 31) + 0$			
$31 = 2\ (\ \ 15) + 1$		$+\ \ 16$	2000
$15 = 2\ (\ \ \ 7) + 1$		$+\ \ \ 8$	2008
$7 = 2\ (\ \ \ 3) + 1$		$+\ \ \ 4$	2012
$3 = 2\ (\ \ \ 1) + 1$		$+\ \ \ 2$	2014
$1 = 2\ (\ \ \ 0) + 1$		$+\ \ \ 1$	2015

$$2015\{10\} = 11\ 111\ 011\ 111\{2\}$$

2. Yes: $83/2 = 41.5$
   $83/3 = 27.666\ 666\ ...$
   $83/5 = 16.6$
   $83/7 = 11.857\ 142\ ...$
   No further values need to be considered because $11 > \sqrt{83}$ or $11^2 = 121 > 83$.

3. $801 = 3*267$
   $802 = 2*401$
   $803 = 11*73$
   $805 = 5*161 = 5*7*23$
   $807 = 3*269$
   809 is prime   (Try 2, 3, 5, 7, 11, 13, 17, 19 and 23... $29^2 = 841$)

© Springer International Publishing AG, part of Springer Nature 2018   483
T. Jenkyns and B. Stephenson, *Fundamentals of Discrete Math for Computer Science: A Problem-Solving Primer*, Undergraduate Topics in Computer Science, https://doi.org/10.1007/978-3-319-70151-6

7. If $t <= \lfloor\sqrt{Q}\rfloor$ then $t <= \sqrt{Q}$ so $t \times t <= Q$ // squaring both sides

   If t is an integer and $t \times t <= Q$ then $t <= \sqrt{Q}$

                                           // taking the positive square root of both sides

   Because t is some integer $<= \sqrt{Q}$, t is less than the largest integer that is
   $<= \sqrt{Q}$; that is, $t <= \lfloor\sqrt{Q}\rfloor$.   ▯

8.

n	$n(n+1)$	$+ 17$
0	0(1)	$+ 17 = 17$ which is prime
1	1(2)	$+ 17 = 19$ which is prime
2	2(3)	$+ 17 = 23$ which is prime
3	3(4)	$+ 17 = 29$ which is prime
4	4(5)	$+ 17 = 37$ which is prime
5	5(6)	$+ 17 = 47$ which is prime
6	6(7)	$+ 17 = 59$ which is prime
7	7(8)	$+ 17 = 73$ which is prime
8	8(9)	$+ 17 = 89$ which is prime
9	9(10)	$+ 17 = 107$ which is prime
10	10(11)	$+ 17 = 127$ which is prime
11	11(12)	$+ 17 = 139$ which is prime
12	12(13)	$+ 17 = 163$ which is prime
13	13(14)	$+ 17 = 189$ which is prime
14	14(15)	$+ 17 = 217$ which is prime
15	15(16)	$+ 17 = 247$ which is prime
16	16(17)	$+ 17 = 279$ $= 17*17$
17	17(18)	$+ 17 = 313$ $= 17*19$
18	18(19)	$+ 17 = 359$ which is prime

14. (a) $\dfrac{|A1 - A|}{|A|} = \dfrac{2.35 - 2.3456}{2.3456} = \dfrac{0.0044}{2.3456} = 0.0018758\ldots \sim 0.19\%$

   (b) $\dfrac{|B1 - B|}{|B|} = \dfrac{2.3541 - 2.3}{2.3541} = \dfrac{0.0541}{2.3541} = 0.0229811\ldots \sim 2.30\%$

   (c) $\dfrac{|(A1 - B1) - (A - B)|}{|A - B|} = \dfrac{|0.05 + 0.0085|}{0.0085} = \dfrac{0.0585}{0.0085} = 6.882352\ldots \sim 688.24\%$

   (d) Because A1 > A and B1 < B and A and B are similar in value, the two
       small errors, which are in opposite directions, are large relative to the
       difference between A and B.

17. GCD(2N+1, 3N+1) = GCD(N, 2N+1) = GCD(1, N) = 1.

A.2 Solutions from Chapter 2

11. (a) $9^4 = 6,561$
    (b) $(9)(8)(7)(6) = 3,024$
    (c) $(1)(8)(7)(6) = 336$
    (d) $\binom{9}{4} = 126$
    (e) Any increasing 4-sequence on X that begins with 3 is a 3 followed by an increasing 3-sequence on $\{4..9\}$, so the number of increasing 4-sequences on X that begin with 3 is the number of increasing 3-sequences on $\{4..9\}$, which is $\binom{6}{3} = 20$.

18. Select the 4 letters and then select the 3 digits.

    (a) Choose 2 positions for the 2 T's
        and choose another letter to place in the left-most free position for a letter
        and choose another letter to place in the right-most free position for a letter
        and choose a digit to place in the left-most position for a digit
        and choose a digit to place in the middle position for a digit.
        Then the number of such license plates $= \binom{4}{2}(25)(25)(10)(10) = 375,000$.
    (b) The number of 4-letter sequences with at least one "T"
        = The number of 4-letter sequences minus the number of 4-letter sequences with no T's
        $= 26^4 - 25^4 = 456,976 - 390,625 = 66,351$
        The number of 3-digit sequences with at least one "4"
        = The number of 3-digit sequences minus the number of 3-digit sequences with no 4's
        $= 10^3 - 9^3 = 1,000 - 729 = 271$
        Then the number of such license plates $= (66,351)(271) = 17,981,121$.

A.3 Solutions from Chapter 3

10. *Proof.*

    Since
    $$\lfloor f \times n \rfloor <= f \times n \quad < \lfloor f \times n \rfloor + 1,$$
    $$n - \lfloor f \times n \rfloor >= n - f \times n \quad > n - \lfloor f \times n \rfloor - 1.$$

Hence $(n - \lfloor f \times n \rfloor) - 1 \; < \; n - f \times n \; <= (n - \lfloor f \times n \rfloor).$ // 2 consecutive integers

so $\lceil n - f \times n \rceil \; = \; n - \lfloor f \times n \rfloor$

and $\lfloor f \times n \rfloor + \lceil (1 - f) \times n \rceil \; = \; n.$ □

11. (a) $n^2 + n + 41 = n(n+1) + 41.$

> If $n = 41$ then $n^2 + n + 41 = 41(41 + 1) + 41 = 41(43)$ which is not prime.
> If $n = 40$ then $n^2 + n + 41 = 40(40 + 1) + 41 = 41(41)$ which is not prime.
>
> If $n = 0$ then $n^2 + n + 41 = 0(1) + 41 = 41 + 0 = 41$ which is prime.
> If $n = 1$ then $n^2 + n + 41 = 1(2) + 41 = 41 + 2 = 43$ which is prime.
> If $n = 2$ then $n^2 + n + 41 = 2(3) + 41 = 41 + 6 = 47$ which is prime.
> If $n = 3$ then $n^2 + n + 41 = 3(4) + 41 = 41 + 12 = 53$ which is prime.
> If $n = 4$ then $n^2 + n + 41 = 4(5) + 41 = 41 + 20 = 61$ which is prime.
> If $n = 5$ then $n^2 + n + 41 = 5(6) + 41 = 41 + 30 = 71$ which is prime.
> If $n = 6$ then $n^2 + n + 41 = 6(7) + 41 = 41 + 42 = 83$ which is prime.
> If $n = 7$ then $n^2 + n + 41 = 7(8) + 41 = 41 + 56 = 97$ which is prime.
> If $n = 8$ then $n^2 + n + 41 = 8(9) + 41 = 41 + 72 = 113$ which is prime.
> If $n = 9$ then $n^2 + n + 41 = 9(10) + 41 = 41 + 90 = 131$ which is prime.
> If $n = 10$ then $n^2 + n + 41 = 10(11) + 41 = 41 + 110 = 151$ which is prime.
> ...
> If $n = 39$ then $n^2 + n + 41 = 39(40) + 41 = 41 + 1560 = 1601$ which is prime.

(b) Let $a = \sqrt{2}$ and let $b = \sqrt{2}$. Then both a and b are irrational // Theorem 3.5.4
   but $a * b = 2$ which is rational.
   Or let $a = \sqrt{2}$ and let $b = \sqrt{8}$. Then both a and b are irrational
                                    // Theorem 3.5.4 plus the fact that $b = 2\sqrt{2}$.
   but $a * b = \sqrt{16} = 4$ which is rational.

(c) Let $a = 0$ and let $b = \sqrt{2}$. Then a is rational and b is irrational
   but $a * b = 0$ which is rational.   // zero is the only exception

14. Let $c = \sqrt{2}^{\sqrt{2}}$.
   If c is rational then take $a = \sqrt{2}$ and $b = \sqrt{2}$. Both are irrational but a^b is
   rational.
   If c is irrational then take $a = c$ and $b = \sqrt{2}$. Both are irrational but $a^b =$
   $$\left\{ (\sqrt{2})^{\sqrt{2}} \right\}^{\sqrt{2}} = (\sqrt{2})^2 \text{ which is rational.}$$

15.

P	Q	R	(P∧Q)	→R	P→	(Q→R)	[P	∧(~R)]	→	(~Q)
T	T	T	T	T	T	T	F	F	T	F
T	T	F	T	F	F	F	T	T	F	F
T	F	T	F	T	T	T	F	F	T	T
T	F	F	F	T	T	T	T	T	T	T
F	T	T	F	T	T	T	F	F	T	F
F	T	F	F	T	T	F	F	T	T	F
F	F	T	F	T	T	T	F	F	T	T
F	F	F	F	T	T	T	F	T	T	T
			↑	↑	↑	↑	↑	↑	↑	↑
			1	2	4	3	6	5	8	7

Columns 2, 4 and 8 are identical, so the three Boolean expressions are equivalent.

26. The answer is "yes".

Theorem: $n^2 < 3^n$ for $\forall$ integers $n \in \mathbf{N}$.

Proof.   // Here $a = 0$ and $P(n)$ is an inequality with a LHS and a RHS
           // **but** the induction step is *easy only* after $k = 2$ (not 0 or 1)

Step 1.   If $n = 0$ then $n^2 = 0 < 1 = 3^n$.   // $P(0)$ is True
            If $n = 1$ then $n^2 = 1 < 3 = 3^n$.   // $P(1)$ is True
            If $n = 2$ then $n^2 = 4 < 9 = 3^n$.   // $P(2)$ is True

Step 2.   Assume $\exists \, k \geq 2$ where $k^2 < 3^k$.   // $P(k)$ is True

Step 3.   If $n = k + 1$ then

$$(k+1)^2 = k^2 + 2k + 1$$
$$< k^2 + k^2 + k^2 \qquad\qquad // \; 1 < 2 <= k < k^2$$
$$= 3k^2$$
$$< 3(3^k) = 3^{k+1}. \qquad\qquad // \text{ by Step2}$$

◻

32. Let r denote the common ratio of this geometric sequence.
If $r = 1$ then $\forall n \in \mathbf{N} \; S_a + S_{a+1} + S_{a+2} + \ldots + S_{a+n} = (n+1)S_a$.
If $r \neq 1$ then $\forall n \in \mathbf{N} \; S_a + S_{a+1} + S_{a+2} + \ldots + S_{a+n} = S_a \times (r^{n+1} - 1)/(r - 1)$.

Proof.
If $r = 1$ then each of the $(n+1)$ consecutive entries is equal to the first,
so $S_a + S_{a+1} + S_{a+2} + \ldots + S_{a+n} = (n+1)S_a$.
If $r \neq 1$ then from Exercise 31 $\forall n \in \{a \ldots\} \, S_n = r^n \times K = r^{n-a} \times S_a$,
so $S_a + S_{a+1} + S_{a+2} + \ldots + S_{a+n} = S_a + rS_a + r^2 S_a + \ldots + r^n S_a$
$$= S_a(1 + r + r^2 + \ldots + r^n)$$
$$= S_a \times (r^{n+1} - 1)/(r - 1). \qquad // \text{ see Theorem 3.6.8}$$

◻

35. *Proof.*   // We will prove (c) by mathematical induction. This also
                              // is also a proof of (a) where $r = 2$ and (b) where $r = 3$.
          // Here $a = 0$ and $P(n)$ is an equation with a LHS and a RHS.

Step 1:   If $n = 0$ then LHS $= (0+1)r^0 = 1$,

$$\text{and} \quad \text{RHS} = \frac{[(r-1)0 + (r-2)]r^{0+1} + 1}{(r-1)^2} = \frac{(r-2)r+1}{(r-1)^2} = 1.$$

                                                                    // $P(1)$ is True

Step 2:   Assume $\exists\ k \in \mathbf{N}$ where $P(k)$ is True.
Step 3:   If $n = k+1$ then   // in the predicate P

$$\text{LHS} = \sum_{j=0}^{k+1}(j+1)r^j = \sum_{j=0}^{k}(j+1)r^j + ([k+1]+1)r^{k+1}$$

$$= \frac{[(r-1)k + (r-2)]r^{k+1} + 1}{(r-1)^2} + (k+2)r^{k+1} \times \frac{(r-1)^2}{(r-1)^2} \qquad \text{// by Step 2}$$

$$= \frac{\{[rk - k + r - 2] + (k+2)[r^2 - 2r + 1]\}r^{k+1} + 1}{(r-1)^2}$$

$$= \frac{\{[r(k+1) - (k+2)] + (k+2)r^2 - (k+2)2r + (k+2)\}r^{k+1} + 1}{(r-1)^2}$$

$$= \frac{\{(k+1) + (k+2)r - (2k+4)\}r \times r^{k+1} + 1}{(r-1)^2}$$

$$= \frac{\{(k+1)r + r - k - 1 - 2\}r^{k+2} + 1}{(r-1)^2}$$

$$= \frac{\{(r-1)(k+1) + (r-2)\}r^{(k+1)+1} + 1}{(r-1)^2}$$

$$= \text{RHS} \qquad\qquad\qquad\qquad\qquad\qquad\qquad \text{// in the predicate } P$$

// Therefore, $\forall n \in \mathbf{N}$, $\displaystyle\sum_{j=0}^{n}(j+1)r^j = \frac{[(r-1)n + (r-2)]r^{n+1} + 1}{(r-1)^2}$.

// Therefore, when r = 2, $\forall n \in \mathbf{N}$, $\displaystyle\sum_{j=0}^{n}(j+1)2^j = \frac{[(2-1)n + (2-2)]2^{n+1} + 1}{(2-1)^2}$

$$= n2^{n+1} + 1$$

// and when r = 3, $\forall n \in \mathbf{N}$, $\displaystyle\sum_{j=0}^{n}(j+1)3^j = \frac{[(3-1)n + (3-2)]3^{n+1} + 1}{(3-1)^2}$

$$= \frac{[2n+1]3^{n+1} + 1}{4}.$$

36. *Proof.*

Step 1: If $n = q$ then LHS $= \binom{q}{q} = 1$, and RHS $= \binom{q+1}{q+1} = 1 =$ LHS.

Step 2: Assume that $\exists$ an integer $k >= q$ where

$$\binom{q}{q} + \binom{q+1}{q} + \binom{q+2}{q} + \ldots + \binom{k}{q} = \binom{k+1}{q+1}$$

Step 3: If $n = q + 1$ then

$$\begin{aligned}
\text{LHS} &= \binom{q}{q} + \binom{q+1}{q} + \binom{q+2}{q} + \ldots + \binom{k}{q} + \binom{k+1}{q} \\
&= \binom{k+1}{q+1} \qquad\qquad\qquad + \binom{k+1}{q} \qquad \text{// from Step 2.} \\
&= \binom{(k+1)+1}{q+1} \qquad\qquad\qquad \text{// the Bad Banana Thm} \\
&= \text{RHS.} \qquad\qquad\qquad\qquad\qquad\qquad\qquad\qquad\qquad \square
\end{aligned}$$

A.4 Solutions from Chapter 4

7. The integer, $k = 2q + r$ where $q \in \mathbf{Z}$ and $r \in \{0, 1\}$.
So $q <= \frac{k}{2} = q + \frac{r}{2} <= q + \frac{1}{2}$ and therefore, $\lfloor k/2 \rfloor = q$.
Also, $q - \frac{1}{2} <= \frac{k-1}{2} = \frac{(2q+r)-1}{2} = q + \frac{r-1}{2} <= q$ and therefore, $\lceil \frac{k-1}{2} \rceil = q$. $\square$

8. Let $k^* = k + 1$.
Then $\lceil k/2 \rceil = \lceil (k^* - 1)/2 \rceil = \lfloor k^*/2 \rfloor$ // from Q7
$= \lfloor (k+1)/2 \rfloor$. $\square$

A.5 Solutions from Chapter 5

39. (a) *Proof:* (using the indirect method)

Assume that d_{ave}, the average degree in G, is $>= 6$. Since $d_{\text{ave}} =$ (the sum of the degrees of all the vertices)/$|V|$, from Eqn. 5.1.1, we get that $2|E| = |V| \times d_{\text{ave}}$. Then $2|E| >= |V| \times 6$ and hence $|V| <= 2|E|/6 = |E|/3$.

Let $d(G)$ be any drawing of G in the plane. Because, G has no bridges, every face of $d(G)$ is bounded by a circuit. Since G has no loops, every face in $d(G)$ has more than 1 edge in its boundary; and, since G has no pair of parallel edges, every face has at least 3 edges. Because every edge is in exactly 2 faces $2|E| >= 3|F|$ and hence $|F| <= 2|E|/3$. Then (from Thm. 5.5.2)

$$2 = |V| - |E| + |F| <= |E|/3 - |E| + 2|E|/3 = 0.$$

Because 2 is not $<= 0$, our assumption must be false, so d_{ave} must be < 6.

$\square$

(b) One example of such a graph is:

40. (a) *Proof:* (using the indirect method)

Assume that f_{ave}, the average face size in some drawing of **G**, is $>= 6$. Since $f_{ave} =$ (the sum of the sizes of all the faces)/$|F|$, from Eqn. 5.5.1, we get that $2|E| = |F| \times f_{ave}$. Then $2|E| >= |F| \times 6$ and hence $|F| <= 2|E|/6 = |E|/3$.

Let $d(G)$ be any drawing of **G** in the plane. Because, **G** has no bridges, every face of $d(G)$ is bounded by a circuit. Since **G** has no vertex of degree < 3, from Equation (5.1.1), we get that $2|E| = |V| \times d_{ave} >= |V| \times 3$ and hence $|V| <= 2|E|/3$. Then (from Thm. 5.5.2)

$$2 = |V| - |E| + |F| <= 2|E|/3 - |E| + |E|/3 = 0.$$

Because 2 is not $<= 0$, our assumption must be false, so f_{ave} must be < 6.

$\square$

(b) One example of such a graph is:

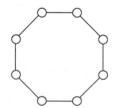

41. (a) Examples of planar drawings are:

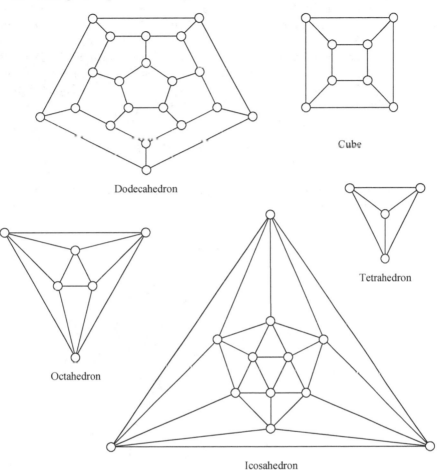

Dodecahedron

Cube

Octahedron

Tetrahedron

Icosahedron

(b) The dual of the *tetrahedron* is the *tetrahedron* itself.
The dual of the *cube* is the *octahedron*.
The dual of the *octahedron* is the *cube*.
The dual of the *dodecahedron* is the *icosahedron*.
The dual of the *icosahedron* is the *dodecahedron*.

(c) *Proof.*

In a 3-dimensional polyhedron, a vertex is a point where 3 or more faces
meet, so each vertex has degree ≥ 3. Also, the planar graph of the surface
of the polyhedron is bridgeless, connected and simple.

If $s = 3$ then all of the faces are triangles. The result given in Q39 gives
us that d must be < 6. The cases where $s = 3$ and $d = 3$, 4 and 5 are: the
tetrahedron, the octahedron, and the icosahedron. No other cases are possible when $s = 3$.

If $d = 3$ then all vertices have degree 3. The result given in Q40 gives us that s must be < 6. The cases where $d = 3$ and $s = 3$, 4 and 5 are: the tetrahedron, the cube, and the dodecahedron. No other cases are possible when $d = 3$.

Assume now that both $d >= 4$ and $s >= 4$. From Equation (5.1.1), we get that $2|E| >= |V| \times 4$ and hence $|V| <= |E|/2$. From Equation (5.5.1), we get that $2|E| >= |F| \times 4$ and hence $|F| <= |E|/2$. Then (from Thm. 5.5.2) we would get

$$2 = |V| - |E| + |F| <= |E|/2 - |E| + |E|/2 = 0.$$

Because 2 is not $<= 0$, our assumption must be false; the only possible Platonic Solids are the 5 listed.   □

42. (a)

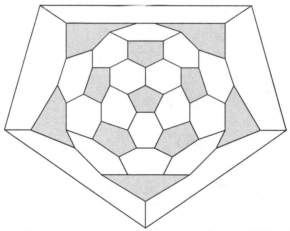

(b)   Each vertex lies on exactly one of the 12 pentagons, so $|V| = (12)(5) = 60$. Each vertex has degree 3, so $2|E| = (60)(3)$. Thus, $|E| = 90$.

A.6 Solutions from Chapter 6

2. (a)   No. See, for example, answer (b).
   (b)

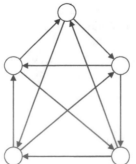

(c) Let $W(x)$ denote the number of wins by player x, and W_{ave} denote the average number of wins over all players. Then

$$6W_{ave} = \text{total \# of wins} = \text{total \# of games} = \binom{6}{2} = 15. \text{ So } W_{ave}$$

$$= 15/6 = 2\frac{1}{2} \text{ and}$$

therefore not all players can have the same number of wins.

(d) Let $L(x)$ denote the number of losses by player x. Then

$$\sum\{[W(x) - L(x)] : x \in V\} = \sum\{W(x) : x \in V\} - \sum\{L(x) : x \in V\}$$
$$= \text{total \# of games} \quad - \text{total \# of games} = 0.$$

So $\sum\{[W(x) - L(x)] : x \in V\backslash\{a\}\} = 0 - [W(a) - L(a)] < 0$ and at least one of the terms in the sum must be negative; i.e. there must be a player b with more losses than wins.

(e) Theorem: <u>Every</u> tournament contains a directed Hamilton Path.

Proof. Let T_n denote any particular tournament resulting from an orientation of K_n.

// We will prove the theorem by Mathematical Induction on n.

Step 1. If $n = 1$ then $V = \{a\}$ and the trivial path (a) contains all the vertices exactly once and is a directed Hamilton Path in T_n.

Step 2. Assume $\exists\ k >= 1$ where every tournament with k players has a directed Hamilton Path.

Step 3. If $n = k + 1$ and T is a tournament with $k + 1$ players, select any player w and remove w, and all arcs to and from w, from T. What remains is a tournament with k players, T_k. By step 2, there is a directed Hamilton Path in T_k,

$$\pi = (v_0, e_1, v_1, e_2, v_2, e_3, \ldots, e_k, v_k)$$

where each e_i is an arc oriented from v_{i-1} to v_i. Denote by α_i the arc joining w and v_i if it is oriented from v_i to w, and by β_i if it is oriented from w to v_i.

If the edge joining w and v_k is oriented from v_k to w, then

$$\pi 1 = (v_0, e_1, v_1, e_2, v_2, e_3, \ldots, e_k, v_k, \alpha_k, w)$$

is a directed Hamilton Path in T.

Otherwise, the edge joining w and v_k is oriented from w to v_k. Now, let q be the smallest index where the edge joining w and v_q is oriented from w to v_q. Then $0 <= q <= k$.

If $q = 0$ then

$$\pi 2 = (w, \beta_0, v_0, e_1, v_1, e_2, v_2, e_3, \ldots, e_k, v_k)$$

is a directed Hamilton Path in T.

If $q > 0$ then the edge joining w and v_{q-1} is oriented from v_{q-1} to w, and

$$\pi 3 = (v_0, e_1, v_1, e_2, \ldots, v_{q-1}, \alpha_{q-1}, w, \beta_q, v_q, \ldots, e_k, v_k)$$

is a directed Hamilton Path in T. □

6. There is only one topological sorting of an acyclic digraph D if and only if D contains a directed Hamilton Path.

7. There are many acyclic orientations – first assign the integers from 1 to $|V|$ to the vertices, then orient the edges so all arcs go from a lower integer to a higher one.

11. First topologically sort the vertices.

Algorithm 6.3.2 yields that the number of dipaths from a to b is 46.

Algorithm 6.3.3 yields that the length of a shortest dipath from a to b is 18.

Algorithm 6.3.4 yields that the length of a longest dipath from a to b is 51.

17. The Ford-Fulkerson Algorithm yields a Max-Flow with value 24.

20. (a) In the construction of the flow network $N(G)$ where the vertices of G are partitioned into two independent sets, L and R: give the arc (s, v) from the new source s to vertex v in L a capacity equal $b(v)$, give the arc (v, t) to the new sink t from a vertex v in R a capacity equal $b(v)$, and give all other arcs a capacity of one unit.

Then any b-*matching* M in G of size k, corresponds to a feasible flow F_M in $N(G)$ with value k; and, any feasible flow F in $N(G)$ with value k corresponds to a b-*matching* M_F in G of size k.

(b) // A maximum cardinality 1-matching in G has size 8.

A maximum cardinality 2-matching in G has size 16.

(c) A maximum cardinality 3-matching in G has size 22.

A.7 Solutions from Chapter 7

6. (a) 7 !R 7
   (b) 6 R 4 and 4 R 5 but 6 !R 5
   (c) 6 R 4 but 4 !R 6
   (d) 3 R 5 and 5 R 3 but 3 $\neq$ 5

8. R is reflexive because the trivial path $\pi = (v)$ joins any vertex v to itself.

R is symmetric because if a path π joins any vertex v to another vertex w, then the path π^R joins w to v.

R is transitive because if $\pi 1$ joins vertex v to vertex w, and $\pi 2$ joins w to another vertex x, then the concatenation of $\pi 1$ and $\pi 2$ joins v to vertex x.

R is an equivalence relation. The equivalence classes of vertices are the vertices in the connected components.

9. R is reflexive because the trivial dipath $\pi = (v)$ joins any vertex v to itself.

R is transitive because if $\pi 1$ joins vertex v to vertex w, and $\pi 2$ joins w to another vertex x, then the concatenation of $\pi 1$ and $\pi 2$ joins v to vertex x.

If D has no cycles, what properties does R have? Recall that a relation R on S is anti-symmetric means whenever a and b are distinct elements of S if $a R b$ then b !R a. If a and b are distinct vertices, and $a R b$ and $b R a$ then

there is a dipath $\pi 1$ from vertex a to vertex b and

there is a dipath $\pi 2$ from vertex a to vertex b.

Hence there is a closed dipath π equal the concatenation of $\pi 1$ and $\pi 2$.
Then there must be a cycle in D.   // by Thm 6.1.1

Thus, if D has no cycles, R is (also) anti-symmetric.

Because R is reflexive, transitive and anti-symmetric it is a Partial Order.

29. (a) $f \sim g$ and $A \in \mathbf{R}^+$

     $\Rightarrow f \ll g$ and $A \in \mathbf{R}^+ \Rightarrow Af \ll g$ // by #3 of Thm 7.5.3

               $A \in \mathbf{R}^+ \Rightarrow f \ll Af$ // by #2 of Thm 7.5.3

     $\Rightarrow g \ll f$ and $f \ll Af \Rightarrow g \ll Af$ // by Thm 7.5.1   □

     $\Rightarrow Af \sim g$

(b) $g \ll (g+f)$ // by #2 of Thm 7.5.3

     $f \ll g \Rightarrow (f+g) \ll g$ // by #4 of Thm 7.5.3   □

     $\Rightarrow (f+g) \sim g$

(c) $f1 \ll g$ and $f2 \sim g$

     $\Rightarrow f1 \ll g$ and $f2 \ll g \Rightarrow (f1+f2) \ll g$ // by #5 of Thm 7.5.3

     $f2 \ll (f2+f1) = (f1+f2)$ // by #1 of Thm 7.5.3

     $g \ll f2$ and $f2 \ll (f2+f1) \Rightarrow g \ll (f2+f1)$

     $\Rightarrow (f1+f2) \sim g$ □

(d) $f1 \ll g1$ and $f2 \ll g2 \Rightarrow (f1+f2) \ll (g1+g2)$ // by #6 of Thm 7.5.3

     $g1 \ll f1$ and $g2 \ll f2 \Rightarrow (g1+g2) \ll (f1+f2)$ // by #6 of Thm 7.5.3

     $\Rightarrow (f1+f2) \sim (g1+g2)$

(e) $f1 \ll g1$ and $f2 \ll g2 \Rightarrow (f1 \times f2) \ll (g1 \times g2)$ // by #7 of Thm 7.5.3

     $g1 \ll f1$ and $g2 \ll f2 \Rightarrow (g1 \times g2) \ll (f1 \times f2)$ // by #7 of Thm 7.5.3

     $\Rightarrow (f1 \times f2) \sim (g1 \times g2)$ □

30. Let $K = 1$ and let $M = 1$. If $n >= M$ then

     $(f1+f2) = n+n^2 <= n^3 + 2n^2 = (g1+g2)$,

so $(f1+f2) = n+n^2 \ll n^3 + 2n^2 = (g1+g2)$.

Suppose that $K \in \mathbf{R}^+$ and $M \in \mathbf{P}$. If $n* = M + \lceil K \rceil$, then $n* \in \mathbf{P}$, $n* >=$
M, $n* >= K$, and when n takes the value $n*$

$$(g1+g2) = n^3 + 2n^2 = n(n^2+2n) >= K(n^2+2n)$$
$$> K(n+n^2) = K(f1+f2).$$

So $(g1+g2) = n^3 + 2n^2 \; !\ll n+n^2 = (f1+f2)$.

Thus $(f1+f2) = n+n^2 \lll n^3 + 2n^2 = (g1+g2)$

and $(f1+f2) = n+n^2 \; !\sim n^3 + 2n^2 = (g1+g2)$.   □

31. Proof.

Anti-symmetry: Suppose that $f \neq g$.

$$f \, S\mathcal{D} \, g \Rightarrow f <<< g \qquad\qquad \text{// Thm. 7.5.7}$$
$$\Rightarrow g \, ! << f$$
$$\Rightarrow g \, ! <<< f \qquad\qquad \text{// Thm. 7.5.5}$$
$$\Rightarrow g \, ! S\mathcal{D} f \qquad\qquad \text{// contra-positive of Thm. 7.5.7}$$

Transitivity: Suppose that $f \, S\mathcal{D} \, g$ and $g \, S\mathcal{D} \, h$.   // Is $f \, S\mathcal{D} \, h$?

$\forall K_1 \in \mathbf{R}^+, \ \exists M_1(K_1) \in \mathbf{P}$ such that if $n > M_1(K_1)$ then $K_1 \times f(n) < g(n)$.
If $K_1 = 1$, $\exists M_1(1) \in \mathbf{P}$ such that if $n > M_1(1)$ then $1 \times f(n) < g(n)$.
$\forall K_2 \in \mathbf{R}^+, \ \exists M_2(K_2) \in \mathbf{P}$ such that if $n > M_2(K_2)$ then $K_2 \times g(n) < h(n)$.
Let K be any element of $\mathbf{R}^+$, and let $M(K) = M_1(1) + M_2(K)$ // then $M(K) \in \mathbf{P}$
If $n > M(K)$ then

$$n > M_1(1) \text{ so } f(n) < g(n) \text{ and } K \times f(n) < K \times g(n),$$

also $n > M_2(K)$ so $K \times g(n) < h(n)$,

and hence $K \times f(n) < h(n)$.

Thus $f \, S\mathcal{D} \, h$.   ☐

32. Tabulating the first few values of these sequences gives:

n	$f(n)$	$g(n)$	
1	1	1	// $n = 1 = 2(0) + 1$ so $r = 0$
2	2	4	
3	6	6	
4	24	96	
5	120	120	

For any $r \in \mathbf{P}$, we have

$$\begin{aligned}
f(2r) \quad &= (2r)! \\
< g(2r) \quad &= (2r)! \, (2r) \\
< f(2r+1) \ &= (2r)! \, (2r+1) \\
= g(2r+1) \ &= (2r)! \, (2r+1) \quad \text{// both } f \text{ and } g \text{ are increasing integer sequences}
\end{aligned}$$

If $n = 2r$ then $g(n) = f(n) \times n$ and if $n = 2r + 1$ then $f(n) = g(n)$.
Let $K = 1$ and let $M = 1$. If $n >= M$ then $f(n) <= K \times g(n)$. Therefore $f << g$.

Recall that $f1 \ !<< f2$ means $\forall K \in \mathbf{R}^+$ and $\forall M \in \mathbf{P}, \ \exists n^* >= M$ where $f1(n^*) > K \times f2(n^*)$.

Let K be any given positive real number and let M be any given positive integer.

Let $r^* = M + \lceil K \rceil$.   // $r^* \in \mathbf{P}$ and $r^* > M$, K
If $n^* = 2r^*$ then $n^* >= M$ and $g(n^*) = f(n^*) \times n^* > K \times f(n^*)$ so $g \mathbin{!} \ll f$.
Therefore $f \lll g$.

Recall that $f1$ $\mathbf{S\!D}$ $f2$ means $\forall K \in \mathbf{R}^+$, $\exists M(K) \in \mathbf{P}$ such that if $n > M(K)$
then $K \times f1(n) < f2(n)$.

Therefore, $f1$ $\mathbf{S\!D}$ $f2$ means $\exists K \in \mathbf{R}^+$ such that $\forall M \in \mathbf{P}$, $\exists n^* > M$ where
$K \times f1(n^*) >= f2(n)$.

Let $K - 1$ and let M be any given positive integer.
If $n^* = 2M + 1$ then $n^* >= M$ and $g(n^*) = f(n^*) >= K \times f(n^*)$.
Therefore $f \mathbin{!}\mathbf{S\!D}$ g. $\square$

Chapter 8

1. Suppose E_n is defined recursively on $\mathbf{P}$ by

$$E_0 = 0, \; E_1 = 2, \text{ and } E_{n+1} = 2n\{E_n + E_{n-1}\} \text{ for all } n >= 1.$$

Determine the value of E_{10}.

$E_2 = E_{1+1} = 2(1)\{E_1 + E_0\} =$	$2\{2+0\} =$	4
$E_3 = E_{2+1} = 2(2)\{E_2 + E_1\} =$	$4\{4+2\} =$	24
$E_4 = E_{3+1} = 2(3)\{E_3 + E_2\} =$	$6\{24+4\} =$	168
$E_5 = E_{4+1} = 2(4)\{E_4 + E_3\} =$	$8\{168+24\} =$	1 536
$E_6 = E_{5+1} = 2(5)\{E_5 + E_4\} =$	$10\{1536 + 168\} =$	17 040
$E_7 = E_{6+1} = 2(6)\{E_6 + E_5\} =$	$12\{17\,040 + 1\,536\} =$	222 912
$E_8 = E_{7+1} = 2(7)\{E_7 + E_6\} =$	$14\{222\,912 + 17\,040\} =$	3 359 328
$E_9 = E_{8+1} = 2(8)\{E_8 + E_7\} =$	$16\{3\,359\,328 + 222\,912\} =$	57 315 840
$E_{10} = E_{9+1} = 2(9)\{E_9 + E_8\} =$	$18\{57\,315\,840 + 3\,359\,328\} =$	1 092 153 024

7. (a) The 15 possible pairings are:
   1. x_1 with x_2 and x_3 with x_4 and x_5 with x_6
   2. x_1 with x_2 and x_3 with x_5 and x_4 with x_6
   3. x_1 with x_2 and x_3 with x_6 and x_4 with x_5
   4. x_1 with x_3 and x_2 with x_4 and x_5 with x_6
   5. x_1 with x_3 and x_2 with x_5 and x_4 with x_6
   6. x_1 with x_3 and x_2 with x_6 and x_4 with x_5
   7. x_1 with x_4 and x_2 with x_3 and x_5 with x_6
   8. x_1 with x_4 and x_2 with x_5 and x_3 with x_6
   9. x_1 with x_4 and x_2 with x_6 and x_3 with x_5
   10. x_1 with x_5 and x_2 with x_3 and x_4 with x_6
   11. x_1 with x_5 and x_2 with x_4 and x_3 with x_6

12. x_1 with x_5 and x_2 with x_6 and x_3 with x_4
13. x_1 with x_6 and x_2 with x_3 and x_4 with x_5
14. x_1 with x_6 and x_2 with x_4 and x_3 with x_5
15. x_1 with x_6 and x_2 with x_5 and x_3 with x_4 // So $P_3 = 15$

(b) Suppose $n >= 2$. Element x_1 may be paired with any of the $(2n - 1)$ other elements in A. This leaves $(2n - 2) = 2(n - 1)$ elements still to be paired, and that can be done in P_{n-1} ways. Thus the number of pairings of $2n$ elements, $P_n = (2n - 1) \times P_{n-1}$.

(c) Theorem: $P_n = (1)(3)(5) \ldots (2n - 1)$
  // product of the first n positive odd integers

Proof. // by mathematical induction

Step 1. $P_1 = 1$ which is the first positive odd integer.

Step 2. Assume $\exists \ k >= 1$ where $P_k = (1)(3)(5) \ldots (2k - 1)$.

Step 3. If $n = k + 1$ then $n >= 2$ and
$$P_{k+1} = (2[k+1]-1)P_k \qquad \text{// using the RE}$$
$$= (2k+1) \times (1)(3)(5) \ldots (2k-1) \qquad \text{// by Step 2}$$
$$= (1)(3)(5) \ldots (2k-1) \times (2[k+1]-1) \qquad \qquad \square$$

8. $y_{n+1} = \dfrac{[n+1]([n+1]-1)}{2} + c = \dfrac{[n+1](n)}{2} + c = \dfrac{n(n-1)+2n}{2} + c = y_n + n.$

9. (a) $f(1) = 11, \quad f(2) = 23, \quad f(3) = 47, \quad f(4) = 95, \quad f(5) = 191,$
  $f(6) = 383, \quad f(7) = 767, \quad f(8) = 1535, \quad f(9) = 3071, \quad f(10) = 6143.$

(b) $f(1) - f(0) = 6$ but $f(2) - f(1) = 12$ so f is not an arithmetic sequence.
  $f(1) / f(0) = 11/5 = 121/55$ but $f(2) / f(1) = 23/11 = 115/55$ so f is not an geometric sequence.

11. (a) $s_1 = (1/5)s_0 - 8 = (1/5)(60) \quad -8 = 12 \quad -8 = \quad +4$
  $s_2 = (1/5)s_1 - 8 = (1/5)(4) \quad -8 = 0.8 \quad -8 = -7.2$
  $s_3 = (1/5)s_2 - 8 = (1/5)(-7.2) \quad -8 = -1.44 \quad -8 = -9.44$

(b) In this RE, $a = 1/5$ and $c = -8$ so $\dfrac{c}{1-a} = \dfrac{-8}{4/5} = -10$ and we have $s_0 = 60$.

  The particular solution is $s_n = a^n \left[I - \dfrac{c}{1-a} \right] + \dfrac{c}{1-a}$
$$= (1/5)^n [60 - (-10)] + (-10)$$
$$= (1/5)^n [70] - 10.$$

(c) Yes. The limit is -10.

(d) No. Because the sequence does not converge to zero.

21.

n	F_n	$T_n = 1 + F_0 + F_1 + \ldots + F_n$
0	1	2
1	1	3
2	2	5
3	3	8
4	5	13
5	8	21

Theorem For $\forall n \in P$, $T_n = 1 + F_0 + F_1 + \ldots + F_n$ equals F_{n+2}.
Proof

Step 1. If $n = 0$ then $T_n = 1 + F_0 = 1 + 1 = 2 = F_2 = F_{0+2}$.
Step 2. Assume that $\exists\ k >= 0$ such that T_k equals F_{k+2}.
Step 3. If $n = k+1$ then $n >= 1$ so

$$
\begin{aligned}
T_{k+1} &= 1 + F_0 + F_1 + \ldots + F_k \quad + F_{k+1} \\
&= T_k \qquad\qquad\qquad\qquad\;\; + F_{k+1} \\
&= F_{k+2} \qquad\qquad\qquad\qquad + F_{k+1} \quad \text{// by Step 2} \\
&= F_{k+3} \qquad\qquad\qquad\qquad\qquad\quad \text{// by the Fibonacci RE} \\
&= F_{(k+1)+2}. \qquad\qquad\qquad\qquad\qquad\qquad\qquad\qquad\qquad \Box
\end{aligned}
$$

23. (a) The General Solution of this Recurrence Equation is

$$
\begin{aligned}
S_n &= A(r)^n + Bn(r)^n \quad \text{// since } r_1 = r_2 = r \\
&= A(11)^n + Bn(11)^n \quad \text{// since } r = 11
\end{aligned}
$$

(b) Find the Particular Solution where $S_0 = 1$ and $S_1 = 5$.

$$
\begin{aligned}
S_0 &= 1 = A(11)^0 + B(0)(11)^0 = A \\
S_1 &= 5 = A(11)^1 + B(1)(11)^1 = 11A + 11B
\end{aligned}
$$

Then $A = 1$
              $B = [5 - 11(1)]/11 = -6/11$
Hence, the Particular Solution where $S_0 = 1$ and $S_1 = 5$ is

$$
\begin{aligned}
S_n &= (1)(11)^n + (-6/11)n(11)^n = 11^n - 6n(11)^{n-1} \\
\text{//} \quad S_0 &= 11^0 - 6(0)(11)^{n-1} \qquad\quad = 1 - 0 = 1 \\
\text{// and} \quad S_1 &= 11^1 - 6(1)(11)^0 \qquad\qquad = 11 - 6 = 5.
\end{aligned}
$$

29. (a) $e^{1/3} = 1.395\,612\,425...$

(b)

j			$\text{term}_j = \frac{1}{j!}\left(\frac{1}{3}\right)^j$	partial sum
0		1	$= 1.000\,000\,000...$	$1.000\,000\,000... = 1$
1		$1/3$	$= 0.333\,333\,333...$	$1.333\,333\,333... = 4/3$
2	$1/(6 \times 3)$	$= 1/18$	$= 0.055\,555\,555...$	$1.388\,888\,888... = 25/18$
3	$1/(9 \times 18)$	$= 1/162$	$= 0.006\,172\,839...$	$1.395\,061\,728... = 226/162$
4	$1/(12 \times 162)$	$= 1/1944$	$= 0.000\,514\,403...$	$1.395\,576\,132... = 2713/1944$
5	$1/(15 \times 1944)$	$= 1/29160$	$= 0.000\,034\,293...$	$1.395\,610\,425... = 40696/29160$

(c) Yes.

(d) $e^{-1/3} = 0.716\,531\,310...$

j			$\text{term}_j = \frac{1}{j!}\left(\frac{-1}{3}\right)^j$	partial sum
0		1	$= +1.000\,000\,000...$	$1.000\,000\,000... = 1$
1		$-1/3$	$= -0.333\,333\,333...$	$0.666\,666\,666... = 2/3$
2	$+1/(6 \times 3)$	$= +1/18$	$= +0.055\,555\,555...$	$0.722\,222\,222... = 13/18$
3	$-1/(9 \times 18)$	$= -1/162$	$= -0.006\,172\,839...$	$0.716\,049\,382... = 116/162$
4	$+1/(12 \times 162)$	$= +1/1944$	$= +0.000\,514\,403...$	$0.716\,563\,786... = 1393/1944$
5	$-1/(15 \times 1944)$	$= -1/29160$	$= -0.000\,034\,293...$	$0.716\,529\,492... = 20894/29160$

Yes. This partial sum also gives 3 decimal places of accuracy.

Chapter 9

4. (a) $[(7)(6)]/[(2)(1)] = 21$

(b) (1, 2, 3, 4, 5)

(c) (3, 4, 5, 6, 7)

(d) (2, 3, **5**, 6, 7)

7. The sequences are (read left to right, top to bottom):

321	421	431	432	521	531	532	541	542	543
621	631	632	641	642	643	651	652	653	654

8. (a) Let $n = 3$, B $= 8$, $V_1 = 10$, $W_1 = 5$, $V_2 = 7$, $W_2 = 4$, $V_3 = 7$, and $W_3 = 4$.
The Greedy Solution is $\{O_1\}$ with total value $= 10$ and total weight $= 5$.
The Optimal Solution is $\{O_2, O_3\}$ with total value $= 14$ and total weight $= 8$.

(b) Let $n = 3$, B $= 8$, $V_1 = 10$, $W_1 = 5$, $\text{Ratio}_1 = 2$
$$V_2 = 7, \ W_2 = 4, \ \text{Ratio}_2 = 1.75$$
$$V_3 = 7, \ W_3 = 4, \ \text{Ratio}_3 = 1.75$$
The Greedy Solution is $\{O_1\}$ with total value $= 10$ and total weight $= 5$.
The Optimal Solution is $\{O_2, O_3\}$ with total value $= 14$ and total weight $= 8$.

9. (a) $\displaystyle\sum_{j=a}^{b}(x_{j-1}-x_j) =$

$$
\begin{aligned}
& x_{a-1} - x_a \\
+\ & x_a - x_{a+1} \\
+\ & x_{a+1} - x_{a+2} \\
& \cdots \\
+\ & x_{b-2} - x_{b-1} \\
+\ & x_{b-1} - x_b \\
=\ & x_{a-1} - x_b. \qquad \text{// All other } x_j\text{'s "cancel out".}
\end{aligned}
$$

(b) $\displaystyle\sum_{j=a}^{b}(x_j - x_{j-1}) = -\sum_{j=a}^{b}(x_{j-1}-x_j) = -(x_{a-1}-x_b) = x_b - x_{a-1}.$

(c) Let $y_j = x_{j+1}$ for each index j. Then

$$\sum_{j=a}^{b}(x_{j+1}-x_j) = \sum_{j=a}^{b}(y_j - y_{j-1}) = y_b - y_{a-1} = x_{b+1} - x_a.$$

(d) Let $x_j = n^j$ for each index j. Then

$$
\sum_{j=a}^{b} n^j(n-1) = \sum_{j=a}^{b}(n^{j+1}-n^j) = \sum_{j=a}^{b}(x_{j+1}-x_j) = x_{b+1} - x_a
$$
$$
= n^{b+1} - n^a.
$$

(e) Let $x_j = n^{k-j}$ for each index j. Then

$$
\sum_{j=a}^{b} n^{k-j}(n-1) = \sum_{j=a}^{b}(n^{k-j+1} - n^{k-j})
$$
$$
= \sum_{j=a}^{b}(x_{j-1}-x_j) = x_{a-1} - x_b \qquad \text{// from part (a)}
$$
$$
= n^{k-(a-1)} - n^{k-b} = n^{k-a+1} - n^{k-b}.
$$

Chapter 10

3. (a) $prob(\mathbf{A}) = 27/80$

   $prob(\mathbf{B}) = 11/80$

   $prob(\mathbf{A}\ \text{and}\ \mathbf{B}) = 4/80$

   $prob(\mathbf{A}\ \text{or}\ \mathbf{B}) = prob(\mathbf{A}) + prob(\mathbf{B}) - prob(\mathbf{A}\ \text{and}\ \mathbf{B})$

   $$= \frac{27}{80} + \frac{11}{80} - \frac{4}{80} = \frac{34}{80}$$

   $prob(\mathbf{A}|\mathbf{B}) = prob(\mathbf{A}\ \text{and}\ \mathbf{B})/prob(\mathbf{B}) = \dfrac{4/80}{11/80} = \dfrac{4}{11}$

   $prob(\mathbf{B}|\mathbf{A}) = prob(\mathbf{B}\ \text{and}\ \mathbf{A})/prob(\mathbf{A}) = \dfrac{4/80}{27/80} = \dfrac{4}{27}$

(b) **A** and **B** are NOT mutually exclusive because $prob(\textbf{A and B}) = 4/80 \neq 0$.

(c) **A** and **B** are NOT independent because

$$prob(\textbf{A and B}) = 4/80 = 320/6400$$
$$\neq prob(\textbf{A}) * prob(\textbf{B}) = (27/80) * (11/80) = 297/6400.$$

5. (a) # delegations that exclude Mike $= \binom{n-1}{k}$

$$prob \text{ (Mike is excluded)} = \binom{n-1}{k} \Big/ \binom{n}{k} = \frac{(n-1)!}{k!(n-k-1)!} \times \frac{k!(n-k)!}{n!} = \frac{n-k}{n}$$

$$prob \text{ (Mike is included)} = 1 - \frac{n-k}{n} = \frac{k}{n}$$

Alternatively…

delegations that include Mike $= \binom{n-1}{k-1}$

$$prob \text{ (Mike is included)} = \binom{n-1}{k-1} \Big/ \binom{n}{k} = \frac{(n-1)!}{(k-1)!(n-k)!} \times \frac{k!(n-k)!}{n!} = \frac{k}{n}$$

(b) $prob$(Flora is included) $= \dfrac{k}{n}$

(c) # delegations that include both Mike and Flora $= \binom{n-2}{k-2}$

$$prob \text{ (Mike and Flora are included)} = \binom{n-2}{k-2} \Big/ \binom{n}{k}$$

$$= \frac{(n-2)!}{(k-2)!(n-k)!} \times \frac{k!(n-k)!}{n!}$$

$$= \frac{k(k-1)}{n(n-1)}.$$

This equals $prob$ (Mike is included) $*$ $prob$ (Flora is included)

$$\Leftrightarrow \quad \frac{k(k-1)}{n(n-1)} = \frac{k(k)}{n(n)}$$

$$\Leftrightarrow \quad \frac{k-1}{n-1} = \frac{k}{n}$$

$$\Leftrightarrow \quad n(k-1) = k(n-1)$$

$$\Leftrightarrow \quad nk - n = kn - k$$

$$\Leftrightarrow \quad -n = -k$$

$$\Leftrightarrow \quad k = n$$

Since $1 < k < n$, the events "Mike is in" and "Flora is in" are NOT independent.

6. $prob(\{X \text{ gets H and you get H}\} \text{ or } \{X \text{ gets T and you get T}\})$

$= prob(X \text{ gets H and you get H}) + prob(X \text{ gets T and you get T})$
$= prob(X \text{ gets H}) * prob(\text{you get H}) + prob(X \text{ gets T}) * prob(\text{you get T})$
$= p * \frac{1}{2} + (1 - p) * (1 - \frac{1}{2})$
$= \frac{1}{2}$

9. (a)

$T =$	Prob	# probes	(#probes) × (Prob)
X_{10}	0.40	1	0.40
X_7	0.22	2	0.44
X_4	0.10	3	0.30
X_9	0.07	4	0.28
X_6	0.06	5	0.30
X_3	0.05	6	0.30
X_1	0.04	7	0.28
X_8	0.03	8	0.24
X_2	0.02	9	0.18
X_5	0.01	10	0.10
	1.00		2.82 = Expected # probes

(b)

# probes	$T =$	Prob	(#probes) × (Prob)
1	X_5	0.01	0.01
2	X_2, X_8	0.05	0.10
3	X_1, X_3, X_6, X_9	0.22	0.66
4	X_4, X_7, X_{10}	0.72	2.88
		1.00	3.65 = Expected # probes

(c)   Yes

10. (a)

	passed test	failed test	
passed course	3230	48	3278
failed course	170	552	722
	3400	600	4000

(b)   P(failed test|passed course) $= \dfrac{48}{3278} = 0.014643075\ldots \sim 1.5\%$

(c)   P(passed test|failed course) $= \dfrac{170}{722} = 0.235457063\ldots \sim 23.5\%$

13. (a) $10^4 = 10\ 000$

(b) $10 * 9 * 8 * 7 = 5\ 040$

(c) The number of passwords with the digit "5" repeated j times is $\binom{4}{j} 9^{4-j}$.

j	$\binom{4}{j} 9^{4-j}$	
0	$(1)9^4=$	6 561
1	$(4)9^3=$	2 916
2	$(6)9^2=$	486
3	$(4)9^1=$	36
4	$(1)9^0=$	1
Total:		$\overline{10\ 000}$

(d)

j	$j \times prob(X{=}j)$
0	$0 \times 6561/10000 = 0.0000$
1	$1 \times 2916/10000 = 0.2916$
2	$2 \times 486/10000 = 0.0972$
3	$3 \times 36/10000 = 0.0108$
4	$4 \times 1/10000 = \underline{0.0004}$

$\qquad\qquad$ E(# of 5's) = 0.4000 $\qquad$ // equals $4 \times (1/10)$

$\qquad\qquad\qquad\qquad\qquad\qquad\qquad\qquad$ // np in some binomial experiment

15. Using a Binomial experiment as a model for this process

where   a trial is a laptop purchase

$\qquad\qquad$ a "success" is *making a claim* on the guarantee

$\qquad\qquad\qquad$ // we're interested in the number of claims made, X

so $\qquad\qquad$ $n = 43$, $p = 1 - 85\% = 15\% = 0.15$ and $q = 0.85$

and $\qquad\quad$ X is the *number of claims* made.

We want $P(X > = 3) = \displaystyle\sum_{k=3}^{43} P(X = k)$ $\qquad\qquad\qquad$ // a very long calculation

$\qquad\qquad\qquad = 1 - \displaystyle\sum_{k=0}^{2} P(X = k)$ $\qquad\qquad\qquad\qquad$ // using the hint

$P(X = 0) = \dbinom{43}{0} p^0 \times q^{43} = 1 \times 1 \times q^{43} = (1)(1)(0.85)^{43} = 0.000\,922\,600\ldots$

$P(X = 1) = \dbinom{43}{1} p^1 \times q^{42} = 43 \times p \times q^{42} = (43)(0.15)(0.85)^{42} = 0.007\,000\,911\ldots$

$P(X = 2) = \dbinom{43}{2} p^2 \times q^{41} = \dfrac{43 \times 42}{2 \times 1}(0.15)^2(0.85)^{41}$

$\qquad\qquad\quad = (903)(0.0225)(0.85)^{41} = 0.025\,944\,554\ldots$

Hence, $P(X < 3) = 0.000\,922\,600\ldots + 0.007\,000\,911\ldots + 0.025\,944\,554\ldots$

$\qquad\qquad\qquad = 0.033\,868\,065\ldots$

and $P(X >= 3) = 1 - 0.033\,868\,065\ldots = 0.966\,131\,934\ldots$

// Also, $\dbinom{43}{0} p^0 \times q^{43} + \dbinom{43}{1} p^1 \times q^{42} + \dbinom{43}{2} p^2 \times q^{41}$

// $= \{1 \times 1 \times q^2 \quad +43 \times p \times q^1 \quad +903 \times p^2\} \times q^{41}$

// $= \{0.7225 \qquad +5.4825 \qquad +20.3175\} \times q^{41}$

// $= \{26.5225\} \times (0.001276956\ldots) = 0.033\,868\,067\ldots$

// We should expect about $np = 43 \times (0.15) = 6.45$ claims.

17. (a) 38%

   (b) Using a Binomial experiment as a model for this
      where   the trials correspond to the top 10 marks

             a "success" is a _woman_ obtaining a top ten mark
      // we're interested in the number of women that obtain a top ten mark, X
      we assume that $p = 38\%$, so $q = 62\%$.

$$P(X = 6) = \dbinom{10}{6} p^6 \times q^4 = \frac{10 \times 9 \times 8 \times 7}{4 \times 3 \times 2 \times 1} (0.38)^6 (0.62)^4$$

$$= (210)(0.003010936\ldots)(0.14776336)$$

$$= 0.093\,430\,276\ldots$$

   (c) $E(X) = np = (10)(0.38) = 3.8$

23. The possible outcomes from the game are summarized in the following table,
    with the amount won shown in brackets:

	Diamond	Spade, Heart or Club
Ace	1 ($10.00)	3 ($5.00)
Jack, Queen or King	3 ($7.00)	9 ($4.50)
2 through 10	9 ($4.50)	27 ($0.00)

Probability Distribution for f

v	$prob(f = v)$	$v \times prob(f = v)$
−$2.50	27/52	−$67.50/52
$2.00	18/52	$36.00/52
$2.50	3/52	$7.50/52
$4.50	3/52	$13.50/52
$7.50	1/52	$7.50/52
	52/52	$E(f) = \quad$ −$ 3.00/52
		$= -\$\,0.057692307\ldots$
		$= -5.7692307\ldots$cents

The average payoff per play is negative because the person operating the game (the "house") needs to cover the cost of operating the game, and wants to make a profit from it. The money needed for these costs and profits comes from having a negative average payoff per play.

Chapter 11

1. (a) (i)

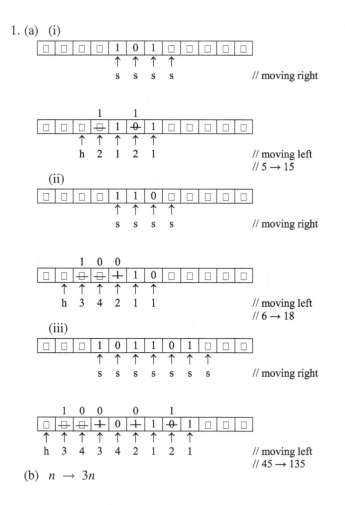

(b) $n \rightarrow 3n$

2. (a)

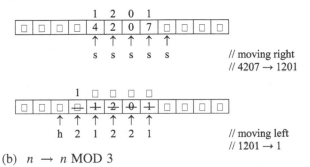

// moving right
// 4207 → 1201

// moving left
// 1201 → 1

(b) $n \rightarrow n$ MOD 3

4. Let L denote the "language" of strings of a's and b's described. Then $L =$ $\{a^n b^n: n \in \mathbf{N}\}$. Furthermore, $w \in L$ if and only if w is the empty string or $w =$ axb where $x \in L$.

(s, a: $\square$, R, 1)   (1, a: a, R, 1)   (2, a: $\square$, L, No)   (3, a: a, L, 3)
(s, b: $\square$, R, No)   (1, b: b, R, 1)   (2, b: $\square$, L, 3)   (3, b: b, L, 3)
(s, $\square$: $\square$, R, Yes) (1, $\square$: $\square$, L, 2) (2, $\square$: $\square$, L, No) (3, $\square$: $\square$, R, s)

Index